A Transition
to
Advanced Mathematics

SEVENTH EDITION

A Transition to Advanced Mathematics

Douglas Smith
University of North Carolina Wilmington

Maurice Eggen
Trinity University

Richard St. Andre
Central Michigan University

BROOKS/COLE
CENGAGE Learning™

Australia • Brazil • Japan • Korea • Mexico • Singapore • Spain • United Kingdom • United States

BROOKS/COLE
CENGAGE Learning™

A Transition to Advanced Mathematics,
7th Edition
Douglas Smith, Maurice Eggen, and
Richard St. Andre

Publisher: Richard Stratton

Senior Sponsoring Editor: Molly Taylor

Associate Editor: Daniel Seibert

Editorial Assistant: Shaylin Walsh

Senior Marketing Manager: Jennifer Pursley Jones

Marketing Communications Manager: Mary Anne Payumo

Marketing Coordinator: Erica O'Connell

Content Project Manager: Alison Eigel Zade

Art Director: Jill Ort

Text Permissions Account Manager: Margaret Chamberlain-Gaston

Photo Permissions Account Manager: Don Schlotman

Senior Print Buyer: Diane Gibbons

Production Service: Elm Street Publishing Services

Compositor: Integra Software Services, Inc.

For product information and technology assistance, contact us at
Cengage Learning Customer & Sales Support, 1-800-354-9706

For permission to use material from this text or product, submit all requests online at **www.cengage.com/permissions**. Further permissions questions can be emailed to **permissionrequest@cengage.com.**

Library of Congress Control Number: 2009939573

Student Edition:
ISBN-13: 978-0-495-56202-3
ISBN-10: 0-495-56202-5

Brooks/Cole
20 Channel Center Street
Boston, MA 02210
USA

Cengage Learning is a leading provider of customized learning solutions with office locations around the globe, including Singapore, the United Kingdom, Australia, Mexico, Brazil and Japan. Locate your local office at **international.cengage.com/region**

Cengage Learning products are represented in Canada by Nelson Education, Ltd.

For your course and learning solutions, visit **www.cengage.com.**

Purchase any of our products at your local college store or at our preferred online store **www.cengagebrain.com.**

Printed in the United States of America
1 2 3 4 5 6 7 14 13 12 11 10

To our wives
Karen, Karen, and Karen

CONTENTS

Preface   viii
Preface to the Student   xii

CHAPTER 1   **Logic and Proofs**   1

   1.1   Propositions and Connectives   1
   1.2   Conditionals and Biconditionals   9
   1.3   Quantifiers   18
   1.4   Basic Proof Methods I   27
   1.5   Basic Proof Methods II   40
   1.6   Proofs Involving Quantifiers   48
   1.7   Additional Examples of Proofs   60

CHAPTER 2   **Set Theory**   71

   2.1   Basic Concepts of Set Theory   71
   2.2   Set Operations   79
   2.3   Extended Set Operations and Indexed Families of Sets   89
   2.4   Mathematical Induction   100
   2.5   Equivalent Forms of Induction   114
   2.6   Principles of Counting   122

CHAPTER 3   **Relations and Partitions**   135

   3.1   Cartesian Products and Relations   135
   3.2   Equivalence Relations   147

3.3   Partitions   157
3.4   Ordering Relations   163
3.5   Graphs   174

CHAPTER   **4   Functions**   **185**

4.1   Functions as Relations   185
4.2   Constructions of Functions   195
4.3   Functions That Are Onto; One-to-One Functions   205
4.4   One-to-One Correspondences and Inverse Functions   213
4.5   Images of Sets   220
4.6   Sequences   225

CHAPTER   **5   Cardinality**   **233**

5.1   Equivalent Sets; Finite Sets   233
5.2   Infinite Sets   242
5.3   Countable Sets   251
5.4   The Ordering of Cardinal Numbers   259
5.5   Comparability of Cardinal Numbers and the Axiom of Choice   267

CHAPTER   **6   Concepts of Algebra**   **275**

6.1   Algebraic Structures   275
6.2   Groups   283
6.3   Subgroups   292
6.4   Operation Preserving Maps   298
6.5   Rings and Fields   306

CHAPTER   **7   Concepts of Analysis**   **315**

7.1   Completeness of the Real Numbers   316
7.2   The Heine–Borel Theorem   324
7.3   The Bolzano–Weierstrass Theorem   336
7.4   The Bounded Monotone Sequence Theorem   341
7.5   Equivalents of Completeness   347

Answers to Selected Exercises   353
Index   393

Excerpts from the Preface to the First Edition

"I understand mathematics but I just can't do proofs."

Our experience has led us to believe that the remark above, though contradictory, expresses the frustration many students feel as they pass from beginning calculus to a more rigorous level of mathematics. This book developed from a series of lecture notes for a course at Central Michigan University that was designed to address this lament. The text is intended to bridge the gap between calculus and advanced courses in at least three ways. First, it provides a firm foundation in the major ideas needed for continued work. Second, it guides students to think and to express themselves mathematically—to analyze a situation, extract pertinent facts, and draw appropriate conclusions. Finally, we present introductions to modern algebra and analysis in sufficient depth to capture some of their spirit and characteristics.

Exercises marked with a solid star (★) have complete answers at the back of the text. Open stars (☆) indicate that a hint or a partial answer is provided. "Proofs to Grade" are a special feature of most of the exercise sets. We present a list of claims with alleged proofs, and the student is asked to assign a letter grade to each "proof" and to justify the grade assigned. Spurious proofs are usually built around a single type of error, which may involve a mistake in logic, a common misunderstanding of the concepts being studied, or an incorrect symbolic argument. Correct proofs may be straightforward, or they may present novel or alternate approaches. We have found these exercises valuable because they reemphasize the theorems and counterexamples in the text and also provide the student with an experience similar to grading papers. Thus the student becomes aware of the variety of possible errors and develops the ability to read proofs critically.

In summary, our main goals in this text are to improve the student's ability to think and write in a mature mathematical fashion and to provide a solid understanding of the material most useful for advanced courses. Student readers, take comfort from the fact that we do not aim to turn you into theorem-proving wizards. Few of you will become research mathematicians. Nevertheless, in almost any mathematically related work you may do, the kind of reasoning you need to be able to do is the same reasoning you use in proving theorems. You must first understand exactly what you want to prove (verify, show, or explain), and you must be familiar with the logical steps that allow you to get from the hypothesis to the conclusion. Moreover, a proof is the ultimate test of your understanding of the subject matter and of mathematical reasoning.

We are grateful to the many students who endured earlier versions of the manuscript and gleefully pointed out misprints. We acknowledge also the helpful comments of Edwin H. Kaufman, Melvin Nyman, Mary R. Wardrop, and especially Douglas W. Nance, who saw the need for a course of this kind at CMU and did a superb job of reviewing the manuscript.

To the Seventh Edition

The seventh edition is based on the same goals and core material as previous editions, but with new organization in several places and many new and revised expositions, examples, and exercises. In the expanded *Preface to the Student*, we have gathered together preliminary ideas that should already be familiar to students (including properties of the number systems, definitions of even, odd and prime numbers, naive notions of sets, and the basic terminology of functions). This arrangement makes the prerequisite material easier to locate and keeps the focus of the text on the use of mathematical reasoning.

The rewritten introduction to concepts of elementary number theory in Section 1.7 is deliberately placed early in the text, before any discussion of inductive proofs and the Well-Ordering Principle, as an opportunity to practice basic proof methods on a coherent set of results about divisibility, the greatest common divisor, and linear combinations. Placing this content here (and accepting the Division Algorithm without proof until inductive proofs are introduced in Chapter 2) allows students to experience significant results that are achieved with relatively simple proof forms. Later, students can observe the power of inductive methods to prove the Division Algorithm and related results.

In Chapter 4 properties of one-to-one and onto functions are now grouped more efficiently and there is a separate section on one-to-one correspondences and permutations of a set. In Section 5.3 on countable sets, the major results (that subsets and unions of countably many countable sets are countable) are moved up to make them more accessible. In Chapter 7, there is even more emphasis on the meaning of the completeness property of the real number system.

Chapter 1 introduces the propositional and predicate logic required by mathematical arguments, not as formal logic, but as tools of reasoning for more complete understanding of concepts (including some ideas of arithmetic, analytic geometry, and calculus with which the student is already familiar). We present methods of proof and carefully analyze examples of each method, giving special attention to the use of definitions and denials. The techniques in this chapter are used and referred to throughout the text. In Chapters 2, 3, and 4 on sets, relations, and functions, we emphasize writing and understanding proofs that require the student to deal precisely with the concepts of set operations, equivalence relations and partitions, and properties of injective and surjective functions.

These first four chapters contain the core material of the text and, in addition, offer the opportunity for further work in several optional sections: basics of number theory (Section 1.7), combinatorial counting (Section 2.6), order relations and graph theory (Sections 3.4 and 3.5), and image sets and sequences (Sections 4.5 and 4.6). See the diagram on the inside front cover for a diagram that highlights the core and shows the prerequisite relationships among sections. For a one-semester course, we recommend the core material along with any one of Chapters 5, 6, or 7, or a selection of optional sections and excursions into one or two of the later chapters—for example, Sections 4.6, 5.1, 5.2, 5.3, 7.1, and 7.2.

Chapters 5, 6, and 7 make use of the skills and concepts the student has acquired in the first four chapters, and thus are a cut above the earlier chapters in terms of level and rigor. Chapter 5 emphasizes a working knowledge of cardinality: finite and infinite sets, denumerable sets and the uncountability of the real numbers, and properties of countable sets. We include sections on the ordering of cardinals and applications of the Cantor–Schröder–Bernstein Theorem and a brief discussion of the Axiom of Choice. In Chapter 6 we consider properties of algebras with a binary operation, groups, substructures, and homomorphisms, and relate these concepts to rings and fields. Chapter 7 considers the completeness property of the real numbers by tracing its consequences: the Heine–Borel Theorem, the Bolzano–Weierstrass Theorem, and the Bounded Monotone Sequence Theorem, and back to completeness.

We sincerely thank our reviewers for the seventh edition: David Bayer, Columbia University; Fernando Burgos, University of South Florida; Yves Nievergelt, Eastern Washington University; and Don Redmond, Southern Illinois University.

We also thank our reviewers of earlier editions: Mangho Ahuja, Southeast Missouri State University; William Ballard, University of Montana; David Barnette, University of California at Davis; Gerald Beer, California State University–Los Angeles; Harry Conce, Mankato State University; Sherralyn Craven, Central Missouri State University; Robert Dean, Stephen F. Austin State University; Ron Dotzel, University of Missouri; Harvey Elder, Murray State University; Michael J. Evans, North Carolina State University; Gerald Farrell. California Polytechnic State University; Benjamin Freed, Clarion University of Pennsylvania; Robert Gamble, Winthrop College; Dennis Garity, Oregon State University; Robert P. Hunter, Pennsylvania State University; Jack Johnson, Brigham Young University–Hawaii; L. Christine Kinsey, Canisuis College; Daniel Kocan, State University of New York, Potsdam; James McKinney, California Polytechnic State University; Blair Madore, The State University of New York at Potsdam; Andrew Martin, Morehead State University; Edward Mosley, Lyon College; Van C. Nall, University of Richmond; Yves Nievergelt, Eastern Washington University; Yewande Olubummo, Spelman College; Hoseph H. Oppenheim, San Francisco State University; John S. Robertson, Georgia College & State University; Victor Schneider, University of Southwestern Louisiana; Dale

Schoenefeld, University of Tulsa; Kenneth Slonnegar, State University of New York at Fredonia; Douglas Smith, University of the Pacific; Joseph Teeters, University of Wisconsin; Mary Treanor, Valparaiso University; and Lawrence Williams, University of Texas, San Antonio.

We also wish to thank Roger Lipsett for his suggestions after proofreading of the final manuscript and the staff at Cengage for their exceptional professional assistance in the development of this edition and previous editions.

Finally, we note that instructors who adopt this text can sign up for online access to complete solutions for all exercises via Cengage's *Solution Builder* service at www.cengage.com/solutionbuilder.

Douglas D. Smith
Richard St. Andre

Welcome to the study of mathematical reasoning. The authors know that many students approach this material with some apprehension and uncertainty. Some students feel that "This isn't like other mathematics courses," or expect that the study of proofs is something they won't really have to do or won't use later. These feelings are natural as you move from calculation-oriented courses where the goals emphasize performing computations or solving certain equations, to more advanced courses where the goal may be to establish whether a mathematical structure has certain properties. This textbook is written to help ease the transition between these courses. Let's consider several questions students commonly have at the beginning of a "transition" course.

Why write proofs?

Mathematicians often collect information and make observations about particular cases or phenomena in an attempt to form a theory (a model) that describes patterns or relationships among quantities and structures. This approach to the development of a theory uses **inductive reasoning**. However, the characteristic thinking of the mathematician is **deductive reasoning**, in which one uses logic to develop and extend a theory by drawing conclusions based on statements accepted as true. Proofs are essential in mathematical reasoning because they demonstrate that the conclusions are true. Generally speaking, a mathematical explanation for a conclusion has no value if the explanation cannot be backed up by an acceptable proof.

Why not just test and repeat enough examples to confirm a theory?

After all, as is typically done in natural and social sciences, the test for truth of a theory is that the results of an experiment conform to predictions, and that when the experiment is repeated under the same circumstances the result is always the same. The difference is that in mathematics we need to know whether a given statement is *always* true, so while the statement may be true for many (even infinitely many) examples, we would never know whether another example might show the statement to be false. By studying examples, we might conclude that the statement

$$\text{``}x^2 - 3x + 43 \text{ is a prime number''}$$

is true for all positive integers x. We could reach this conclusion testing the first 10 or 20 or even the first 42 integers $1, 2, 3, \ldots, 42$. In each of these cases and others, such as 44, 45, 47, 48, 49, 50 and more, $x^2 - 3x + 43$ *is* a prime number. But the statement is not always true because $43^2 - 3(43) + 43 = 1763$, which is $41 \cdot 43$. Checking examples is helpful in gaining insight for understanding concepts and relationships in mathematics, but is not a valid proof technique unless we can somehow check all examples.

Why not just rely on proofs that someone else has done?

One answer follows from the statement above that deductive reasoning characterizes the way mathematicians think. In the sciences, a new observation may force a complete rethinking of what was thought to be true; in mathematics what we know to be true (by proof) is true forever unless there was a flaw in the reasoning. By learning the techniques of reasoning and proof, you are learning the tools of the trade.

The first goal of this text is to examine standard proof techniques, especially concentrating on how to get started on a proof, and how to construct correct proofs using those techniques. You will discover how the logical form of a statement can serve as a guide to the structure of a proof of the statement. As you study more advanced courses, it will become apparent that the material in this book is indeed fundamental and the knowledge gained will help you succeed in those courses. Moreover, many of the techniques of reasoning and proof that may seem so difficult at first will become completely natural with practice. In fact, the reasoning that you will study is the essence of advanced mathematics and the ability to reason abstractly is a primary reason why applicants trained in mathematics are valuable to employers.

What am I supposed to know before beginning Chapter 1?

The usual prerequisite for a transition course is at least one semester of calculus. We will sometimes refer to topics that come from calculus and earlier courses (for example, differentiable functions or the graph of a parabola), but we won't be solving equations or finding derivatives.

You will need a good understanding of the basic concepts and notations from earlier courses. The list of definitions and relationships below includes the main things **you will need to have ready for immediate use** at any point in the text.

Be aware that definitions in mathematics, however, are not like definitions in ordinary English, which are based on how words are typically used. For example, the ordinary English word "cool" came to mean something good or popular when many people used it that way, not because it has to have that meaning. If people stop using the word that way, this meaning of the word will change. Definitions in mathematics have precise, fixed meanings. When we say that an integer is odd, we do not mean that it's strange or unusual. Our definition below tells you exactly what odd means. You may form a concept or a mental image that you may use to help understand (such as "ends in 1, 3, 5, 7, or 9"), but the mental image you form is not what has been defined. For this reason, definitions are usually stated with the "if and only if" connective because they describe exactly—no more, no less—the condition(s) to meet the definition.

Sets

A **set** is a collection of objects, called the **elements**, or members of the set. When the object x is in the set A, we write $x \in A$; otherwise $x \notin A$. The set $K = \{6, 7, 8, 9\}$ has four elements; we see that $7 \in K$ but $3 \notin K$. We may use set-builder notation to write the set K as

$$\{x: x \text{ is an integer greater than 5 and less than 10}\},$$

which we read as "the set of x such that x is . . ." Observe that the set whose only element is 5 is not the same as the number 5; that is, $\{5\} \neq 5$. The **empty set** $\varnothing$ is a set with no elements.

We say that A is a **subset** of B, and write $A \subseteq B$, if and only if every element of A is an element of B. If sets A and B have exactly the same elements, we say they are **equal** and write $A = B$.

We use these notations for the number systems:

$\mathbb{N} = \{1, 2, 3, \ldots\}$ is the set of natural numbers.
$\mathbb{Z} = \{\ldots -3, -2, -1, 0, 1, 2, \ldots\}$ is the set of integers.
$\mathbb{Q}$ is the set of all rational numbers.
$\mathbb{R}$ is the set of all real numbers.
$\mathbb{C}$ is the set of all complex numbers.

A set is **finite** if it is empty or if it has n elements for some natural number n. Otherwise it is **infinite**. Thus the set $\{6, 7, 8, 9\}$ is finite. All the number systems listed above are infinite.

The Natural Numbers

The properties below describe the basic arithmetical and ordering structure of the set $\mathbb{N}$.

1. *Successor properties*

    1 is a natural number.
    Every natural number x has a unique successor $x + 1$.
    1 is not the successor of any natural number.

2. *Closure properties*
   The sum of two natural numbers is a natural number.
   The product of two natural numbers is a natural number.

3. *Associativity properties*
   For all $x, y, z \in \mathbb{N}$, $x + (y + z) = (x + y) + z$.
   For all $x, y, z \in \mathbb{N}$, $x(yz) = (xy)z$.

4. *Commutativity properties*
   For all $x, y \in \mathbb{N}$, $x + y = y + x$.
   For all $x, y \in \mathbb{N}$, $xy = yx$.

5. *Distributivity properties*
   For all $x, y, z \in \mathbb{N}$, $x(y + z) = xy + xz$.
   For all $x, y, z \in \mathbb{N}$, $(y + z)x = yx + zx$.

6. *Cancellation properties*
   For all $x, y, z \in \mathbb{N}$, if $x + z = y + z$, then $x = y$.
   For all $x, y, z \in \mathbb{N}$, if $xz = yz$, then $x = y$.

For natural numbers a and b we say a **divides** b (or a is a **divisor** of b, or b is a **multiple** of a) if and only if there is a natural number k such that $b = ak$. For example, 7 divides 56 because there is a natural number (namely 8) such that $56 = 7 \cdot 8$.

A natural number p is **prime** if and only if p is greater than 1 and the only natural numbers that divide p are 1 and p. A **composite** is a natural number that is neither 1 nor prime.

The Fundamental Theorem of Arithmetic:

Every natural number larger than 1 is prime or can be expressed uniquely as a product of primes. For example, 440 can be expressed as $440 = 2^3 \cdot 5 \cdot 11$. If we list the prime factors in increasing order, then there is only one prime factorization: the primes and their exponents are uniquely determined.

The Integers

The integers share properties 2 through 6 listed above for $\mathbb{N}$ (with the exception that we can't cancel $z = 0$ from the product $xz = yz$). Other important properties are:

For all x in $\mathbb{Z}$, $x + 0 = 0$, $x \cdot 0 = 0$ and $x + (-x) = 0$.
For all x, y, z in $\mathbb{Z}$, if $x < y$ and $z > 0$, $xy < yz$.
The product of two positive or two negative integers is positive; the product of a positive and a negative is negative.

The natural numbers and integers provide excellent settings for developing an understanding of the structure of a correct proof, so we will use the following definitions extensively in early examples of proof writing. In those proofs we make use of the properties of number systems and the fact that every integer is either even or odd, but not both.

An integer x is **even** if and only if there is an integer k such that $x = 2k$. An integer x is **odd** if and only if there is an integer j such that $x = 2j + 1$. For integers a and b with $a \neq 0$ we say a **divides** b if and only if there is an integer k such that $b = ak$.

Real and Rational Numbers

We think of the real numbers as being all the numbers along the number line. Each real number can be represented as an integer together with a finite or infinite decimal part. We use the standard notations for intervals on the number line. For real numbers a and b with $a < b$:

$(a, b) = \{x: x \in \mathbb{R}$ and $a < x < b\}$ is the **open interval from a to b**.

$[a, b] = \{x: x \in \mathbb{R}$ and $a \leq x \leq b\}$ is the **closed interval from a to b**.

$(a, \infty) = \{x: x \in \mathbb{R}$ and $a < x\}$ and $(-\infty, b) = \{x: x \in \mathbb{R}$ and $x < b\}$ are **open rays.**

$[a, \infty) = \{x: x \in \mathbb{R}$ and $a \leq x\}$ and $(-\infty, b] = \{x: x \in \mathbb{R}$ and $x \leq b\}$ are **closed rays.**

Note that the infinity symbol "∞" is simply a notational convenience and does not represent any real number. Also, one should be careful not to confuse $(1, 6)$ with $\{2, 3, 4, 5\}$, since $(1, 6)$ is the set of all real numbers between 1 and 6 and contains, for example, $2, \pi, \sqrt{13}$, and $\frac{27}{5}$.

The real number x is **rational** if and only if there are integers p and q, with $q \neq 0$, such that $x = p/q$.

The rationals are exactly the numbers along the number line that have terminating or repeating decimal expressions. All other real numbers are **irrational**. In Chapter 1 we will see a proof that $\sqrt{2}$ is irrational. The number systems $\mathbb{R}$ and $\mathbb{Q}$ share many of the arithmetic and ordering properties of the naturals and integers, along with a new property:

Every number x except 0 has a multiplicative inverse; that is, there is a number y such that $xy = 1$.

Complex Numbers

A complex number has the form $a + bi$, where a and b are real numbers and $i = \sqrt{-1}$. The **conjugate** of $a + bi$ is $a - bi$ and $(a + bi)(a - bi) = a^2 + b^2$. The set of reals is a subset of the complex numbers because any real number x may be written as $x + 0i$. Complex numbers do not share the ordering properties of the reals.

Functions

A **function** (or a **mapping**) is a rule of correspondence that associates to each element in a set A a unique element in a second set B. No restriction is placed on the sets A and B, which may be sets of numbers, or functions, or vegetables. To denote that f is a function from A to B, we write

$$f: A \rightarrow B$$

and say "f maps A to B." If $a \in A$ and the corresponding element of B is b, we write

$$f(a) = b.$$

The elements of A are sometimes called the **arguments** or **inputs** of the function. If $f(a) = b$ we say that b is the **image** of a, or b is the **value** of the function f at a. We also say that a is a **pre-image** of b.

For example, $f: \mathbb{R} \to \mathbb{R}$ given by $f(x) = x^2 + 1$ represents the correspondence that assigns to each real number x the number that is one more than the square of x. The image of the real number 2 is 5 and -3 is a pre-image of 10.

The features that make f a function from A to B are that every element of A must have an image, that image must be in B, and most importantly, that no element of A has more than one image. It is this **single-valued** property that make functions so useful.

If $f: A \to B$, the set A is the **domain** of f, denoted $\mathrm{Dom}\,(f)$, and B is the **codomain** of f. The set

$$\mathrm{Rng}\,(f) = \{f(x): x \in A\}$$

of all images under the function f is called the **range** of f. The range of the function $f: \mathbb{R} \to \mathbb{R}$ given by $f(x) = x^2 + 1$ is $[1, \infty)$.

It is sometimes convenient to describe a function by giving only a domain and a rule. For functions whose domains and codomains are subsets of $\mathbb{R}$, the domain is sometimes left unspecified and assumed to be the largest possible subset of $\mathbb{R}$ for which image values may be obtained. With this assumption, the domain of $g(x) = \sqrt{x + 1}$ is $[-1, \infty)$, because this is the largest set of real numbers for which $\sqrt{x + 1}$ may be calculated.

When we say that $f: A \to B$, it is required that $\mathrm{Rng}\,(f) \subseteq B$. However, it may be that some elements of the codomain are not images under the function f; that is, the set $\mathrm{Rng}\,(f)$ may not be equal to B. In the special case when the range of f is equal to B, we say f maps A **onto** B. It may also be that two different elements of A have the same image in B. In the special case when any two different arguments have different images, we say that f is **one-to-one**. Because the range of $f(x) = x^2 + 1$ is $[1, \infty)$, f is not onto $\mathbb{R}$. Since $f(3)$ and $f(-3)$ have value 10, f is not one-to-one.

What am I allowed to assume for a proof?

You may be given specific instructions for some proof writing exercises, but generally the idea is that you may use what someone studying the topic of your proof would know. That is, when we prove something about intersecting lines we might use facts about the slope of a line, but we probably would not use properties of derivatives. This really is not much of a problem, except for our first proof examples, which deal with elementary concepts such as even and odd (because they provide meaningful examples and a familiar context in which to study logic and reasoning). For these proofs we are allowed to use the properties of integers and natural numbers that we already know *except* what we already know about evenness and oddness.

Remember, we don't expect you to become an expert at proving theorems overnight. With practice—studying lots of examples and exercises—the skills will come. Our goal is to help you write and think as mathematicians do, and to present a solid foundation in material that is useful in advanced courses. We hope you enjoy it.

Douglas D. Smith
Richard St. Andre

Logic and Proofs

We recommend that you read the *Preface to the Student* before beginning this first chapter. Most of the terms and concepts in that *Preface* should be familiar to you, but it is well worth making sure you know the terminology and notations we will use throughout the book. It is especially important that you know precisely the definitions of such terms as: "divides," "prime," "rational," and "even" and "odd."

As described in the *Preface*, mathematics is concerned with the formation of a **theory** (collection of true statements) that describes patterns or relationships among quantities and structures. It is characterized by **deductive reasoning**, in which one uses logic to develop and extend a theory by drawing conclusions based on statements accepted as true. We give **proofs** to demonstrate that our conclusions are true. This chapter will provide a working knowledge of the basics of logic and how to construct a proof.

1.1   Propositions and Connectives

Our goal in this section is to understand truth values of propositions and how propositions can be combined using logical connectives.

Most sentences, such as "$\pi > 3$" and "Earth is the closest planet to the sun," have a truth value. That is, they are either true or false. We call these sentences propositions. Other sentences, such as "What time is it?" and "Look out!" are interrogatory or exclamatory; they express complete thoughts but have no truth value.

> **DEFINITION**   A **proposition** is a sentence that has exactly one truth value: true, which we denote by T, or false, which we denote by F.

Some propositions, such as "$7^2 = 60$," have easily determined truth values. It will take years to determine the truth value of the proposition "The North Pacific right whale will be an extinct species before the year 2525." Other statements, such as "Euclid was left-handed," are propositions whose truth values may never be known.

Sentences like "She lives in New York City" and "$x^2 = 36$" are not propositions because each could be true or false depending upon the person to whom "she" refers and what numerical value is assigned to x. We will deal with sentences like these in Section 1.3.

The statement "This sentence is false" is not a proposition because it is neither true nor false. It is an example of a **paradox**—a situation in which, from premises that look reasonable, one uses apparently acceptable reasoning to derive a conclusion that seems to be contradictory. If the statement "This sentence is false" is true, then by its meaning it must be false. On the other hand, if the given statement is false, then what it claims is false, so it must be true. The study of paradoxes such as this has played a key role in the development of modern mathematical logic. A famous example of a paradox formulated in 1901 by Bertand Russell* is discussed in Section 2.1.

By applying logical connectives to propositions, we can form new propositions.

DEFINITION   The **negation** of a proposition P, denoted $\sim P$, is the proposition "not P." The proposition $\sim P$ is true exactly when P is false.

The truth value of the negation of a proposition is the opposite of the truth value of the proposition. For example, the negation of the false proposition "7 is divisible by 2" is the true statement "It is not the case that 7 is divisible by 2," or "7 is not divisible by 2."

DEFINITIONS   Given propositions P and Q, the **conjunction** of P and Q, denoted $P \wedge Q$, is the proposition "P and Q." $P \wedge Q$ is true exactly when *both* P and Q are true.

The **disjunction** of P and Q, denoted $P \vee Q$, is the proposition "P or Q." $P \vee Q$ is true exactly when *at least one* of P or Q is true.

Examples. If C is the proposition "19 is composite" and M is "45 is a multiple of 3," we know C is false and M is true. Thus "19 is composite and 45 is a multiple of 3," written using logical connectives as $C \wedge M$, is a false proposition, while "19 is composite or 45 is a multiple of 3," which has form $C \vee M$, is true. The false proposition "Either 19 is composite or 45 is not a multiple of 3" has the form $C \vee \sim M$.

The English words *but*, *while*, and *although* are usually translated symbolically with the conjunction connective, because they have the same meaning as *and*. For

---

* Bertrand Russell (1872–1970) was a British philosopher, mathematician, and advocate for social reform. He was a strong voice for precision and clarity of arguments in mathematics and logic. He coauthored *Principia Mathematica* (1910–1913), a monumental effort to derive all of mathematics from a specific set of axioms and well-defined rules of inference.

example, we would write "19 is not composite, but 45 is a multiple of 3" in symbolic form as: $(\sim C) \wedge M$.

An important distinction must be made between a statement and the *form* of a statement. In the previous example "19 is composite and 45 is a multiple of 3" is a proposition with truth value F. We used the form $C \wedge M$ to represent this proposition, but *the form $C \wedge M$ itself has no truth value* unless C and M are assigned to be specific propositions. If we let C be "Copenhagen is the capital of Denmark" and M be "Madrid is the capital of Spain," then $C \wedge M$ would have the value T.

To repeat: a propositional form does not have *a* truth value. Instead, each form has a *list* of truth values that depend on the values assigned to its components. This list is displayed by presenting all possible combinations for the truth values of its components in a truth table. Since the connectives $\wedge$ and $\vee$ involve two components, their truth tables must list the four possible combinations of the truth values of those components:

P	Q	$P \wedge Q$
T	T	T
F	T	F
T	F	F
F	F	F

P	Q	$P \vee Q$
T	T	T
F	T	T
T	F	T
F	F	F

Since the value of $\sim P$ depends only on the two possible values for P, its truth table is

P	$\sim P$
T	F
F	T

Frequently you will encounter compound propositions formed from more than two propositional variables. The propositional form $(P \wedge Q) \vee \sim R$ has three variables P, Q, and R; it follows that there are $2^3 = 8$ possible combinations of truth values. The two main components are $P \wedge Q$ and $\sim R$. We make truth tables for these and combine them by using the truth table for $\vee$.

P	Q	R	$P \wedge Q$	$\sim R$	$(P \wedge Q) \vee \sim R$
T	T	T	T	F	T
F	T	T	F	F	F
T	F	T	F	F	F
F	F	T	F	F	F
T	T	F	T	T	T
F	T	F	F	T	T
T	F	F	F	T	T
F	F	F	F	T	T

The statement "Either 7 is prime and 9 is even or else 11 is not less than 3" may be symbolized by $(P \wedge Q) \vee \sim R$, where P is "7 is prime," Q is "9 is even," and R

is "11 is less than 3." We know P is true, Q is false and R is false. Therefore, $(P \wedge Q)$ is false and $\sim R$ is true. Thus $(P \wedge Q) \vee \sim R$ is true, in agreement with line 7 of the table. Thus the proposition "Either 7 is prime and 9 is even or else 11 is not less than 3" is a true statement.

Some compound forms always yield the value true just because of the way they are formed; others are always false.

DEFINITIONS   A **tautology** is a propositional form that is true for every assignment of truth values to its components.

A **contradiction** is a propositional form that is false for every assignment of truth values to its components.

, For example, the *Law of Excluded Middle*, $P \vee \sim P$, is a tautology because $P \vee \sim P$ is true when P is true and true when P is false. We know that a statement like "The absolute value function is continuous or it is not continuous" must be true because it has the form of the Law of Excluded Middle.

Example.   Show that $(P \vee Q) \vee (\sim P \wedge \sim Q)$ is a tautology.

The truth table for this propositional form is

P	Q	$P \vee Q$	$\sim P$	$\sim Q$	$\sim P \wedge \sim Q$	$(P \vee Q) \vee (\sim P \wedge \sim Q)$
T	T	T	F	F	F	T
F	T	T	T	F	F	T
T	F	T	F	T	F	T
F	F	F	T	T	T	T

Since the last column is all true, $(P \vee Q) \vee (\sim P \wedge \sim Q)$ is a tautology.

Both $\sim(P \vee \sim P)$ and $Q \wedge \sim Q$ are examples of contradictions. The negation of a contradiction is, of course, a tautology.

Writing a proof requires the ability to connect statements so that the truth of any given statement in the proof follows logically from previous statements in the proof, from known results, or from basic assumptions. Particularly important is the ability to recognize or write a statement equivalent to another. Sometimes, it is the *form* of a compound statement that may be used to find a useful equivalent.

DEFINITION   Two propositional forms are **equivalent** if and only if they have the same truth tables.

Example. The propositional forms P and $\sim(\sim P)$ are equivalent. The truth tables for these forms may be combined in one table to show that they are the same:

P	$\sim P$	$\sim(\sim P)$
T	F	T
F	T	F

The fact that P and $\sim(\sim P)$ have the same truth value for each line of the truth table means that whatever proposition we choose for P, the truth value of P and $\sim(\sim P)$ are identical.

Some of the most commonly used equivalent forms are presented in the following theorem.

Theorem 1.1.1 For propositions P, Q, and R, the following are equivalent:

(a)	P	and	$\sim(\sim P)$	Double Negation Law
(b)	$P \vee Q$	and	$Q \vee P$	Commutative Laws
(c)	$P \wedge Q$	and	$Q \wedge P$	
(d)	$P \vee (Q \vee R)$	and	$(P \vee Q) \vee R$	Associative Laws
(e)	$P \wedge (Q \wedge R)$	and	$(P \wedge Q) \wedge R$	
(f)	$P \wedge (Q \vee R)$	and	$(P \wedge Q) \vee (P \wedge R)$	Distributive Laws
(g)	$P \vee (Q \wedge R)$	and	$(P \vee Q) \wedge (P \vee R)$	
(h)	$\sim(P \wedge Q)$	and	$\sim P \vee \sim Q$	DeMorgan's* Laws
(i)	$\sim(P \vee Q)$	and	$\sim P \wedge \sim Q$	

Proof.
(a) See the discussion above.
(h) By examining the fourth and seventh columns of their combined truth tables as shown here,

P	Q	$P \wedge Q$	$\sim(P \wedge Q)$	$\sim P$	$\sim Q$	$\sim P \vee \sim Q$
T	T	T	F	F	F	F
F	T	F	T	T	F	T
T	F	F	T	F	T	T
F	F	F	T	T	T	T

we see that the truth tables for $\sim(P \wedge Q)$ and $\sim P \vee \sim Q$ are identical. Thus $\sim(P \wedge Q)$ and $\sim P \vee \sim Q$ are equivalent propositional forms.

Proofs of the remaining parts are left as exercises. ∎

In addition to making tables to verify the remaining parts of Theorem 1.1.1, you should also think about why two propositional forms are equivalent by looking

---

* Augustus DeMorgan (1806–1871) was an English logician and mathematician whose contributions include his notational system for symbolic logic. He also introduced the term "mathematical induction" (see Section 2.4) and developed a rigorous foundation for that proof technique.

at their meanings. For part (h), negation is applied to a conjunction. The form $\sim(P \wedge Q)$ is true precisely when $P \wedge Q$ is false. This happens when one of P or Q is false, or in other words, when one of $\sim P$ or $\sim Q$ is true. Thus, $\sim(P \wedge Q)$ is equivalent to $\sim P \vee \sim Q$. That is, to say "You don't have both P and Q" is the same as saying "You don't have P or you don't have Q."

As an example of how this theorem might be useful in dealing with statements, suppose we are told that the statement "The function f is increasing and concave upward" is false. The statement has the form $P \wedge Q$, where P is the statement "f is increasing" and Q is the statement "f is concave upward." The negation $\sim(P \wedge Q)$ is "It is not the case that f is increasing and f is concave upward." By part (h) above, this is equivalent to $\sim P \vee \sim Q$, which is

"It is not the case that f is increasing or it is not the case that f is concave upward."

An easier way to say this is

"f is not increasing or f is not concave upward."

A **denial** of a proposition P is any proposition equivalent to $\sim P$. A proposition has only one negation, $\sim P$, but always has many denials, including $\sim P$, $\sim\sim\sim P$, $\sim\sim\sim\sim\sim P$, etc. DeMorgan's Laws provide others ways to construct useful denials.

Example.   A denial of "Either Miss Scarlet is not guilty or the crime did not take place in the ballroom" is "The crime took place in the ballroom and Miss Scarlet is guilty." This can be verified by writing the two propositions symbolically as $(\sim S) \vee (\sim B)$ and $B \wedge S$, respectively, and checking that their truth tables have exactly opposite values. We could also observe that $B \wedge S$ is equivalent to $S \wedge B$ so a denial of $B \wedge S$ is equivalent to $\sim(S \wedge B)$, which we know by DeMorgan's Laws is equivalent to $(\sim S) \vee (\sim B)$.

Example.   The statement "Line L_1 has slope 3/5 or line L_2 does not have slope -4" may be symbolized using the form $P \vee \sim Q$, so its negation is $\sim(P \vee \sim Q)$. We can write a simpler denial $(\sim P) \wedge Q$ by applying DeMorgan's Laws and the Double Negation Law. The simplified denial says "Line L_1 does not have slope 3/5 and line L_2 has slope -4."

Notice that someone might read the negation $\sim(P \vee \sim Q)$ as "It is not the case that L_1 has slope 3/5 or line L_2 does not have slope -4." This sentence is ambiguous because without some further explanation, it is not clear if the phrase "It is not the case" refers to the entire remainder of the sentence or to just "L_1 has slope 3/5."

Ambiguities like the one above are sometimes allowable in English but can cause trouble in mathematics. To avoid ambiguities, you should use delimiters, such as parentheses ( ), square brackets [ ], and braces { }.

To avoid writing large numbers of delimiters, we use the following rules, which we refer to as the *hierarchy of connectives*.

First, $\sim$ always is applied to the smallest proposition following it.
Then, $\wedge$ always connects the smallest propositions surrounding it.
Finally, $\vee$ connects the smallest propositions surrounding it.

Thus, $\sim P \vee Q$ is an abbreviation for $(\sim P) \vee Q$, but $\sim(P \vee Q)$ is the only way to write the negation of $P \vee Q$. Here are some other examples:

$P \vee Q \wedge R$ abbreviates $P \vee (Q \wedge R)$.
$P \wedge \sim Q \vee \sim R$ abbreviates $[P \wedge (\sim Q)] \vee (\sim R)$.
$\sim P \vee \sim Q$ abbreviates $(\sim P) \vee (\sim Q)$.
$\sim P \wedge \sim R \vee \sim P \wedge R$ abbreviates $[(\sim P) \wedge (\sim R)] \vee [(\sim P) \wedge R]$.

When the same connective is used several times in succession, parentheses may be omitted. We reinsert parentheses from the left, so that $P \vee Q \vee R$ is really $(P \vee Q) \vee R$. For example, $R \wedge P \wedge \sim P \wedge Q$ abbreviates $[(R \wedge P) \wedge (\sim P)] \wedge Q$, whereas $R \vee P \wedge \sim P \vee Q$, which does not use the same connective consecutively, abbreviates $(R \vee [P \wedge (\sim P)]) \vee Q$. Leaving out parentheses is not required; some propositional forms are much easier to read with a few well-chosen "unnecessary" parentheses.

Exercises 1.1

1. Use your knowledge of number systems to determine whether each is true or false.
   (a) 11 is a rational number.
   ★ (b) 5π is a rational number.
   (c) There are exactly 3 prime numbers between 40 and 50.
   (d) There are exactly 5 prime numbers less than 10.
   (e) 29 is a composite number.
   (f) 0 is a natural number.
   ★ (g) $(5 + 2i)(5 - 2i)$ is a real number.
   (h) 18 is a multiple of 12.

2. Which of the following are propositions? Give the truth value of each proposition.
   (a) What time is dinner?
   (b) It is not the case that $5 + \pi$ is not a rational number.
   ★ (c) $x/2$ is a rational number.
   (d) $2x + 3y$ is a real number.
   (e) Either $3 + \pi$ is rational or $3 - \pi$ is rational.
   ★ (f) Either 2 is rational and π is irrational, or 2π is rational.
   (g) Either 5π is rational and 4.9 is rational, or 3π is rational.
   (h) $-\frac{1}{2}$ is rational, and either $3\pi < 10$ or $3\pi > 15$.
   (i) It is not the case that 39 is prime, or that 64 is a power of 2.
   (j) There are more than three false statements in this book and this statement is one of them.

3. Make truth tables for each of the following propositional forms.
   ★ (a) $P \wedge \sim P$.   (b) $P \vee \sim P$.
   ★ (c) $P \wedge \sim Q$.   (d) $P \wedge (Q \vee \sim Q)$.
   ★ (e) $(P \wedge Q) \vee \sim Q$.   (f) $\sim(P \wedge Q)$.
   (g) $(P \vee \sim Q) \wedge R$.   (h) $\sim P \wedge \sim Q$.

⋆ **(i)** $P \wedge (Q \vee R)$.   **(j)** $(P \wedge Q) \vee (P \wedge R)$.

(k) $P \wedge P$.   **(l)** $(P \wedge Q) \vee (R \wedge \sim S)$.

4. If P, Q, and R are true while S and T are false, which of the following are true?

⋆ **(a)** $Q \wedge (R \wedge S)$.   **(b)** $Q \vee (R \wedge S)$.

⋆ **(c)** $(P \vee Q) \wedge (R \vee S)$.   **(d)** $(\sim P \vee \sim Q) \vee (\sim R \vee \sim S)$.

(e) $\sim P \vee \sim Q$.   ⋆ **(f)** $(\sim Q \vee S) \wedge (Q \vee S)$.

⋆ **(g)** $(P \vee S) \wedge (P \vee T)$.

5. Use truth tables to prove the remaining parts of Theorem 1.1.1.

6. Which of the following pairs of propositional forms are equivalent?

⋆ **(a)** $P \wedge P, P$.   **(b)** $P \vee P, P$.

⋆ **(c)** $P \wedge Q, Q \wedge P$.   **(d)** $(\sim P) \vee (\sim Q), \sim (P \vee \sim Q)$.

⋆ **(e)** $\sim P \wedge \sim Q, \sim (P \wedge \sim Q)$.   **(f)** $\sim (P \wedge Q), \sim P \wedge \sim Q$.

⋆ **(g)** $(P \wedge Q) \vee R, P \wedge (Q \vee R)$.   **(h)** $(P \wedge Q) \vee R, P \vee (Q \wedge R)$.

7. Determine the propositional form and truth value for each of the following:

(a) It is not the case that 2 is odd.

(b) $f(x) = e^x$ is increasing and concave up.

(c) Both 7 and 5 are factors of 70.

(d) Perth or Panama City or Pisa is located in Europe.

8. P, Q, and R are propositional forms, and P is equivalent to Q, and Q is equivalent to R. Prove that

⋆ **(a)** Q is equivalent to P.

(b) P is equivalent to R.

(c) $\sim Q$ is equivalent to $\sim P$.

9. Use a truth table to determine whether each of the following is a tautology, a contradiction, or neither.

(a) $(P \wedge Q) \vee (\sim P \wedge \sim Q)$.

(b) $\sim (P \wedge \sim P)$.

⋆ **(c)** $(P \wedge Q) \vee (\sim P \vee \sim Q)$.

(d) $(A \wedge B) \vee (A \wedge \sim B) \vee (\sim A \wedge B) \vee (\sim A \wedge \sim B)$.

(e) $(Q \wedge \sim P) \wedge \sim (P \wedge R)$.

(f) $P \vee [(\sim Q \wedge P) \wedge (R \vee Q)]$.

10. Suppose A is a tautology and B is a contradiction. Are the following tautologies, contradictions, or neither?

⋆ **(a)** $A \wedge B$.   **(b)** $A \wedge \sim B$.

⋆ **(c)** $A \vee B$.   **(d)** $\sim (\sim A \wedge B)$.

11. Give a useful denial of each statement.

⋆ **(a)** x is a positive integer. (Assume that x is some fixed integer.)

(b) Cleveland will win the first game or the second game.

⋆ **(c)** $5 \geq 3$.

(d) 641,371 is a composite integer.

⋆ **(e)** Roses are red and violets are blue.

(f) T is not bounded or T is compact. (Assume that T is a fixed object.)

(g) M is odd and one-to-one. (Assume that M is some fixed function.)

(h)   The function f has positive first and second derivatives at x_0. (Assume that f is a fixed function and x_0 is a fixed real number.)

(i)   The function g has a relative maximum at $x = 2$ or $x = 4$ and a relative minimum at $x = 3$. (Assume that g is a fixed function.)

(j)   Neither $z < s$ nor $z \leq t$ is true. (Assume that z, s, and t are fixed real numbers.)

(k) R is transitive but not reflexive. (Assume that R is a fixed object.)

12.   Restore parentheses to these abbreviated propositional forms.

(a) $\sim\sim P \vee \sim Q \wedge \sim S$.

(b) $Q \wedge \sim S \vee \sim(\sim P \wedge Q)$.

(c) $P \wedge \sim Q \ \vee \sim P \wedge \sim R \vee \sim P \wedge S$.

(d) $\sim P \vee Q \wedge \sim\sim P \wedge Q \vee R$.

13.   Other logical connectives between two propositions P and Q are possible.

(a)   The word *or* is used in two different ways in English. We have presented the truth table for $\vee$, the **inclusive or,** whose meaning is "one or the other or both." The **exclusive or,** meaning "one or the other but not both" and denoted $\varovee$, has its uses in English, as in "She will marry Heckle or she will marry Jeckle." The "inclusive or" is much more useful in mathematics and is the accepted meaning unless there is a statement to the contrary.

★   (i)   Make a truth table for the "exclusive or" connective $\varovee$.

(ii)   Show that $A \varovee B$ is equivalent to $(A \vee B) \wedge \sim(A \wedge B)$.

(b)   "NAND" and "NOR" circuits are commonly used as a basis for flash memory chips. A NAND B is defined to be the negation of "A and B." A NOR B is defined to be the negation of "A or B."

(i)   Write truth tables for NAND and NOR connectives.

(ii)   Show that $(A$ NAND $B) \vee (A$ NOR $B)$ is equivalent to $(A$ NAND $B)$.

(iii)   Show that $(A$ NAND $B) \wedge (A$ NOR $B)$ is equivalent to $(A$ NOR $B)$.

1.2   Conditionals and Biconditionals

Sentences of the form "If P, then Q" are the most important kind of propositions in mathematics. You have seen many examples of such statements in mathematics courses: from precalculus, "If two lines in a plane have the same slope, then the lines are parallel"; from trigonometry, "If $\sec \theta = \frac{5}{3}$, then $\sin \theta = \frac{4}{5}$."; from calculus, "If f is differentiable at x_0 and $f(x_0)$ is a relative minimum for f, then $f'(x_0) = 0$."

DEFINITIONS   For propositions P and Q, the **conditional sentence** $P \Rightarrow Q$ is the proposition "If P, then Q." Proposition P is called the **antecedent** and Q is the **consequent.** The conditional sentence $P \Rightarrow Q$ is true if and only if P is false or Q is true.

The truth table for $P \Rightarrow Q$ is

P	Q	$P \Rightarrow Q$
T	T	T
F	T	T
T	F	F
F	F	T

According to this table, there is only one way that $P \Rightarrow Q$ can be false: when P is true and Q is false. Thus, this truth table agrees with the way we understand promises: the only situation where a promise is broken is when the antecedent is true but the person making the promise fails to make the consequent true.

Example. Suppose someone says to a friend "If the concert is sold out, I'll take you sailing." This promise is broken (the conditional sentence is false) only when the concert was sold out (the antecedent is true) and the person who made the promise did not take the other person sailing (the consequent is false). This is line 3 of the truth table. In all other situations, the promise is true. If there were tickets left (lines 2 and 4 of the table), we don't say the promise was broken, regardless of whether the friends decided to go sailing. The promise is also kept in the situation where the concert is sold out and the friends went sailing, which is line 1 of the table.

One curious consequence of the truth table for $P \Rightarrow Q$ is that a conditional sentence may be true even when there is no connection between the antecedent and the consequent. The reason for this is that the truth value of $P \Rightarrow Q$ depends *only* on the truth value of components P and Q, not on their interpretation. For this reason all of the following are true:

"If $\sin \pi = 1$, then 6 is prime." (line 4 of the truth table)
"$13 > 7 \Rightarrow 2 + 3 = 5$." (line 1 of the truth table)
"$\pi = 3 \Rightarrow$ Paris is the capital of France." (line 2 of the truth table)

and both of these are false by line 3 of the truth table:

"If Saturn has rings, then $(2 + 3)^2 = 2^2 + 3^2$."
"If $4\pi > 10$, then 1 is a prime number."

Other consequences of the truth table for $P \Rightarrow Q$ are worth noting. When P is false, it doesn't matter what truth value Q has: $P \Rightarrow Q$ will be true by lines 2 and 4. When Q is true, it doesn't matter what truth value P has: $P \Rightarrow Q$ will be true by lines 1 and 2. Finally, when P and $P \Rightarrow Q$ are both true (on line 1), Q must also be true.

Example. Both propositions

"If Isaac Newton was born in 1642, then $3 \cdot 5 = 15$"
"If Isaac Newton was born in 1643, then $3 \cdot 5 = 15$"

are true because the consequent "$3 \cdot 5 = 15$" is true.

Our truth table definition for $P \Rightarrow Q$ captures the same meaning for "If ...,
then ..." that you have always used in mathematics. For example, if we think of x
as some fixed real number, we all know that

"If $x > 8$, then $x > 5$"

is a true statement, no matter what number x we have in mind. Let's examine why
we say this sentence is true for some specific values of x, where the antecedent P is
"$x > 8$" and the consequent Q is "$x > 5$."

In the case $x = 11$, both P and Q are true, as in line 1 of the truth table. The case
$x = 7$ corresponds to the second line of the table, and for $x = 3$ we have the situation in
line 4. There is no case corresponding to line 3 because $P \Rightarrow Q$ is true. Note that when
we say "If P, then Q" is true, we don't claim that either P or Q is true. What we do say
is that no matter what number we think of, *if* it's larger than 8, it's also larger than 5.

Two propositions closely related to $P \Rightarrow Q$ are its converse and contrapositive.

DEFINITION   Let P and Q be propositions.
The **converse** of $P \Rightarrow Q$ is $Q \Rightarrow P$.
The **contrapositive** of $P \Rightarrow Q$ is $(\sim Q) \Rightarrow (\sim P)$.

For the conditional sentence "If π is an integer, then 14 is even," the converse
of the sentence is "If 14 is even, then π is an integer" and the contrapositive is "If
14 is not even, then π is not an integer." The converse is false, but the sentence and
its contrapositive are true.

For the sentence "If $1 + 1 = 2$, then $\sqrt{10} > 3$," the converse and contrapositive are, respectively, "If $\sqrt{10} > 3$, then $1 + 1 = 2$ " and "If $\sqrt{10}$ is not greater
than 3, then $1 + 1$ is not equal to 2." In this example, all three sentences are true.

The previous two examples suggest that the truth values of a conditional sentence and its contrapositive are related, but there seems to be little connection
between the truth values of $P \Rightarrow Q$ and its converse. We describe the relationships
in the following theorem.

Theorem 1.2.1

For propositions P and Q,

(a) $P \Rightarrow Q$ is equivalent to its contrapositive $(\sim Q) \Rightarrow (\sim P)$.
(b) $P \Rightarrow Q$ is *not* equivalent to its converse $Q \Rightarrow P$.

Proof.   The proofs are carried out by examination of the truth tables.

P	Q	$P \Rightarrow Q$	$\sim P$	$\sim Q$	$(\sim Q) \Rightarrow (\sim P)$	$Q \Rightarrow P$
T	T	T	F	F	T	T
F	T	T	T	F	T	F
T	F	F	F	T	F	T
F	F	T	T	T	T	T

(a) $P \Rightarrow Q$ is equivalent to $(\sim Q) \Rightarrow (\sim P)$ because the third column in the truth table is identical to the sixth column in the table.

(b) $P \Rightarrow Q$ is not equivalent to $Q \Rightarrow P$ because column 3 in the truth table differs from column 7 in rows 2 and 3.   ∎

We have seen cases where a conditional sentence and its converse have the same truth value. Theorem 1.2.1(b) simply says that this need not always be the case—the truth values of $P \Rightarrow Q$ cannot be inferred from its converse $Q \Rightarrow P$.

The next connective we need is the biconditional connective $\Leftrightarrow$. The double arrow $\Leftrightarrow$ reminds one of both $\Leftarrow$ and $\Rightarrow$, and this is no accident, because $P \Leftrightarrow Q$ is equivalent to $(P \Rightarrow Q) \wedge (Q \Rightarrow P)$.

DEFINITION   For propositions P and Q, the **biconditional sentence** $P \Leftrightarrow Q$ is the proposition "P if and only if Q." $P \Leftrightarrow Q$ is true exactly when P and Q have the same truth values. We also write P iff Q to abbreviate P if and only if Q.

The truth table for $P \Leftrightarrow Q$ is

P	Q	$P \Leftrightarrow Q$
T	T	T
F	T	F
T	F	F
F	F	T

Examples.   The proposition "$2^3 = 8$ iff 49 is a perfect square" is true because both components are true. The proposition "$\pi = 22/7$ iff $\sqrt{2}$ is a rational number" is true because both components are false. The proposition "$6 + 1 = 7$ iff Lake Michigan is in Kansas" is false because the truth values of the components differ.

Definitions, fully stated with the "if and only if" connective, are important examples of biconditional sentences because they describe exactly the condition(s) to meet the definition. Although sometimes a definition does not explicitly use the iff wording, biconditionality does provide a good test of whether a statement could serve as a definition or just a description.

Example.   The statement "Vertical lines have undefined slope" could be used as a definition because a line is vertical iff its slope is undefined. However, "A zebra is a striped animal" is not a definition, because the sentence "An animal is a zebra iff the animal is striped" is false.

Because the biconditional sentence $P \Leftrightarrow Q$ is true exactly when the truth values of P and Q agree, we can use the biconditional connective to restate the meaning of equivalent propositional forms:

The propositional forms P and Q are equivalent precisely when P $\Leftrightarrow$ Q is a tautology.

Thus each statement in Theorem 1.1.1 may be restated using the $\Leftrightarrow$ connective. For example, DeMorgan's Laws are:

$$\sim(P \wedge Q) \Leftrightarrow (\sim P \vee \sim Q) \text{ and}$$
$$\sim(P \vee Q) \Leftrightarrow (\sim P \wedge \sim Q).$$

All of the statements in Theorem 1.1.1 are used regularly in proofs. The next theorem contains several additional important pairs of equivalent propositional forms that involve implication. They, too, will be used often.

Theorem 1.2.2   For propositions P, Q, and R,

    (a) $P \Rightarrow Q$ is equivalent to $\sim P \vee Q$.
    (b) $P \Leftrightarrow Q$ is equivalent to $(P \Rightarrow Q) \wedge (Q \Rightarrow P)$.
    (c) $\sim(P \Rightarrow Q)$ is equivalent to $P \wedge \sim Q$.
    (d) $\sim(P \wedge Q)$ is equivalent to $P \Rightarrow \sim Q$ and to $Q \Rightarrow \sim P$.
    (e) $P \Rightarrow (Q \Rightarrow R)$ is equivalent to $(P \wedge Q) \Rightarrow R$.
    (f) $P \Rightarrow (Q \wedge R)$ is equivalent to $(P \Rightarrow Q) \wedge (P \Rightarrow R)$.
    (g) $(P \vee Q) \Rightarrow R$ is equivalent to $(P \Rightarrow R) \wedge (Q \Rightarrow R)$.

Exercise 8 asks you to prove each part of Theorem 1.2.2. The natural way to proceed is by constructing and then comparing truth tables, but you should also think about the meaning of both sides of each statement of equivalence. With part (a), for example, we reason as follows: $P \Rightarrow Q$ is false exactly when P is true and Q is false, which happens exactly when both $\sim P$ and Q are false. Since this happens exactly when $\sim P \vee Q$ is false, the truth tables for $P \Rightarrow Q$ and $\sim P \vee Q$ are identical.

Note that many of the statements in Theorems 1.1.1 and 1.2.2 are related. For example, once we have established Theorem 1.1.1 and 1.2.2(a), we reason that part (c) is correct as follows:

$\sim(P \Rightarrow Q)$ is equivalent, by part (a), to
$\sim(\sim P \vee Q)$, which is equivalent, by Theorem 1.1.1(i), to
$\sim(\sim P) \wedge \sim Q$, which is equivalent, by Theorem 1.1.1(a), to
$P \wedge \sim Q$.

Recognizing the structure of a sentence and translating the sentence into symbolic form using logical connectives are aids in determining its truth or falsity. The translation of sentences into propositional symbols is sometimes very complicated because some natural languages such as English are rich and powerful with many nuances. The ambiguities that we tolerate in English would destroy structure and usefulness if we allowed them in mathematics.

Even the translations of simple sentences can present special problems. Suppose a teacher says to a student

"If you score 74% or higher on the next test, you will pass this course."

This sentence clearly has the form of a conditional sentence, although almost everyone will interpret the meaning as a biconditional.

Contrast this with the situation in mathematics where "If $x = 2$, then x is a solution to $x^2 = 2x$" must have only the meaning of the connective $\Rightarrow$, because $x^2 = 2x$ does not imply $x = 2$.

Shown below are some phrases in English that are ordinarily translated by using the connectives $\Rightarrow$ or $\Leftrightarrow$. In the accompanying examples, think of a and t as fixed real numbers.

Use $P \Rightarrow Q$ to translate: Examples:

If P, then Q.	If $a > 5$, then $a > 3$.
P implies Q.	$a > 5$ implies $a > 3$.
P is sufficient for Q.	$a > 5$ is sufficient for $a > 3$.
P only if Q.	$a > 5$ only if $a > 3$.
Q, if P.	$a > 3$, if $a > 5$.
Q whenever P.	$a > 3$ whenever $a > 5$.
Q is necessary for P.	$a > 3$ is necessary for $a > 5$.
Q, when P.	$a > 3$, when $a > 5$.

Use $P \Leftrightarrow Q$ to translate: Examples:

P if and only if Q.	$\lvert t \rvert = 2$ if and only if $t^2 = 4$.
P if, but only if, Q.	$\lvert t \rvert = 2$ if, but only if, $t^2 = 4$.
P is equivalent to Q.	$\lvert t \rvert = 2$ is equivalent to $t^2 = 4$.
P is necessary and sufficient for Q.	$\lvert t \rvert = 2$ is necessary and sufficient for $t^2 = 4$.

The word *unless* is one of those connective words in English that poses special problems because it has so many different interpretations. See Exercise 11.

Examples.   In these sentence translations, we assume that S, G, and e have been specified. It is not necessary to know the meanings of all the words because the form of the sentence is sufficient to determine the correct translation.

"S is compact is sufficient for S to be bounded" is translated

$$S \text{ is compact} \Rightarrow S \text{ is bounded.}$$

"A necessary condition for a group G to be cyclic is that G is abelian" is translated

$$G \text{ is cyclic} \Rightarrow G \text{ is abelian.}$$

"A set S is infinite if S has an uncountable subset" is translated

$$S \text{ has an uncountable subset} \Rightarrow S \text{ is infinite.}$$

"A necessary and sufficient condition for the graph G to be a tree is that G is connected and every edge of G is a bridge" is translated

$$G \text{ is a tree} \Leftrightarrow (G \text{ is connected} \wedge \text{ every edge of } G \text{ is a bridge}).$$

Example.   If we let P denote the proposition "Roses are red" and Q denote the proposition "Violets are blue," we can translate the sentence "It is not the case that roses are red, nor that violets are blue" in at least two ways: $\sim(P \vee Q)$ or $(\sim P) \wedge (\sim Q)$. Fortunately, these are equivalent by Theorem 1.1.1(h). Note that the proposition "Violets are purple" requires a new symbol, say R, since it expresses a new idea that cannot be formed from the components P and Q.

The sentence "17 and 35 have no common divisors" shows that the meaning, and not just the form of the sentence, must be considered in translating; it cannot be broken up into the two propositions: "17 has no common divisors" and "35 has no common divisors." Compare this with the proposition "17 and 35 have digits totaling 8," which can be written as a conjunction.

Example. Suppose b is a fixed real number. The form of the sentence "If b is an integer, then b is either even or odd" is $P \Rightarrow (Q \vee R)$, where P is "b is an integer," Q is "b is even," and R is "b is odd."

Example. Suppose a, b, and p are fixed integers. "If p is a prime number that divides ab, then p divides a or b" has the form $(P \wedge Q) \Rightarrow (R \vee S)$, where P is "p is a prime," Q is "p divides ab," R is "p divides a," and S is "p divides b."

The hierarchy of connectives in Section 1.1 that governs the use of parentheses for propositional forms can be extended to the connectives $\Rightarrow$ and $\Leftrightarrow$:

The connectives $\sim$, $\wedge$, $\vee$, $\Rightarrow$, and $\Leftrightarrow$ are always applied in the order listed.

Thus, $\sim$ applies to the smallest possible proposition, then $\wedge$ is applied with the next smallest scope, and so forth. For example,

$$P \Rightarrow \sim Q \vee R \Leftrightarrow S \text{ is an abbreviation for } (P \Rightarrow [(\sim Q) \vee R]) \Leftrightarrow S,$$
$$P \vee \sim Q \Leftrightarrow R \Rightarrow S \text{ is an abbreviation for } [P \vee (\sim Q)] \Leftrightarrow (R \Rightarrow S),$$

and

$$P \Rightarrow Q \Rightarrow R \text{ is an abbreviation for } (P \Rightarrow Q) \Rightarrow R.$$

Exercises 1.2

1. Identify the antecedent and the consequent for each of the following conditional sentences. Assume that a, b, and f represent some fixed sequence, integer, or function, respectively.
   ⋆ **(a)** If squares have three sides, then triangles have four sides.
   (b) If the moon is made of cheese, then 8 is an irrational number.
   (c) b divides 3 only if b divides 9.
   ⋆ **(d)** The differentiability of f is sufficient for f to be continuous.
   (e) A sequence a is bounded whenever a is convergent.
   ⋆ **(f)** A function f is bounded if f is integrable.
   (g) $1 + 2 = 3$ is necessary for $1 + 1 = 2$.

  (h)   The fish bite only when the moon is full.
★ (i)   A time of 3 minutes, 48 seconds or less is necessary to qualify for the Olympic team.

☆ **2.** Write the converse and contrapositive of each conditional sentence in Exercise 1.

3. What can be said about the truth value of Q when
  (a) P is false and $P \Rightarrow Q$ is true?   (b) P is true and $P \Rightarrow Q$ is true?
  (c) P is true and $P \Rightarrow Q$ is false?   (d) P is false and $P \Leftrightarrow Q$ is true?
  (e) P is true and $P \Leftrightarrow Q$ is false?

4. Identify the antecedent and consequent for each conditional sentence in the following statements from this book.
  (a) Theorem 1.3.1(a)   (b) Exercise 3 of Section 1.6
  (c) Theorem 2.1.4   (d) The PMI, Section 2.4
  (e) Theorem 2.6.4   (f) Theorem 3.4.2
  (g) Theorem 4.2.2   (h) Theorem 5.1.7(a)

5. Which of the following conditional sentences are true?
★ (a) If triangles have three sides, then squares have four sides.
  (b) If a hexagon has six sides, then the moon is made of cheese.
★ (c) If $7 + 6 = 14$, then $5 + 5 = 10$.
  (d) If $5 < 2$, then $10 < 7$.
★ (e) If one interior angle of a right triangle is $92°$, then the other interior angle is $88°$.
  (f) If Euclid's birthday was April 2, then rectangles have four sides.
  (g) 5 is prime if $\sqrt{2}$ is not irrational.
  (h) $1 + 1 = 2$ is sufficient for $3 > 6$.

6. Which of the following are true?
★ (a) Triangles have three sides iff squares have four sides.
  (b) $7 + 5 = 12$ iff $1 + 1 = 2$.
★ (c) b is even iff $b + 1$ is odd. (Assume that b is some fixed integer.)
  (d) m is odd iff m^2 is odd. (Assume that m is some fixed integer.)
  (e) $5 + 6 = 6 + 5$ iff $7 + 1 = 10$.
  (f) A parallelogram has three sides iff 27 is prime.
  (g) The Eiffel Tower is in Paris if and only if the chemical symbol for helium is H.
  (h) $\sqrt{10} + \sqrt{13} < \sqrt{11} + \sqrt{12}$ iff $\sqrt{13} - \sqrt{12} < \sqrt{11} - \sqrt{10}$.
  (i) $x^2 \geq 0$ iff $x \geq 0$. (Assume that x is a fixed real number.)
  (j) $x^2 - y^2 = 0$ iff $(x - y)(x + y) = 0$. (Assume that x and y are fixed real numbers.)
  (k) $x^2 + y^2 = 50$ iff $(x + y)^2 = 50$. (Assume that x and y are fixed real numbers.)

7. Make truth tables for these propositional forms.
  (a) $P \Rightarrow (Q \wedge P)$.   ★ (b) $(\sim P \Rightarrow Q) \vee (Q \Leftrightarrow P)$.
★ (c) $\sim Q \Rightarrow (Q \Leftrightarrow P)$.   (d) $(P \vee Q) \Rightarrow (P \wedge Q)$.
  (e) $(P \wedge Q) \vee (Q \wedge R) \Rightarrow P \vee R$.
  (f) $[(Q \Rightarrow S) \wedge (Q \Rightarrow R)] \Rightarrow [(P \vee Q) \Rightarrow (S \vee R)]$.

8. Prove Theorem 1.2.2 by constructing truth tables for each equivalence.

9. Determine whether each statement qualifies as a definition.
   (a) $y = f(x)$ is a linear function when its graph is a straight line.
   (b) $y = f(x)$ is a quadratic function when it contains an x^2 term.
   (c) m is a perfect square when $m = n^2$ for some integer n.
   (d) A triangle is a right triangle when the sum of two of its interior angles is $90°$.
   (e) Two lines are parallel when their slopes are the same number.
   (f) A sundial is an instrument for measuring time.

10. Rewrite each of the following sentences using logical connectives. Assume that each symbol f, x_0, n, x, S, **B** represents some fixed object.
   ⋆ (a) If f has a relative minimum at x_0 and if f is differentiable at x_0, then $f'(x_0) = 0$.
   (b) If n is prime, then $n = 2$ or n is odd.
   (c) A number x is real and not rational whenever x is irrational.
   ⋆ (d) If $x = 1$ or $x = -1$, then $|x| = 1$.
   ⋆ (e) f has a critical point at x_0 iff $f'(x_0) = 0$ or $f'(x_0)$ does not exist.
   (f) S is compact iff S is closed and bounded.
   (g) **B** is invertible is a necessary and sufficient condition for det **B** $\neq 0$.
   (h) $6 \geq n - 3$ only if $n > 4$ or $n > 10$.
   (i) x is Cauchy implies x is convergent.
   (j) f is continuous at x_0 whenever $\lim\limits_{x \to x_0} f(x) = f(x_0)$.
   (k) If f is differentiable at x_0 and f is strictly increasing at x_0, then $f'(x_0) > 0$.

11. Dictionaries indicate that the conditional meaning of *unless* is preferred, but there are other interpretations as a converse or a biconditional. Discuss the translation of each sentence.
   (a) I will go to the store unless it is raining.
   ⋆ (b) The Dolphins will not make the playoffs unless the Bears win all the rest of their games.
   (c) You cannot go to the game unless you do your homework first.
   (d) You won't win the lottery unless you buy a ticket.

12. Show that the following pairs of statements are equivalent.
   (a) $(P \vee Q) \Rightarrow R$ and $\sim R \Rightarrow (\sim P \wedge \sim Q)$.
   ⋆ (b) $(P \wedge Q) \Rightarrow R$ and $(P \wedge \sim R) \Rightarrow \sim Q$.
   (c) $P \Rightarrow (Q \wedge R)$ and $(\sim Q \vee \sim R) \Rightarrow \sim P$.
   (d) $P \Rightarrow (Q \vee R)$ and $(P \wedge \sim R) \Rightarrow Q$.
   (e) $(P \Rightarrow Q) \Rightarrow R$ and $(P \wedge \sim Q) \vee R$.
   (f) $P \Leftrightarrow Q$ and $(\sim P \vee Q) \wedge (\sim Q \vee P)$.

13. Give, if possible, an example of a true conditional sentence for which
   ⋆ (a) the converse is true.   (b) the converse is false.
   ⋆ (c) the contrapositive is false.   (d) the contrapositive is true.

14. Give, if possible, an example of a false conditional sentence for which
   (a) the converse is true.   (b) the converse is false.
   (c) the contrapositive is false.   (d) the contrapositive is true.

15. Give the converse and contrapositive of each sentence of Exercises 10(a), (b), (c), and (d). Tell whether each converse and contrapositive is true or false.

16. Determine whether each of the following is a tautology, a contradiction, or neither.
  * **(a)** $[(P \Rightarrow Q) \Rightarrow P] \Rightarrow P$.
  (b) $P \Leftrightarrow P \wedge (P \vee Q)$.
  (c) $P \Rightarrow Q \Leftrightarrow P \wedge \sim Q$.
  * **(d)** $P \Rightarrow [P \Rightarrow (P \Rightarrow Q)]$.
  (e) $P \wedge (Q \vee \sim Q) \Leftrightarrow P$.
  (f) $[Q \wedge (P \Rightarrow Q)] \Rightarrow P$.
  (g) $(P \Leftrightarrow Q) \Leftrightarrow \sim(\sim P \vee Q) \vee (\sim P \wedge Q)$.
  (h) $[P \Rightarrow (Q \vee R)] \Rightarrow [(Q \Rightarrow R) \vee (R \Rightarrow P)]$.
  (i) $P \wedge (P \Leftrightarrow Q) \wedge \sim Q$.
  (j) $(P \vee Q) \Rightarrow Q \Rightarrow P$.
  (k) $[P \Rightarrow (Q \wedge R)] \Rightarrow [R \Rightarrow (P \Rightarrow Q)]$.
  (l) $[P \Rightarrow (Q \wedge R)] \Rightarrow R \Rightarrow (P \Rightarrow Q)$.

17. The **inverse**, or **opposite**, of the conditional sentence $P \Rightarrow Q$ is $\sim P \Rightarrow \sim Q$.
  (a) Show that $P \Rightarrow Q$ and its inverse are not equivalent forms.
  (b) For what values of the propositions P and Q are $P \Rightarrow Q$ and its inverse both true?
  (c) Which is equivalent to the converse of a conditional sentence, the contrapositive of its inverse, or the inverse of its contrapositive?

1.3   Quantifiers

Unless there has been a prior agreement about the value of x, the statement "$x \geq 3$" is neither true nor false. A sentence that contains variables is called an **open sentence** or **predicate**, and becomes a proposition only when its variables are assigned specific values. For example, "$x \geq 3$" is true when x is given the value 7 and false when $x = 2$.

When P is an open sentence with a variable x, the sentence is symbolized by $P(x)$. Likewise, if P has variables $x_1, x_2, x_3, \ldots, x_n$, the sentence may be denoted by $P(x_1, x_2, x_3, \ldots, x_n)$. For example, for the sentence "$x + y = 3z$" we write $P(x, y, z)$, and we see that $P(4, 5, 3)$ is true because $4 + 5 = 3(3)$, while $P(1, 2, 4)$ is false.

The collection of objects that may be substituted to make an open sentence a true proposition is called the **truth set** of the sentence. Before a truth set can be determined, we must be given or must decide what objects are available for consideration; that is, we must have specified a **universe of discourse.** In many cases the universe will be understood from the context. For a sentence such as "x likes chocolate," the universe is presumably the set of all people. We will often use the number systems $\mathbb{N}, \mathbb{Z}, \mathbb{Q}, \mathbb{R}$, and $\mathbb{C}$ as our universes. (See the *Preface to the Student*.)

Example. The truth set of the open sentence "$x^2 < 5$" depends upon the collection of objects we choose for the universe of discourse. With the universe specified as the set $\mathbb{N}$, the truth set is $\{1, 2\}$. For the universe $\mathbb{Z}$, the truth set is $\{-2, -1, 0, 1, 2\}$. When the universe is $\mathbb{R}$, the truth set is the open interval $(-\sqrt{5}, \sqrt{5})$.

> **DEFINITION**   With a universe specified, two open sentences $P(x)$ and $Q(x)$ are **equivalent** iff they have the same truth set.

Examples.   The sentences "$3x + 2 = 20$" and "$x = 6$" are equivalent open sentences in any of the number systems we have named. On the other hand, "$x^2 = 4$" and "$x = 2$" are *not* equivalent when the universe is $\mathbb{R}$. They *are* equivalent when the universe is $\mathbb{N}$.

The notions of truth set, universe, and equivalent open sentences should not be new concepts for you. Solving an equation such as $(x^2 + 1)(x - 3) = 0$ is a matter of determining what objects x make the open sentence "$(x^2 + 1)(x - 3) = 0$" true. For the universe $\mathbb{R}$, the only solution is $x = 3$ and thus the truth set is $\{3\}$. But if we choose the universe to be $\mathbb{C}$, the equation may be replaced by the equivalent open sentence $(x + i)(x - i)(x - 3) = 0$, which has truth set (solutions) $\{3, i, -i\}$.

A sentence such as

"There is a prime number between 5060 and 5090"

is treated differently from the propositions we considered earlier. To determine whether this sentence is true in the universe $\mathbb{N}$, we might try to individually examine every natural number, checking whether it is a prime and between 5060 and 5090, until we eventually find any *one* of the primes 5077, 5081, or 5087 and conclude that the sentence is true. (A quicker way is to search through a complete list of the first thousand primes.) The key idea here is that although the open sentence "x is a prime number between 5060 and 5090" is not a proposition, the sentence

"There is a number x such that x is a prime number between 5060 and 5090"

does have a truth value. This sentence is formed from the original open sentence by applying a quantifier.

> **DEFINITION**   For an open sentence $P(x)$, the sentence $(\exists x)P(x)$ is read "There exists x such that $P(x)$" or "For some x, $P(x)$." The sentence $(\exists x)P(x)$ is true iff the truth set of $P(x)$ is nonempty. The symbol $\exists$ is called the **existential quantifier**.

An open sentence $P(x)$ does not have a truth value, but the quantified sentence $(\exists x)P(x)$ does. One way to show that $(\exists x)P(x)$ is true for a particular universe is to identify an object a in the universe such that the proposition $P(a)$ is true. To show $(\exists x)P(x)$ is false, we must show that the truth set of $P(x)$ is empty.

Examples.   Let's examine the truth values of these statements for the universe $\mathbb{R}$:

(a) $(\exists x)(x \geq 3)$
(b) $(\exists x)(x^2 = 0)$
(c) $(\exists x)(x^2 = -1)$

Statement (a) is true because the truth set of $x \geq 3$ contains 3, 7.02, and many other real numbers. Thus the truth set contains at least one real number. Statement (b) is true because the truth set of $x^2 = 0$ is precisely $\{0\}$ and thus is nonempty. Since the open sentence $x^2 = -1$ is never true for real numbers, the truth set of $x^2 = -1$ is empty. Statement (c) is false.

In the universe $\mathbb{N}$, only statement (a) is true. The three statements are all true in the universe $\{0, 5, i\}$ and all three statements are false in the universe $\{1, 2\}$.

Sometimes we can say $(\exists x) P(x)$ is true even when we do not know a specific object in the universe in the truth set of $P(x)$, only that there (at least) is one.

Example.   Show that $(\exists x)(x^7 - 12x^3 + 16x - 3 = 0)$ is true in the universe of real numbers.

For the polynomial $f(x) = x^7 - 12x^3 + 16x - 3$, $f(0) = -3$ and $f(1) = 2$. From calculus, we know that f is continuous on $[0, 1]$. The Intermediate Value Theorem tells us there is a zero for f between 0 and 1. Even if we don't know the exact value of the zero, we know it exists. Therefore, the truth set of $x^7 - 12x^3 + 16x - 3 = 0$ is nonempty. Hence $(\exists x)(x^7 - 12x^3 + 16x - 3 = 0)$ is true.

The sentence "The square of every number is greater than 3" uses a different quantifier for the open sentence "$x^2 > 3$." To decide the truth value of the given sentence in the universe $\mathbb{N}$ it is not enough to observe that $3^2 > 3$, $4^2 > 3$, and so on. In fact, the sentence is false in $\mathbb{N}$ because 1 is in the universe but not in the truth set. The sentence is true, however, in the universe $[1.74, \infty)$ because with this universe the truth set for $x^2 > 3$ is the same as the universe.

DEFINITION   For an open sentence $P(x)$, the sentence $(\forall x) P(x)$ is read "For all x, $P(x)$" and is true iff the truth set of $P(x)$ is the *entire* universe. The symbol $\forall$ is called the **universal quantifier**.

Examples.   For the universe of all real numbers,

$(\forall x)(x + 2 > x)$ is true.
$(\forall x)(x > 0 \vee x = 0 \vee x < 0)$ is true. That is, every real number is positive, zero or negative.
$(\forall x)(x \geq 3)$ is false because there are (many) real numbers x for which $x \geq 3$ is false.
$(\forall x)(|x| > 0)$ is false, because 0 is not in the truth set.

There are many ways to express a quantified sentence in English. Look for key words such as "for all," "for every," "for each," or similar words that require universal quantifiers. Look for phrases such as "some," "at least one," "there exist(s)," "there is (are)," and others that indicate existential quantifiers.

You should also be alert for hidden quantifiers because natural languages allow for imprecise quantified statements where the words "for all" and "there exists" are not present. Someone who says "Polynomial functions are continuous" means that "All polynomial functions are continuous," but someone who says "Rational functions have vertical asymptotes" must mean "Some rational functions have vertical asymptotes."

We agree that "All apples have spots" is quantified with ∀, but what form does it have? If we limit the universe to just apples, a correct symbolization would be $(\forall x)(x$ has spots$)$. But if the universe is all fruits, we need to be more careful. Let $A(x)$ be "x is an apple" and $S(x)$ be "x has spots." Should we write the sentence as $(\forall x)[A(x) \wedge S(x)]$ or $(\forall x)[A(x) \Rightarrow S(x)]$?

The first quantified form, $(\forall x)[A(x) \wedge S(x)]$, says "For all objects x in the universe, x is an apple and x has spots." Since we don't really intend to say that all fruits are spotted apples, this is not the meaning we want. Our other choice, $(\forall x)[A(x) \Rightarrow S(x)]$, is the correct one because it says "For all objects x in the universe, if x is an apple then x has spots." In other words, "If a fruit is an apple, then it has spots."

Now consider "Some apples have spots." Should this be $(\exists x)[A(x) \wedge S(x)]$ or $(\exists x)[A(x) \Rightarrow S(x)]$? The first form says "There is an object x such that it is an apple and it has spots," which is correct. On the other hand, $(\exists x)[A(x) \Rightarrow S(x)]$ reads "There is an object x such that, if it is an apple, then it has spots," which does *not* ensure the existence of apples with spots. The sentence $(\exists x)[A(x) \Rightarrow S(x)]$ is true in every universe for which there is an object x such that either x is not an apple or x has spots, which is not the meaning we want.

In general, a sentence of the form "All $P(x)$ are $Q(x)$" should be symbolized $(\forall x)[P(x) \Rightarrow Q(x)]$. And, *in general,* a sentence of the form "Some $P(x)$ are $Q(x)$" should be symbolized $(\exists x)[P(x) \wedge Q(x)]$.

Examples.   The sentence "For every odd prime x less than 10, $x^2 + 4$ is prime" means that if x is prime, and odd, and less than 10, then $x^2 + 4$ is prime. It is written symbolically as

$$(\forall x)(x \text{ is prime} \wedge x \text{ is odd} \wedge x < 10 \Rightarrow x^2 + 4 \text{ is prime}).$$

The sentence "Some functions defined at 0 are not continuous at 0" can be written symbolically as $(\exists f)(f$ is defined at $0 \wedge f$ is not continuous at 0$)$.

Example.   The sentence "Some real numbers have a multiplicative inverse" could be symbolized

$$(\exists x)(x \text{ is a real number} \wedge x \text{ has a real multiplicative inverse}).$$

However, "x has an inverse" means there is some number that is an inverse for x (hidden quantifier), so a more complete symbolic translation is

$$(\exists x)[x \text{ is a real number} \wedge (\exists y)(y \text{ is a real number} \wedge xy = 1)].$$

Example.   One correct translation of "Some integers are even and some integers are odd" is

$$(\exists x)(x \text{ is even}) \wedge (\exists x)(x \text{ is odd})$$

because the first quantifier ($\exists x$) extends only as far as the word "even." After that, any variable (even x again) may be used to express "some are odd." It would be equally correct and sometimes preferable to write

$$(\exists x)(x \text{ is even}) \wedge (\exists y)(y \text{ is odd}),$$

but it would be wrong to write

$$(\exists x)(x \text{ is even} \wedge x \text{ is odd}),$$

because there is no integer that is both even and odd.

Several of our essential definitions given in the *Preface to the Student* are in fact quantified statements. For example, the definition of a rational number may be symbolized:

$$r \text{ is a rational number iff } (\exists p)(\exists q)(p \in \mathbb{Z} \wedge q \in \mathbb{Z} \wedge q \neq 0 \wedge r = \tfrac{p}{q})$$

Statements of the form "Every element of the set A has the property P" and "Some element of the set A has property P" occur so frequently that abbreviated symbolic forms are desirable. "Every element of the set A has the property P" could be restated as "If $x \in A$, then . . ." and symbolized by

$$(\forall x \in A)P(x).$$

"Some element of the set A has property P" is abbreviated by

$$(\exists x \in A)P(x).$$

Examples.   The definition of a rational number given above may be written as

$$r \text{ is a rational number iff } (\exists p \in \mathbb{Z})(\exists q \in \mathbb{Z})(q \neq 0 \wedge r = \tfrac{p}{q}).$$

The statement "For every rational number there is a larger integer" may be symbolized by

$$(\forall x)[x \in \mathbb{Q} \Rightarrow (\exists z)(z \in \mathbb{Z} \text{ and } z > x)]$$

or

$$(\forall x \in \mathbb{Q})(\exists z \in \mathbb{Z})(z > x).$$

> **DEFINITION**   Two quantified sentences are **equivalent in a given universe** iff they have the same truth value in that universe. Two quantified sentences are **equivalent** iff they are equivalent in every universe.

Example. $(\forall x)(x > 3)$ and $(\forall x)(x \geq 4)$ are equivalent in the universe of integers (because both are false), the universe of natural numbers greater than 10 (because both are true), and in many other universes. However, if we chose a number between 3 and 4, say 3.7, and let U be the universe of real numbers larger than 3.7, then $(\forall x)(x > 3)$ is true and $(\forall x)(x \geq 4)$ is false in U. The sentences are not equivalent in this universe, so they are not equivalent sentences.

As was noted with propositional forms, it is necessary to make a distinction between a quantified sentence and its logical form. With the universe all integers, the sentence "All integers are odd" is an instance of the logical form $(\forall x)P(x)$, where $P(x)$ is "x is odd." The form itself, $(\forall x)P(x)$, is neither true nor false, but becomes false when "x is odd" is substituted for $P(x)$ and the universe is all integers.

The pair of quantified forms $(\exists x)([P(x) \land Q(x)]$ and $(\exists x)([Q(x) \land P(x)]$ are equivalent because for any choices of P and Q, $P \land Q$ and $Q \land P$ are equivalent propositional forms. Another pair of equivalent sentences is $(\forall x)[P(x) \Rightarrow Q(x)]$ and $(\forall x)[\sim Q(x) \Rightarrow \sim P(x)]$.

The next two equivalences are essential for reasoning about quantifiers.

Theorem 1.3.1   If $A(x)$ is an open sentence with variable x, then

(a) $\sim (\forall x)A(x)$ is equivalent to $(\exists x) \sim A(x)$.
(b) $\sim (\exists x)A(x)$ is equivalent to $(\forall x) \sim A(x)$.

Proof.
(a)   Let U be any universe.
　　　The sentence $\sim (\forall x)A(x)$ is true in U
　　　　　iff $(\forall x)A(x)$ is false in U
　　　　　iff the truth set of $A(x)$ is not the universe
　　　　　iff the truth set of $\sim A(x)$ is nonempty
　　　　　iff $(\exists x) \sim A(x)$ is true in U.

(b)   The proof of this part is Exercise 7.   ∎

Theorem 1.3.1 is helpful for finding useful denials (that is, simplified forms of negations) of quantified sentences. For example, in the universe of natural numbers, the sentence "All primes are odd" is symbolized $(\forall x)(x$ is prime $\Rightarrow x$ is odd). The negation is $\sim (\forall x)(x$ is prime $\Rightarrow x$ is odd). By applying Theorem 1.3.1(a), this becomes $(\exists x)[\sim (x$ is prime $\Rightarrow x$ is odd)]. By Theorem 1.2.2(c) this is equivalent to $(\exists x)[x$ is prime $\land \sim (x$ is odd)]. We read this last statement as "There exists a number that is prime and is not odd" or "Some prime number is even."

Example.   A simplified denial of $(\forall x)(\exists y)(\exists z)(\forall u)(\exists v)(x + y + z > 2u + v)$ begins with its negation

$$\sim (\forall x)(\exists y)(\exists z)(\forall u)(\exists v)(x + y + z > 2u + v).$$

After 5 applications of Theorem 1.3.1, beginning with the outermost quantifier $(\forall x)$, we arrive at the simplified form

$$(\exists x)(\forall y)(\forall z)(\exists u)(\forall v)(x + y + z \leq 2u + v).$$

Example.   For the universe of all real numbers, find a denial of "Every positive real number has a multiplicative inverse."

The sentence is symbolized $(\forall x)[x > 0 \Rightarrow (\exists y)(xy = 1)]$. The negation and successively rewritten equivalents are:

$$\sim(\forall x)[x > 0 \Rightarrow (\exists y)(xy = 1)]$$

$$(\exists x) \sim [x > 0 \Rightarrow (\exists y)(xy = 1)]$$

$$(\exists x)[x > 0 \wedge \sim(\exists y)(xy = 1)]$$

$$(\exists x)[x > 0 \wedge (\forall y) \sim(xy = 1)]$$

$$(\exists x)[x > 0 \wedge (\forall y)(xy \neq 1)]$$

This last sentence may be translated as "There is a positive real number that has no multiplicative inverse."

Example.   For the universe of living things, find a denial of "Some children do not like clowns."

The sentence is $(\exists x)$ [x is a child $\wedge$ $(\forall y)(y$ is a clown $\Rightarrow x$ does not like $y)$]. Its negation and several equivalents are:

$\sim(\exists x)$ [x is a child $\wedge$ $(\forall y)(y$ is a clown $\Rightarrow x$ does not like y)]

$(\forall x) \sim$ [x is a child $\wedge$ $(\forall y)(y$ is a clown $\Rightarrow x$ does not like y)]

$(\forall x)$ [x is a child $\Rightarrow \sim(\forall y)(y$ is a clown $\Rightarrow x$ does not like y)]

$(\forall x)$ [x is a child $\Rightarrow (\exists y) \sim(y$ is a clown $\Rightarrow x$ does not like y)]

$(\forall x)$ [x is a child $\Rightarrow (\exists y)(y$ is a clown $\wedge \sim x$ does not like y)]

$(\forall x)$ [x is a child $\Rightarrow (\exists y)(y$ is a clown $\wedge x$ likes y)]

The denial we seek is "Every child has some clown that he/she likes."

We sometimes hear statements like the complaint one fan had after a great Little League baseball game. "The game was fine," he said, "but everybody didn't get to play." We easily understand that the fan did not mean this literally, because otherwise there would have been no game. The meaning we understand is "Not everyone got to play" or "Some team members did not play." Such misuse of quantifiers, while tolerated in casual conversations, is always to be avoided in mathematics.

The $\exists!$ quantifier, defined next, is a special case of the existential quantifier.

DEFINITION   For an open sentence $P(x)$, the proposition $(\exists!x) P(x)$ is read "**there exists a unique x such that $P(x)$**" and is true iff the truth set of $P(x)$ has *exactly one element*. The symbol $\exists!$ is called the **unique existential quantifier.**

Recall that for $(\exists x)P(x)$ to be true it is unimportant how many elements are in the truth set of $P(x)$, as long as there is at least one. For $(\exists!x)P(x)$ to be true, the number of elements in the truth set of $P(x)$ is crucial—there must be exactly one.

In the universe of natural numbers, $(\exists!x)$ (x is even and x is prime) is true because the truth set of "x is even and x is prime" contains only the number 2. The sentence $(\exists!x)(x^2 = 4)$ is true in the universe of natural numbers, but false in the universe of all integers.

Theorem 1.3.2   If $A(x)$ is an open sentence with variable x, then

(a) $(\exists!x)A(x) \Rightarrow (\exists x)A(x)$.

(b) $(\exists!x)A(x)$ is equivalent to $(\exists x)A(x) \wedge (\forall y)(\forall z)(A(y) \wedge A(z) \Rightarrow y = z)$.

Part (a) of Theorem 1.3.2 says that $\exists!$ is indeed a special case of the quantifier $\exists$. Part (b) says that "There exists a unique x such that $A(x)$" is equivalent to "There is an x such that $A(x)$ and if both $A(y)$ and $A(z)$, then $y = z$." The proofs are left to Exercise 11.

Exercises 1.3

1. Translate the following English sentences into symbolic sentences with quantifiers. The universe for each is given in parentheses.
   ★ **(a)**   Not all precious stones are beautiful. (All stones)
   ☆ **(b)**   All precious stones are not beautiful. (All stones)
   (c)   Some isosceles triangle is a right triangle. (All triangles)
   (d)   No right triangle is isosceles. (All triangles)
   (e)   All people are honest or no one is honest. (All people)
   (f)   Some people are honest and some people are not honest. (All people)
   (g)   Every nonzero real number is positive or negative. (Real numbers)
   ★ **(h)**   Every integer is greater than -4 or less than 6. (Real numbers)
   (i)   Every integer is greater than some integer. (Integers)
   ★ **(j)**   No integer is greater than every other integer. (Integers)
   (k)   Between any integer and any larger integer, there is a real number. (Real numbers)
   ★ **(l)**   There is a smallest positive integer. (Real numbers)
   ★ **(m)**   No one loves everybody. (All people)
   (n)   Everybody loves someone. (All people)
   (o)   For every positive real number x, there is a unique real number y such that $2^y = x$. (Real numbers)

   ☆ 2. For each of the propositions in Exercise 1, write a useful denial, and give a translation into ordinary English.

   3. Translate these definitions from the *Preface to the Student* into quantified sentences.
   (a)   The integer x is **even**.
   (b)   The integer x is **odd**.

   (c)   The integer a **divides** the integer b.
   (d)   The natural number n is **prime**.
   (e)   The natural number n is **composite**.

4.   Translate these definitions in this text into quantified sentences. You need not know the specifics of the terms and symbols to complete this exercise.
   (a)   The relation R is **symmetric**. (See page 147.)
   (b)   The relation R is **transitive**. (See page 147.)
   (c)   The function f is **one-to-one**. (See page 208.)
   (d)   The operation $*$ is **commutative**. (See page 277.)

5.   The sentence "People dislike taxes" might be interpreted to mean "All people dislike all taxes," "All people dislike some taxes," "Some people dislike all taxes," or "Some people dislike some taxes." Give a symbolic translation for each of these interpretations.

6.   Let $T = \{17\}$, $U = \{6\}$, $V = \{24\}$, and $W = \{2, 3, 7, 26\}$. In which of these four different universes is the statement true?
   ★   a) $(\exists x)(x \text{ is odd} \Rightarrow x > 8)$.
       b) $(\exists x)(x \text{ is odd} \wedge x > 8)$.
       c) $(\forall x)(x \text{ is odd} \Rightarrow x > 8)$.
       d) $(\forall x)(x \text{ is odd} \wedge x > 8)$.

7.   (a)   Complete this proof of Theorem 1.3.1(b):
       Proof: Let U be any universe.
       The sentence $\sim(\exists x)A(x)$ is true in U
       iff ...
       iff $(\forall x)\sim A(x)$ is true in U.
   ☆   (b)   Give a proof of part (b) of Theorem 1.3.1 that uses part (a).

8.   Which of the following are true? The universe for each statement is given in parentheses.
   (a) $(\forall x)(x + x \geq x)$. ($\mathbb{R}$)
   ★   (b) $(\forall x)(x + x \geq x)$. ($\mathbb{N}$)
   (c) $(\exists x)(2x + 3 = 6x + 7)$. ($\mathbb{N}$)
   (d) $(\exists x)(3^x = x^2)$. ($\mathbb{R}$)
   ★   (e) $(\exists x)(3^x = x)$. ($\mathbb{R}$)
   (f) $(\exists x)(3(2 - x) = 5 + 8(1 - x))$. ($\mathbb{R}$)
   (g) $(\forall x)(x^2 + 6x + 5 \geq 0)$. ($\mathbb{R}$)
   ★   (h) $(\forall x)(x^2 + 4x + 5 \geq 0)$. ($\mathbb{R}$)
   (i) $(\exists x)(x^2 - x + 41 \text{ is prime})$. ($\mathbb{N}$)
   (j) $(\forall x)(x^2 - x + 41 \text{ is prime})$. ($\mathbb{N}$)
   (k) $(\forall x)(x^3 + 17x^2 + 6x + 100 \geq 0)$. ($\mathbb{R}$)
   (l) $(\forall x)(\forall y)[x < y \Rightarrow (\exists w)(x < w < y)]$. ($\mathbb{Q}$)

9.   Give an English translation for each. The universe is given in parentheses.
   (a) $(\forall x)(x \geq 1)$. ($\mathbb{N}$)
   ★   (b) $(\exists! x)(x \geq 0 \wedge x \leq 0)$. ($\mathbb{R}$)
   (c) $(\forall x)(x \text{ is prime} \wedge x \neq 2 \Rightarrow x \text{ is odd})$. ($\mathbb{N}$)
   ★   (d) $(\exists! x)(\log_e x = 1)$. ($\mathbb{R}$)

     (e) $\sim(\exists x)(x^2 < 0)$. ($\mathbb{R}$)

     (f) $(\exists!x)(x^2 = 0)$. ($\mathbb{R}$) ?

     (g) $(\forall x)(x$ is odd $\Rightarrow x^2$ is odd). ($\mathbb{N}$)

10. Which of the following are true in the universe of all real numbers?

★ **(a)** $(\forall x)(\exists y)(x + y = 0)$.

     (b) $(\exists x)(\forall y)(x + y = 0)$.

     (c) $(\exists x)(\exists y)(x^2 + y^2 = -1)$.

★ **(d)** $(\forall x)[x > 0 \Rightarrow (\exists y)(y < 0 \wedge xy > 0)]$.

     (e) $(\forall y)(\exists x)(\forall z)(xy = xz)$.

★ **(f)** $(\exists x)(\forall y)(x \leq y)$.

     (g) $(\forall y)(\exists x)(x \leq y)$.

     (h) $(\exists!y)(y < 0 \wedge y + 3 > 0)$.

★ **(i)** $(\exists!x)(\forall y)(x = y^2)$.

     (j) $(\forall y)(\exists!x)(x = y^2)$.

     (k) $(\exists!x)(\exists!y)(\forall w)(w^2 > x - y)$.

11. Let $A(x)$ be an open sentence with variable x.

☆ **(a)** Prove Theorem 1.3.2 (a).

☆ **(b)** Show that the converse of Theorem 1.3.2 (a) is false.

     (c) Prove Theorem 1.3.2 (b).

     (d) Prove that $(\exists!x)A(x)$ is equivalent to $(\exists x)[A(x) \wedge (\forall y)(A(y) \Rightarrow x = y)]$.

★ **(e)** Find a useful denial for $(\exists!x)A(x)$.

12. **(a)** Write the symbolic form for the definition of "f is continuous at a."

     (b) Write the symbolic form of the statement of the Mean Value Theorem.

     (c) Write the symbolic form for the definition of "$\lim\limits_{x \to a} f(x) = L$."

     (d) Write a useful denial of each sentence in parts (a), (b), and (c).

13. Which of the following are denials of $(\exists!x)P(x)$?

     (a) $(\forall x)P(x) \vee (\forall x)\sim P(x)$.

     (b) $(\forall x) \sim P(x) \vee (\exists y)(\exists z)(y \neq z \wedge P(y) \wedge P(z))$.

     (c) $(\forall x)[P(x) \Rightarrow (\exists y)(P(y) \wedge x \neq y)]$.

★ **(d)** $\sim(\forall x)(\forall y)[(P(x) \wedge P(y)) \Rightarrow x = y]$.

★ **14.** *Riddle:* What is the English translation of the symbolic statement $\forall\exists\exists\forall$?

              all for one & one for all

1.4   Basic Proof Methods I

In mathematics, a **theorem** is a statement that describes a pattern or relationship among quantities or structures and a **proof** is a justification of the truth of a theorem. Before beginning to examine valid proof techniques it is recommended that you review the comments about proofs and the definitions in the *Preface to the Student*.

    We cannot define all terms nor prove all statements from previous ones. We begin with an initial set of statements, called **axioms** (or **postulates**), that are *assumed to be true*. We then derive theorems that are true in any situation where the axioms are true. The Pythagorean* Theorem, for example, is a theorem whose proof is ultimately based on the five axioms of Euclidean[†] geometry. In a situation where the Euclidean axioms are not all true (which can happen), the Pythagorean Theorem may not be true.

There must also be an initial set of **undefined terms**—concepts fundamental to the context of study. In geometry, the concept of a point is an undefined term. In this text the real numbers are not formally defined. Instead, they are described in the *Preface to the Student* as the decimal numbers along the number line. While a precise definition of a real number could be given[‡], doing so would take us far from our intended goals.

From the axioms and undefined terms, new concepts (new **definitions**) can be introduced. And finally, new theorems can be proved. The structure of a proof for a particular theorem depends greatly on the logical form of the theorem. Proofs may require some ingenuity or insightfulness to put together the right statements to build the justification. Nevertheless, much can be gained in the beginning by studying the fundamental components found in proofs and examples that exhibit them. The four rules that follow provide guidance about what statements are allowed in a proof, and when.

Some steps in a proof may be statements of axioms of the basic theory upon which the discussion rests. Other steps may be previously proved results. Still other steps may be assumptions you wish to introduce. In any proof you may

At any time state an assumption, an axiom, or a previously proved result.

The statement of an assumption generally takes the form "Assume P" to alert the reader that the statement is not derived from a previous step or steps. We must be careful about making assumptions, because we can only be certain that what we proved will be true *when all the assumptions are true*. The most common assumptions are hypotheses given as components in the statement of the theorem to be proved. We will discuss assumptions in more detail later in this section.

The statement of an axiom is usually easily identified as such by the reader because it is a statement about a very fundamental fact assumed about the theory. Sometimes the axiom is so well known that its statement is omitted from proofs, but there are cases (such as the Axiom of Choice in Chapter 5) for which it is prudent to mention the axiom in every proof employing it.

Proof steps that use previously proven results help build a rich theory from the basic assumptions. In calculus, for example, before one proves that the derivative of $\sin x$ is $\cos x$, there is a proof of the separate result that $\lim\limits_{\Delta x \to 0} \dfrac{\sin \Delta x}{\Delta x} = 1$. It is easier to prove this result first, then cite the result in the proof of the fact that the derivative of $\sin x$ is $\cos x$.

---

* Pythagoras, latter half of the 6th century, B.C.E., was a Greek mathematician and philosopher who founded a secretive religious society based on mathematical and metaphysical thought. Although Pythagoras is regularly given credit for the theorem named for him, the result was known to Babylonian and Indian mathematicians centuries earlier.

† Euclid of Alexandria, circa 300 B.C.E., made his immortal contribution to mathematics with his famous text on geometry and number theory. His *Elements* sets forth a small number of axioms from which additional definitions and many familiar geometric results were developed in a rigorous way. Other geometries, based on different sets of axioms, did not begin to appear until the 1800s.

‡ See the references cited in Section 7.5.

An important skill for proof writing is the ability to rewrite a complex statement in an equivalent form that is more useful or helps to clarify its meaning. You may:

At any time state a sentence equivalent to any statement earlier in the proof.

This **replacement rule** is often used in combination with the equivalences of Theorems 1.1.1 and 1.2.2 to rewrite a statement involving logical connectives. For example, suppose we have been able to establish the step

"It is not the case that x is even and prime."

Because the form of this statement is $\sim(P \wedge Q)$, where P is "x is even" and Q is "x is prime," we may deduce that

"x is not even or x is not prime,"

which has form $\sim P \vee \sim Q$. We have applied the replacement rule, using one of De Morgan's Laws. A working knowledge of the equivalences of Theorems 1.1.1 and 1.2.2 is essential.

The replacement rule allows you to use definitions in two ways. First, if you are told or have shown that x is odd, then you can correctly state that for some natural number k, $x = 2k + 1$. You now have an equation to use. Second, if you need to prove that x is odd, then the definition gives you something equivalent to work toward: It suffices to show that x can be expressed as $x = 2k + 1$, for some natural number k. You'll find it useful in writing proofs to keep in mind these two ways we use definitions.

Example. If a proof contains the line "The product of real numbers a and b is zero," we could assert that "Either $a = 0$ or $b = 0$." In this example, the equivalence of the two statements comes from our knowledge of the real numbers that $(ab = 0) \Leftrightarrow (a = 0 \text{ or } b = 0)$.

Tautologies are important both because a statement that has the form of a tautology may be used as a step in a proof, and because tautologies are used to create rules for making deductions in a proof. The **tautology rule** says that you may:

At any time state a sentence whose symbolic translation is a tautology.

For example, if a proof involves a real number x, you may at any time assert "Either $x > 0$ or $x \leq 0$," since this is an instance of the tautology $P \vee \sim P$.

The rules above allow us to reword a statement or say something that's always true or is assumed to be true. The next rule is the one that allows us to make a connection so that we can get from statement P to a *different* statement Q.

The most fundamental rule of reasoning is **modus ponens**, which is based on the tautology $[P \wedge (P \Rightarrow Q)] \Rightarrow Q$. As we have seen in Section 1.2, what this means is that when P and $P \Rightarrow Q$ are both true, we may deduce that Q must also be true. The **modus ponens rule** says you may:

At any time after P and $P \Rightarrow Q$ appear in a proof, state that Q is true.

Example. From calculus we know that if a function f is differentiable on an interval (a, b), then f is continuous on the interval (a, b). A proof writer who had already written:

$$f \text{ is differentiable on the interval } (a, b)$$

could use modus ponens to write as a subsequent step:

$$\text{Therefore } f \text{ is continuous on the interval } (a, b).$$

This deduction uses the statements D, $D \Rightarrow C$, and C, where D is the statement "f is differentiable on interval (a, b)" and C is "f is continuous on the interval (a, b)."

Notice that in this example it would make the proof shorter and easier to read if we didn't write out the sentence $D \Rightarrow C$ in the proof. This is because the connection between differentiability and continuity is a well-known theorem, which the proof writer may assume that the reader knows.

When we use modus ponens to deduce statement Q from P and $P \Rightarrow Q$, the statement P could be an instance of a tautology, a simple or compound proposition whose components are either hypotheses, axioms, earlier statements deduced in the proof, or statements of previously proved theorems. Likewise, $P \Rightarrow Q$ may have been deduced earlier in the proof or may be a previous theorem, axiom, or tautology.

Example. You are at a crime scene and have established the following facts:

(1) If the crime did not take place in the billiard room, then Colonel Mustard is guilty.

(2) The lead pipe is not the weapon.

(3) Either Colonel Mustard is not guilty or the weapon used was a lead pipe.

From these facts and modus ponens, you may construct a proof that shows the crime took place in the billiard room:

Proof.

Statement (1)	$\sim B \Rightarrow M$
Statement (2)	$\sim L$
Statement (3)	$\sim M \vee L$
Statements (1) and (2) and (3)	$(\sim B \Rightarrow M) \wedge \sim L \wedge (\sim M \vee L)$
Statements (1), (2), and (3)	$[(\sim B \Rightarrow M) \wedge \sim L \wedge (\sim M \vee L)] \Rightarrow B$
imply the crime took place	is a tautology (see Exercise 2).
in the billiard room.	
Therefore, the crime took place	B
in the billiard room.	∎

The last three statements above are an application of the modus ponens rule: We deduced Q from the statements P and $P \Rightarrow Q$, where Q is B and P is $(\sim B \Rightarrow M) \wedge \sim L \wedge (\sim M \vee L)$.

The previous example shows the power of pure reasoning: It is the *forms* of the propositions and not their meanings that allowed us to make the deductions.

Because our proofs are always about mathematical phenomena, we also need to understand the subject matter of the proof—the concepts involved and how they are related. Therefore, when you develop a strategy to construct a proof, keep in mind both the logical form of the theorem's statement and the mathematical concepts involved.

You won't find truth tables displayed or referred to in proofs that you encounter in mathematics: It is expected that readers are familiar with the rules of logic and correct forms of proof. As a general rule, when you write a step in a proof, ask yourself if deducing that step is valid in the sense that it uses one of the four rules above. If the step follows as a result of the use of a tautology, it is not necessary to cite the tautology in your proof. In fact, with practice you should eventually come to write proofs without purposefully thinking about tautologies. What *is* necessary is that every step be justifiable.

The first—and most important—proof method is the **direct proof** of statement of the form $P \Rightarrow Q$, which proceeds in a step by step fashion from the antecedent P to the consequent Q. Since $P \Rightarrow Q$ is false only when P is true and Q is false, it suffices to show that this situation cannot happen. The direct way to proceed is to assume that P is true and show (deduce) that Q is also true. A direct proof of $P \Rightarrow Q$ will have the following form:

DIRECT PROOF OF $P \Rightarrow Q$
Proof.
Assume P.
$$\vdots$$
Therefore, Q.
Thus, $P \Rightarrow Q$. ∎

Some of the examples that follow actually involve quantified sentences. Since we won't consider proofs with quantifiers until Section 1.6, you should imagine for now that a variable represents some fixed object. Out first example proves the familiar fact that "If x is odd, then $x + 1$ is even." You should think of x as being some particular integer.

Example. Let x be an integer. Prove that if x is odd, then $x + 1$ is even.

Proof. *⟨The theorem has the form $P \Rightarrow Q$, where P is "x is odd" and Q is "$x + 1$ is even."⟩* Let x be an integer. *⟨We may assume this hypothesis since it is given in the statement of the theorem.⟩* Suppose x is odd. *⟨We assume that the antecedent P is true. The goal is to derive the consequent Q as our last step.⟩* From the definition of odd, $x = 2k + 1$ for some integer k. *⟨This deduction is the replacement* *of P by an equivalent statement—the definition of "odd." We now have an equation to use.*) Then $x + 1 = (2k + 1) + 1$ for some integer k. *(This is another replacement using an algebraic property of* $\mathbb{N}$.*)* Since $(2k + 1) + 1 = 2k + 2 = 2(k + 1)$, $x + 1$ is the product of 2 and an integer. *(Another equivalent using algebra.)* Thus $x + 1$ is even. *(We have deduced Q.)*

Therefore, if x is an odd integer, then $x + 1$ is even. *(We conclude that $P \Rightarrow Q$.)* ∎

In this example, we did not worry about what would happen if x were not odd. Remember that it is appropriate to assume P is true when giving a direct proof of $P \Rightarrow Q$. (If P is false, it does not matter what the truth of Q is; the statement we are trying to prove, $P \Rightarrow Q$, will be true.) The process of assuming that the antecedent is true and proceeding step by step to show the consequent is true is what makes this type of proof direct.

This example also includes parenthetical comments offset by $\langle \ldots \rangle$ and in italics to explain how and why a proof is proceeding as it is. Such comments are not a requisite part of the proof, but are inserted to help clarify the workings of the proof. The proof above would stand alone as correct with all the comments deleted, or it could be written in shorter form, as follows.

Proof.   Let x be an integer. Suppose x is odd. Then $x = 2k + 1$ for some integer k. Then $x + 1 = (2k + 1) + 1 = 2k + 2 = 2(k + 1)$. Since $k + 1$ is an integer and $x + 1 = 2(k + 1)$, $x + 1$ is even.

Therefore, if x is an odd integer, then $x + 1$ is even. ∎

Great latitude is allowed for differences in taste and style among proof writers. Generally, in advanced mathematics, only the minimum amount of explanation is included in a proof. The reader is expected to know the definitions and previous results and be able to fill in computations and deductions as necessary. In this text, we shall include parenthetical comments for more complete explanations.

Example.   Suppose a, b, and c are integers. Prove that if a divides b and b divides c, then a divides c.

Proof.   Let a, b, and c be integers. *(We start by assuming that the hypothesis is true.)* Suppose a divides b and b divides c. *(The antecedent is the compound sentence "a divides b and b divides c.")* Then $b = ak$ for some integer k and $c = bm$ for some integer m. *(We replaced the assumptions by equivalents using the definition of "divides." Notice that we did not assume that k and m are the same integer.)* *(To show that a divides c, we must write c as a multiple of a.)* Therefore, $c = bm = (ak)m = a(km)$. Then c is a multiple of a. *(We use the fact that if k and m are integers, then km is an integer.)*

Therefore, if a divides b and b divides c, then a divides c. ∎

Both of the above examples and many more to follow use the following strategy for developing a direct proof of a conditional sentence:

1. Determine precisely the hypotheses (if any) and the antecedent and consequent.
2. Replace (if necessary) the antecedent with a more usable equivalent.
3. Replace (if necessary) the consequent by something equivalent and more readily shown.
4. Beginning with the assumption of the antecedent, develop a chain of statements that leads to the consequent. Each statement in the chain must be deducible from its predecessors or other known results.

As you write a proof, be sure it is not just a string of symbols. Every step of your proof should express a complete sentence. Be sure to include important connective words.

Example. Suppose a, b, and c are integers. Prove that if a divides b and a divides c, then a divides $b - c$.

Proof. Suppose a, b, and c are integers and a divides b and a divides c. ⟨*Now use the definition of divides.*⟩ Then $b = an$ for some integer n and $c = am$ for some integer m. Thus, $b - c = an - am = a(n - m)$. Since $n - m$ is an integer ⟨*using the fact that the difference of two integers is an integer* ⟩, a divides $b - c$. ∎

Our next example of a direct proof, which comes from an exercise in precalculus mathematics, involves a point (x, y) in the Cartesian plane (Figure 1.4.1). It uses algebraic properties available to students in such a class.

Example. Prove that if $x < -4$ and $y > 2$, then the distance from (x, y) to $(1, -2)$ is at least 6.

Proof. Assume that $x < -4$ and $y > 2$. Then $x - 1 < -5$, so $(x - 1)^2 > 25$. Also $y + 2 > 4$, so $(y + 2)^2 > 16$. Therefore,

$$\sqrt{(x - 1)^2 + (y + 2)^2} > \sqrt{25 + 16} > \sqrt{36},$$

so the distance from (x, y) to $(1, -2)$ is at least 6. ∎

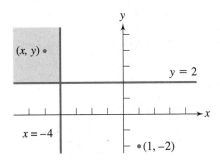

Figure 1.4.1

To get a sense of how a proof of $P \Rightarrow Q$ should proceed, it is sometimes useful to "work backward" from what is to be proved: To show that a consequent is true, decide what statement could be used to prove it, another statement that could be used to prove that one, and so forth. Continue until you reach a hypothesis, the antecedent, or a fact known to be true. After doing such preliminary work, construct a proof "forward" so that your conclusion is the consequent.

Example. Let a and b be positive real numbers. Prove that if $a < b$, then $b^2 - a^2 > 0$.

Proof. ⟨*Working backward, rewrite $b^2 - a^2 > 0$ as $(b - a)(b + a) > 0$. This inequality will be true when both $b - a > 0$ and $b + a > 0$. The first inequality $b - a > 0$ will be true because we will assume the antecedent $a < b$. The second inequality $b + a > 0$ is true because of our hypothesis that a and b are positive. We now proceed with the direct proof.*⟩ Assume a and b are positive real numbers and that $a < b$. Since both a and b are positive, $b + a > 0$. Since $a < b$, $b - a > 0$. Because the product of two positive real numbers is positive, $(b - a)(b + a) > 0$. Therefore $b^2 - a^2 > 0$.   ∎

It is often helpful to work both ways—backward from what is to be proved and forward from the hypothesis—until you reach a common statement from each direction.

Example.   Prove that if $x^2 \leq 1$, then $x^2 - 7x > -10$.
Working backward from $x^2 - 7x > -10$, we note that this can be deduced from $x^2 - 7x + 10 > 0$. This can be deduced from $(x - 5)(x - 2) > 0$, which could be concluded if we knew that $x - 5$ and $x - 2$ were both positive or both negative.
Working forward from $x^2 \leq 1$, we have $-1 \leq x \leq 1$, so $x \leq 1$. Therefore, $x < 5$ and $x < 2$, from which we can conclude that $x - 5 < 0$ and $x - 2 < 0$, which is exactly what we need.

Proof.   Assume that $x^2 \leq 1$. Then $-1 \leq x \leq 1$. Therefore $x \leq 1$. Thus $x < 5$ and $x < 2$, and so we have $x - 5 < 0$ and $x - 2 < 0$. Therefore, $(x - 5)(x - 2) > 0$. Thus $x^2 - 7x + 10 > 0$. Hence $x^2 - 7x > -10$.   ∎

We now consider direct proofs of statements of the form $P \Rightarrow Q$ when either P or Q is itself a compound proposition. We have in fact already constructed proofs of statements of the form $(P \wedge Q) \Rightarrow R$. When we give a direct proof of a statement of this form, we have the advantage of assuming both P and Q at the beginning of the proof, as we did in the proof (above) that if a divides b and a divides c, then a divides $b - c$.

A proof of a statement symbolized by $P \Rightarrow (Q \wedge R)$ would probably have two parts. In one part we prove $P \Rightarrow Q$ and in the other part we prove $P \Rightarrow R$. We would use this method to prove the statement "If two parallel lines are cut by a transversal, then corresponding angles are equal and corresponding lines are equal."

To prove a conditional sentence whose consequent is a disjunction, that is, a sentence of the form $P \Rightarrow (Q \lor R)$, one often proves either the equivalent $P \land \sim Q \Rightarrow R$ or the equivalent $P \land \sim R \Rightarrow Q$. For instance, to prove "If the polynomial f has degree 4, then f has a real zero or f can be written as the product of two irreducible quadratics," we would prove "If f has degree 4 and no real zeros, then f can be written as the product of two irreducible quadratics."

A statement of the form $(P \lor Q) \Rightarrow R$ has the meaning: "If either P is true or Q is true, then R is true," or "In case either P or Q is true, R must be true." A natural way to prove such a statement is by cases, so the proof outline would have the form:

Case 1. Assume P. . . . Therefore R.
Case 2. Assume Q. . . . Therefore R.

This method is valid because of the tautology

$$[(P \lor Q) \Rightarrow R] \Leftrightarrow [(P \Rightarrow R) \land (Q \Rightarrow R)].$$

The statement "If a quadrilateral has opposite sides equal or opposite angles equal, then it is a parallelogram" is proved by showing both "A quadrilateral with opposite sides equal is a parallelogram" and "A quadrilateral with opposite angles equal is a parallelogram."

The two similar statement forms $(P \Rightarrow Q) \Rightarrow R$ and $P \Rightarrow (Q \Rightarrow R)$ have remarkably dissimilar direct proof outlines. For $(P \Rightarrow Q) \Rightarrow R$, we assume $P \Rightarrow Q$ and deduce R. We cannot assume P; we must assume $P \Rightarrow Q$. On the other hand, in a direct proof of $P \Rightarrow (Q \Rightarrow R)$, we do assume P and show $Q \Rightarrow R$. Furthermore, after the assumption of P, a direct proof of $Q \Rightarrow R$ begins by assuming Q is true as well. This is not surprising since $P \Rightarrow (Q \Rightarrow R)$ is equivalent to $(P \land Q) \Rightarrow R$.

The main lesson to be learned from this discussion is that the method of proof you choose will depend on the form of the statement to be proved. The outlines we have given are the most natural, but not the only ways, to construct correct proofs. Of course constructing a proof also requires knowledge of the subject matter.

Example. Suppose n is an odd integer. Then $n = 4j + 1$ for some integer j, or $n = 4i - 1$ for some integer i.

Proof. Suppose n is odd. Then $n = 2m + 1$ for some integer m. $\langle$*A little experimentation shows that when m is even, for example when n is $2(-2) + 1$, $2(0) + 1, 2(2) + 1, 2(4) + 1$, etc., n has the form $4j + 1$; otherwise n has the form $4i - 1$. We now show that $(P \lor Q) \Rightarrow (R_1 \lor R_2)$, where P is "m is even," Q is "m is odd," R_1 is "$n = 4j + 1$ for some integer j," and R_2 is "$n = 4i - 1$ for some integer i." The method we choose is to show that $P \Rightarrow R_1$ and $Q \Rightarrow R_2$.*$\rangle$

Case 1. If m is even, then $m = 2j$ for some integer j, and so $n = 2(2j) + 1 = 4j + 1$.
Case 2. If m is odd, then $m = 2k + 1$ for some integer k. In this case, $n = 2(2k + 1) + 1 = 4k + 3 = 4(k + 1) - 1$. Choosing i to be the integer $k + 1$, we have $n = 4i - 1$. ∎

The form of proof known as **proof by exhaustion** consists of an examination of every possible case. The statement to be proved may have any form P. For example, to prove that every number x in the closed interval $[0, 5]$ has a certain property, we might consider the cases $x = 0$, $0 < x < 5$, and $x = 5$. The exhaustive method was our method in the example above, and in the proof of Theorem 1.1.1, where we examined all four combinations of truth values for two propositions. Naturally, the idea of proof by exhaustion is appealing only when the number of cases is small, or when large numbers of cases can be systematically handled. Care must be taken to ensure that all possible cases have been considered.

Example. Let x be a real number. Prove that $-|x| \leq x \leq |x|$.

Proof. ⟨*Since the absolute value of x is defined by cases ($|x| = x$ if $x \geq 0$; $|x| = -x$ if $x < 0$) this proof will proceed by cases.*⟩

Case 1. Suppose $x \geq 0$. Then $|x| = x$. Since $x \geq 0$, we have $-x \leq x$. Hence, $-x \leq x \leq x$, which is $-|x| \leq x \leq |x|$ in this case.
Case 2. Suppose $x < 0$. Then $|x| = -x$. Since $x < 0$, $x \leq -x$. Hence, we have $x \leq x \leq -x$, or $-(-x) \leq x \leq -x$, which is $-|x| \leq x \leq |x|$.

Thus, in all cases we have $-|x| \leq x \leq |x|$. ∎

There have been instances of truly exhausting proofs involving great numbers of cases. In 1976, Kenneth Appel and Wolfgang Haken of the University of Illinois announced a proof of the Four-Color Theorem. The original version of their proof of the famous Four-Color Conjecture contains 1,879 cases and took $3\frac{1}{2}$ years to develop.*

Finally, there are proofs by exhaustion with cases so similar in reasoning that we may simply present a single case and alert the reader with the phrase "without loss of generality" that this case represents the essence of arguments for the other cases. Here is an example.

Example. Prove that for the integers m and n, one of which is even and the other odd, $m^2 + n^2$ has the form $4k + 1$ for some integer k.

Proof. Let m and n be integers. Without loss of generality, we may assume that m is even and n is odd. ⟨*The case where m is odd and n is even is similar.*⟩ Then there exist integers s and t such that $m = 2s$ and $n = 2t + 1$. Therefore, $m^2 + n^2 = (2s)^2 + (2t + 1)^2 = 4s^2 + 4t^2 + 4t + 1 = 4(s^2 + t^2 + t) + 1$. Since $s^2 + t^2 + t$ is an integer, $m^2 + n^2$ has the form $4k + 1$ for some integer k. ∎

---

* The Four-Color Theorem involves coloring regions or countries on a map in such a way that no two adjacent countries have the same color. It states that four colors are sufficient, no matter how intertwined the countries may be. The fact that the proof depended so heavily on the computer for checking cases raised questions about the nature of proof. Verifying the 1,879 cases required more than 10 billion calculations. Many people wondered whether there might have been at least one error in a process so lengthy that it could not be carried out by one human being in a lifetime. Haken and Appel's proof has since been improved, and the Four-Color Theorem is accepted; but the debate about the role of computers in proof continues.

Exercises 1.4

1. Analyze the logical form of each of the following statements and construct just the outline of a proof. Since the statements may contain terms with which you are not familiar, you should not (and perhaps could not) provide any details of the proof.

   ⋆ **(a)** Outline a direct proof that if $(G, *)$ is a cyclic group, then $(G, *)$ is abelian.

   (b) Outline a direct proof that if **B** is a nonsingular matrix, then the determinant of **B** is not zero.

   (c) Suppose A, B, and C are sets. Outline a direct proof that if A is a subset of B and B is a subset of C, then A is a subset of C.

   (d) Outline a direct proof that if the maximum value of the differentiable function f on the closed interval $[a, b]$ occurs at x_0, then either $x_0 = a$ or $x_0 = b$ or $f'(x_0) = 0$.

   (e) Outline a direct proof that if **A** is a diagonal matrix, then **A** is invertible whenever all its diagonal entries are nonzero.

2. A theorem of linear algebra states that if **A** and **B** are invertible matrices, then the product **AB** is invertible. As in Exercise 1, outline

   (a) a direct proof of the theorem.

   (b) a direct proof of the converse of the theorem.

3. Verify that $[(\sim B \Rightarrow M) \wedge \sim L \wedge (\sim M \vee L)] \Rightarrow B$ is a tautology. See the example on page 30.

4. These facts have been established at a crime scene.

   (i) If Professor Plum is not guilty, then the crime took place in the kitchen.

   (ii) If the crime took place at midnight, Professor Plum is guilty.

   (iii) Miss Scarlet is innocent if and only if the weapon was not the candlestick.

   (iv) Either the weapon was the candlestick or the crime took place in the library.

   (v) Either Miss Scarlet or Professor Plum is guilty.

   Use the above and the additional fact(s) below to solve the case. Explain your answer.

   ⋆ **(a)** The crime lab determines that the crime took place in the library.

   (b) The crime lab determines that the crime did not take place in the library.

   (c) The crime lab determines that the crime was committed at noon with the revolver.

   (d) The crime took place at midnight in the conservatory. (Give a complete answer.)

5. Let x and y be integers. Prove that

   (a) if x and y are even, then $x + y$ is even.

   (b) if x is even, then xy is even.

   (c) if x and y are even, then xy is divisible by 4.

   (d) if x and y are even, then $3x - 5y$ is even.

   (e) if x and y are odd, then $x + y$ is even.

    (f)   if x and y are odd, then $3x - 5y$ is even.

    (g)   if x and y are odd, then xy is odd.

★ **(h)**   if x is even and y is odd, then $x + y$ is odd.

    (i)   if x is even and y is odd, then xy is even.

6. Let a and b be real numbers. Prove that

    (a) $|ab| = |a||b|$.

    (b) $|a - b| = |b - a|$.

    (c) $\left|\dfrac{a}{b}\right| = \dfrac{|a|}{|b|}$, for $b \neq 0$.

☆ **(d)** $|a + b| \leq |a| + |b|$.

    (e)   if $|a| \leq b$, then $-b \leq a \leq b$.

    (f)   if $-b \leq a \leq b$, then $|a| \leq b$.

7. Suppose a, b, c, and d are integers. Prove that

    (a) $2a - 1$ is odd.

★ **(b)**   if a is even, then $a + 1$ is odd.

    (c)   if a is odd, then $a + 2$ is odd.

☆ **(d)** $a(a + 1)$ is even.

    (e)   1 divides a.

    (f) a divides a.

★ **(g)**   if a and b are positive and a divides b, then $a \leq b$.

    (h)   if a divides b, then a divides bc.

★ **(i)**   if a and b are positive and $ab = 1$, then $a = b = 1$.

    (j)   if a and b are positive, a divides b and b divides a, then $a = b$.

    (k)   if a divides b and c divides d, then ac divides bd.

    (l)   if ab divides c, then a divides c.

    (m) if ac divides bc, then a divides b.

8. Give two proofs that if n is a natural number, then $n^2 + n + 3$ is odd.

    (a)   Use two cases.

    (b)   Use Exercises 7(d) and 5(h).

9. Let a, b, and c be integers and x, y, and z be real numbers. Use the technique of working backward from the desired conclusion to prove that

    (a)   if x and y are nonnegative, then $\dfrac{x + y}{2} \geq \sqrt{xy}$.

           Where in the proof do we use the fact that x and y are nonnegative?

    (b)   if a divides b and a divides $b + c$, then a divides $3c$.

    (c)   if $ab > 0$ and $bc < 0$, then $ax^2 + bx + c = 0$ has two real solutions.

    (d)   if $x^3 + 2x^2 < 0$, then $2x + 5 < 11$.

    (e)   if an isosceles triangle has sides of length x, y, and z, where $x = y$ and $z = \sqrt{2xy}$, then it is a right triangle.

10. Recall that except for degenerate cases, the graph of $Ax^2 + Bxy + Cy^2 + Dx + Ey + F = 0$ is

$$\text{an ellipse iff } B^2 - 4AC < 0,$$
$$\text{a parabola iff } B^2 - 4AC = 0,$$
$$\text{a hyperbola iff } B^2 - 4AC > 0.$$

★ **(a)** Prove that the graph of the equation is an ellipse whenever $A > C > B > 0$.

(b) Prove that the graph of the equation is a hyperbola if $AC < 0$ or $B < C < 4A < 0$.

(c) Prove that if the graph is a parabola, then $BC = 0$ or $A = B^2/(4C)$.

Proofs to Grade   **11.** Exercises throughout the text with this title ask you to examine "Proofs to Grade." These are allegedly true claims and supposed "proofs" of the claims. You should decide the merit of the claim and the validity of the proof and then assign a grade of

A (correct), if the claim and proof are correct, even if the proof is not the simplest or the proof you would have given.

C (partially correct), if the claim is correct *and* the proof is largely correct. The proof may contain one or two incorrect statements or justifications, but the errors are easily correctable.

F (failure), if the claim is incorrect, or the main idea of the proof is incorrect, or there are too many errors.

You must justify assignments of grades other than A and if the proof is incorrect, explain what is incorrect and why.

★ **(a)** Suppose a is an integer.
Claim. If a is odd then $a^2 + 1$ is even.
"Proof." Let a. Then, by squaring an odd we get an odd. An odd plus odd is even. So $a^2 + 1$ is even. ∎

(b) Suppose a, b, and c are integers.
Claim. If a divides b and a divides c, then a divides $b + c$.
"Proof." Suppose a divides b and a divides c. Then for some integer q, $b = aq$, and for some integer q, $c = aq$. Then $b + c = aq + aq = 2aq = a(2q)$, so a divides $b + c$. ∎

★ **(c)** Suppose x is a positive real number.
Claim. The sum of x and its reciprocal is greater than or equal to 2. That is,

$$x + \frac{1}{x} \geq 2.$$

"Proof." Multiplying by x, we get $x^2 + 1 \geq 2x$. By algebra, $x^2 - 2x + 1 \geq 0$. Thus, $(x - 1)^2 \geq 0$. Any real number squared is greater than or equal to zero, so $x + \frac{1}{x} \geq 2$ is true. ∎

★ **(d)** Suppose m is an integer.
Claim. If m^2 is odd, then m is odd.
"Proof." Assume m is odd. Then $m = 2k + 1$ for some integer k. Therefore, $m^2 = (2k + 1)^2 = 4k^2 + 4k + 1 = 2(2k^2 + 2k) + 1$, which is odd. Therefore, if m^2 is odd, then m is odd. ∎

(e) Suppose a is an integer.
Claim. $a^3 + a^2$ is even.
"Proof." $a^3 + a^2 = a^2(a + 1)$, which is always an odd number times an even number. Therefore, $a^3 + a^2$ is even. ∎

In the last section, we saw that the method of direct proof for $P \Rightarrow Q$ proceeds as a chain of statements from the antecedent to the consequent. This is the most basic form of proof and is the foundation for several other proof techniques. The techniques in this section are based on tautologies that replace the statement to be proved by an equivalent statement or statements. We call these **indirect proofs.**

A **proof by contraposition** or **contrapositive proof** for a conditional sentence $P \Rightarrow Q$ makes use of the tautology $(P \Rightarrow Q) \Leftrightarrow (\sim Q \Rightarrow \sim P)$. Since $P \Rightarrow Q$ and $\sim Q \Rightarrow \sim P$ are equivalent statements, we first give a proof of $\sim Q \Rightarrow \sim P$ and then conclude by replacement that $P \Rightarrow Q$.

PROOF BY CONTRAPOSITION OF $P \Rightarrow Q$
Proof.
Assume $\sim Q$.
$\vdots$
Therefore, $\sim P$.
Thus, $\sim Q \Rightarrow \sim P$
Therefore, $P \Rightarrow Q$.   ∎

This method can work well when the connection between denials of P and Q are easier to understand than the connection between P and Q themselves, or when the statement of either P and Q is itself a negation.

In the following examples of proof by contraposition we use familiar properties of inequalities and the property that every integer is either even or odd, but not both. As in the last section, we assume that variables represent fixed quantities.

Example. Let m be an integer. Prove that if m^2 is even, then m is even.

Proof. ⟨*The antecedent is P, "m^2 is even" and the consequent is Q, "m is even."*⟩
Suppose that the integer m is not even. ⟨*Suppose $\sim Q$.*⟩ Then m is odd so $m = 2k + 1$ for some integer k. Then

$$m^2 = (2k + 1)^2 = 4k^2 + 4k + 1 = 2(2k^2 + 2k) + 1.$$

Since m^2 is twice an integer, plus 1, m^2 is odd. ⟨*Since k is an integer, $2k^2 + 2k$ is an integer.*⟩ Therefore, m^2 is not even. ⟨*We have concluded $\sim P$.*⟩

Thus, if m is not even, then m^2 is not even. By contraposition, if m^2 is even, then m is even.   ∎

Example. Let x and y be real numbers such that $x < 2y$. Prove that if $7xy \le 3x^2 + 2y^2$, then $3x \le y$.

Proof.   Suppose x and y are real numbers and $x < 2y$. ⟨*Let P be* $7xy \leq 3x^2 + 2y^2$ *and Q be* $3x \leq y$.⟩ Suppose $3x > y$. ⟨*We assume* $\sim Q$.⟩ Then $2y - x > 0$ and $3x - y > 0$. Therefore, $(2y - x)(3x - y) = 7xy - 3x^2 - 2y^2 > 0$. Hence, $7xy > 3x^2 + 2y^2$.

We have shown that if $3x > y$, then $7xy > 3x^2 + 2y^2$. Therefore, by contraposition, if $7xy \leq 3x^2 + 2y^2$, then $3x \leq y$.   ■

Another indirect proof technique is **proof by contradiction.** The logic behind such a proof is that if a statement cannot be false, then it must be true. Thus, to prove by contradiction that a statement P is true, we temporarily assume that P is false and then see what would happen. If what happens is an impossibility—that is, a contradiction—then we know that P must be true. Here is an example of a proof by contradiction.

Example.   Prove that the graphs of $y = x^2 + x + 2$ and $y = x - 2$ do not intersect.

Proof.   Suppose the graphs of $y = x^2 + x + 2$ and $y = x - 2$ do intersect at some point (a, b). ⟨*Suppose* $\sim P$.⟩ Since (a, b) is a point on both graphs, $b = a^2 + a + 2$ and $b = a - 2$. Therefore, $a - 2 = a^2 + a + 2$, so $a^2 = -4$. Thus, $a^2 < 0$. But a is a real number, so $a^2 \geq 0$. This is impossible. ⟨*The statement* $a^2 < 0 \wedge a^2 \geq 0$ *is a contradiction.*⟩ Therefore, the graphs do not intersect.   ■

A proof by contradiction is based on the tautology $P \Leftrightarrow [(\sim P) \Rightarrow (Q \wedge \sim Q)]$. That is, to prove a proposition P, we prove $(\sim P) \Rightarrow (Q \wedge \sim Q)$ for some proposition Q. In the example above, Q is the statement $a^2 < 0$. A proof by contradiction has the following form:

PROOF OF P BY CONTRADICTION
Proof.
Suppose $\sim P$.
$\vdots$
Therefore, Q.
$\vdots$
Therefore, $\sim Q$.
Hence, $Q \wedge \sim Q$ a contradiction.
Thus, P.   ■

Two aspects about proofs by contradiction are especially noteworthy. First, this method of proof can be applied to any proposition P, whereas direct proofs and proofs by contraposition can be used only for conditional sentences. Second, the proposition Q does not appear on the left side of the tautology. The strategy of proving P by proving $\sim P \Rightarrow (Q \wedge \sim Q)$, then, has an advantage and a disadvantage. We don't know what proposition to use for Q, but any proposition that will do the job is a good one. This means a proof by contradiction may require a spark of insight to determine a useful Q.

The next proof by contradiction is a classical result whose proof can be traced back to Hippasus, a disciple of Pythagoras, circa 500 B.C.E. One of several legends has it that Hippasus proved that $\sqrt{2}$ is not a rational number while traveling by ship with his Pythagorean colleagues. The Pythagoreans, steadfast believers that all numbers are rational*, supposedly threw him into the sea to drown.

The proof relies on the definition of a rational number: r is rational iff $r = \frac{a}{b}$ for some integers a and b, with $b \neq 0$. We may assume that that a and b have no common factors, because otherwise we would simply reduce $\frac{a}{b}$ by cancelling any common factors.

Example. $\sqrt{2}$ is an irrational number.

Proof.   Assume that $\sqrt{2}$ is a rational number. ⟨*We assume* ∼*P.*⟩ Then $\sqrt{2} = \frac{a}{b}$ for some integers a and b, where $b \neq 0$ and a and b have no common factors. ⟨*The statement Q is "a and b have no common factors."*⟩ From $\sqrt{2} = \frac{a}{b}$ we have $2 = \frac{a^2}{b^2}$, which implies that $2b^2 = a^2$. Therefore a^2 is even and so a is even. (Recall the example we proved on page 40.) It follows that there exists an integer k such that $a = 2k$ and therefore

$$2b^2 = a^2$$
$$= (2k)^2$$
$$= 4k^2.$$

Thus $b^2 = 2k^2$, which shows b^2 is even. Therefore b is even. Since both a and b are even, a and b do have a common factor of 2. ⟨*We have deduced the statement* ∼*Q.*⟩ This is a contradiction. We conclude that $\sqrt{2}$ is irrational.   ■

Recall that a natural number greater than 1 is prime iff its only positive divisors are 1 and itself. The next proof by contradiction, attributed to Euclid, shows that there are infinitely many primes. By this we mean that it is impossible to list all of the prime numbers from the first to the kth (last) one, where k is a natural number. It uses the fundamental result that every natural number greater than 1 has a prime divisor.

Example.   The set of primes is infinite.

Proof.   Suppose the set of primes is finite. ⟨*Suppose* ∼*P. This means that the set of primes has k elements for some natural number k. Then the set of all primes can be listed, from the first one to the kth (last) one.*⟩ Let $p_1, p_2, p_3, \ldots, p_k$ be all those primes. Let n be one more than the product of all of them: $n = (p_1 p_2 p_3 \ldots p_k) + 1$. ⟨*We made up a number n which will not have any of the p_i as prime factors.*⟩ Then n is a natural number, so n has a prime divisor q. Since q is prime, $q > 1$. ⟨*The Q statement is*

---

* You may wonder why $\sqrt{2}$ is important or why it should be the first number to be proved irrational. The ancient Greeks geometers constructed numbers (lengths of line segments) using only a compass and a straightedge. It's easy to construct a square with sides of length 1, for which the length of a diagonal is $\sqrt{2}$. The fundamental Pythagorean belief that all numbers that arise in nature are either integers or ratios of integers is disproved by the irrationality of $\sqrt{2}$.

"$q > 1$.") Since q is a prime and $p_1, p_2, p_3, \ldots, p_k$ are *all* the primes, q is one of the p_i in the list. Thus, q divides the product $p_1 p_2 p_3 \ldots p_k$. Since q divides n, q divides the difference $n - (p_1 p_2 p_3 \ldots p_k)$. But this difference is 1, so $q = 1$. ⟨*This is* $\sim Q$.⟩ From the contradiction, $q > 1$ and $q = 1$, we conclude that the assumption that the set of primes is finite is false. Therefore, the set of primes is infinite. ■

Example. Prove the square shown in Figure 1.5.1(a) cannot be completed to form a "magic square" whose rows, columns, and diagonals all sum to the same number.

1	2	3	
	4	5	6
7		8	
	9		10

(a)

1	2	3	a
b	4	5	6
7	c	8	d
e	9	f	10

(b)

Figure 1.5.1

Proof. Suppose the square can be completed with entries a, b, c, d, e, f, as shown in Figure 1.5.1(b). Since the sums of the second row and second column are the same, $b + 15 = c + 15$. Thus, $b = c$. Comparing the sums of the first column and the lower-left to upper-right diagonal, $1 + b + 7 + e = e + c + 5 + a$. Thus, $a = 3$ and the first row sums to 9. Thus, the "magic sum" is 9. ⟨*This is our Q statement.*⟩ But the main diagonal sum ($1 + 4 + 8 + 10 = 23$) is not 9. ⟨*This is our* $\sim Q$ *statement.*⟩ This is a contradiction. We conclude that the square cannot be completed. ■

Proofs of biconditional sentences $P \Leftrightarrow Q$ often make use of the tautology $(P \Leftrightarrow Q) \Leftrightarrow (P \Rightarrow Q) \wedge (Q \Rightarrow P)$. Proofs of $P \Leftrightarrow Q$ generally have the following two-part form:

TWO-PART PROOF OF $P \Leftrightarrow Q$
Proof.
  (i) Show $P \Rightarrow Q$.
  (ii) Show $Q \Rightarrow P$.
Therefore, $P \Leftrightarrow Q$. ■

The separate proofs of parts (i) and (ii) may use different methods. Often the proof of one part is easier than the other. This is true, for example, of the proof that

"The natural number p is prime iff there is no positive integer greater than 1 and less than or equal to $\sqrt{x}$ that divides x."

It immediately follows from the definition of prime that "x is prime" implies "there is no positive integer greater than 1 and less than or equal to $\sqrt{x}$ that divides x." The converse requires more thought and is an exercise in the next section.

The **parity** of an integer is the attribute of being either odd or even. The integer 31 has odd parity while 42 has even parity. The integers 12 and 15 have opposite parity. The next example is a proof of a biconditional statement about parity with a two part proof. Both parts of the proof have two cases. The proof we give is not the shortest possible, but it does illustrate the two part approach to proving a biconditional statement.

Example.   Let m and n be integers. Then m and n have the same parity iff $m^2 + n^2$ is even.

Proof.

(i)   Suppose m and n have the same parity. We have two cases.
   (a)   If both m and n are even then $m = 2k$ and $n = 2j$ for some integers k and j. Then $m^2 + n^2 = (2k)^2 + (2j)^2 = 2(2k^2 + 2j^2)$, which is even.
   (b)   If both m and n are odd then $m = 2k + 1$ and $n = 2j + 1$ for some integers k and j. Then $m^2 + n^2 = (2k + 1)^2 + (2j + 1)^2 = 2(2k^2 + 2k + 2j^2 + 2j + 1)$, which is even.
   In both cases $m^2 + n^2$ is even.

(ii)   Suppose $m^2 + n^2$ is even. ⟨*To show that n has the same parity as m, we use some previous examples and exercises about even and odd integers.*⟩ Again we have two cases.
   (a)   If m is even, then m^2 is even. Therefore, since $m^2 + n^2$ is even and m^2 is even, $n^2 = (m^2 + n^2) - m^2$ is even. From n^2 is even, we conclude that n is even.
   (b)   If m is odd, then m^2 is odd. Therefore, since $m^2 + n^2$ is even and m^2 is odd, $n^2 = (m^2 + n^2) - m^2$ is odd. From n^2 is odd, we conclude that n is odd.
   Hence, if m is even, then n is even, and if m is odd, then n is odd. Therefore, m and n have the same parity.   ■

In some cases it is possible to prove a biconditional sentence $P \Leftrightarrow Q$ that uses the "iff" connective throughout. This amounts to starting with P and then replacing it with a sequence of equivalent statements, the last one being Q. With n intermediate statements $R_1, R_2, \ldots, R_n$, a biconditional proof of $P \Leftrightarrow Q$ has the form:

BICONDITIONAL PROOF OF $P \Leftrightarrow Q$
Proof.
P iff R_1
   iff R_2
   . . .
   iff R_n
   iff Q.   ■

Example. The triangle in Figure 1.5.2 has sides of length a, b, and c. Use the Law of Cosines to prove that the triangle is a right triangle with hypotenuse c if and only if $a^2 + b^2 = c^2$.

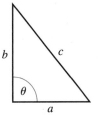

Figure 1.5.2

Proof. By the Law of Cosines, $a^2 + b^2 = c^2 - 2ab \cos \theta$, where θ is the angle between the sides of length a and b. Therefore,

$$a^2 + b^2 = c^2 \quad \text{iff} \quad 2ab \cos \theta = 0$$
$$\text{iff} \quad \cos \theta = 0$$
$$\text{iff} \quad \theta = 90°.$$

Thus, $a^2 + b^2 = c^2$ iff the triangle is a right triangle with hypotenuse c. ∎

As the following example shows, many theorems are amenable to more than one proof technique. Two of the proofs below will use the fact that if a prime (2 in our case) divides the product of two integers, then it must divide at least one of the integers. This property, known as Euclid's Lemma, will be proved in Section 1.7.

Example. For given integers x and y, give a direct proof, a proof by contraposition, and a proof by contradiction of the following statement: If x and y are odd integers, then xy is odd.

Direct Proof. Assume x is odd and y is odd. Then integers m and n exist so that $x = 2m + 1$ and $y = 2n + 1$. Thus,

$$xy = (2m + 1)(2n + 1)$$
$$= 4mn + 2m + 2n + 1$$
$$= 2(2mn + m + n) + 1.$$

Thus xy is an odd integer. ∎

Proof by Contraposition. ⟨*The contrapositive of x is odd $\wedge$ y is odd $\Rightarrow$ xy is odd is the statement xy is even $\Rightarrow$ ~(x is odd $\wedge$ y is odd), or equivalently,*

$$xy \text{ is even} \Rightarrow (x \text{ is even} \vee y \text{ is even}).⟩$$

Assume xy is even. Thus, 2 is a factor of xy. But since 2 is a prime number and 2 divides the product xy, then either 2 divides x or 2 divides y by Euclid's Lemma. We have shown that if xy is even, then either x or y is even. Thus, if x and y are odd, then xy is odd. ∎

Proof by Contradiction.   Suppose that the statement "If x and y are odd integers, then xy is odd" is false. Then x is odd and y is odd, and xy is not odd. Since xy is not odd, xy is even. Therefore 2 divides xy. Then by Euclid's Lemma, 2 divides x or 2 divides y. Thus either x is even or y is even. But x is odd and y is odd. This is a contradiction. We conclude that if x and y are odd integers, then xy is odd.   ■

By now you may have the impression that, given a set of axioms and definitions of a mathematical system, any properly stated proposition in that system can be proved true or proved false. This is not the case. There are important examples in mathematics of **consistent axiom systems** (so that there exist structures satisfying all the axioms) for which there are statements such that neither the statement nor its negation can be proved. It is not a matter of these statements being difficult to prove or that no one has yet been clever enough to devise a proof; it has been proved that there can be no proof of either the statement or its negation within the system. Such statements are called **undecidable** in the system because their truth is independent of the truth of the axioms.

The classic case of an undecidable statement involves the fifth of five postulates that Euclid set forth as his basis for plane geometry: "Given a line and a point not on that line, exactly one line can be drawn through the point parallel to the line." For centuries, some thought Euclid's axioms were not independent, believing that the fifth postulate could be proved from the other four. It was not until the 19th century that it became clear that the fifth postulate was undecidable. There are now theories of Euclidean geometry for which the fifth postulate is assumed true and non-Euclidean geometries for which it is assumed false. Both are perfectly reasonable subjects for mathematical study and application.

Exercises 1.5

1. Analyze the logical form of each of the following statements and construct just the outline of a proof by the given method. Since the statements may contain terms with which you are not familiar, you should not (and perhaps could not) provide any details of the proof.

   ★ (a)   Outline a proof by contraposition that if $(G, *)$ is a cyclic group, then $(G, *)$ is abelian.

   (b)   Outline a proof by contraposition that if **B** is a nonsingular matrix, then the determinant of **B** is not zero.

   ★ (c)   Outline a proof by contradiction that the set of natural numbers is not finite.

   (d)   Outline a proof by contradiction that the multiplicative inverse of a nonzero real number x is unique.

   ★ (e)   Outline a two-part proof that the inverse of the function f from A to B is a function from B to A if and only if f is one-to-one and onto B.

   (f)   Outline a two-part proof that a subset A of the real numbers is compact if and only if A is closed and bounded.

2. A theorem of linear algebra states that if **A** and **B** are invertible matrices, then the product **AB** is invertible. As in Exercise 1,
    (a) outline a proof of the theorem by contraposition.
    (b) outline a proof of the converse of the theorem by contraposition.
    (c) outline a proof of the theorem by contradiction.
    (d) outline a proof of the converse of the theorem by contradiction.
    (e) outline a two-part proof that **A** and **B** are invertible matrices if and only if the product **AB** is invertible.

3. Let x, y, and z be integers. Write a proof by contraposition to show that
    ★ (a) if x is even, then $x + 1$ is odd.
    (b) if x is odd, then $x + 2$ is odd.
    (c) if x^2 is not divisible by 4, then x is odd.
    (d) if xy is even, then either x or y is even.
    ☆ (e) if $x + y$ is even, then x and y have the same parity.
    (f) if xy is odd, then both x and y are odd.
    (g) if 8 does not divide $x^2 - 1$, then x is even.
    (h) if x does not divide yz, then x does not divide z.

4. Write a proof by contraposition to show that for any real number x,
    (a) if $x^2 + 2x < 0$, then $x < 0$.
    ★ (b) if $x^2 - 5x + 6 < 0$, then $2 < x < 3$.
    (c) if $x^3 + x > 0$, then $x > 0$.

5. A circle has center (2, 4).
    (a) Prove that $(-1, 5)$ and $(5, 1)$ are not both on the circle.
    (b) Prove that if the radius is less than 5, then the circle does not intersect the line $y = x - 6$.
    (c) Prove that if (0, 3) is not inside the circle, then (3, 1) is not inside the circle.

6. Suppose a and b are positive integers. Write a proof by contradiction to show that
    (a) if a divides b, then $a \le b$.
    ★ (b) if ab is odd, then both a and b are odd.
    (c) if a is odd, then $a + 1$ is even.
    (d) if $a - b$ is odd, then $a + b$ is odd.
    (e) if $a < b$ and $ab < 3$, then $a = 1$.

7. Suppose a, b, c, and d are positive integers. Write a proof of each biconditional statement.
    (a) ac divides bc if and only if a divides b.
    (b) $a + 1$ divides b and b divides $b + 3$ if and only if $a = 2$ and $b = 3$.
    (c) a is odd if and only if $a + 1$ is even.
    (d) $a + c = b$ and $2b - a = d$ if and only if $a = b - c$ and $b + c = d$.

8. Let m and n be integers. Then prove that m and n have different parity iff $m^2 - n^2$ is odd.

9. Prove by contradiction that if n is a natural number, then
$$\frac{n}{n+1} > \frac{n}{n+2}.$$

10. Prove that $\sqrt{5}$ is not a rational number.

11. Three real numbers, x, y, and z, are chosen between 0 and 1 with $0 < x < y < z < 1$. Prove that at least two of the numbers x, y, and z are within $\frac{1}{2}$ unit from one another.

Proofs to Grade

12. Assign a grade of A (correct), C (partially correct), or F (failure) to each. Justify assignments of grades other than A.

   (a) Suppose m is an integer. $\sqrt{}$ m is odd, M^2 is odd.
       Claim. If m^2 is odd, then m is odd.
       "Proof." Assume that m^2 is not odd. Then m^2 is even and $m^2 = 2k$ for some integer k. Thus $2k$ is a perfect square; that is, $\sqrt{2k}$ is an integer. If $\sqrt{2k}$ is odd, then $\sqrt{2k} = 2n + 1$ for some integer n, which means $m^2 = 2k = (2n + 1)^2 = 4n^2 + 4n + 1 = 2(2n^2 + 2n) + 1$. Thus m^2 is odd, contrary to our assumption. Therefore $\sqrt{2k} = m$ must be even. Thus if m^2 is not odd, then m is not odd. Hence if m^2 is odd, then m is odd. ∎

   ⋆ (b) Suppose t is a real number.
       Claim. If t is irrational, then $5t$ is irrational.
       "Proof." Suppose $5t$ is rational. Then $5t = p/q$, where p and q are integers and $q \neq 0$. Therefore, $t = p/(5q)$, where p and $5q$ are integers and $5q \neq 0$, so t is rational. Therefore, if t is irrational, then $5t$ is irrational. ∎

   (c) Suppose x and y are integers.
       Claim. If x and y are even then $x + y$ is even.
       "Proof." Suppose x and y are even but $x + y$ is odd. Then, for some integer k, $x + y = 2k + 1$. Therefore, $x + y + (-2)k = 1$. The left side of the equation is even because it is the sum of even numbers. However, the right side, 1, is odd. Since an even cannot equal an odd, we have a contradiction. Therefore, $x + y$ is even. ∎

   (d) Suppose a, b, and c are integers.
       Claim. If a divides both b and c, then a divides $b + c$.
       "Proof." Assume that a does not divide $b + c$. Then there is no integer k such that $ak = b + c$. However, a divides b, so $am = b$ for some integer m; and a divides c, so $an = c$ for some integer n. Thus $am + an = a(m + n) = b + c$. Therefore $k = m + n$ is an integer satisfying $ak = b + c$. Thus the assumption that a does not divide $b + c$ is false, and a does divide $b + c$. ∎

1.6   Proofs Involving Quantifiers

Recall that in our first example of a direct proof in Section 1.4 we proved the statement "If x is odd then $x + 1$ is even." That statement has the meaning "For every integer x, if x is odd then $x + 1$ is even." We dealt with the quantifier in that example by asking you to think of the variable x as some fixed integer. This section discusses specifically the proof methods for statements with quantifiers.

To prove a proposition of the form $(\forall x)P(x)$, we must show that $P(x)$ is true for every object x in the universe. A direct proof is begun by letting x represent an arbitrary object in the universe, and then showing that $P(x)$ is true for that object. In the proof we may use only properties of x that are shared by every element of the universe. Then, since x is arbitrary, we can conclude that $(\forall x)P(x)$ is true.

Thus a **direct proof** of $(\forall x)P(x)$ has the following form:

DIRECT PROOF OF $(\forall x)P(x)$
Proof.
Let x be an arbitrary object in the universe. (The universe should be named or its objects described.)
$\vdots$
Hence $P(x)$ is true.
Since x is arbitrary, $(\forall x)P(x)$ is true. ∎

A review of the proof examples in Sections 1.4 and 1.5 shows that whenever the statement was universally quantified, the proof given had the form of a complete proof, because each begins with an assumption such as "Let x be an integer" or "Let x and y be real numbers."

Example.   Prove that for every natural number n, $4n^2 - 6.8n + 2.88 > 0$.

Proof.   ⟨*The statement has the form $(\forall x)P(x)$, where the universe is $\mathbb{N}$ and $P(x)$ is* "$4n^2 - 6.8n + 2.88 > 0$."⟩ Let n be a natural number. Then $n \geq 1$, so $n - .8$ and $n - .9$ are both positive. Therefore $4(n - .8)(n - .9) = 4n^2 - 6.8n + 2.88$ is positive. We conclude that $4n^2 - 6.8n + 2.88 > 0$ for all natural numbers n.   ∎

Since the open sentence $P(x)$ in $(\forall x)P(x)$ will often be a combination of other open sentences joined by the logical connectives, the selection of an appropriate proof technique will depend on the logical form of $P(x)$. In the next example $P(x)$ has the form of a conditional sentence.

Example.   If x is an even integer, then x^2 is an even integer.

Proof.   ⟨*The statement has the form $(\forall x)(A(x) \Rightarrow B(x))$, where the universe is $\mathbb{Z}$, $A(x)$ is* "x is even," *and $B(x)$ is* "x^2 is even."⟩ Let $x \in \mathbb{Z}$. ⟨*We give a direct proof of $A(x) \Rightarrow B(x)$, which we begin by assuming $A(x)$.*⟩ Assume x is even. Then $x = 2k$ for some integer k. Thus $x^2 = (2k)^2 = 2(2k^2)$. Since $2k^2$ is an integer, x^2 is even. Since x is arbitrary, we have that for all $x \in \mathbb{Z}$, if x is even, then x^2 is even.   ∎

It is essential in a direct proof of $(\forall x)P(x)$ that the first step assume nothing about x other than it is an object in the universe. In the example above there are *two* assumptions about the variable x — for two very different reasons. The assumption "Let $x \in \mathbb{Z}$" appears first because we are assuming x is an object in the universe.

We make the statement "Assume x is even" because we are initiating a direct proof of a conditional sentence, which starts by assuming the antecedent.

It is a mistake to give an example (or several examples) of the statement "If x is even, then x^2 is even" and then claim that the statement has been proved for all natural numbers n. Examples may sometimes help decide whether a statement is true. Examples can also help guide our thinking about how to proceed with a proof. However, we cannot prove that a universally quantified statement is true by showing that it's true for selected values of the variable.

The next example involves two quantifiers.

Example.   For all rational numbers x and y, $\dfrac{x+y}{2}$ is a rational number.

Proof.   ⟨*The statement has the form* $(\forall x)(\forall y) P(x, y)$, *where the universe is* $\mathbb{Q}$ *and* $P(x, y)$ *is* " $\dfrac{x+y}{2}$ *is rational.*"⟩ Let x and y be rational numbers. Then

$$\frac{x+y}{2} = \frac{1}{2}\left(\frac{p}{q} + \frac{s}{t}\right) = \frac{1}{2}\left(\frac{pt+qs}{qt}\right) = \frac{pt+qs}{2qt}.$$

Both $pt + qs$ and $2qt$ are integers and $2qt \neq 0$. ⟨*The sums and products of integers are integers. The product of three nonzero numbers is not zero.*⟩ Therefore, $\dfrac{x+y}{2}$ is a rational number.   ■

The method of proof by contradiction is often used to prove statements of the form $(\forall x) P(x)$. Since $\sim(\forall x) P(x)$ is equivalent to $(\exists x) \sim P(x)$, the form of the proof is as follows:

PROOF OF $(\forall x) P(x)$ BY CONTRADICTION
Proof.
Suppose $\sim(\forall x) P(x)$.
Then $(\exists x) \sim P(x)$.
Let t be an object such that $\sim P(t)$.
$\vdots$
Therefore $Q \wedge \sim Q$.
Thus $(\exists x) \sim P(x)$ is false, so $(\forall x) P(x)$ is true.   ■

The following example of a proof by contradiction comes from an exercise in a trigonometry class. It uses algebraic and trigonometric properties available to students in the class.

Example.   Prove that for all $x \in \left(0, \frac{\pi}{2}\right), \sin x + \cos x > 1$.

Proof.   ⟨*The statement has the form* $(\forall x) P(x)$, *where the universe is the open interval* $\left(0, \frac{\pi}{2}\right)$ *and* $P(x)$ *is* "$\sin x + \cos x > 1$."⟩ Suppose that the statement is false. Then there exists a real number t, with $0 < t < \frac{\pi}{2}$, such that $\sin t + \cos t \leq 1$. ⟨*We have deduced* $(\exists t) \sim P(t)$.⟩ Since the functions $\sin x$ and $\cos x$ are positive for every $x \in \left(0, \frac{\pi}{2}\right)$ $\sin t > 0$ and $\cos t > 0$. Therefore,

$$0 < \sin t + \cos t \leq 1$$
$$0 < (\sin t + \cos t)^2 \leq 1^2 = 1$$
$$0 < \sin^2 t + 2\sin t \cos t + \cos^2 t \leq 1$$
$$0 < 1 + 2\sin t \cos t \leq 1$$
$$-1 < 2\sin t \cos t \leq 0.$$

⟨*We use the identity* $\sin^2 t + \cos^2 t = 1$.⟩ But $2\sin t \cos t \leq 0$ is impossible since both $\sin t$ and $\cos t$ are positive. Therefore, if $0 < x < \frac{\pi}{2}$, then $\sin x + \cos x > 1$. ∎

Notice the different roles that the symbols "x" and "t" play in the above example. The variable x is used to express the statement of the theorem and also appears as the independent variable in the sine and cosine functions. The symbol t represents some fixed value in $\left(0, \frac{\pi}{2}\right)$ with the property that $\sin t + \cos t \leq 1$.

There are several ways to prove existence theorems—that is, propositions of the form $(\exists x)P(x)$. In a **constructive proof** we actually name an object a in the universe such that $P(a)$ is true, which directly verifies that the truth set of $P(x)$ is nonempty. Some constructive proofs are quite easy to devise. For example, to prove that "There is an even prime natural number," we simply observe that 2 is prime and 2 is even.

Other constructive proofs have eluded mathematicians for centuries. The question of whether any nth power is a sum of fewer than n nth powers was raised by Leonard Euler* in the mid 1700s. A computer search in 1968 discovered a fifth power that was the sum of four fifth powers. Here is an example for fourth powers.

Example. Prove that there exists a natural number whose fourth power is the sum of three other fourth powers.

Proof. 20,615,673 is one such number because

$$20615673^4 = 2682440^4 + 1536539^4 + 18796760^4.$$ ∎

Another strategy to prove $(\exists x)P(x)$ is to show that there must be some object for which $P(x)$ is true, without ever actually producing a particular object. Both Rolle's Theorem and the Mean Value Theorem from calculus are good examples of this. Here is another.

---

* Leonard Euler (1707–1783) was a brilliant Swiss mathematician who spent much of his career at the Imperial Russian Academy of Sciences in St. Petersburg and the Berlin Academy. He made profound contributions to calculus, number theory, and graph theory as well as physics and astronomy. He was the first to introduce the idea of function and the familiar $f(x)$ notation.

Example.   Prove that the polynomial

$$r(x) = x^{71} - 2x^{39} + 5x - 0.3$$

has a real zero.

Proof.   ⟨*The universe is* $\mathbb{R}$. *The statement has the form* $(\exists t)(r(t) = 0).$⟩ By the Fundamental Theorem of Algebra[†], $r(x)$ has 71 zeros that are either real or complex. Since the polynomial has real coefficients, its nonreal zeros come in pairs ⟨*by the Complex Root Theorem*⟩. Hence the number of nonreal zeros is even, and that leaves an odd number of real zeros. Therefore, $r(x)$ has at least one real zero. ∎

Existence theorems may also be proved by **contradiction.** The proof technique has the following form:

PROOF OF $(\exists x) P(x)$ **BY CONTRADICTION**
Proof.
Suppose $\sim (\exists x) P(x)$.
Then $(\forall x) \sim P(x)$
$\vdots$

Therefore, $\sim Q \wedge Q$, a contradiction.
Thus $\sim (\exists x) P(x)$ is false.
Therefore $(\exists x) P(x)$ is true.   ∎

The core of a proof of $(\exists x) P(x)$ by contradiction involves making deductions from the statement $(\forall x) \sim P(x)$.

Example.   Starting at 9 a.m. on Monday a hiker walked from a base camp up a mountain trail and reached the summit at exactly 3 p.m. The hiker camped for the night and then hiked back down the same trail, again starting at 9 a.m. On this second walk the hiker stopped to look at a scenic overlook, but walked faster on other parts of the trail and returned to the starting point in exactly six hours. Prove that there is some point on the trail that the hiker passed at the identical time on the two days.

Proof.   Clearly, the point on the trail is not at the base camp or summit. ⟨*The universe is the open interval* (0, 6), *representing the time between* $t = 0$ (9 *a.m.*) *and* $t = 6$ (3 *p.m.*) *along the trail. The statement has the form* $\exists t \in (0, 6)$ (*the point on the trail at time t on Monday is the same as the point on the trail at time t on Tuesday*).⟩ Suppose there is no such point along the trail. Then for every time $t \in (0, 6)$, the point where the hiker is at time t on Monday is different from the point where the hiker is at time t on Tuesday. Have two other people simultaneously walk the trail, starting at 9 a.m. One goes up the trail at exactly the pace set by the hiker on

---

[†] The Fundamental Theorem of Algebra says that every polynomial in one variable with complex coefficients and degree $n > 0$ has exactly n zeros, counting multiplicities.

Monday and the other walks down the trail at exactly the pace set by the hiker on Tuesday. Since these two people are at different points at every time between 9 a.m. and noon, they will never meet. But they must meet at some point on the trail. This is a contradiction. Therefore there is some point on the trail that the hiker passed at the same time on the two days. ∎

Sometimes a statement to be proved has the form $(\exists x)P(x) \Rightarrow Q$. As a first step, we assume $(\exists x)P(x)$. However, the fact that *some* object x in the universe has the property $P(x)$ does not give us much to work with. A useful next step is to name some particular object that has the property and use the property of the object to derive Q.

Example. The graph of $x^2 + y^2 = r^2$, with $r > 0$, is a circle with center $(0, 0)$ and radius r. Prove that if one of the x-intercepts of the circle has rational coordinates, then all four intercepts have rational coordinates.

Proof. Suppose an x-intercept $(a, 0)$ of the circle has rational coordinates. Then a is a rational number and $a^2 + 0^2 = r^2$, so $a^2 = r^2$ and $a = \pm r$. Then the other x-intercept is $(-a, 0)$. To find the y-intercepts, we solve $0^2 + y^2 = r^2$ and find $y = \pm r = \pm a$. Therefore, the four intercepts are $(a, 0)$, $(-a, 0)$, $(0, a)$, and $(0, -a)$, all of which have rational coordinates. ∎

Many statements have more than one quantifier. We must deal with each in succession, starting from the left.

Example. Between any two rational numbers x and y, where $x < y$, there is always another rational number z.

Proof. ⟨*The statement may be symbolized* $(\forall x \in \mathbb{Q})(\forall y \in \mathbb{Q})[x < y \Rightarrow (\exists z \in \mathbb{Q})(x < z < y)]$. *We begin with the two universal quantifers.*⟩ Suppose x and y are rational numbers. Assume that $x < y$. ⟨*Now we must prove the existence of a rational number z with the given property.*⟩ We choose $z = \dfrac{x + y}{2}$. By a previous example, z is a rational number. Furthermore,

$$x = \frac{x + x}{2} < \frac{x + y}{2} < \frac{y + y}{2} = y.$$

Therefore $x < z < y$. ∎

Example. Prove that for every natural number n, there is a natural number M such that for all natural numbers $m > M$,

$$\frac{1}{m} < \frac{1}{3n}.$$

Proof. ⟨*The statement may be symbolized by*

$$(\forall n \in \mathbb{N})(\exists M \in \mathbb{N})(\forall m \in \mathbb{N}) \left(m > M \Rightarrow \frac{1}{m} < \frac{1}{3n} \right).$$

We begin with the universal quantifier on the left.⟩ Let n be a natural number. ⟨*We must prove the existence of a natural number M with the given property.⟩* Choose M to be $3n$. Let m be a natural number, and suppose $m > M$. Then $m > 3n$, and $3mn > 0$, so dividing by $3mn$ we have $\dfrac{1}{m} < \dfrac{1}{3n}$. ⟨*The choice of 3n for M is the result of some scratchwork, working backward from the intended conclusion* $\dfrac{1}{m} < \dfrac{1}{3n}$.⟩   ∎

Example. There is a real number with the property that for any two larger numbers there is another real number that is larger than the sum of the two numbers and less than their product.

Proof. ⟨*The universe is* $\mathbb{R}$. *A symbolic form of the statement is*

$$(\exists z)(\forall x)(\forall y)[(x > z \wedge y > z) \Rightarrow (\exists w)(x + y < w < xy)].$$

We must choose z so that the statement

$$(x > z \wedge y > z) \Rightarrow (\exists w)(x + y < w < xy)$$

will be true for all x and y.⟩ We chose $z = 2$. ⟨*To understand this choice for z, first notice that* $x + y$ *is not always less than* xy. *For example, let* $x = 1.6$ *and* $y = 1.4$.⟩ Let x and y be real numbers such that $x > z$ and $y > z$. Without loss of generality, we may assume that $y \geq x$. ⟨*Otherwise, we could rename x and y.*⟩ Then

$$x + y \leq 2y < xy.$$

Now choose w to be the midpoint between $x + y$ and xy, so $w = \dfrac{(x+y) + xy}{2}$. We have $x + y < w < xy$.   ∎

A proof of a statement about unique existence always involves multiple quantifiers. The standard technique for proving a proposition of the form $(\exists! x)P(x)$ is based on proving the equivalent statement: $(\exists x)P(x) \wedge (\forall y)(\forall z)$ $[P(y) \wedge P(z) \Rightarrow y = z]$. Since the main connective is a conjunction, the method will have two parts:

PROOF OF $(\exists! x)P(x)$
Proof.
 (i) Prove that $(\exists x)P(x)$ is true. Use any method.
 (ii) Prove that $(\forall y)(\forall z)[P(y) \wedge P(z) \Rightarrow y = z]$.
   Assume that y and z are objects in the universe such that $P(y)$ and $P(z)$ are true.

   $\vdots$

   Therefore, $y = z$.
 From (i) and (ii) conclude that $(\exists! x)P(x)$ is true.   ∎

Example. Every nonzero real number has a unique multiplicative inverse.

Proof. ⟨*The statement has the form* $(\forall x \in \mathbb{R})(x \neq 0 \Rightarrow (\exists! y \in \mathbb{R})(xy = 1))$.⟩ Let $x \neq 0$. ⟨*We show there is a unique real number y such that $xy = 1$ in two steps: First show that such a number y exists, and then show that x cannot have two different inverses.*⟩

(i) ⟨*This part is a constructive proof.*⟩ Let $y = \dfrac{1}{x}$. Since $x \neq 0$, y is a real number. Then $xy = x\left(\dfrac{1}{x}\right) = 1$. Therefore, x has a multiplicative inverse.

(ii) Now suppose that y and z are multiplicative inverses for x. ⟨*We do not assume that this y is the same as the y in part* (i).⟩ Then $xy = 1$ and $xz = 1$, so

$$xy = xz$$
$$xy - xz = 0$$
$$x(y - z) = 0.$$

Since $x \neq 0$, $y - z = 0$. Therefore $y = z$. ∎

Great care must be taken in proofs that contain expressions involving more than one quantifier. Here are some manipulations of quantifiers that permit valid deductions.

1. $(\forall x)(\forall y)P(x, y) \Leftrightarrow (\forall y)(\forall x)P(x, y)$.
2. $(\exists x)(\exists y)P(x, y) \Leftrightarrow (\exists y)(\exists x)P(x, y)$.
3. $[(\forall x)P(x) \vee (\forall x)Q(x)] \Rightarrow (\forall x)[P(x) \vee Q(x)]$.
4. $(\forall x)[P(x) \Rightarrow Q(x)] \Rightarrow [(\forall x)P(x) \Rightarrow (\forall x)Q(x)]$.
5. $(\forall x)[P(x) \wedge Q(x)] \Leftrightarrow [(\forall x)P(x) \wedge (\forall x)Q(x)]$.
6. $(\exists x)(\forall y)P(x, y) \Rightarrow (\forall y)(\exists x)P(x, y)$.

You should convince yourself that each of these is a logically valid conditional or biconditional. For example, the last on the list is always true because if $(\exists x)(\forall y)P(x, y)$ is true, then there is (at least) one x that makes $P(x, y)$ true no matter what y is. Therefore, for any y, $(\exists x)P(x, y)$ is true because this particular x exists.

It is important to be aware of the most common *incorrect deductions* making use of quantifiers. We list four here and show by example that each is not valid. Notice that statements 2, 3, and 4 in the following list are the converses, respectively, of valid deductions of statements 3, 4, and 6 above.

1. $(\exists x)P(x) \Rightarrow (\forall x)P(x)$ *is not valid.*
   The implication says that if some object has property P, then all objects have property P. If the universe is all integers and $P(x)$ is the sentence "x is odd," then $P(5)$ is true and $P(8)$ is false. Thus, $(\exists x)P(x)$ is true and $(\forall x)P(x)$ is false, so the implication fails.
2. $(\forall x)[P(x) \vee Q(x)] \Rightarrow [(\forall x)P(x) \vee (\forall x)Q(x)]$ *is not valid.*
   This implication says that if every object has one of two properties, then either every object has the first property or every object has the second property.

Suppose the universe is the integers, $P(x)$ is "x is odd" and $Q(x)$ is "x is even." Then it is true that "All integers are either odd or even" but false that "Either all integers are odd or all integers are even."

3. $[(\forall x) P(x) \Rightarrow (\forall x) Q(x)] \Rightarrow (\forall x)[P(x) \Rightarrow Q(x)]$ *is not valid.*

   The implication says that if every object has property P implies every object has property Q, then every object that has property P must also have property Q. Again, let the universe be the integers and let $P(x)$ be "x is odd" and $Q(x)$ be "x is even." Because $(\forall x) P(x)$ is false, $(\forall x) P(x) \Rightarrow (\forall x) Q(x)$ is true. However, $(\forall x)[P(x) \Rightarrow Q(x)]$ is false.

4. $(\forall y)(\exists x) P(x, y) \Rightarrow (\exists x)(\forall y) P(x, y)$ *is not valid.*

   This is probably the most troublesome of all the possibilities for dealing with quantifiers. The implication says that if for every y there is some x that satisfies P, then there is an x that works with every y to satisfy P. Let the universe be the set of all married people and $P(x, y)$ be the sentence "x is married to y." Then $(\forall y)(\exists x) P(x, y)$ is true, since everyone is married to someone. But $(\exists x)(\forall y) P(x, y)$ would be translated as "There is some married person who is married to every married person," which is clearly false.

There are times when we will want to prove a quantified statement is *false*. We know that $(\forall x) P(x)$ is false precisely when $\sim(\forall x) P(x)$ is true and $\sim(\forall x) P(x)$ is equivalent to $(\exists x) \sim P(x)$. Therefore, one way to prove $(\forall x) P(x)$ is false is to prove $(\exists x) \sim P(x)$ is true.

A constructive proof of $(\exists x)(\sim P(x))$ names an object a in the universe such that $P(a)$ is false. The object a is called a **counterexample** to $(\forall x) P(x)$. The number 2 is a counterexample to the statement "All primes are odd." The function $f(x) = |x|$ is a counterexample to "Every function that is continuous at 0 is differentiable at 0."

Example. Some beginning algebra students believe that $(x + y)^2 = x^2 + y^2$. In symbolic terms, they believe that $(\forall x)(\forall y)[(x + y)^2 = x^2 + y^2]$ is true in the universe of real numbers. This mistake could be corrected by providing a counterexample—for instance, $x = 3$ and $y = 4$.

Our last example in this section is a proof of a statement of the form $\sim(\exists x) P(x)$, which means it is also an example of a proof of an equivalent statement of the form $(\forall x) \sim P(x)$. We proved in Section 1.4 that every odd integer can be written in the form $4j - 1$ or $4k + 1$. We now show that there does not exist an integer that can be written in both of these forms. The proof is by contradiction.

Example. There is no odd integer that can be expressed in the form $4j - 1$ and in the form $4k + 1$ for integers j and k.

Proof. Suppose n is an odd integer, and suppose $n = 4j - 1$ and $n = 4k + 1$ for integers j and k. Then $4j - 1 = 4k + 1$, so $4j - 4k = 2$. Therefore, $2j - 2k = 1$. The left side of this equation is $2(j - k)$, which is even, but 1 is odd. This is a contradiction. ∎

Exercises 1.6

1. Prove that
   ⋆ (a) there exist integers m and n such that $2m + 7n = 1$.
   (b) there exist integers m and n such that $15m + 12n = 3$.
   ⋆ (c) there do not exist integers m and n such that $2m + 4n = 7$.
   (d) there do not exist integers m and n such that $12m + 15n = 1$.
   (e) for every integer t, if there exist integers m and n such that $15m + 16n = t$, then there exist integers r and s such that $3r + 8s = t$.
   ☆ (f) if there exist integers m and n such that $12m + 15n = 1$, then m and n are both positive.
   (g) for every odd integer m, if m has the form $4k + 1$ for some integer k, then $m + 2$ has the form $4j - 1$ for some integer j.
   (h) for every odd integer m, $m^2 = 8k + 1$ for some integer k. (*Hint:* Use the fact that $k(k + 1)$ is an even integer for every integer k.)
   (i) for all odd integers m and n, if $mn = 4k - 1$ for some integer k, then m or n is of the form $4j - 1$ for some integer j.

2. Prove that for all integers a, b, and c,
   (a) if c divides a and c divides b, then for all integers x and y, c divides $ax + by$.
   ⋆ (b) if a divides $b - 1$ and a divides $c - 1$, then a divides $bc - 1$.
   (c) if a divides b, then for all natural numbers n, a^n divides b^n.
   (d) if a is odd, $c > 0$, c divides a and e divides $a + 2$, then $c = 1$.
   (e) if there exist integers m and n such that $am + bn = 1$ and $c \neq \pm 1$, then c does not divide a or c does not divide b.

3. Prove that if every even natural number greater than 2 is the sum of two primes,* then every odd natural number greater than 5 is the sum of three primes.

4. Provide either a proof or a counterexample for each of these statements.
   (a) For all positive integers x, $x^2 + x + 41$ is a prime.
   (b) $(\forall x)(\exists y)(x + y = 0)$. (Universe of all reals)
   (c) $(\forall x)(\forall y)(x > 1 \wedge y > 0 \Rightarrow y^x > x)$. (Universe of all reals)
   (d) For integers a, b, c, if a divides bc, then either a divides b or a divides c.
   (e) For integers a, b, c, and d, if a divides $b - c$ and a divides $c - d$, then a divides $b - d$.
   (f) For all positive real numbers x, $x^2 - x \geq 0$.
   (g) For all positive real numbers x, $2^x > x + 1$.
   ☆ (h) For every positive real number x, there is a positive real number y less than x with the property that for all positive real numbers z, $yz \geq z$.
   ☆ (i) For every positive real number x, there is a positive real number y with the property that if $y < x$, then for all positive real numbers z, $yz \geq z$.

5. (a) Prove that the natural number x is prime iff $x > 1$ and there is no positive integer greater than 1 and less than or equal to $\sqrt{x}$ that divides x.

---

* No one knows whether every even number greater than 2 is the sum of two prime numbers. This is the famous Goldbach Conjecture, proposed by the Prussian mathematician Christian Goldbach in 1742. You should search the Web to learn about the million dollar prize (never claimed) for proving Goldbach's Conjecture. Fortunately, you don't have to prove Goldbach's Conjecture to do this exercise.

(b) Prove that if p is a prime number and $p \neq 3$, then 3 divides $p^2 + 2$. (*Hint:* When p is divided by 3, the remainder is either 0, 1, or 2. That is, for some integer k, $p = 3k$ or $p = 3k + 1$ or $p = 3k + 2$.)

6. Prove that

(a) for every natural number n, $\dfrac{1}{n} \leq 1$. (*Hint:* Use the fact that $n \geq 1$ and divide by the positive number n.)

(b) there is a natural number M such that for all natural numbers $n > M$, $\dfrac{1}{n} < 0.13$.

★ **(c)** for every natural number n, there is a natural number M such that $2n < M$.

(d) there is a natural number M such that for every natural number n, $\dfrac{1}{n} < M$.

(e) there is no largest natural number.

(f) there is no smallest positive real number.

★ **(g)** for every real number $\varepsilon > 0$, there is a natural number M such that for all natural numbers $n > M$, $\dfrac{1}{n} < \varepsilon$.

☆ **(h)** for every real number $\varepsilon > 0$, there is a natural number M such that if $m > n > M$, then $\dfrac{1}{n} - \dfrac{1}{m} < \varepsilon$.

(i) there is a natural number K such that $\dfrac{1}{r^2} < 0.01$ whenever r is a real number larger than K.

(j) there exist integers L and G such that $L < G$ and for every real number x, if $L < x < G$, then $40 > 10 - 2x > 12$.

(k) there exists an odd integer M such that for all real numbers r larger than M, $\dfrac{1}{2r} < 0.01$.

(l) for every natural number x there is an integer k such that $3.3x + k < 50$.

(m) there exist integers $x < 100$ and $y < 30$ such that $x + y < 128$ and for all real numbers r and s, if $r > x$ and $s > y$, then $(r - 50)(s - 20) > 390$.

(n) for every pair of positive real numbers x and y where $x < y$, there exists a natural number M such that if n is a natural number and $n > M$, then $\dfrac{1}{n} < (y - x)$.

Proofs to Grade

7. Assign a grade of A (correct), C (partially correct), or F (failure) to each. Justify assignments of grades other than A.

★ **(a)** **Claim.** Every polynomial of degree 3 with real coefficients has a real zero.

"*Proof.*" The polynomial $p(x) = x^3 - 8$ has degree 3, real coefficients, and a real zero ($x = 2$). Thus the statement "Every polynomial of degree 3 with real coefficients does not have a real zero" is false, and hence its denial, "Every polynomial of degree 3 with real coefficients has a real zero," is true. ∎

★ **(b)** **Claim.** There is a unique polynomial whose first derivative is $2x + 3$ and which has a zero at $x = 1$.

"*Proof.*" The antiderivative of $2x + 3$ is $x^2 + 3x + C$. If we let $p(x) = x^2 + 3x - 4$, then $p'(x) = 2x + 3$ and $p(1) = 0$. So $p(x)$ is the desired polynomial. ∎

(c) $^\complement$ **Claim.**   Every prime number greater than 2 is odd.
   "Proof."   The prime numbers greater than 2 are 3, 5, 7, 11, 13, 17, 19, ... None of these are even, so all of them are odd.   ■

★ (d) ^{A} **Claim.**   There exists an irrational number r such that $r^{\sqrt{2}}$ is rational.
   "Proof."   If $\sqrt{3}^{\sqrt{2}}$ is rational, then $r = \sqrt{3}$ is the desired example. Otherwise, $\sqrt{3}^{\sqrt{2}}$ is irrational and $(\sqrt{3}^{\sqrt{2}})^{\sqrt{2}} = (\sqrt{3})^2 = 3$, which is rational. Therefore either $\sqrt{3}$ or $\sqrt{3}^{\sqrt{2}}$ is an irrational number r such that $r^{\sqrt{2}}$ is rational.   ■

(e) $^\complement$ **Claim.**   For every real number x, $|x| \geq 0$.
   "Proof."   We proceed by three cases: $x > 0$, $x = 0$, and $x < 0$.
   Case 1. $x > 0$. Choose, for example, $x = 4$. Then $|4| = 4$. Thus $|x| \geq 0$.
   Case 2. $x = 0$. Then $|0| = 0$. Thus, $|x| \geq 0$.
   Case 3. $x < 0$. Choose, for example, $x = -5$. Then $|-5| = 5$. Thus $|x| \geq 0$.   ■

(f) ^{A} **Claim.**   If x is prime, then $x + 7$ is composite.
   "Proof."   Let x be a prime number. If $x = 2$, then $x + 7 = 9$, which is composite. If $x \neq 2$, then x is odd, so $x + 7$ is even and greater than 2. In this case, too, $x + 7$ is composite. Therefore, if x is prime, then $x + 7$ is composite.   ■

(g) $^\ltimes$ **Claim.**   For all irrational numbers t, $t - 8$ is irrational.
   "Proof."   Suppose there exists an irrational number t such that $t - 8$ is rational. Then $t - 8 = \frac{p}{q}$, where p and q are integers and $q \neq 0$. Then $t = \frac{p}{q} + 8 = \dfrac{p + 8q}{q}$, with $p + 8q$ and q integers and $q \neq 0$. This is a contradiction because t is irrational. Therefore, for all irrational numbers t, $t - 8$ is irrational.   ■

(h) $^\complement$ **Claim.**   For real numbers x and y, if $xy = 0$ then $x = 0$ or $y = 0$.
   "Proof."

   Case 1.   If $x = 0$, then $xy = 0y = 0$.
   Case 2.   If $y = 0$, then $xy = x0 = 0$.

   In either case, $xy = 0$.   ■

☆ (i) $^\epsilon$ **Claim.**   For every real number $\varepsilon > 0$, there is a natural number K such that for all real numbers $x > K$, $\dfrac{1}{4x} < \varepsilon$.
   "Proof."   Let $\varepsilon > 0$ be a real number. Let K be $\dfrac{1}{2\varepsilon}$. Assume x is a real number and $x > K$. Then $x > \dfrac{1}{2\varepsilon}$, so $x > \dfrac{1}{4\varepsilon}$. Therefore, $4x\varepsilon > 1$, so $\dfrac{1}{4x} < \varepsilon$.   ■

(j) **Claim.**   For every natural number n, $n \leq n^2$.
   "Proof."   Let n be a natural number. Since n is a natural number, $1 \leq n$. Since n is positive, $n \cdot 1 \leq n \cdot n$. Therefore, $n \leq n^2$ for all natural numbers n.   ■

1.7   Additional Examples of Proofs

This section contains no new proof techniques but does offer pointers about how to begin a proof and how the form of the statement to be proved usually suggests a method for proving it. These discussions include references to exercises that have complete solutions in the Answers to Selected Exercises section. And because the subject provides an excellent setting for examples of proofs, we conclude this section with additional concepts from number theory.

Here are some strategies to consider when you begin to write a proof.

1.   **Make a start**. For most people, the hardest part of writing a proof is knowing where or how to start. *The most important step is to make a start—almost any start.* Once you've begun you may get stuck and need to begin again with a different approach, but often the first attempt will give you some ideas that can be useful in a new approach. Writing a proof is not done by staring at a statement to be proved until a full-blown proof pops into your head. It is done step by step, piecing together facts, definitions, and previous results, and building toward the statement to be proved.

2.   **Identify the assumption(s) and conclusion**. Most theorems can be stated in the form of a conditional sentence. The antecedent gives your hypotheses; the consequent is your goal. Look for known facts and previous results that might connect the antecedent with the consequent. For example, later in this section we shall see that theorems about the greatest common divisor use the Division Algorithm.

3.   **Try working backwards and/or fill in the "middle" of the proof**. Once the hypotheses and conclusion have been identified, write your assumptions, leave some space, and write the conclusion as the last line. Try to deduce statements from the hypothesis that are more useful. Rewrite the conclusion or find a suitable statement from which the conclusion follows. The idea is to try to reason forward from your assumption and backward from your conclusion until you join them. At the middle you will have steps that follow from the hypotheses and from which the conclusion follows. This makes a complete proof. See Exercise 1(a).

4.   **Understand the concepts**. Make sure you know the definitions of any technical terms that appear in the statement to be proved. Often the terms are defined by equations or formulas that can be manipulated for use in the next steps of the proof, as we did in previous sections with the definitions of even and odd integers, rational numbers, and other terms. See Exercise 1(b).

5.   **Determine the logical form of the statement**. It is important to be able to write (or at least visualize) the complete symbolic translation, with quantifiers, of the statement to be proved, because *the logical form of the statement will usually offer you insight into how to proceed*. Don't be overly concerned with naming different types of proofs and devising "formulas" for writing proofs of a given type. It's not true that if a statement has a certain form you must always use a certain proof technique. However, for each logical form there is always at least one natural outline for its proof, as described in the following examples.

"If *P*, then *Q*." First consider a direct proof. Begin with the first step, "Assume *P*." *P* may be a conjunction of several statements, so we may assume all of these statements are true. See Exercise 5(a). When a direct proof fails, consider a proof by contrapositive, especially when *Q* has the form of a negation. See Exercise 3(a). If direct proofs of $P \Rightarrow Q$ and $\sim Q \Rightarrow \sim P$ both fail, try the method of proof by contradiction. Assuming *P* and $\sim Q$ gives you more hypotheses to work with as you aim for some contradiction. See Exercise 5(b).

"If $P_1 \vee P_2$, then *Q*." It is usually best to first try to prove this by cases because $(P_1 \vee P_2) \Rightarrow Q$ is equivalent to $(P_1 \Rightarrow Q) \wedge (P_2 \Rightarrow Q)$. That is, try to show that (i) $P_1 \Rightarrow Q$ and (ii) $P_2 \Rightarrow Q$. See Exercise 7(b). Any proof of $P \Rightarrow Q$ that is done by considering cases has this form. See Exercise 1(h).

"If *P*, then $Q_1 \vee Q_2$." A good first step is to try to prove the equivalent form $(P \wedge \sim Q_1) \Rightarrow Q_2$. This method has that advantage of assuming that both *P* and $\sim Q_1$ are true, giving you more hypotheses to utilize. Or if you prefer, you could assume both *P* and $\sim Q_2$, and deduce Q_1. See Exercises 1(d) and (e).

"*P* iff *Q*." What you should hope for is an "iff" proof in which you construct a list of equivalent statements linking *P* and *Q*. But usually, and especially when *P* and *Q* are complicated, you will need to prove $P \Rightarrow Q$ and $Q \Rightarrow P$ separately. Rather than worrying about which proof form to use, a good strategy is to begin by proving either of the two implications and then checking to see whether each step can be reversed so that (by modifying the words that connect statements) the proof can be converted to an "iff" proof. See Exercise 1(c).

Here are some strategies for writing proofs of quantified sentences.

"$(\forall x) P(x)$." Usually there will be one or more universal quantifiers, which may be hidden. Your first sentence will almost always have the form "Let *x* be an object in _____" or "Suppose *x* is in _____," where we specify the universe. See Exercises 1(f) and 2(c). Proofs by contradiction of universally quantified statements are not so common. See the comments below on the form $\sim (\exists x) P(x)$, and Exercise 1(g).

"$(\exists x) P(x)$." You may be able to construct or guess an object that has the desired property. See Exercise 6(b). If not, you may be able to still prove existence without producing an actual object, perhaps by contradiction. See Exercise 4(b).

"$\sim (\exists x) P(x)$." You have two options, and the one you choose will depend on the form of $P(x)$. You might first try a direct proof of the equivalent statement $(\forall x) \sim P(x)$. The alternative is to assume $(\exists x) P(x)$ and find a contradiction. (This amounts to proving $(\forall x) \sim P(x)$ by contradiction.) See Exercise 1(g).

"$(\exists! x) P(x)$." First prove $(\exists x) P(x)$ as described above. To prove uniqueness, you may choose any one of several approaches. You may (1) prove that any two objects with the property must be equal, (2) derive a contradiction from the assumption that two different objects have the property, or (3) prove that every object with the property is identical to some specific object. See Exercises 3(c) and 4(d).

The remainder of this section is devoted to examples of proofs from elementary number theory—that branch of mathematics concerned with the integers and questions about divisibility, primes, and factorizations. The term "elementary" is used, not because the subject is low level, but because no methods from other fields of mathematics are used. Some of the most simply stated, yet still unsolved problems in mathematics come from elementary number theory.

Our proof examples are all concerned with the greatest common divisor (gcd) of two integers; a concept that is probably already familiar to you. We can't rely on just a general idea of gcd to prove theorems: It's not enough just to be able to find the gcd of 12 and 15. As you gain experience you will find that writing good proofs requires that we understand and use concepts precisely. By precisely, we mean *as specified by the definition*.

The most fundamental theorem about the integers is the Division Algorithm, which we state here without proof. In Chapter 2 the Division Algorithm will be presented as Theorem 2.5.1 and proved using a technique that will be introduced in Section 2.5.

The Division Algorithm (See Theorem 2.5.1)

For all integers a and b, with $a \neq 0$, there exist unique integers q and r such that

$$b = aq + r \text{ and } 0 \leq r < |a|.$$

The integer a is the **divisor**, q is the **quotient**, and r is the **remainder**. For example, 23 divided by 4 gives a quotient of 5 and remainder 3, because $23 = 4 \cdot 5 + 3$. Note, however, it would be incorrect to say that -23 divided by 4 has quotient -5 and remainder -3, even though $-23 = 4(-5) + (-3)$. Remainders can't be negative, so when we divide by 4 the only possible remainders are 0, 1, 2, and 3. Thus when -23 is divided by 4 the quotient is -6 and the remainder is 1.

It is the fact that the remainder must be nonnegative and as small as possible that makes the quotient and remainder unique. Notice that dividing b by a produces a remainder of 0 exactly when there is an integer q such that $b = aq + 0$, which happens exactly when a divides b.

One of the most useful concepts regarding integers is that of the greatest common divisor.

DEFINITIONS   Let a, b, c, and d be nonzero integers.

We say c is a **common divisor** of a and b iff c divides a and c divides b.
We say d is the **greatest common divisor** of a and b, and write $d = gcd(a, b)$, iff
    (i) d is a common divisor of a and b, and
    (ii) every common divisor c of a and b is less than or equal to d.

For example, the common divisors of 18 and 24 are $-6, -3, -2, -1, 1, 2, 3$, and 6, so $\gcd(18, 24) = 6$. There is no requirement that a and b must be positive. For example $\gcd(-5, 20) = 5$, $\gcd(21, -35) = 7$, and $\gcd(-9, -27) = 9$. The integers 24 and 35 have no positive common divisors except 1, so $\gcd(24, 35) = 1$. Since $\gcd(a, b)$ is greater than or equal to any common divisor of nonzero integers a and b, $\gcd(a, b)$ is always a positive integer.

An integer of the form $ax + by$, for integers x and y, is called a **linear combination** of a and b. For example, some linear combinations of 3 and 7 are:

$$1 = 3 \cdot (-2) + 7 \cdot 1 \qquad\qquad -2 = 3 \cdot 4 + 7 \cdot (-2)$$

$$58 = 3 \cdot 10 + 7 \cdot 4 \qquad\qquad -7 = 3 \cdot 0 + 7 \cdot (-1).$$

You could experiment with different values of x and y to find that every integer multiple of 3 is a linear combination of 12 and 15. For example,

$$0 = 12 \cdot 0 + 15 \cdot 0 \qquad\qquad 3 = 12 \cdot (-1) + 15 \cdot 1$$

$$-3 = 12 \cdot (-4) + 15 \cdot 3 \qquad\qquad 6 = 12 \cdot 8 + 15 \cdot (-6).$$

An exercise in the previous section established an interesting result about linear combinations: For all integers x and y, if c divides both a and b, then c also divides $ax + by$. This fact (Exercise 2(a)) can be restated in our new terminology as:

Theorem 1.7.1   Let a and b be integers. If c is a common divisor of a and b, then c divides every linear combination of a and b. In particular, $\gcd(a, b)$ divides every linear combination of a and b.

There is much more to be said about linear combinations. Whereas we look for the *greatest* common divisor of a and b, we look for the *smallest* positive linear combination of a and b. We see from the example above that 1 is a linear combination of 3 and 7, and so 1 must be the smallest positive linear combination. We also see above that 3 is a linear combination of 12 and 15, so the smallest positive linear combination of 12 and 15 must be 1, 2, or 3. But we can see that 1 is not a linear combination of 12 and 15 (See Exercise 1(d) of Section 1.6), and we can show in the same way that 2 is not a linear combination. Therefore, 3 is the smallest positive linear combination of 12 and 15.

It's a natural question to ask whether there is, for every pair a, b of nonzero integers, a smallest positive linear combination of a and b. There is, but once again we simply state the result here and wait until we have the tools in Chapter 2 to give the proof. See Theorem 2.5.2. Still, it's not too soon to see how we can use this result and basic proof techniques to understand the essential connection between the gcd and linear combinations.

Lemma 1.7.2   Let a and b be nonzero integers. Then the smallest positive linear combination of a and b is a common divisor of a and b.

Proof.   Let $d = as + bt$ be the smallest positive linear combination of a and b. ⟨*We need to show that d divides a and d divides b.*⟩ By the Division Algorithm there exist integers q and r such that $a = dq + r$, where $0 \leq r < d$. Then

$$r = a - dq$$
$$= a - (as + bt)q$$
$$= a - as - btq$$
$$= a(1 - s) + b(-tq),$$

which is a linear combination of a and b. But $0 \leq r < d$ and d is the smallest positive linear combination. We conclude that $r = 0$, so d divides a. In the same way, d divides b. Thus d is a common divisor of a and b.   ∎

Theorem 1.7.3   Let a and b be nonzero integers. The gcd of a and b is the smallest positive linear combination of a and b.

Proof.   Let $d = as + bt$ be the smallest positive linear combination of a and b. By Lemma 1.7.2, d is a common divisor of a and b. We must now show that every common divisor of a and b is less than or equal to d.

⟨*To show that every common divisor is less than or equal to d, we first prove that if c is any common divisor of a and b, then c divides d.*⟩ Suppose c is a common divisor of a and b. Then for some integers n and m, $a = cn$ and $b = cm$. Then

$$d = as + bt$$
$$= (cn)s + (cm)t$$
$$= c(ns + mt).$$

Therefore c divides d. We conclude that $c \leq d$. ⟨*We have used Exercise 7(g) of Section 1.4.*⟩ Therefore d is the greatest common divisor of a and b.   ∎

Now we know that gcd(a, b) is a linear combination of a and b, in fact the smallest linear combination, and it divides every linear combination. These facts are useful in many important applications, from coding theory to the solution of equations with integer coefficients. One immediate application is in establishing divisibility relationships among integers. For example, if we know that we can write 1 as a linear combination of two integers, then the only common divisors of those integers are 1 and -1.

DEFINITION   We say nonzero integers a and b are **relatively prime**, or **coprime**, iff gcd$(a, b) = 1$.

The numbers 12 and 35 are relatively prime. The numbers 15 and 36 are not, because $\gcd(15, 36) = 3$. The integer 2 is coprime with every odd integer.

Theorem 1.7.4

Let a and b be nonzero integers that are relatively prime, and let c be an integer. Then the equation $ax + by = c$ has an integer solution.

Proof.   See Exercise 18.   ∎

The next result, which is found in Euclid's *Elements*, makes use of the concepts of gcd and relatively prime.

Lemma 1.7.5

Euclid's Lemma. Let a, b, and p be integers. If p is a prime and p divides ab, then p divides a or p divides b.

Proof.   Suppose p is prime and p divides ab. Assume that p does not divide a. ⟨*We must show that p divides b.*⟩ Since p does not divide a, p and a are relatively prime, so there exist integers s and t such that $as + pt = 1$. Then $b = abs + bpt$. Since p divides abs and bpt, it divides their sum, so p divides b. We conclude that p divides a or p divides b.   ∎

Euclid's Lemma is frequently used in one of its equivalent forms:

> if p divides ab and p does not divide a, then p must divide b,

or

> if p does not divide a and p does not divide b, then p does not divide ab.

Exercises 1.7

1.

★ **(a)**   Prove that if n is an integer and $3n + 1$ is odd, then $2n + 8$ is divisible by 4.

★ **(b)**   Assume $a \neq 3$. Prove that if a is a solution to $x^2 - x - 6 = 0$, then a is a solution to $x^3 + 2x^2 + x + 3 = 0$.

★ **(c)**   Assume $a \neq 3$. Prove that a is a solution to $x^2 - x - 6 = 0$ iff a is a solution to $x^3 + 2x^2 + x + 3 = 0$.

★ **(d)**   Let x be a real number. Prove that if $x^2 = 2x + 15$, then $x \leq 2$ or $\dfrac{(x-4)}{(x-3)} > 0$.

★ **(e)**   Let x and y be real numbers. Prove that if $x + y$ is irrational, then either x or y is irrational.

★ **(f)**   Prove that if two nonvertical lines are perpendicular, then the product of their slopes is -1. (Recall that nonvertical lines are those lines in the plane that have slope.)

★ **(g)**   No point inside the circle $(x - 3)^2 + y^2 = 6$ is on the line $y = x + 1$.

★ **(h)**   Prove that for all real numbers $x \geq 1$, $\dfrac{3|x - 2|}{x} \leq 4$.

2. Prove that
   (a) for all integers n, $5n^2 + 3n + 4$ is even.
   (b) for all odd integers n, $2n^2 + 3n + 4$ is odd.
   (c) the sum of 5 consecutive integers is always divisible by 5.
   (d) if two nonvertical lines have slopes whose product is -1, then the lines are perpendicular.
   ☆ (e) for all integers n, $n^3 - n$ is divisible by 6.
   (f) for all integers n, $(n^3 - n)(n + 2)$ is divisible by 12.

3. Let L be the line $2x + ky = 3k$. Prove that
   (a) if $k \neq -6$, then L does not have slope $\frac{1}{3}$.
   (b) for every real number k, L is not parallel to the x-axis.
   (c) there is a unique real number k such that L passes through $(1, 4)$.

4. (a) Prove that if x is rational and y is irrational, then $x + y$ is irrational.
   ☆ (b) Prove that there exist irrational numbers x and y such that $x + y$ is rational.
   (c) Prove that for every rational number z, there exist irrational numbers x and y such that $x + y = z$.
   (d) Prove that for every rational number z and every irrational number x, there exists a unique irrational number y such that $x + y = z$.

5. (a) Prove that except for two points on the circle, if (x, y) is on the circle with center at the origin and radius r, then the line passing through (x, y) and $(r, 0)$ is perpendicular to the line passing through (x, y) and $(-r, 0)$. Which two points are the exceptions?
   (b) Let (x, y) be a point inside the circle with center at the origin and radius r. Prove that the line passing through (x, y) and $(r, 0)$ is not perpendicular to the line passing through (x, y) and $(-r, 0)$.

6. Prove that
   (a) every point on the line $y = 6 - x$ is outside the circle with radius 4 and center $(-3, 1)$.
   (b) Prove that there exists a three-digit natural number less than 400 with distinct digits, such that the sum of the digits is 17 and the product of the digits is 108.
   (c) Use the Extreme Value Theorem to prove that if f does not have a maximum value on the interval $[5, 7]$, then f is not differentiable on $[5, 7]$.
   (d) Use Rolle's Theorem to show that $x^3 + 6x - 1 = 0$ does not have more than one real solution.

7. Prove that for all real numbers x,
   (a) if $x > 0$, then $\dfrac{|2x - 1|}{x + 1} \leq 2$.

   (b) if $-2 < x < 1$ or $x > 3$, then $\dfrac{(x - 1)(x + 2)}{(x - 3)(x + 4)} > 0$.

8. Prove or disprove:
   ★ (a) Every point inside the circle $(x - 3)^2 + (y - 2)^2 = 4$ is inside the circle $x^2 + y^2 = 41$.
   (b) If (x, y) is inside the circle $(x - 3)^2 + (y - 2)^2 = 4$, then $x - 6 < 3y$.
   (c) Every point inside the circle $(x - 3)^2 + (y - 2)^2 = 4$ is inside the circle $(x - 5)^2 + (y + 1)^2 = 25$.

9. For each given pair a, b of integers, find the unique quotient and remainder when b is divided by a.
   (a) $a = 8, b = 310$
   (b) $a = 5, b = 36$
   (c) $a = -5, b = 36$
   ★ (d) $a = 5, b = -36$
   (e) $a = 7, b = 44$
   (f) $a = -8, b = -52$

10. (a) Let a and b be integers and $a > b$. Prove that if $b \geq 0$, then when b is divided by a, the quotient is 0.
    (b) Let a and b be integers and $a > b$. Prove that if the quotient is 0 when b is divided by a, then $b \geq 0$.

11. For each pair of integers, list all positive and negative common divisors, and find gcd (a, b).
    (a) $a = 8, b = 310$
    (b) $a = -5, b = 36$
    (c) $a = 18, b = -54$
    (d) $a = -8, b = -52$

12. (a) Write 2 in two different ways as a linear combination of 12 and 22.
    (b) Write -4 in two different ways as a linear combination of 12 and 22.
    (c) What is the set of all linear combinations of 12 and 22?

13. Find $d = \gcd(a, b)$ and integers x and y such that $d = ax + by$.
    (a) $a = 13, b = 15$
    (b) $a = 26, b = 32$
    (c) $a = 9, b = 30$.

14. Let a, b, and c be natural numbers and $\gcd(a, b) = d$. Prove that
    ☆ (a) if c divides a and c divides b, then c divides d.
    (b) a divides b iff $d = a$.
    ☆ (c) if a divides bc and $d = 1$, then a divides c.
    (d) if c divides a and c divides b, then $\gcd\left(\frac{a}{c}, \frac{b}{c}\right) = \frac{d}{c}$. In particular, $\gcd\left(\frac{a}{d}, \frac{b}{d}\right) = 1$.
    (e) for every natural number n, $\gcd(an, bn) = dn$.

15. Which elements of the set $\{3, 6, 10, 63\}$ are relatively prime to 7? to 21? to 30?

16. Prove that for every prime p and for all natural numbers a,
    ★ (a) $\gcd(p, a) = p$ iff p divides a.
    (b) $\gcd(p, a) = 1$ iff p does not divide a.

17. Let q be a natural number greater than 1 with the property that q divides a or q divides b whenever q divides ab. Prove that q is prime.

18. Let a and b be nonzero integers that are relatively prime, and let c be an integer. Prove that the equation $ax + by = c$ has an integer solution. (Theorem 1.7.4.) *Hint:* Use the fact that 1 is a linear combination of a and b.

19. Let a and b be nonzero integers and $d = \gcd(a, b)$. Let $m = \frac{b}{d}$ and $n = \frac{a}{d}$. Show that if $x = s$ and $y = t$ is a solution to $ax + by = c$, then so is $x = s + km$ and $y = t - kn$ for every integer k. (This shows how linear combinations help to describe solutions to equations.)

20. For nonzero integers a and b, the integer n is a **common multiple of a and b** iff a divides n and b divides n. We say the positive integer m is the **least common multiple of a and b**, written as **lcm(a, b)**, iff
   (i) m is a common multiple of a and b, and
   (ii) if n is a positive common multiple of a and b, then $m \le n$.
Find lcm(a, b) for
  ⋆ **(a)** $a = 6, b = 14$
      (b) $a = 10, b = 35$
      (c) $a = 21, b = 39$
      (d) $a = 12, b = 48$

21. Let a, b, and c be natural numbers, $\gcd(a, b) = d$ and lcm(a, b) $= m$. Prove that
  ☆ **(a)** a divides b iff $m = b$.
      (b) $m \le ab$.
  ☆ **(c)**   if $d = 1$, then $m = ab$.
      (d)   if c divides a and c divides b, then lcm$\left(\frac{a}{c}, \frac{b}{c}\right) = \frac{m}{c}$.
      (e)   for every natural number n, lcm(an, bn) $= mn$.
  ☆ **(f)** $\gcd(a, b) \cdot$ lcm(a, b) $= ab$.

22. Let a and b be integers, and let $m = $ lcm(a, b). Use the Division Algorithm to prove that if c is a common multiple of a and b, then m divides c.

Proofs to Grade   **23.** Assign a grade of A (correct), C (partially correct), or F (failure) to each. Justify assignments of grades other than A.
      (a) **Claim.**   There is a unique 3-digit number whose digits have sum 8 and product 10.
            "*Proof.*"   Let x, y, and z be the digits. Then $x + y + z = 8$ and $xyz = 10$. The only factors of 10 are 1, 2, 5, and 10, but since 10 is not a digit, the digits must be 1, 2, and 5. The sum of these digits is 8. Therefore, 125 is the only 3-digit number whose digits have sum 8 and product 10.   ■
  ⋆ **(b)** **Claim.**   There is a unique set of three consecutive odd numbers that are all prime.
            "*Proof.*"   The consecutive odd numbers 3, 5, and 7 are all prime. Suppose x, y, and z are consecutive odd numbers, all prime, and $x \ne 3$. Then $y = x + 2$ and $z = x + 4$. Since x is prime, when x is divided by 3, the remainder is 1 or 2. In case the remainder is 1, then $x = 3k + 1$ for some

integer $k \geq 1$. But then $y = x + 2 = 3k + 3 = 3(k + 1)$, so y is not prime. In case the remainder is 2, then $x = 3k + 2$ for some integer $k \geq 1$. But then $z = x + 4 = 3k + 2 + 4 = 3(k + 2)$, so z is not prime. In either case we reach the contradiction that y or z is not prime. Thus $x = 3$ and $y = 5$, $z = 7$. Therefore, the only three consecutive odd primes are 3, 5, and 7. ∎

(c) **Claim.** If x is any real number, then either $\pi - x$ is irrational or $\pi + x$ is irrational.

"***Proof.***" It is known that π is an irrational number; that is, π cannot be written in the form $\frac{a}{b}$ for integers a and b. Consider $x = \pi$. Then $\pi - x = 0$, which is rational, but $\pi + x = 2\pi$. If 2π were rational, then $2\pi = \frac{a}{b}$ for some integers a and b. Then $\pi = \frac{a}{2b}$, so π is rational. This is impossible, so 2π is irrational. Therefore either $\pi - x$ or $\pi + x$ is irrational. ∎

(d) **Claim.** If x is any real number, then either $\pi - x$ is irrational or $\pi + x$ is irrational.

"***Proof.***" It is known that π is an irrational number; that is, π cannot be written in the form $\frac{a}{b}$ for integers a and b. Let x be any real number. Suppose both $\pi - x$ and $\pi + x$ are rational. Then since the sum of two rational numbers is always rational, $(\pi - x) + (\pi + x) = 2\pi$ is rational. Then $2\pi = \frac{a}{b}$ for some integers a and b. Then $\pi = \frac{a}{2b}$, so π is rational. This is impossible. Therefore, at least one of $\pi - x$ or $\pi + x$ is irrational. ∎

(e) **Claim.** For all natural numbers n, $\gcd(n, n + 1) = 1$.

"***Proof.***" (i) 1 divides n and 1 divides $n + 1$. (ii) Suppose c divides n and c divides $n + 1$. Then 1 divides c. Therefore, $\gcd(n, n + 1) = 1$. ∎

(f) **Claim.** For all natural numbers n, $\gcd(2n - 1, 2n + 1) = 1$.

"***Proof.***" Obviously 1 divides both $2n - 1$ and $2n + 1$. Suppose c divides $2n - 1$ and $2n + 1$. Then c divides their sum, $4n$, so c also divides $4n^2$. Furthermore, c divides their product, $4n^2 - 1$. Since c divides $4n^2$ and $4n^2 - 1$, c divides $4n^2 - (4n^2 - 1) = 1$. Therefore, $c \leq 1$. Thus 1 is the greatest common divisor. ∎

Set Theory

Starting from the theory of sets, one can construct all the number systems, functions, calculus, and other areas of mathematics. Thus, the study of sets is the foundation for the entire structure of mathematics.

This chapter does not develop these constructions but does provide some set-theoretic concepts used throughout the text and advanced mathematics. Sections 2.1 and 2.2 provide precise definitions for familiar concepts such as union and intersection. In Section 2.3 we extend the union and intersection operations to collections of sets and discuss how to use indices to organize a family of sets. Proofs methods using forms of mathematical induction are discussed in Sections 2.4 and 2.5. Basic methods for counting the elements in a finite set appear in the optional Section 2.6.

2.1   Basic Concepts of Set Theory

We assume that you have had some experience with sets, set notations, and common sets of numbers such as the integers and real numbers as described in the *Preface to the Student*. In general, capital letters will be used to denote sets and lowercase letters to denote the elements in sets. To designate a set, we use the notation

$$\{x \colon P(x)\},$$

where $P(x)$ is a one-variable open sentence description of the property that defines the set. For example, the set $A = \{1, 3, 5, 7, 9, 11, 13\}$ may be written as

$$\{x \colon x \in \mathbb{N}, x \text{ is odd, and } x < 14\}.$$

The set of all integer multiples of 3 is the set $3\mathbb{Z} = \{3z \colon z \in \mathbb{Z}\}$, and this set contains $0, 3, -3, 6, -6, 9, -9$, etc.

A word of caution: Some sentences $P(x)$ may not be used to define a set. In 1902, when the theory of sets was new, Bertrand Russell* and others pointed out flaws in the then common assumption that for every open sentence $P(x)$, there corresponds a set $\{x\colon P(x)\}$. See Exercise 3 for a version of the Russell paradox.

The resolution of Russell's and other paradoxes involved making a distinction between sets and arbitrary collections of objects. Sets may be defined within a system of axioms for set theory, first developed by Ernst Zermelo[†] and Abraham Fraenkel.** Their axioms assert, for example, that a collection of two sets constitutes a set (Axiom of Pairing) and that the collection of all subsets of a set is a set (Axiom of Powers). Under their system, known paradoxes such as Russell's are avoided.

It is not our purpose here to carry out a formal study of axiomatic set theory.[††] However, all of our discussions of sets are consistent with the Zermelo–Fraenkel system of axiomatic set theory.

A second word of caution: Recall that the universe of discourse is a collection of objects understood from the context or specified at the outset of a discussion and that all objects under consideration must belong to the universe. Some ambiguity may arise unless the universe is known. For example, membership in the set $A = \{x\colon x^2 - 6x = 0\}$ depends on an agreed upon universe. For the universe of real numbers A is $\{0, 6\}$, but A is $\{6\}$ for the universe of natural numbers.

DEFINITION   Let $\varnothing = \{x\colon x \neq x\}$. Then $\varnothing$ is a set, called the **empty set** or **null set**.

It is an *axiom* that $\varnothing$ is a set. Since for every object x in every universe, x is equal (identical) to x, there are no elements in the collection $\varnothing$. That is, the statement $x \in \varnothing$ is *false* for every object x. We could define other empty collections, such as $B = \{x\colon x \in \mathbb{R} \text{ and } x^2 < 0\}$, but we will soon prove that all such collections are equal, so there really is just one empty set.

In the *Preface to the Student* we said A is a **subset** of B and wrote $A \subseteq B$ if and only if every element of A is an element of B. If A is not a subset of B we write $A \nsubseteq B$. For $X = \{2, 4\}$, $Y = \{2, 3, 4, 5\}$, and $Z = \{2, 3, 6\}$, $X \subseteq Y$ and $X \nsubseteq Z$.

In symbols, we write the definition of $A \subseteq B$ as

$$A \subseteq B \Leftrightarrow (\forall x)(x \in A \Rightarrow x \in B).$$

---

* Bertrand Russell (1872–1970) was a British philosopher and mathematician and strong proponent for social reform. He coauthored *Principia Mathematica* (1910–1913), a monumental effort to derive all of mathematics from a specific set of axioms and a well defined set of rules of inference.

[†] Ernst Zermelo (1871–1953) was a German mathematician whose work on the axioms of set theory has profoundly influenced the foundations of mathematics. In 1905 he discovered a paradox similar to the Russell paradox. He developed a theory of sets based on seven axioms, but was unable to prove that no new paradoxes could arise in his system.

** Abraham Fraenkel (1891–1965), born in Germany, spent much of his career in Israel. In the 1920s he made attempts to improve the set theoretic axioms of Zermelo to eliminate paradoxes. Within his system of ten axioms he proved the independence of the Axiom of Choice. (See Section 5.5.)

[††] A complete study of the foundations of set theory from the Zermelo–Fraenkel axioms may be found in *Notes on Set Theory* by Y. N. Moschovakis (Springer-Verlag, Berlin, 1994). The study of set theory is still active today, with many unsolved problems.

Therefore, a proof of the statement $A \subseteq B$ is often a direct proof, taking the form:

DIRECT PROOF OF $A \subseteq B$
Proof.
Let x be any object.
Suppose $x \in A$.
$\vdots$

Thus $x \in B$.
Therefore $A \subseteq B$.   ∎

Example. Let $A = \{2, -3\}$ and $B = \{x \in \mathbb{R}: x^3 + 3x^2 - 4x - 12 = 0\}$. Prove that $A \subseteq B$.

Proof. Suppose $x \in A$. ⟨*We show $A \subseteq B$ by individually checking each element of A.*⟩ Then $x = 2$ or $x = -3$. For $x = 2$, $2^3 + 3(2^2) - 4(2) - 12 = 8 + 12 - 8 - 12 = 0$. For $x = -3$, $(-3)^3 + 3(-3)^2 - 4(-3) - 12 = -27 + 27 + 12 - 12 = 0$. In both cases, $x \in B$. Thus, $A \subseteq B$.   ∎

Example. Let a and b be natural numbers, and let $a\mathbb{Z}$ and $b\mathbb{Z}$ be the sets of all integer multiples of a and b, respectively. Prove that if a divides b, then $b\mathbb{Z} \subseteq a\mathbb{Z}$.

Proof. Suppose that a divides b. Then there exists an integer c such that $b = ac$. ⟨*To show $b\mathbb{Z} \subseteq a\mathbb{Z}$, we start with an element from $b\mathbb{Z}$.*⟩ Let $x \in b\mathbb{Z}$. Then x is a multiple of b, so there exists an integer d such that $x = bd$. But then $x = bd = (ac)d = a(cd)$. Therefore x is a multiple of a, so $x \in a\mathbb{Z}$.   ∎

Theorem 2.1.1

(a)   For every set A, $\emptyset \subseteq A$.
(b)   For every set A, $A \subseteq A$.
(c)   For all sets A, B, and C, if $A \subseteq B$ and $B \subseteq C$, then $A \subseteq C$.

Proof.

(a)   Let A be any set. Let x be any object. Because the antecedent is false, the sentence $x \in \emptyset \Rightarrow x \in A$ is true. Therefore, $\emptyset \subseteq A$.
(b)   Let A be any set. ⟨*To prove $A \subseteq A$, we must show that for all objects x, if $x \in A$ then $x \in A$.*⟩ Let x be any object. Then $x \in A \Rightarrow x \in A$ is true. ⟨*Here we use the tautology $P \Rightarrow P$.*⟩ Therefore, $(\forall x)(x \in A \Rightarrow x \in A)$ and so $A \subseteq A$.
(c)   See Exercise 8.   ∎

Recall that sets A and B are **equal** iff they have exactly the same elements; that is,

$$A = B \text{ iff } (\forall x)(x \in A \Leftrightarrow x \in B).$$

Thus, one method to prove $A = B$ is to give a sequence of equivalent statements starting with the statement $x \in A$ and ending with $x \in B$. However, since $x \in A \Leftrightarrow x \in B$ is equivalent to $(x \in A \Rightarrow x \in B) \wedge (x \in B \Rightarrow x \in A)$, we may also say

$$A = B \text{ iff } A \subseteq B \text{ and } B \subseteq A.$$

For this reason, a proof that $A = B$ will typically have the form:

> **TWO PART PROOF OF $A = B$**
> **Proof.**
>   (i) Prove that $A \subseteq B$ (by any method).
>   (ii) Prove that $B \subseteq A$ (by any method).
>   (iii) Therefore $A = B$.   ∎

Example.   Prove that $X = Y$ where $X = \{x \in \mathbb{R}: x^2 - 1 = 0\}$ and $Y = \{-1, 1\}$.

Proof.

   (i)   We show $Y \subseteq X$ by individually checking each element of Y. By substitution, we see that both 1 and -1 are solutions to $x^2 - 1 = 0$. Thus $Y \subseteq X$.

   (ii)   Next, we must show $X \subseteq Y$. Let $t \in X$. Then, by definition of X, t is a solution to $x^2 - 1 = 0$. Thus $t^2 - 1 = 0$. Factoring, we have $(t - 1)(t + 1) = 0$. This product is 0 exactly when $t - 1 = 0$ or $t + 1 = 0$. Therefore, $t = 1$ or $t = -1$. Thus if t is a solution, then $t = 1$ or $t = -1$; so $t \in Y$. This proves $X \subseteq Y$.

   (iii) By (i) and (ii), $X = Y$.   ∎

The set B is a **proper** subset of the set A iff $B \subseteq A$ and $A \neq B$. To denote that B is a proper subset of A, some authors write $B \subset A$ and others write $B \subsetneq A$. The only improper subset of A is the set A itself.

We are now in a position to prove that there is only one empty set, in the sense that any two empty sets are equal.

Theorem 2.1.2   If A and B are sets with no elements, then $A = B$.

Proof.   Since A has no elements, the sentence $(\forall x)(x \in A \Rightarrow x \in B)$ is true. Therefore, $A \subseteq B$. Similarly, $(\forall x)(x \in B \Rightarrow x \in A)$ is true, so $B \subseteq A$. Therefore, by definition of set equality, $A = B$.   ∎

Theorem 2.1.3   For any sets A and B, if $A \subseteq B$ and $A \neq \emptyset$, then $B \neq \emptyset$.

Proof.   Suppose $A \subseteq B$ and $A \neq \emptyset$. Since A is nonempty, there is an object t such that $t \in A$. Since $t \in A$, $t \in B$. Therefore, $B \neq \emptyset$.   ∎

We sometimes use Venn* diagrams to display simple relationships among sets. For example, suppose we want to find nonempty sets A, B, and C such that $A \subseteq B$, $A \neq B$, $C \subseteq A$, and $A \nsubseteq C$. We begin with three overlapping sets that represent the sets A, B, and C in Figure 2.1.1(a). Since $A \subseteq B$, there are no elements in the two regions of A that are outside B. Since $C \subseteq A$, there are no elements in the two regions of C that are outside A. These four regions are shaded in Figure 2.1.1(b). Since A is not a subset of C, there is some element x in the remaining region of A that does not overlap C, and since $A \neq B$ there is some element y in B that is not in A. Finally, C is required to be nonempty, so there is an element z in C. There may be other elements in these sets, but the solution we have found is $A = \{x, z\}$, $B = \{x, y, z\}$ and $C = \{z\}$. (See Figure 2.1.1(c).)

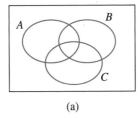
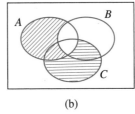
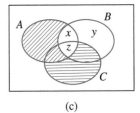

(a)   (b)   (c)

Figure 2.1.1

One of the axioms of set theory asserts that for every set A, the collection of all subsets of A is also a set.

DEFINITION   Let A be a set. The **power set** of A is the set whose elements are the subsets of A and is denoted $\mathscr{P}(A)$. Thus

$$\mathscr{P}(A) = \{B : B \subseteq A\}.$$

Notice that the power set of a set A is a set whose elements are themselves sets, specifically the subsets of A. For example, if $A = \{a, b, c, d\}$, then the power set of A is

$$\mathscr{P}(A) = \{\varnothing, \{a\}, \{b\}, \{c\}, \{d\}, \{a, b\}, \{a, c\}, \{a, d\}, \{b, c\}, \{b, d\}, \{c, d\},$$
$$\{a, b, c\}, \{a, b, d\}, \{a, c, d\}, \{b, c, d\}, A\}.$$

When we work with sets whose elements are sets, it is important to recognize the distinction between "is an element of" and "is a subset of." To use $A \in B$ correctly, we must consider whether the object A (which happens to be a set) is an element of the set B, whereas $A \subseteq B$ requires determining whether all objects in the set A are also in B. If $x \in A$ and $B \subseteq A$, the correct terminology is that A *contains* x and A *includes* B.

---

* John Venn (1834–1923) was a British philosopher and logician best known for his diagrams to describe relationships.

Example. Let $X = \{\{1, 2, 3\}, \{4, 5\}, 6\}$. Then X is a set with three elements, namely, the set $\{1, 2, 3\}$, the set $\{4, 5\}$, and the number 6. $\mathcal{P}(X) = \{\varnothing, \{\{1, 2, 3\}\}, \{\{4, 5\}\}, \{6\}, \{\{1, 2, 3\}, \{4, 5\}\}, \{\{1, 2, 3\}, 6\}, \{\{4, 5\}, 6\}, X\}$. The set $\{\{4, 5\}\}$ has one element; it is $\{4, 5\}$. For the set X:

$$6 \in X \quad \text{and} \quad \{4, 5\} \in X, \quad \text{but} \quad 4 \notin X.$$
$$\{4\} \notin \{4, 5\} \quad \text{but} \quad \{4\} \subseteq \{4, 5\}.$$
$$\{4, 5\} \not\subseteq X \quad \text{because} \quad 5 \notin X.$$
$$\{6\} \subseteq X \quad \text{but} \quad \{6\} \notin X.$$
$$\{6\} \in \mathcal{P}(X) \quad \text{but} \quad \{6\} \not\subseteq \mathcal{P}(X).$$
$$\{\{4, 5\}\} \subseteq X \quad \text{because} \quad \{4, 5\} \in X.$$
$$\{4, 5\} \notin \mathcal{P}(X) \quad \text{but} \quad \{\{4, 5\}\} \in \mathcal{P}(X).$$
$$\varnothing \subseteq X, \quad \text{so} \quad \varnothing \in \mathcal{P}(X), \quad \text{and} \quad \{\varnothing\} \subseteq \mathcal{P}(X).$$

Notice that for the set $A = \{a, b, c, d\}$, which has four elements, $\mathcal{P}(A)$ has $16 = 2^4$ elements and for the set X above with three elements, $\mathcal{P}(X)$ has $8 = 2^3$ elements. These observations illustrate the next theorem.

Theorem 2.1.4   If A is a set with n elements, then $\mathcal{P}(A)$ is a set with 2^n elements.*

Proof. ⟨*The number of elements in $\mathcal{P}(A)$ is the number of subsets of A. Thus to prove this result, we must count all the subsets of A.*⟩ If $n = 0$, that is, if A is the empty set, then $\mathcal{P}(\varnothing) = \{\varnothing\}$, which is a set with $2^0 = 1$ elements. Thus the theorem is true for $n = 0$.

Suppose A has n elements, for $n \geq 1$. We may write A as $A = \{x_1, x_2, \ldots, x_n\}$. To describe a subset B of A, we need to know for each $x_i \in A$ whether the element is in B. For each x_i, there are two possibilities ($x_i \in B$ or $x_i \notin B$), so there are $2 \cdot 2 \cdot 2 \cdots \cdots 2$ (n factors) different ways of making a subset of A. Therefore $\mathcal{P}(A)$ has 2^n elements. ⟨*The counting rule used here is called the Product Rule. See Theorem 2.6.5 and the discussion following that theorem.*⟩ ∎

The next theorem is a good example of a biconditional statement for which a two-part proof is easier than an iff proof.

Theorem 2.1.5   Let A and B be sets. Then $A \subseteq B$ iff $\mathcal{P}(A) \subseteq \mathcal{P}(B)$.

Proof.

(i)   We must show that $A \subseteq B$ implies $\mathcal{P}(A) \subseteq \mathcal{P}(B)$. Assume that $A \subseteq B$ and suppose $X \in \mathcal{P}(A)$. We must show that $X \in \mathcal{P}(B)$. But $X \in \mathcal{P}(A)$ implies $X \subseteq A$. Since $X \subseteq A$ and $A \subseteq B$, then $X \subseteq B$ by Theorem 2.1.1. But $X \subseteq B$ implies $X \in \mathcal{P}(B)$. Therefore, $X \in \mathcal{P}(A)$ implies $X \in \mathcal{P}(B)$. Thus $\mathcal{P}(A) \subseteq \mathcal{P}(B)$.

(ii)   We must show that $\mathcal{P}(A) \subseteq \mathcal{P}(B)$ implies $A \subseteq B$. Assume that $\mathcal{P}(A) \subseteq \mathcal{P}(B)$. By Theorem 2.1.1 $A \subseteq A$; so $A \in \mathcal{P}(A)$. Since $\mathcal{P}(A) \subseteq \mathcal{P}(B)$, $A \in \mathcal{P}(B)$. Therefore $A \subseteq B$. ∎

---

* This theorem is the reason that some mathematicians use 2^A to denote the power set of A.

The second half of the proof of Theorem 2.1.5 could have been done differently. We could have shown that $A \subseteq B$ by giving a direct proof that $x \in A$ implies $x \in B$. A proof that consists of a series of steps beginning with "Assume $x \in A$" and leads to a conclusion that "Therefore $x \in B$" is often called an *element-chasing proof,* and is the natural way to prove the most basic facts about sets. As we build our knowledge of set properties in the next section, we may use theorems already proved, as we did above by using Theorem 2.1.1, to shorten our proof of Theorem 2.1.5. An element-chasing proof of part (ii) would be just as correct, but most people prefer a shorter, more elegant proof. When you write proofs, you may choose one method of proof over another because it is shorter, is easier to understand, or for any other reason.

Exercises 2.1

1. Write the following sets by using the set notation $\{x: P(x)\}$.
   ⋆ **(a)**   The set of natural numbers strictly less than 6
   (b)   The set of integers whose square is less than 17
   ⋆ **(c)** $[2, 6]$
   (d) $(-1, 9]$
   (e) $[-5, -1)$
   (f)   The set of rational numbers less than -1

2. Let $X = \{x: P(x)\}$. Are the following statements true or false?
   (a)   If $a \in X$, then $P(a)$.
   (b)   If $P(a)$, then $a \in X$.
   (c)   If $\sim P(a)$, then $a \notin X$.

3. ⋆ **(a)**   (*Russell paradox*)   A logical difficulty arises from the idea, which at first appears natural, of calling any collection of objects a set. Let's say that set B is **ordinary** if $B \notin B$. For example, if B is the set of all chairs, then $B \notin B$, because B is not a chair. It is only in the case of very unusual collections that we are tempted to say that a set is a member of itself. (The collection of all abstract ideas certainly is an abstract idea.) Let $X = \{x: x \text{ is an ordinary set}\}$. Is $X \in X$? Is $X \notin X$? What should we say about the collection of all ordinary sets?
   (b)   In the town of Seville, the (male) barber shaves all the men, and only the men, who do not shave themselves. Let A be the set of all men in the town who do not shave themselves. Who shaves the barber? (That is, is the barber an element of A? Is he not an element of A?)

4. True or false?
   ⋆ **(a)** $\mathbb{N} \subseteq \mathbb{Q}$.
   (b) $\mathbb{Q} \subseteq \mathbb{Z}$.
   ⋆ **(c)** $\mathbb{N} \subseteq \mathbb{R}$.
   (d) $\left[\frac{1}{2}, \frac{5}{2}\right] \subseteq \mathbb{Q}$.
   ⋆ **(e)** $\left[\frac{1}{2}, \frac{5}{2}\right] \subseteq \left(\frac{1}{2}, \frac{5}{2}\right)$.
   (f) $\mathbb{R} \subseteq \mathbb{Q}$.
   ⋆ **(g)** $[7, 10] \subseteq \mathbb{R}$.
   (h) $[2, 5] = \{2, 3, 4, 5\}$.
   ⋆ **(i)** $[7, 10) \subseteq \{7, 8, 9, 10\}$.
   (j) $(6, 9] \subseteq [6, 10)$.

5. True or false?
   ★ (a) $\varnothing \in \{\varnothing, \{\varnothing\}\}$.
   ★ (b) $\varnothing \subseteq \{\varnothing, \{\varnothing\}\}$.
   ★ (c) $\{\varnothing\} \in \{\varnothing, \{\varnothing\}\}$.
   (d) $\{\varnothing\} \subseteq \{\varnothing, \{\varnothing\}\}$.
   ★ (e) $\{\{\varnothing\}\} \in \{\varnothing, \{\varnothing\}\}$.
   (f) $\{\{\varnothing\}\} \subseteq \{\varnothing, \{\varnothing\}\}$.
   ★ (g) For every set A, $\varnothing \in A$.
   (h) For every set A, $\{\varnothing\} \subseteq A$.
   ★ (i) $\{\varnothing, \{\varnothing\}\} \subseteq \{\{\varnothing, \{\varnothing\}\}\}$.
   (j) $\{1, 2\} \in \{\{1, 2, 3\}, \{1, 3\}, 1, 2\}$.
   ★ (k) $\{1, 2, 3\} \subseteq \{1, 2, 3, \{4\}\}$.
   (l) $\{\{4\}\} \subseteq \{1, 2, 3, \{4\}\}$.

6. Give an example, if there is one, of sets A, B, and C such that the following are true. If there is no example, write "Not possible."
   ★ (a) $A \subseteq B$, $B \nsubseteq C$, and $A \subseteq C$.
   (b) $A \subseteq B$, $B \subseteq C$, and $C \subseteq A$.
   ★ (c) $A \nsubseteq B$, $B \nsubseteq C$, and $A \subseteq C$.
   (d) $A \subseteq B$, $B \nsubseteq C$, and $A \nsubseteq C$.

7. Prove that if $x \notin B$ and $A \subseteq B$, then $x \notin A$.

☆ 8. Prove part (c) of Theorem 2.1.1: For all sets A, B, and C, if $A \subseteq B$ and $B \subseteq C$, then $A \subseteq C$.

☆ 9. Prove that if $A \subseteq B$, $B \subseteq C$, and $C \subseteq A$, then $A = B$ and $B = C$.

10. Suppose that $X = \{x : x \in \mathbb{R}$ and x is a solution to $x^2 - 7x + 12 = 0\}$ and $Y = \{3, 4\}$. Prove that $X = Y$.

11. Let $X = \{x \in \mathbb{Z} : |x| \leq 3\}$ and $Y = \{-3, -2, -1, 0, 1, 2, 3\}$. Prove that $X = Y$.

12. Prove that $X = Y$, where $X = \{x \in \mathbb{N} : x^2 < 30\}$ and $Y = \{1, 2, 3, 4, 5\}$.

13. For a natural number a, let $a\mathbb{Z}$ be the set of all integer multiples of a. Prove that for all $a, b \in \mathbb{N}$, $a = b$ iff $a\mathbb{Z} = b\mathbb{Z}$.

14. Write the power set, $\mathcal{P}(X)$, for each of the following sets.
   ★ (a) $X = \{0, \triangle, \square\}$
   (b) $X = \{S, \{S\}\}$
   ★ (c) $X = \{\varnothing, \{a\}, \{b\}, \{a, b\}\}$
   (d) $X = \{1, \{\varnothing\}, \{2, \{3\}\}\}$

15. Let A, B, and C be sets and x and y be any objects. True or false?
   ★ (a) If $x \in A$, then $x \in \mathcal{P}(A)$.
   (b) If $x \in A$, then $\{x\} \in \mathcal{P}(A)$.
   (c) If $x \in A$, then $\{x\} \subseteq \mathcal{P}(A)$.
   (d) If $\{x, y\} \in \mathcal{P}(A)$, then $x \in A$ and $y \in A$.
   ★ (e) If $B \subseteq A$, then $\{B\} \in \mathcal{P}(A)$.
   (f) If $B \subseteq A$, then $B \in \mathcal{P}(A)$.
   ★ (g) If $B \in \mathcal{P}(A)$, then $B \subseteq A$.
   (h) If $C \subseteq B$ and $B \in \mathcal{P}(A)$, then $C \in \mathcal{P}(A)$.

16. List all of the proper subsets for each of the following sets.
   ★ (a) $\varnothing$
   (b) $\{\varnothing, \{\varnothing\}\}$
   ★ (c) $\{1, 2\}$
   (d) $\{0, \triangle, \square\}$

17. True or false?
   ★ (a) $\varnothing \in \mathcal{P}(\{\varnothing, \{\varnothing\}\})$.
   (b) $\{\varnothing\} \in \mathcal{P}(\{\varnothing, \{\varnothing\}\})$.
   ★ (c) $\{\{\varnothing\}\} \in \mathcal{P}(\{\varnothing, \{\varnothing\}\})$.
   (d) $\varnothing \subseteq \mathcal{P}(\{\varnothing, \{\varnothing\}\})$.
   ★ (e) $\{\varnothing\} \subseteq \mathcal{P}(\{\varnothing, \{\varnothing\}\})$.
   (f) $\{\{\varnothing\}\} \subseteq \mathcal{P}(\{\varnothing, \{\varnothing\}\})$.

★ **(g)** $3 \in \mathbb{Q}$.   **(h)** $\{3\} \subseteq \mathcal{P}(\mathbb{Q})$.

★ **(i)** $\{3\} \in \mathcal{P}(\mathbb{Q})$.   **(j)** $\{\{3\}\} \subseteq \mathcal{P}(\mathbb{Q})$.

★ **(k)** $\{3\} \subseteq \mathbb{Q}$.   **(l)** $\{\{3\}\} \in \mathcal{P}(\mathbb{Q})$.

18. Let A and B be sets. Prove that $A = B$ iff $\mathcal{P}(A) = \mathcal{P}(B)$.

Proofs to Grade   **19.** Assign a grade A (correct), C (partially correct), or F (failure) to each. Justify assignments of grades other than A.

(a) **Claim.** If $X = \{x \in \mathbb{N}: x^2 < 14\}$ and $Y = \{1, 2, 3\}$, then $X = Y$.

   "Proof." Since $1^2 = 1 < 14$, $2^2 = 4 < 14$, and $3^2 = 9 < 14$, $X = Y$.   ■

(b) **Claim.** If A, B, and C are sets, and $A \subseteq B$ and $B \subseteq C$, the $A \subseteq C$.

   "Proof." Let $A = \{1, 5, 8\}$, $B = \{1, 4, 5, 8, 10\}$, and $C = \{1, 2, 4, 5, 6, 8, 10\}$. Then $A \subseteq B$, and $B \subseteq C$, and $A \subseteq C$.   ■

★ **(c)** **Claim.** If A, B, and C are sets, and $A \subseteq B$ and $B \subseteq C$, then $A \subseteq C$.

   "Proof." Suppose x is any object. If $x \in A$, then $x \in B$, since $A \subseteq B$. If $x \in B$, then $x \in C$, since $B \subseteq C$. Thus, $x \in C$. Therefore, $A \subseteq C$.   ■

(d) **Claim.** If A, B, and C are sets, and $A \subseteq B$ and $B \subseteq C$, then $A \subseteq C$.

   "Proof." If $x \in C$, then, since $B \subseteq C$, $x \in B$. Since $A \subseteq B$ and $x \in B$, it follows that $x \in A$. Thus $x \in C$ implies $x \in A$. Therefore, $A \subseteq C$.   ■

★ **(e)** **Claim.** If A, B, and C are sets, and $A \subseteq B$ and $B \subseteq C$, then $A \subseteq C$.

   "Proof." Suppose $A \subseteq B$ and $B \subseteq C$. Then $x \in A$ and $x \in B$, because $A \subseteq B$. Then $x \in B$ and $x \in C$, because $B \subseteq C$. Therefore, $x \in A$ and $x \in C$, so $A \subseteq C$.   ■

(f) **Claim.** If A is a set, $A \subseteq \mathcal{P}(A)$.

   "Proof." Assume A is a set. Suppose $x \in A$. Then $x \subseteq A$. Thus $x \in \mathcal{P}(A)$. Therefore $A \subseteq \mathcal{P}(A)$.   ■

(g) **Claim.** If A is a set, $A \subseteq \mathcal{P}(A)$.

   "Proof." Assume A is a set. Suppose $x \in A$. Then $\{x\} \subseteq A$. Thus $\{x\} \in \mathcal{P}(A)$. Therefore, $A \subseteq \mathcal{P}(A)$.   ■

★ **(h)** **Claim.** If A and B are sets and $\mathcal{P}(A) \subseteq \mathcal{P}(B)$, then $A \subseteq B$.

   "Proof." $x \in A \Rightarrow \{x\} \subseteq A$

$\Rightarrow \{x\} \in \mathcal{P}(A)$

$\Rightarrow \{x\} \in \mathcal{P}(B)$

$\Rightarrow \{x\} \subseteq B$

$\Rightarrow x \in B$.

Therefore, $x \in A \Rightarrow x \in B$. Thus $A \subseteq B$.   ■

★ **(i)** **Claim.** If $A \subseteq B$ and $B \nsubseteq C$, then $A \nsubseteq C$.

   "Proof." Suppose $A \subseteq B$ and $B \nsubseteq C$. Then there exists $x \in B$ such that $x \notin C$. Since $x \in B$, $x \in A$ by definition of subset. Thus $x \in A$ and $x \notin C$. Therefore $A \nsubseteq C$.   ■

2.2   Set Operations

In this section we give precise definitions and prove some well-known properties of familiar operations on sets. Set union, interesection, and difference are called *binary* operations because each combines two sets to produce another set.

> **DEFINITIONS**   Let A and B be sets.
>
> The **union of A and B** is the set $A \cup B = \{x : x \in A \text{ or } x \in B\}$.
> The **intersection of A and B** is the set $A \cap B = \{x : x \in A \text{ and } x \in B\}$.
> The **difference of A and B** is the set $A - B = \{x : x \in A \text{ and } x \notin B\}$.

The set $A \cup B$ is a set formed from A and B by choosing as elements the objects contained in at least one of A or B; $A \cap B$ consists of all objects that appear in both A and B; and $A - B$ contains exactly those elements of A that are not in B. The shaded areas in the first three Venn diagrams of Figure 2.2.1 represent, respectively, the result of forming the union, intersection, and difference of two sets. These visual representations are often useful for understanding relationships among sets. However, when there are more than three sets involved, it is difficult or impossible to use Venn diagrams.

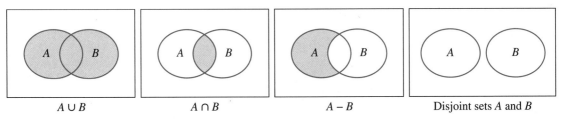

| $A \cup B$ | $A \cap B$ | $A - B$ | Disjoint sets A and B |

Figure 2.2.1

Examples.   For $A = \{1, 2, 4, 5, 7\}$ and $B = \{1, 3, 5, 9\}$,

$$A \cup B = \{1, 2, 3, 4, 5, 7, 9\},$$
$$A \cap B = \{1, 5\},$$
$$A - B = \{2, 4, 7\},$$
$$B - A = \{3, 9\}.$$

Examples.   For intervals of real numbers, we have

$$[3, 6] \cup [4, 8) = [3, 8)$$
$$[3, 6] \cap [4, 8) = [4, 6]$$
$$[3, 6] - [4, 8) = [3, 4)$$
$$[4, 8) - [3, 6] = (6, 8)$$
$$[4, 8) - (5, 6] = [4, 5] \cup (6, 8).$$

Two sets are said to be disjoint if they have no elements in common.

> **DEFINITION**   Sets A and B are **disjoint** iff $A \cap B = \varnothing$.

As shown in the last Venn diagram of Figure 2.2.1, when sets A and B are known to be disjoint we represent them as non-overlapping regions.

Examples. The sets $\{1, 2, b\}$ and $\{-1, t, \nu, 8\}$ are disjoint. The set of even natural numbers and the set of odd natural numbers are disjoint. The intervals $(0, 1)$ and $[1, 2]$ are disjoint, but $(0, 1]$ and $[1, 2]$ are not disjoint because they both contain the element 1.

The set operations of union, intersection, and difference obey certain rules that allow us to simplify our work or replace an expression with an equivalent one. Some of the 18 relationships in the next theorem seem to be obviously true, especially if you compare sets using Venn diagrams. For example, the Venn diagrams for $A \cap B$ and $B \cap A$ are exactly the same (see part (h)). However, simply drawing a Venn diagram does not constitute a proof. Each statement requires a confirmation of the relationship between the sets by using the set operation definitions. We prove parts (b), (f), (h), (m), and (p) and leave the others as exercises.

Theorem 2.2.1   For all sets A, B, and C,

  (a) $A \subseteq A \cup B$.
  (b) $A \cap B \subseteq A$.
  (c) $A \cap \varnothing = \varnothing$.
  (d) $A \cup \varnothing = A$.
  (e) $A \cap A = A$.
  (f) $A \cup A = A$.
  (g) $A \cup B = B \cup A$. $\left.\right\}$ Commutative Laws
  (h) $A \cap B = B \cap A$.
  (i) $A - \varnothing = A$.
  (j) $\varnothing - A = \varnothing$.
  (k) $A \cup (B \cup C) = (A \cup B) \cup C$. $\left.\right\}$ Associative Laws
  (l) $A \cap (B \cap C) = (A \cap B) \cap C$.
  (m) $A \cap (B \cup C) = (A \cap B) \cup (A \cap C)$. $\left.\right\}$ Distributive Laws
  (n) $A \cup (B \cap C) = (A \cup B) \cap (A \cup C)$.
  (o) $A \subseteq B$ iff $A \cup B = B$.
  (p) $A \subseteq B$ iff $A \cap B = A$.
  (q) If $A \subseteq B$, then $A \cup C \subseteq B \cup C$.
  (r) If $A \subseteq B$, then $A \cap C \subseteq B \cap C$.

Proof.

  (b) ⟨*We must show that, if $x \in A \cap B$, then $x \in A$*⟩. Suppose $x \in A \cap B$. Then $x \in A$ and $x \in B$. Therefore $x \in A$. ⟨*We used the tautology $P \wedge Q \Rightarrow Q$.*⟩
  (f) ⟨*We must show that $x \in A \cup A$ iff $x \in A$*⟩. By the definition of union, $x \in A \cup A$ iff $x \in A$ or $x \in A$. This is equivalent to $x \in A$. Therefore, $A \cup A = A$.

(h)   ⟨*This biconditional proof uses the definition of intersection and the equivalence of $P \wedge Q$ and $Q \wedge P$.*⟩

$x \in A \cap B$ iff $x \in A$ and $x \in B$
    iff $x \in B$ and $x \in A$
    iff $x \in B \cap A$.

(m)   ⟨*As you read this proof, watch for the steps in which the definitions of union and intersection are used (two for each). Watch also for the use of the equivalence from Theorem 1.1.1 (f).*⟩

$x \in A \cap (B \cup C)$ iff $x \in A$ and $x \in B \cup C$
    iff $x \in A$ and $(x \in B$ or $x \in C)$
    iff $(x \in A$ and $x \in B)$ or $(x \in A$ and $x \in C)$
    iff $x \in A \cap B$ or $x \in A \cap C$
    iff $x \in (A \cap B) \cup (A \cap C)$.

Therefore $A \cap (B \cup C) = (A \cap B) \cup (A \cap C)$.

(p)   ⟨*We give separate proofs for each implication, making use of earlier parts of this theorem.*⟩ First, assume that $A \subseteq B$. We must show that $A \cap B = A$. Suppose $x \in A$. Then from the hypothesis $A \subseteq B$, we have $x \in B$. Therefore $x \in A$ and $x \in B$, so $x \in A \cap B$. This shows that $A \subseteq A \cap B$, which, combined with $A \cap B \subseteq A$ from part (b) of this theorem, gives $A \cap B = A$.

Second, assume that $A \cap B = A$. We must show that $A \subseteq B$. By parts (b) and (h) of this theorem, we have $B \cap A \subseteq B$ and $B \cap A = A \cap B$. Therefore, $A \cap B \subseteq B$. By hypothesis, $A \cap B = A$, so $A \subseteq B$.   ∎

When you suspect that a relationship among sets is not always true, try to construct a counterexample. To find a counterexample for $(A \cup B) \cap C = A \cup (B \cap C)$ we need sets such that the shaded regions of Figures 2.2.2 (a) and (b) have different elements. That is, we find sets A, B, and C such that A contains at least one element that is not in C. One counterexample is $B = \{3, 5\}$, $C = \{4, 5, 6\}$, and $A = \{2, 3, 4\}$. Then $(A \cup B) \cap C = \{4, 5\}$ while $A \cup (B \cap C) = \{2, 3, 4, 5\}$.

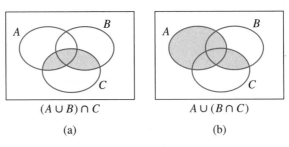

$(A \cup B) \cap C$ $A \cup (B \cap C)$

(a)   (b)

Figure 2.2.2

Recall that the universe of discourse is a collection of objects understood from the context or specified at the outset of a discussion and that all objects under consideration must belong to the universe.

DEFINITION   Let U be the universe and $A \subseteq U$. The **complement** of A is the set $A^c = U - A$.

The set A^c is the set of all elements of the universe that are not in A. (See Figure 2.2.3.)

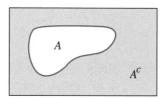

Figure 2.2.3

For the set $A = \{2, 4, 6, 8\}$, we have $A^c = \{10, 12, 14, 16, \ldots\}$ if the universe is all even natural numbers, while $A^c = \{1, 3, 5, 7, 9, 10, 11, 12, 13, \ldots\}$ if the universe is $\mathbb{N}$. For the universe $\mathbb{R}$, if $B = (0, \infty)$, then $B^c = (-\infty, 0]$. If $D = \{5\}$ then $D^c = (-\infty, 5) \cup (5, \infty)$.

Since the universe is fixed throughout a discussion, finding the complement may be thought of as a *unary* operation—it applies to a single set. The next theorem includes several results about the relationships between complementation and the other set operations.

Theorem 2.2.2   Let U be the universe, and let A and B be subsets of U. Then

(a) $(A^c)^c = A$.
(b) $A \cup A^c = U$.
(c) $A \cap A^c = \varnothing$.
(d) $A - B = A \cap B^c$.
(e) $A \subseteq B$ iff $B^c \subseteq A^c$.
(f) $(A \cup B)^c = A^c \cap B^c$. $\left.\begin{array}{l}\\ \\\end{array}\right\}$ De Morgan's Laws
(g) $(A \cap B)^c = A^c \cup B^c$.
(h) $A \cap B = \varnothing$ iff $A \subseteq B^c$.

Proof.

(a)   By definition of the complement $x \in (A^c)^c$ iff $x \notin A^c$ iff $x \in A$. Therefore $(A^c)^c = A$.

(e)   To demonstrate different styles, we give two separate proofs.

First proof.   ⟨*This is a two-part proof. The first part is an element chasing proof. The second part is proved using the first part.*⟩

(1)   ⟨*Show that if* $A \subseteq B$ *then* $B^c \subseteq A^c$.⟩ Assume that $A \subseteq B$. Suppose $x \in B^c$. Then $x \notin B$. Since $A \subseteq B$ and $x \notin B$, we have $x \notin A$. Therefore, $x \in A^c$. Thus, $B^c \subseteq A^c$.

(2)   ⟨*Show that if* $B^c \subseteq A^c$, *then* $A \subseteq B$.⟩ Assume that $B^c \subseteq A^c$. Then by part (1), $(A^c)^c \subseteq (B^c)^c$. Therefore, using part (a), $A \subseteq B$.

By parts (1) and (2), we conclude that $A \subseteq B$ iff $B^c \subseteq A^c$.

Second proof.   ⟨*This iff proof makes use of the fact that a conditional sentence is equivalent to its contrapositive.*⟩

$A \subseteq B$ iff for all x, if $x \in A$ then $x \in B$
iff for all x, if $x \notin B$ then $x \notin A$
iff for all x, if $x \in B^c$ then $x \in A^c$
iff $B^c \subseteq A^c$.

(f) $x \in (A \cup B)^c$ iff $x \notin A \cup B$
iff it is not the case that $x \in A$ or $x \in B$
iff $x \notin A$ and $x \notin B$
iff $x \in A^c$ and $x \in B^c$
iff $x \in A^c \cap B^c$.

The proofs of the remaining parts are left as Exercise 8.   ■

The **ordered pair** formed from two entities a and b is the object (a, b). Ordered pairs have the property that if either of the **coordinates** a or b is changed, the ordered pair changes. That is, two ordered pairs (a, b) and (c, d) are **equal** iff $a = c$ and $b = d$. Thus, $(3, 7) \neq (7, 3)$ even though the sets $\{3, 7\}$ and $\{7, 3\}$ are equal. A more rigorous definition of an ordered pair as a set is given in Exercise 17.

In previous study you have dealt with the ambiguity of using the same notation $(3, 7)$ for the ordered pair that represents a point in the plane and also for the open interval of real numbers with endpoints 3 and 7. The context in which $(3, 7)$ appears should always make the meaning clear.

We also say the **ordered n-tuples** $(a_1, a_2, \ldots, a_n)$ and $(c_1, c_2, \ldots, c_n)$ are equal iff $a_i = c_i$ for $i = 1, 2, \ldots, n$. Thus the ordered 5-tuples $(4, 9, 5, 0, 1)$, $(5, 4, 9, 0, 1)$ and $(0, 1, 4, 5, 9)$ are all different.

DEFINITION   Let A and B be sets. The **product** (or **cross product**) of A and B is

$$A \times B = \{(a, b) : a \in A \text{ and } b \in B\}.$$

We read $A \times B$ as "A cross B."

The set $A \times B$ is the set of all ordered pairs having first coordinate in A and second coordinate in B. The cross product is sometimes called the *Cartesian product* of A and B, in honor of René Descartes.*

Example.   If $A = \{1, 2\}$ and $B = \{2, 3, 4\}$, then

$$A \times B = \{(1, 2), (1, 3), (1, 4), (2, 2), (2, 3), (2, 4)\}.$$

Thus $(1, 2) \in A \times B$, $(2, 1) \notin A \times B$, and $\{(1, 3), (2, 2)\} \subseteq A \times B$. In this example, $A \times B \neq B \times A$ since

$$B \times A = \{(2, 1), (2, 2), (3, 1), (3, 2), (4, 1), (4, 2)\}.$$

The product of three or more sets is defined similarly. For example, for sets A, B, and C, $A \times B \times C = \{(a, b, c): a \in A, b \in B, \text{ and } c \in C\}$.

Some useful relationships between the cross product of sets and the other set operations are presented in the next theorem.

Theorem 2.2.3   If, A, B, C, and D are sets, then

(a) $A \times (B \cup C) = (A \times B) \cup (A \times C)$.
(b) $A \times (B \cap C) = (A \times B) \cap (A \times C)$.
(c) $A \times \emptyset = \emptyset$.
(d) $(A \times B) \cap (C \times D) = (A \cap C) \times (B \cap D)$.
(e) $(A \times B) \cup (C \times D) \subseteq (A \cup C) \times (B \cup D)$.
(f) $(A \times B) \cap (B \times A) = (A \cap B) \times (A \cap B)$.

Proof.

(a)   ⟨*Since both $A \times (B \cup C)$ and $(A \times B) \cup (A \times C)$ are sets of ordered pairs, their elements have the form (x, y). To show that each set is a subset of the other we use an "iff argument."*⟩

$(x, y) \in A \times (B \cup C)$ iff $x \in A$ and $y \in B \cup C$
                iff $x \in A$ and $(y \in B$ or $y \in C)$
                iff $(x \in A$ and $y \in B)$ or $(x \in A$ and $y \in C)$
                iff $(x, y) \in A \times B$ or $(x, y) \in A \times C$
                iff $(x, y) \in (A \times B) \cup (A \times C)$.
Therefore, $A \times (B \cup C) = (A \times B) \cup (A \times C)$.

(e)   If $(x, y) \in (A \times B) \cup (C \times D)$, then $(x, y) \in A \times B$ or $(x, y) \in C \times D$. If $(x, y) \in A \times B$, then $x \in A$ and $y \in B$. Thus $x \in A \cup C$ and $y \in B \cup D$. ⟨*Because $A \subseteq A \cup C$ and $B \subseteq B \cup D$.*⟩ Thus, $(x, y) \in (A \cup C) \times (B \cup D)$.

---

* René Descartes (1596–1650) was a French mathematician, philosopher, and scientist. His work *Discours de la méthode* defined analytical geometry by combining the geometric notions of curves and areas with algebraic equations and computations. He was the first person to use superscripts to indicate exponential powers of a quantity.

If $(x, y) \in C \times D$, a similar argument shows $(x, y) \in (A \cup C) \times (B \cup D)$. This shows that $(A \times B) \cup (C \times D) \subseteq (A \cup C) \times (B \cup D)$. ∎

Parts (b), (c), (d), and (f) are proved in Exercise 15. Part (e) of Theorem 2.2.3 cannot be sharpened to equality. See Exercise 16(a).

Exercises 2.2

1. Let $A = \{1, 3, 5, 7, 9\}$, $B = \{0, 2, 4, 6, 8\}$, $C = \{1, 2, 4, 5, 7, 8\}$, and $D = \{1, 2, 3, 5, 6, 7, 8, 9, 10\}$. Find
    ★ (a) $A \cup B$.
    (b) $A \cap B$.
    ★ (c) $A - B$.
    (d) $A - (B - C)$.
    ★ (e) $(A - B) - C$.
    (f) $A \cup (C \cap D)$.
    ★ (g) $(A \cap C) \cap D$.
    (h) $A \cap (B \cup C)$.
    ★ (i) $(A \cap B) \cup (A \cap C)$.
    (j) $(A \cup B) - (C \cap D)$.

2. Let the universe be all real numbers. Let $A = [3, 8)$, $B = [2, 6]$, $C = (1, 4)$, and $D = (5, \infty)$. Find
    (a) $A \cup B$.
    ★ (b) $A \cup C$.
    (c) $A \cap B$.
    ★ (d) $B \cap C$.
    (e) $A - B$.
    (f) $B - D$.
    (g) $D - A$.
    ★ (h) A^c.
    (i) $B - (A \cup C)$.
    (j) $(A \cup C) - (B \cap D)$.

3. Let the universe be the set $\mathbb{Z}$. Let E, D, $\mathbb{Z}^+$, and $\mathbb{Z}^-$ be the sets of all even, odd, positive, and negative integers, respectively. Find
    ★ (a) $E - \mathbb{Z}^+$.
    (b) $\mathbb{Z}^+ - E$.
    (c) $D - E$.
    ★ (d) $(\mathbb{Z}^+)^c$.
    (e) $\mathbb{Z}^+ - \mathbb{Z}^-$.
    (f) E^c.
    ★ (g) $E - \mathbb{Z}^-$.
    (h) $(E \cap \mathbb{Z}^-)^c$.
    (i) $\varnothing^c$.

★ 4. Let A, B, C, and D be as in Exercise 1. Which pairs of these four sets are disjoint?

5. Let A, B, C, and D be as in Exercise 2. Which pairs of these four sets are disjoint?

6. Give an example of nonempty sets A, B and C such that
    ★ (a) $C \subseteq A \cup B$ and $A \cap B \not\subseteq C$.
    (b) $A \subseteq B$ and $C \subseteq A \cap B$.
    (c) $A \cup B \subseteq C$ and $C \not\subseteq B$.
    (d) $A \not\subseteq B \cup C$, $B \not\subseteq A \cup C$, and $C \subseteq A \cup B$.
    (e) $A \subseteq B \cup C$, $B \subseteq A \cup C$, $C \subseteq A \cup B$, $A \cap B = A \cap C$, and $A \neq B$.
    (f) $A \cap B \subseteq C$, $A \cap C \subseteq B$, $B \cap C \subseteq A$, and $A = B \cup C$.

7. Prove the remaining parts of Theorem 2.2.1.

8. Prove the remaining parts of Theorem 2.2.2.

9. Let A, B, and C be sets. Prove that
    ☆ (a) $A \subseteq B$ iff $A - B = \varnothing$.

    (b) if $A \subseteq B \cup C$ and $A \cap B = \varnothing$, then $A \subseteq C$.

★ **(c)** $C \subseteq A \cap B$ iff $C \subseteq A$ and $C \subseteq B$.

    (d) if $A \subseteq B$, then $A - C \subseteq B - C$.

    (e) $(A - B) - C = (A - C) - (B - C)$.

    (f) if $A \subseteq C$ and $B \subseteq C$, then $A \cup B \subseteq C$.

    (g) $(A \cup B) \cap C \subseteq A \cup (B \cap C)$.

    (h) $A - B$ and B are disjoint.

10. Let A, B, C, and D be sets. Prove that

    (a) if $C \subseteq A$ and $D \subseteq B$, then $C \cap D \subseteq A \cap B$.

    (b) if $C \subseteq A$ and $D \subseteq B$, then $C \cup D \subseteq A \cup B$.

★ **(c)** if $C \subseteq A$, $D \subseteq B$, and A and B are disjoint, then C and D are disjoint.

    (d) if $C \subseteq A$ and $D \subseteq B$, then $D - A \subseteq B - C$.

    (e) if $A \cup B \subseteq C \cup D$, $A \cap B = \varnothing$, and $C \subseteq A$, then $B \subseteq D$.

11. Provide counterexamples for each of the following.

★ **(a)** If $A \cup C \subseteq B \cup C$, then $A \subseteq B$.

    (b) If $A \cap C \subseteq B \cap C$, then $A \subseteq B$.

★ **(c)** If $(A - B) \cap (A - C) = \varnothing$, then $B \cap C = \varnothing$.

    (d) $\mathscr{P}(A) - \mathscr{P}(B) \subseteq \mathscr{P}(A - B)$.

★ **(e)** $A - (B - C) = (A - B) - (A - C)$.

    (f) $A - (B - C) = (A - B) - C$.

12. Let A and B be sets.

★ **(a)** Prove that $\mathscr{P}(A \cap B) = \mathscr{P}(A) \cap \mathscr{P}(B)$. You may use Exercise 9(c).

    (b) Prove that $\mathscr{P}(A) \cup \mathscr{P}(B) \subseteq \mathscr{P}(A \cup B)$.

    (c) Show by example that set equality need not be the case in part (b). Under what conditions on A and B is $\mathscr{P}(A \cup B) = \mathscr{P}(A) \cup \mathscr{P}(B)$?

★ **(d)** Show that there are no sets A and B such that $\mathscr{P}(A - B) = \mathscr{P}(A) - \mathscr{P}(B)$.

13. List the ordered pairs in $A \times B$ and $B \times A$ in each case:

    (a) $A = \{1, 3, 5\}$, $B = \{a, e, k, n, r\}$.

★ **(b)** $A = \{1, 2, \{1, 2\}\}$, $B = \{q, \{t\}, \pi\}$.

    (c) $A = \{\varnothing, \{\varnothing\}, \{\varnothing, \{\varnothing\}\}\}$, $B = \{(\varnothing, \{\varnothing\}), \{\varnothing\}, (\{\varnothing\}, \varnothing)\}$.

    (d) $A = \{(2, 4), (3, 1)\}$, $B = \{(4, 1), (2, 3)\}$.

14. Let A and B be nonempty sets. Prove that $A \times B = B \times A$ iff $A = B$.

15. Complete the proof of Theorem 2.2.3 by proving

★ **(a)** $A \times (B \cap C) = (A \times B) \cap (A \times C)$.

    (b) $A \times \varnothing = \varnothing$.

    (c) $(A \times B) \cap (C \times D) = (A \cap C) \times (B \cap D)$.

    (d) $(A \times B) \cap (B \times A) = (A \cap B) \times (A \cap B)$.

16. Give an example of nonempty sets A, B, C, and D such that

    (a) $(A \times B) \cup (C \times D) \neq (A \cup C) \times (B \cup D)$.

    (b) $(C \times C) - (A \times B) \neq (C - A) \times (C - B)$.

    (c) $A \times (B \times C) \neq (A \times B) \times C$.

17. One way to define an ordered pair in terms of sets is to say $(a, b) = \{\{a\}, \{a, b\}\}$. Using this definition, prove that $(a, b) = (x, y)$ iff $a = x$ and $b = y$.

18. Let A and B be sets. Define the **symmetric difference** of A and B to be $A \triangle B = (A - B) \cup (B - A)$. Prove that
    (a) $A \triangle B = B \triangle A$.
    (b) $A \triangle B = (A \cup B) - (A \cap B)$.
    (c) $A \triangle A = \varnothing$.
    (d) $A \triangle \varnothing = A$.

Proofs to Grade

19. Assign a grade of A (correct), C (partially correct), or F (failure) to each. Justify assignments of grades other than A.
    (a) **Claim.** If $A \subseteq B$, then $A - C \subseteq B - C$.
    "Proof." Assume $A \subseteq B$. Suppose $x \in A$. Then $x \in B$, since $A \subseteq B$. Let C be any set. Then $x \in A$ and $x \notin C$. Then $x \in B$ and $x \notin C$. Thus $x \in A - C$ and $x \in B - C$. Therefore, $A - C \subseteq B - C$. ∎

    (b) **Claim.** If $A \subseteq B$, then $A - C \subseteq B - C$.
    "Proof." Assume $A \subseteq B$. Suppose $A - C$. Then $x \in A$ and $x \notin C$. Then $x \in B$, because $A \subseteq B$. Since $x \in B$ and $x \notin C$, $B - C$. Therefore, $A - C \subseteq B - C$. ∎

    (c) **Claim.** If $A \subseteq B$, then $A - C \subseteq B - C$.
    "Proof." Assume $A \subseteq B$. Then $x \in A$ and $x \in B$. Suppose $x \in A - C$. Then $x \in A$ and $x \notin C$. Since $x \in B$ and $x \notin C$, $x \in B - C$. Therefore, $A - C \subseteq B - C$. ∎

    (d) **Claim.** $A \subseteq B$ iff $A \cap B = A$.
    "Proof." Assume that $A \subseteq B$. Suppose $x \in A \cap B$. Then $x \in A$ and $x \in B$, so $x \in A$. This shows that $A \cap B = A$. Now assume that $A \cap B = A$. Suppose $x \in A$. Then $x \in A \cap B$, since $A = A \cap B$; and, therefore, $x \in B$. This shows that $x \in A$ implies $x \in B$, and so $A \subseteq B$. ∎

    (e) **Claim.** $A \cap \varnothing = A$.
    "Proof." We know that $x \in A \cap \varnothing$ iff $x \in A$ and $x \in \varnothing$. Since $x \in \varnothing$ is false, $x \in A$ and $x \in \varnothing$ iff $x \in A$. Therefore, $x \in A \cap \varnothing$ iff $x \in A$; that is, $A \cap \varnothing = A$. ∎

    (f) **Claim.** If $A \cap B \neq \varnothing$ and $B \cap C \neq \varnothing$, then $A \cap C \neq \varnothing$.
    "Proof." Assume $A \cap B \neq \varnothing$ and $B \cap C \neq \varnothing$. Since $A \cap B \neq \varnothing$, there exists x such that $x \in A \cap B$; thus $x \in A$. Since $B \cap C \neq \varnothing$, there exists $x \in B \cap C$; thus $x \in C$. Hence $x \in A$ and $x \in C$. Therefore, $x \in A \cap C$, which show $A \cap C \neq \varnothing$. ∎

    (g) **Claim.** $A \cap A^c = \varnothing$.
    "Proof." ⟨We show each side is a subset of the other.⟩ By Theorem 2.1.1, $\varnothing \subseteq A \cap A^c$. Now suppose $x \in A \cap A^c$. Then $x \in A$ and $x \in A^c$. Thus $x \in A$ and $x \notin A$. Therefore, $x \neq x$. Hence, by the definition of $\varnothing$, $x \in \varnothing$. Therefore, $A \cap A^c \subseteq \varnothing$. ∎

    (h) **Claim.** $\mathscr{P}(A - B) - \{\varnothing\} \subseteq \mathscr{P}(A) - \mathscr{P}(B)$.
    "Proof." Suppose $x \in \mathscr{P}(A - B) - \{\varnothing\}$. Then $x \in \mathscr{P}(A - B)$ and $x \neq \varnothing$. Since $\varnothing \in \mathscr{P}(A)$ and $\varnothing \in \mathscr{P}(B)$, $\varnothing \notin \mathscr{P}(A) - \mathscr{P}(B)$. Since $x \in \mathscr{P}(A - B)$, $x \in \mathscr{P}(A) - \mathscr{P}(B)$. Therefore, we can conclude that $\mathscr{P}(A - B) - \{\varnothing\} \subseteq \mathscr{P}(A) - \mathscr{P}(B)$. ∎

$\star$ **(i)**   **Claim.**   If $A \subseteq B$, then $A \cup B = B$.
"***Proof.***"   Let $A \subseteq B$. Then A and B are related as in this figure.

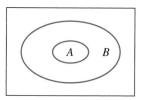

Since $A \cup B$ is the set of elements in either of the sets A or B, $A \cup B$ is the shaded area in this figure.

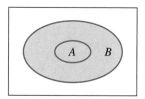

Since this is B, $A \cup B = B$.   ∎

2.3   Extended Set Operations and Indexed Families of Sets

A set of sets is often called a **family** or a **collection** of sets. In this section we extend the definitions of union and intersection to families of sets and prove generalizations of parts of Theorem 2.2.1.

Throughout this section we will use script letters, $\mathcal{A}, \mathcal{B}, \mathcal{C}, \ldots$ to denote families of sets. For example,

$$\mathcal{A} = \{\{1, 2, 3\}, \{3, 4, 5\}, \{3, 6\}, \{2, 3, 6, 7, 9, 10\}\}$$

is a family consisting of four sets. Notice that $5 \in \{3, 4, 5\}$ and $\{3, 4, 5\} \in \mathcal{A}$, but $5 \notin \mathcal{A}$. The set $\mathcal{B} = \{(-x, x): x \in \mathbb{R} \text{ and } x > 0\}$ is a family of open intervals. The sets $(-1, 1)$, $(-\sqrt{2}, \sqrt{2})$, and $(-5, 5)$ are elements of $\mathcal{B}$. See Figure 2.3.1.

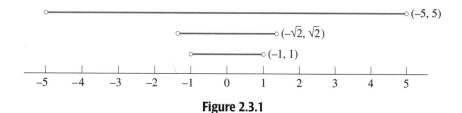

Figure 2.3.1

DEFINITION Let $\mathcal{A}$ be a family of sets. The **union over** $\mathcal{A}$ is

$$\bigcup_{A \in \mathcal{A}} A = \{x: x \in A \text{ for some } A \in \mathcal{A}\}.$$

Using this definition, for any object x we may write:

$$x \in \bigcup_{A \in \mathcal{A}} A \text{ iff } (\exists A \in \mathcal{A})(x \in A).$$

This symbolic statement expresses the direct relationship between the union over a family and the existential quantifier $\exists$. To show that an object is in the union of a family, we must show the existence of at least one set in the family that contains the object. Figure 2.3.2(a) is a Venn diagram showing the union over the family $\mathcal{M} = \{R, S, T\}$.

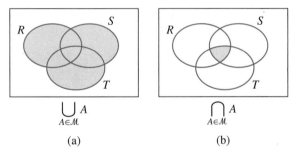

$$\bigcup_{A \in \mathcal{M}} A \qquad\qquad \bigcap_{A \in \mathcal{M}} A$$

(a)   (b)

Figure 2.3.2

For the family $\mathcal{A}$ of four sets given above, $\bigcup_{A \in \mathcal{A}} A = \{1, 2, 3, 4, 5, 6, 7, 9, 10\}$.
The union of the family $\mathcal{B} = \{(-x, x): x \in \mathbb{R} \text{ and } x > 0\}$ is the set of all real numbers because every real number b is an element of the open interval $(-|b| - 1, |b| + 1)$.

DEFINITION Let $\mathcal{A}$ be a family of sets. The **intersection over** $\mathcal{A}$ is

$$\bigcap_{A \in \mathcal{A}} A = \{x: x \in A \text{ for every } A \in \mathcal{A}\}.$$

For the intersection over a family $\mathcal{A}$, we write

$$x \in \bigcap_{A \in \mathcal{A}} A \text{ iff } (\forall A \in \mathcal{A})(x \in A).$$

Figure 2.3.2(b) shows the Venn diagram for the intersection over the family $\mathcal{M} = \{R, S, T\}$. Using the family $\mathcal{A}$ above again as an example, $\bigcap\limits_{A \in \mathcal{A}} A = \{3\}$ because 3 is the only object contained in all four sets in $\mathcal{A}$. The intersection of the family $\mathcal{B} = \{(-x, x): x \in \mathbb{R} \text{ and } x > 0\}$ is the set $\{0\}$ because 0 is the only number in every set in $\mathcal{B}$.

Example. For the family $\mathcal{A} = \{\{r, k, s, t, a\}, \{k, d, s\}\}$, $\bigcup\limits_{A \in \mathcal{A}} A = \{r, k, s, t, a, d\}$ and $\bigcap\limits_{A \in \mathcal{A}} A = \{k, s\}$. If there are only two sets in the family, the union and intersection over the family are the same as the union and intersection defined in Section 2.2.

Theorem 2.3.1 For every set B in a family $\mathcal{A}$ of sets,

(a) $\bigcap\limits_{A \in \mathcal{A}} A \subseteq B.$

(b) $B \subseteq \bigcup\limits_{A \in \mathcal{A}} A.$

(c)   If the family $\mathcal{A}$ is nonempty, then $\bigcap\limits_{A \in \mathcal{A}} A \subseteq \bigcup\limits_{A \in \mathcal{A}} A.$

Proof.

(a)   Let $\mathcal{A}$ be a family of sets and $B \in \mathcal{A}$. Suppose $x \in \bigcap\limits_{A \in \mathcal{A}} A$. Then $x \in A$ for every $A \in \mathcal{A}$. ⟨*Notice that the set A in the last sentence is a dummy symbol. It stands for any set in the family. The set B is in the family.*⟩ In particular, $x \in B$. Therefore $\bigcap\limits_{A \in \mathcal{A}} A \subseteq B$.

(b)   The proof that B is a subset of the union over $\mathcal{A}$ is left as Exercise 3.

(c)   Let $\mathcal{A}$ be a nonempty family. Choose any set $C \in \mathcal{A}$. By parts (a) and (b) $\bigcap\limits_{A \in \mathcal{A}} A \subseteq C \subseteq \bigcup\limits_{A \in \mathcal{A}} A$ and therefore, $\bigcap\limits_{A \in \mathcal{A}} A \subseteq \bigcup\limits_{A \in \mathcal{A}} A$.   ∎

It was necessary in part (c) that the family $\mathcal{A}$ be nonempty. If $\mathcal{A}$ is the empty family, then intersection is equal to the universe of discourse. (See Exercise 4.) This observation is a reason to be cautious about dealing with the empty family of sets.

Example. Let $\mathcal{B}$ be the collection $\{B_n: n \in \mathbb{N}\}$, where $B_n = \{0, 1, 2, \ldots, n\}$. Members of $\mathcal{B}$ include $B_2 = \{0, 1, 2\}$ and $B_6 = \{0, 1, 2, 3, 4, 5, 6\}$. Then $\bigcup\limits_{B \in \mathcal{B}} B = \mathbb{N} \cup \{0\}$ and $\bigcap\limits_{B \in \mathcal{B}} B = \{0, 1\}$.

It is often helpful to associate an identifying tag, or index, with each set in a family of sets. In the example above, each natural number n corresponds to a set B_n. By specifying the index, as we did when we selected $n = 2$ or 6, we specified the corresponding set. By specifying a set of indices, we can specify the family of sets we want to consider.

> **DEFINITIONS**   Let Δ be a nonempty set such that for each $\alpha \in \Delta$ there is a corresponding set A_α. The family $\{A_\alpha : \alpha \in \Delta\}$ is an **indexed family of sets**. The set Δ is called the **indexing set** and each $\alpha \in \Delta$ is an **index**.

Indexing is a common phenomenon in everyday life. Suppose an apartment building has six rental units, labeled A through F. At any given time, for each apartment, there is a set of people residing in that apartment. These sets may be indexed by $\Delta = \{A, B, C, D, E, F\}$. Let P_k be the set of people living in apartment k. Then $\mathcal{P} = \{P_k : k \in \Delta\}$ is an indexed family of sets. An index is simply a label that provides a convenient way to refer to a certain set.

Example.   For all $n \in \mathbb{N}$, let $A_n = \{n, n + 1, 2n\}$. Then $A_1 = \{1, 2\}$, $A_2 = \{2, 3, 4\}$, $A_3 = \{3, 4, 6\}$, and so forth. The set with index 10 is $A_{10} = \{10, 11, 20\}$. Except for the set A_1, every set in the family $\{A_i : i \in \mathbb{N}\}$ has 3 elements. To form the family of sets that contains only A_2, A_3, A_{10}, and A_{15}, we change the index set as follows: $\{A_2, A_3, A_5, A_{10}\} = \{A_i : i \in \{2, 3, 10, 15\}\}$.

There is no real difference between a family of sets and an indexed family. Every family of sets could be indexed by finding a large enough set of indices to label each set in the family.

Example.   For the sets $A_1 = \{1, 2, 4, 5\}$, $A_2 = \{2, 3, 5, 6\}$, and $A_3 = \{3, 4, 5, 6\}$, the index set has been chosen to be $\Delta = \{1, 2, 3\}$. The family $\mathcal{A}$ indexed by Δ is $\mathcal{A} = \{A_1, A_2, A_3\} = \{A_i : i \in \Delta\}$. The family $\mathcal{A}$ could be indexed by another set. For instance, if $\Gamma = \{10, 21, \pi\}$, and $A_{10} = \{1, 2, 4, 5\}$, $A_{21} = \{2, 3, 5, 6\}$, and $A_\pi = \{3, 4, 5, 6\}$, then $\{A_i : i \in \Delta\} = \{A_i : i \in \Gamma\}$.

Example.   Let $\Delta = \{0, 1, 2, 3, 4\}$ and let $A_x = \{2x + 4, 8, 12 - 2x\}$ for each $x \in \Delta$. Then $A_0 = \{4, 8, 12\}$, $A_1 = \{6, 8, 10\}$, $A_2 = \{8\}$, $A_3 = \{6, 8, 10\}$, and $A_4 = \{4, 8, 12\}$. The indexing set has five elements but the indexed family $\mathcal{A} = \{A_x : x \in \Delta\}$ has only three members, since $A_1 = A_3$ and $A_0 = A_4$.

As the above examples demonstrate, an indexing family may be finite or infinite, the number of elements in the member sets do not have to be the same, and different indices need not correspond to different sets in the family.

The operations of union and intersection over families of sets apply to indexed families, although the notation is slightly different. For a family $\mathcal{A} = \{A_\alpha : \alpha \in \Delta\}$, the notations for unions and intersection are:

$$\bigcup_{\alpha \in \Delta} A_\alpha = \bigcup_{A \in \mathcal{A}} A \quad \text{and} \quad x \in \bigcup_{\alpha \in \Delta} A_\alpha \text{ iff } (\exists \alpha \in \Delta)(x \in A_\alpha).$$

$$\bigcap_{\alpha \in \Delta} A_\alpha = \bigcap_{A \in \mathcal{A}} A \quad \text{and} \quad x \in \bigcap_{\alpha \in \Delta} A_\alpha \text{ iff } (\forall \alpha \in \Delta)(x \in A_\alpha).$$

In the previous example with $\Delta = \{0, 1, 2, 3, 4\}$, $\bigcup_{\alpha \in \Delta} A_\alpha = \{4, 6, 8, 10, 12\}$ and $\bigcap_{\alpha \in \Delta} A_\alpha = \{8\}$.

Example. For $n \in \mathbb{N}$, let $A_n = \{n, n + 1, 2n\}$. Then $\bigcup_{n \in \mathbb{N}} A_n = \mathbb{N}$. To show, for example, that $27 \in \bigcup_{n \in \mathbb{N}} A_n$, we need only point out some index n such that $27 \in A_n$. Either index 26 or 27 will do. Since there is no number x such that $x \in C_n$ for all $n \in \mathbb{N}$, $\bigcap_{n \in \mathbb{N}} A_n = \varnothing$.

Example. For each real number x, define $B_x = [x^2, x^2 + 1]$. Then $B_{-1/2} = \left[\frac{1}{4}, \frac{5}{4}\right]$, $B_0 = [0, 1]$, and $B_{10} = [100, 101]$. This is another example in which we have different indices representing the same set. For example, $B_{-2} = B_2 = [4, 5]$. Here the index set is $\mathbb{R}$, $\bigcap_{x \in \mathbb{R}} B_x = \varnothing$, and $\bigcup_{x \in \mathbb{R}} B_x = [0, \infty)$.

There is a convenient variation on the notation for union and intersection when the index set is the natural numbers. For an indexed family $\mathscr{A} = \{A_n : n \in \mathbb{N}\}$, we can write $\bigcup_{i=1}^{\infty} A_i$ instead of $\bigcup_{n \in \mathbb{N}} A_n$. The intersection over $\mathscr{A}$ is written $\bigcap_{i=1}^{\infty} A_i$. Also, $A_2 \cup A_3 \cup A_4 = \bigcup_{i=2}^{4} A_i$ and $A_{11} \cap A_{12} \cap A_{13} \cap A_{14} \cap A_{15} = \bigcap_{i=11}^{15} A_i$.

Example. For each $n \in \mathbb{N}$, let $A_n = \{n, n + 1, n^2\}$. For $\mathscr{A} = \{A_n : n \in \mathbb{N}\}$

$$\bigcap_{i=1}^{\infty} A_i = \varnothing \qquad\qquad \bigcup_{i=4}^{6} A_i = \{4, 5, 6, 7, 16, 25, 36\}$$

$$\bigcup_{i=1}^{\infty} A_i = \mathbb{N} \qquad\qquad \bigcap_{i=2}^{4} A_i = \{4\}$$

$$\bigcup_{i=1}^{3} A_i = \{1, 2, 3, 4, 9\} \qquad\qquad \bigcap_{i=8}^{10} A_i = \varnothing$$

The next theorem restates Theorem 2.3.1 for indexed families and gives a version of De Morgan's Laws for indexed families.

Theorem 2.3.2   Let $\mathscr{A} = \{A_\alpha : \alpha \in \Delta\}$ be an indexed collection of sets. Then

(a) $\bigcap_{\alpha \in \Delta} A_\alpha \subseteq A_\beta$ for each $\beta \in \Delta$.

(b) $A_\beta \subseteq \bigcup_{\alpha \in \Delta} A_\alpha$ for each $\beta \in \Delta$.

(c) $\left(\bigcap_{\alpha \in \Delta} A_\alpha \right)^c = \bigcup_{\alpha \in \Delta} A_\alpha^c$

(d) $\left(\bigcup_{\alpha \in \Delta} A_\alpha \right)^c = \bigcap_{\alpha \in \Delta} A_\alpha^c$

$\Bigg\}$ De Morgan's Laws

Proof.   The proofs of parts (a) and (b) are similar to those for Theorem 2.3.1 and are left for Exercise 5(a).

(c) $x \in \left(\bigcap_{\alpha \in \Delta} A_\alpha \right)^c$

iff $x \notin \bigcap_{\alpha \in \Delta} A_\alpha$

iff it is not the case that for every $\alpha \in \Delta$, $x \in A_\alpha$

iff for some $\beta \in \Delta$, $x \notin A_\beta$

iff for some $\beta \in \Delta$, $x \in A_\beta^c$

iff $x \in \bigcup_{\alpha \in \Delta} A_\alpha^c$.

Therefore, $\left(\bigcap_{\alpha \in \Delta} A_\alpha \right)^c = \bigcup_{\alpha \in \Delta} A_\alpha^c$.

(d)   ⟨*One proof of part* (d) *is very similar to that given for part* (c) *and is left as Exercise 5(b). However, since part* (c) *has been proved, it is permissible to use it. We also use (twice) the fact that* $(A^c)^c = A$.⟩

$$\left(\bigcup_{\alpha \in \Delta} A_\alpha \right)^c = \left(\bigcup_{\alpha \in \Delta} (A_\alpha^c)^c \right)^c$$

$$= \left(\left(\bigcap_{\alpha \in \Delta} A_\alpha^c \right)^c \right)^c$$

$$= \bigcap_{\alpha \in \Delta} A_\alpha^c.$$

∎

DEFINITION   The indexed family $\mathscr{A} = \{A_\alpha : \alpha \in \Delta\}$ of sets is **pairwise disjoint** iff for all α and β in Δ, either $A_\alpha = A_\beta$ or $A_\alpha \cap A_\beta = \varnothing$.

The family $\{A_1, A_2, A_3\}$ in Figure 2.3.3(a) is pairwise disjoint. However, the family $\{B_1, B_2, B_3\}$ in Figure 2.3.3(b) is not pairwise disjoint. Although $B_1 \cap B_2 = \varnothing$, the sets B_1 and B_3 are neither identical nor disjoint.

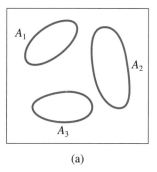

(a)

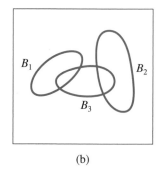

(b)

Figure 2.3.3

Two questions are commonly asked about the concept of pairwise disjoint families. The first is about why we bother with such a definition when we could simply say a family is disjoint if and only if $\bigcap_{\alpha \in \Delta} A_\alpha = \varnothing$. Having an empty intersection is not the same as being pairwise disjoint and not nearly as useful (see Section 3.3). The family $\{C_1, C_2, C_3\}$ with $C_1 = \{a, b\}$, $C_2 = \{b, c\}$, and $C_3 = \{a, c\}$ is not pairwise disjoint, even though $\bigcap_{i=1}^{3} C_i = \varnothing$.

The second common question asks why the definition says "either $A_\alpha = A_\beta$ or $A_\alpha \cap A_\beta = \varnothing$" instead of "$A_\alpha \cap A_\beta = \varnothing$ whenever $\alpha \neq \beta$." That is, why not say "if $\alpha \neq \beta$ then A_α and A_β are disjoint?" The family $\{[n, n + 1): n \in \mathbb{N}\}$ is pairwise disjoint because whenever $n \neq m$ we have $[n, n + 1) \cap [m, m + 1) = \varnothing$. However, in some families of sets it happens that different indices correspond to the same set, so the definition allows for this possibility.

Example. Suppose $\mathscr{B} = \{B_1, B_2, B_3, B_4, B_5, B_6\}$, where

$$B_1 = \{a, c, e\} \qquad\qquad B_2 = \{d, g\}$$
$$B_3 = \{d, g\} \qquad\qquad B_4 = \{b, f, h\}$$
$$B_5 = \{a, c, e\} \qquad\qquad B_6 = \{a, c, e\}.$$

The family $\mathscr{B}$ is pairwise disjoint. Note that $B_1 = B_5 = B_6$ and $B_2 = B_3$, so $\mathscr{B} = \{B_1, B_2, B_4\}$.

Example. Suppose $A_x = \{-x, x\}$ for every x in $\mathbb{R}$ and $\mathscr{D} = \{A_x : x \in \mathbb{R}\}$. Then $A_3 = \{-3, 3\} = A_{-3}$ and $A_{-7} = \{7, -7\} = A_7$. The family $\mathscr{D}$ is pairwise disjoint because $A_x = A_y$ whenever $|x| = |y|$ and $A_x \cap A_y = \varnothing$ whenever $|x| \neq |y|$.

Exercises 2.3

1. Find the union and intersection of each of the following families or indexed collections.

   ★ **(a)** Let $\mathscr{A} = \{\{1, 2, 3, 4, 5\}, \{2, 3, 4, 5, 6\}, \{3, 4, 5, 6, 7\}, \{4, 5, 6, 7, 8\}\}$.

   (b) Let $\mathscr{A} = \{\{1, 3, 5\}, \{2, 4, 6\}, \{7, 9, 11, 13\}, \{8, 10, 12\}\}$.

   ★ **(c)** For each natural number n, let $A_n = \{5n, 5n + 1, 5n + 2 \ldots, 6n\}$, and let $\mathscr{A} = \{A_n : n \in \mathbb{N}\}$.

   (d) For each natural number n, let $B_n = \mathbb{N} - \{1, 2, 3, \ldots, n\}$ and let $\mathscr{B} = \{B_n : n \in \mathbb{N}\}$.

   ★ **(e)** Let $\mathscr{A}$ be the set of all sets of integers that contain 10.

   (f) Let $A_1 = \{1\}, A_2 = \{2, 3\}, A_3 = \{3, 4, 5\}, \ldots, A_{10} = \{10, 11, \ldots, 19\}$, and let $\mathscr{A} = \{A_n : n \in \{1, 2, 3, \ldots, 10\}\}$.

   ★ **(g)** For each natural number, let $A_n = \left(0, \frac{1}{n}\right)$, and let $\mathscr{A} = \{A_n : n \in \mathbb{N}\}$.

   (h) For $r \in (0, \infty)$, let $A_r = [-\pi, r)$, and let $\mathscr{A} = \{A_r : r \in (0, \infty)\}$.

   ★ **(i)** For each real number r, let $A_r = [|r|, 2|r| + 1]$, and let $\mathscr{A} = \{A_r : r \in \mathbb{R}\}$.

   (j) For each $n \in \mathbb{N}$, let $M_n = \{\ldots, -3n, -2n, -n, 0, n, 2n, 3n, \ldots\}$, and let $\mathscr{M} = \{M_n : n \in \mathbb{N}\}$.

   (k) For each natural number $n \geq 3$, let $A_n = \left[\frac{1}{n}, 2 + \frac{1}{n}\right]$ and $\mathscr{A} = \{A_n : n \geq 3\}$.

   (l) For each $n \in \mathbb{Z}$, let $C_n = [n, n + 1)$ and let $\mathscr{C} = \{C_n : n \in \mathbb{Z}\}$.

   (m) For each $n \in \mathbb{Z}$, let $A_n = (n, n + 1)$ and $\mathscr{A} = \{A_n : n \in \mathbb{Z}\}$.

   (n) For each $n \in \mathbb{Z}$, let $D_n = \left(-n, \frac{1}{n}\right)$ and $\mathscr{D} = \{D_n : n \in \mathbb{N}\}$.

   (o) For each prime number p, let $p\mathbb{N} = \{np : n \in \mathbb{N}\}$ and $\mathscr{A}$ be the family $\{p\mathbb{N} : n \in \mathbb{N}$ and p is prime$\}$.

   ★ **(p)** For each $n \in \mathbb{Z}$, let $T_n = \{(x, y) \in \mathbb{R} \times \mathbb{R} : 0 \leq x \leq 1, 0 \leq y \leq x^n\}$ and $\mathscr{T} = \{T_n : n \in \mathbb{N}\}$.

   (q) For each $n \in \mathbb{Z}$, let $V_n = \{(x, y) \in \mathbb{R} \times \mathbb{R} : 0 \leq x \leq 1, x^n \leq y \leq \sqrt[n]{x}\}$ and $\mathscr{V} = \{V_n : n \in \mathbb{N}\}$.

☆ 2. Which families in Exercise 1 are pairwise disjoint?

3. Prove part (b) of Theorem 2.3.1.

4. Let the universe of discourse be the set $\mathbb{R}$ of real numbers, and let $\mathscr{A}$ be the empty family of subsets of $\mathbb{R}$.

   ☆ **(a)** Show that $\bigcap_{A \in \mathscr{A}} A = \mathbb{R}$.

   ☆ **(b)** Show that $\bigcup_{A \in \mathscr{A}} A = \varnothing$.

   (c) Conclude that $\bigcap_{A \in \mathscr{A}} A \subseteq \bigcup_{A \in \mathscr{A}} A$ is false in this example.

5. ☆ **(a)**   Prove parts (a) and (b) of Theorem 2.3.2.

(b)   Give a direct proof of part (d) of Theorem 2.3.2 that does not use part (c).

6.   Let $\mathcal{A} = \{A_\alpha : \alpha \in \Delta\}$ be a family of sets and let B be a set. Prove that

★ **(a)** $B \cap \bigcup\limits_{\alpha \in \Delta} A_\alpha = \bigcup\limits_{\alpha \in \Delta} (B \cap A_\alpha)$.

(b) $B \cup \bigcap\limits_{\alpha \in \Delta} A_\alpha = \bigcap\limits_{\alpha \in \Delta} (B \cup A_\alpha)$.

7.   Let $\mathcal{A} = \{A_\alpha : \alpha \in \Delta\}$ and let $\mathcal{B} = \{B_\beta : \beta \in \Gamma\}$. Use Exercise 6 to write

★ **(a)** $\left(\bigcup\limits_{\alpha \in \Delta} A_\alpha \right) \cap \left(\bigcup\limits_{\beta \in \Gamma} B_\beta \right)$ as a union of intersections.

(b) $\left(\bigcap\limits_{\alpha \in \Delta} A_\alpha \right) \cup \left(\bigcap\limits_{\beta \in \Gamma} B_\beta \right)$ as an intersection of unions.

8.   Let $\mathcal{A} = \{A_\alpha : \alpha \in \Delta\}$ be a family of sets, $\Delta \neq \varnothing$, and B be a set. For each part either prove the statement is true or give a counterexample.

(a) $B - \left(\bigcap\limits_{\alpha \in \Delta} A_\alpha \right) = \bigcap\limits_{\alpha \in \Delta} (B - A_\alpha)$. f

(b) $B - \left(\bigcup\limits_{\alpha \in \Delta} A_\alpha \right) = \bigcup\limits_{\alpha \in \Delta} (B - A_\alpha)$. f

☆ **(c)** $\left(\bigcap\limits_{\alpha \in \Delta} A_\alpha \right) - B = \bigcap\limits_{\alpha \in \Delta} (A_\alpha - B)$. T

(d) $\left(\bigcup\limits_{\alpha \in \Delta} A_\alpha \right) - B = \bigcup\limits_{\alpha \in \Delta} (A_\alpha - B)$. T

9.   If $\mathcal{A} = \{A_\alpha : \alpha \in \Delta\}$ is a family of sets and $\Gamma \subseteq \Delta$, prove that

★ **(a)** $\bigcup\limits_{\alpha \in \Gamma} A_\alpha \subseteq \bigcup\limits_{\alpha \in \Delta} A_\alpha$.

(b) $\bigcap\limits_{\alpha \in \Delta} A_\alpha \subseteq \bigcap\limits_{\alpha \in \Gamma} A_\alpha$.

10.   Let $\mathcal{A}$ be a nonempty family of sets.

★ **(a)**   Suppose $B \subseteq A$ for every $A \in \mathcal{A}$. Prove that $B \subseteq \bigcap\limits_{A \in \mathcal{A}} A$.

★ **(b)**   What is the largest set X such that $X \subseteq A$ for all $A \in \mathcal{A}$? That is, find the set X such that (i) $X \subseteq A$ for all $A \in \mathcal{A}$; and (ii) if $V \subseteq A$ for all $A \in \mathcal{A}$, then $V \subseteq X$.

(c)   Suppose $A \subseteq D$ for every $A \in \mathcal{A}$. Prove that $\bigcup\limits_{A \in \mathcal{A}} A \subseteq D$.

(d)   What is the smallest set Y such that $A \subseteq Y$ for all $A \in \mathcal{A}$? That is, find the set Y such that (i) $A \subseteq Y$ for all $A \in \mathcal{A}$; and (ii) if $A \subseteq W$ for all $A \in \mathcal{A}$, then $Y \subseteq W$.

11. Let $X = \{1, 2, 3, 4, \ldots, 20\}$. Give an example of each of the following:

    (a) a family $\mathscr{A}$ of subsets of X such that $\bigcap_{A \in \mathscr{A}} A = \{1\}$ and $\bigcup_{A \in \mathscr{A}} A = X$.

    (b) a family $\mathscr{B}$ of four pairwise disjoint subsets of X such that $\bigcup_{B \in \mathscr{B}} B = X$.

    (c) a family $\mathscr{C}$ of twenty pairwise disjoint subsets of X such that $\bigcup_{C \in \mathscr{C}} C = X$.

12. Give an example of an indexed collection of sets $\{A_n : n \in \mathbb{N}\}$ such that each $A_n \subseteq (0, 1)$, and for all $m, n \in \mathbb{N}$, $A_m \cap A_n \neq \varnothing$ but $\bigcap_{n \in \mathbb{N}} A_n = \varnothing$.

13. Let $\mathscr{A}$ be a family of pairwise disjoint sets. Prove that if $\mathscr{B} \subseteq \mathscr{A}$, then $\mathscr{B}$ is a family of pairwise disjoint sets.

14. Let $\mathscr{A}$ and $\mathscr{B}$ be two pairwise disjoint families of sets. Let $\mathscr{C} = \mathscr{A} \cap \mathscr{B}$ and $\mathscr{D} = \mathscr{A} \cup \mathscr{B}$.

    (a) Prove that $\mathscr{C}$ is a family of pairwise disjoint sets.

    (b) Give an example to show that $\mathscr{D}$ need not be pairwise disjoint.

    (c) Prove that if $\bigcup_{A \in \mathscr{A}} A$ and $\bigcup_{B \in \mathscr{B}} B$ are disjoint, then $\mathscr{D}$ is pairwise disjoint.

15. Let $\mathscr{A} = \{A_i : i \in \mathbb{N}\}$ be a family of sets and k, m be natural numbers with $k \leq m$. Prove that

    (a) $\displaystyle\bigcup_{i=1}^{k+1} A_i = \bigcup_{i=1}^{k} A_i \cup A_{k+1}$.

    (b) $\displaystyle\bigcap_{i=1}^{k+1} A_i = \bigcap_{i=1}^{k} A_i \cap A_{k+1}$.

    ★ **(c)** $\displaystyle\bigcup_{i=k}^{m} A_i \subseteq \bigcup_{i=1}^{\infty} A_i$.

    (d) $\displaystyle\bigcap_{i=1}^{\infty} A_i \subseteq \bigcap_{i=k}^{m} A_i$.

    (e) $\displaystyle\bigcup_{i=1}^{k} A_i \subseteq \bigcup_{i=1}^{m} A_i$.

    (f) $\displaystyle\bigcap_{i=1}^{m} A_i \subseteq \bigcap_{i=1}^{k} A_i$.

16. Suppose $\mathscr{A} = \{A_i : i \in \mathbb{N}\}$ is a family of sets such that for all $i, j \in \mathbb{N}$, if $i \leq j$, then $A_j \subseteq A_i$. (Such a family is called a **decreasing nested family** of sets.)

    (a) Prove that for every $k \in \mathbb{N}$, $\displaystyle\bigcap_{i=1}^{k} A_i = A_k$.

    (b) Prove that $\displaystyle\bigcup_{i=1}^{\infty} A_i = A_1$.

17. Give an example of a decreasing nested family $\{A_i : i \in \mathbb{N}\}$ (see Exercise 16) for each condition.

    ★ **(a)** $\displaystyle\bigcap_{i=1}^{\infty} A_i = [0, 1]$.

    (b) $\displaystyle\bigcap_{i=1}^{\infty} A_i = (0, 1)$.

(c) $\displaystyle\bigcap_{i=1}^{\infty} A_i = \{0, 1\}$.

(d) $\displaystyle\bigcap_{i=1}^{\infty} A_i = \varnothing$.

Proofs to Grade

18. Assign a grade of A (correct), C (partially correct), or F (failure) to each. Justify assignments of grades other than A.

⋆ C(a) **Claim.** For every indexed family $\{A_\alpha : \alpha \in \Delta\}$, $\displaystyle\bigcap_{\alpha \in \Delta} A_\alpha \subseteq \bigcup_{\alpha \in \Delta} A_\alpha$.

"*Proof.*" Choose any $A_\beta \in \{A_\alpha : \alpha \in \Delta\}$. Then by Theorem 2.3.2,

$$\bigcap_{\alpha \in \Delta} A_\alpha \subseteq A_\beta \quad \text{and} \quad A_\beta \subseteq \bigcup_{\alpha \in \Delta} A_\alpha.$$

Therefore, by transitivity of set inclusion, $\displaystyle\bigcap_{\alpha \in \Delta} A_\alpha \subseteq \bigcup_{\alpha \in \Delta} A_\alpha$. ∎

⋆ C(b) **Claim.** If $A_\alpha \subseteq B$ for all $\alpha \in \Delta$, then $\displaystyle\bigcup_{\alpha \in \Delta} A_\alpha \subseteq B$.

"*Proof.*" Suppose $x \in \displaystyle\bigcup_{\alpha \in \Delta} A_\alpha$. Then, since $A_\alpha \subseteq B$ for all $\alpha \in \Delta$, $x \in B$. Therefore, $\displaystyle\bigcup_{\alpha \in \Delta} A \subseteq B$. ∎

(c) **Claim.** For every indexed family $\{A_\alpha : \alpha \in \Delta\}$, $\displaystyle\bigcap_{\alpha \in \Delta} A_\alpha \subseteq \bigcup_{\alpha \in \Delta} A_\alpha$.

"*Proof.*" Let $\Delta = \{r, s, t\}$, $A_r = \{a, b, c, d\}$, $A_s = \{b, c, d, e\}$, $A_t = \{c, d, e, f\}$. Then $\displaystyle\bigcap_{\alpha \in \Delta} A_\alpha = \{c, d\} \subseteq \{a, b, c, d, e, f\} = \bigcup_{\alpha \in \Delta} A_\alpha$. ∎

(d) **Claim.** For every indexed family $\{A_\alpha : \alpha \in \Delta\}$, $\displaystyle\bigcap_{\alpha \in \Delta} A_\alpha \subseteq \bigcup_{\alpha \in \Delta} A_\alpha$.

"*Proof.*" Assume $\displaystyle\bigcap_{\alpha \in \Delta} A_\alpha \nsubseteq \bigcup_{\alpha \in \Delta} A_\alpha$. Then for some $x \in \displaystyle\bigcap_{\alpha \in \Delta} A_\alpha$, $x \notin \displaystyle\bigcup_{\alpha \in \Delta} A_\alpha$. Since $x \notin \displaystyle\bigcup_{\alpha \in \Delta} A_\alpha$, it is not the case that $x \in A_\alpha$ for some $\alpha \in \Delta$. Therefore, $x \notin A_\alpha$ for every $\alpha \in \Delta$. But since $x \in \displaystyle\bigcap_{\alpha \in \Delta} A_\alpha$, $x \in A_\alpha$ for every $\alpha \in \Delta$. This is a contradiction, so we conclude

$$\bigcap_{\alpha \in \Delta} A_\alpha \subseteq \bigcup_{\alpha \in \Delta} A_\alpha.$$ ∎

⋆ F(e) **Claim.** $\displaystyle\bigcup_{n=1}^{\infty} [n, n+1) = \mathbb{R}$.

"*Proof.*" Let $x \in \mathbb{R}$. Choose a natural number y such that $y \le x < y + 1$. Thus $x \in [y, y + 1)$. Therefore, x is an element of $\displaystyle\bigcup_{n=1}^{\infty} [n, n+1)$. Since $[n, n+1) \subseteq \mathbb{R}$ for all $n \in \mathbb{N}$, $\displaystyle\bigcup_{n=1}^{\infty} [n, n+1) \subseteq \mathbb{R}$.

Therefore, $\displaystyle\bigcup_{n=1}^{\infty} [n, n+1) = \mathbb{R}$. ∎

2.4   Mathematical Induction

In 1889 Giuseppe Peano* set forth five axioms that provided a complete and rigorous definition of the natural numbers based on the notion of successors. The axioms assert that

  (i)   1 is a natural number,

 (ii)   every natural number has a unique successor, which is a natural number,

(iii)   no two natural numbers have the same successor,

(iv)   1 is not a successor for any natural number,

 (v)   if a property is possessed by 1 and possessed by the successor of every natural number that possesses it, then the property is possessed by all natural numbers.

These axioms are sufficient to derive all the familiar arithmetic and order properties of $\mathbb{N}$ that are listed in the *Preface to the Student*. The development of all these properties as consequences of Peano's axioms is certainly a worthy activity, but it would take more time than we can devote to the topic here. Instead, we focus our attention on the *inductive property* of $\mathbb{N}$ given in the fifth axiom. Peano's fifth axiom can be restated as a property of sets of natural numbers.

Principle of Mathematical Induction (PMI)

Let S be a subset of $\mathbb{N}$ with these two properties:

 (i) $1 \in S$,

(ii)   for all $n \in \mathbb{N}$, if $n \in S$, then $n + 1 \in S$.

Then $S = \mathbb{N}$.

A set S of natural numbers is called an **inductive set** iff it has the property that whenever $n \in S$, then $n + 1 \in S$. The set $\{5, 6, 7, 8, \ldots\}$ is inductive, as is the set $\{100, 101, 102, 103, \ldots\}$. We leave it as an exercise to show that $\mathbb{N}$ and $\varnothing$ are inductive sets. The set $\{1, 3, 5, 7, 9, \ldots\}$ is not inductive because, for example, 7 is a member but 8 is not. Many sets of natural numbers have the inductive property, but only one set is inductive and contains 1. By the Principle of Mathematical Induction, that set is $\mathbb{N}$.

An important use of the PMI is to make inductive definitions. These definitions follow the form of the PMI: We define a first object, and then the $(n + 1)$st object is defined in terms of the nth object. The PMI ensures that the set of all n for which the corresponding object is defined is $\mathbb{N}$.

Example.   The noninductive definition of the **factorial** of a natural number n is

$$n! = n \cdot (n - 1) \cdot \cdots \cdot 3 \cdot 2 \cdot 1.$$

* Giuseppe Peano (1858–1932) was an Italian mathematician who made many contributions to mathematical logic and set theory, especially its language and symbolism. He was the first to use the modern symbols for union and intersection. His "Formulario Mathematico" manuscript (1908) contains 4,200 precisely stated mathematical formulae and theorems. Other contributions include his "space filling curve" counterexample, a forerunner of fractal images.

For example, $5! = 5 \cdot 4 \cdot 3 \cdot 2 \cdot 1 = 120$. The inductive definition of $n!$ is

(i) $1! = 1$.
(ii)   For $n \in \mathbb{N}$, $(n + 1)! = (n + 1)n!$.

To show that the inductive definition defines $n!$ for all natural numbers n, we let S be the set of n for which $n!$ is defined. First, $1 \in S$ because of part (i). Second, S is inductive because if $n \in S$, then $n!$ is defined and hence, by part (ii), $(n + 1)!$ is also defined. Thus $n + 1 \in S$. By the PMI, $S = \mathbb{N}$. In other words, the set of numbers for which the factorial is defined is $\mathbb{N}$, so $n!$ has been defined for all natural numbers.

The inductive definition makes clear the relationship between the factorial of a number and the factorial of the next number; if you happen to know that $11! = 39{,}916{,}800$, then you compute $12! = 12 \cdot 11! = 479{,}001{,}600$.

Example.   Sets may be defined inductively. Suppose we let T be the set of integers defined by

(i) $5 \in T$,
(ii)   if $x \in T$, then $x + 4 \in T$.

The set $T = \{5, 9, 13, 17, \dots\}$, which may also be defined using the non-inductive definition $T = \{4k + 1 : k \in \mathbb{N}\}$.

The real power of the Principle of Mathematical Induction is as a method for proving statements that are true for all natural numbers. For example, we note that the sum of the first three odd numbers is $1 + 3 + 5 = 9$, which happens to be 3^2, the sum of the first four odd numbers is $1 + 3 + 5 + 7 = 16$, which is 4^2, the sum of the first five odd numbers is $25 = 5^2$, and the sum of the first 6 odd numbers is 6^2. This pattern leads to the conjecture that

$$\text{for all } n \in \mathbb{N}, \; 1 + 3 + 5 + \cdots + (2n - 1) = n^2.$$

We could never verify this statement by checking all possible values for n, but we can prove it using the PMI.

Example.   Prove that for every natural number n,

$$1 + 3 + 5 + \cdots + (2n - 1) = n^2.$$

Proof.   Let $S = \{n \in \mathbb{N} : 1 + 3 + 5 + \cdots + (2n - 1) = n^2\}$. *(We have defined S to be the set of natural numbers for which the statement is true. We show the statement is true for all natural numbers by showing that $S = \mathbb{N}$.)*

(i) $1 = 1^2$, so $1 \in S$.
(ii)   Let n be a natural number such that $n \in S$. Then

$$1 + 3 + 5 + \cdots + (2n - 1) = n^2.$$

⟨*We have assumed that some n is in S. From this assumption we will show that $n + 1 \in S$. Showing $n + 1 \in S$ is accomplished by verifying that*

$$1 + 3 + 5 + \cdots + [2(n + 1) - 1] = (n + 1)^2.$$

At this point in the proof it is essential to compare the statements for n and for $n + 1$. Notice that the left-hand sides of the two equations are almost identical, but the statement about $n + 1$ has one more term.⟩ By adding $2(n + 1) - 1$ to both sides of $1 + 3 + 5 + \cdots + (2n - 1) = n^2$, we have

$$1 + 3 + 5 + \cdots + (2n - 1) + [2(n + 1) - 1] = n^2 + [2(n + 1) - 1]$$
$$= n^2 + 2n + 1$$
$$= (n + 1)^2.$$

This shows that if $n \in S$, then $n + 1 \in S$.

(iii)   By the PMI, $S = \mathbb{N}$. That is, $1 + 3 + 5 + \cdots + (2n - 1) = n^2$ for every natural number n.   ■

The first key step in the proof above was to define the set S as the set of all natural numbers for which $1 + 3 + 5 + \cdots + (2n - 1) = n^2$ is true. In general, for an open sentence $P(n)$, the statement $(\forall n \in \mathbb{N})P(n)$ is true iff the set of numbers for which $P(n)$ is true equals $\mathbb{N}$.

The second key step in the proof above was to assume that *some* natural number n is in S. This assumption is called the **hypothesis of induction**. Notice that we must not assume that $n \in S$ for all $n \in \mathbb{N}$, because that would be assuming what we want to prove. For a direct proof of the inductive step, we start from the assumption that $n \in S$ (for some natural number n) and deduce that $n + 1 \in S$.

The third key step in the proof above was to compare the statement about n with the statement about $n + 1$. *Every good proof by induction will use the hypothesis of induction* to show that $n + 1 \in S$. Finding the connection between these statements is the heart of a proof by induction.

Thus, a proof of $(\forall n \in \mathbb{N})P(n)$ using the PMI may take the form:

PROOF OF $(\forall n \in \mathbb{N})P(n)$ USING THE PMI
Proof.
Let $S = \{n \in \mathbb{N}: P(n) \text{ is true}\}$.
   (i)   (Basis step) Show that $1 \in S$.
   (ii)   (Inductive step) Show that for all $n \in \mathbb{N}$, if $n \in S$ then $n + 1 \in S$.
   (iii)   Therefore, by the PMI, $S = \mathbb{N}$. Thus $(\forall n \in \mathbb{N})P(n)$ is true.   ■

In actual practice, very few induction proofs start by defining the set S. Since "$1 \in S$" is equivalent to "$P(1)$ is true," the basis step is a determination that $P(1)$

is true. Since "$n \in S$ implies $n + 1 \in S$" is equivalent to "$P(n)$ implies $P(n+1)$," the inductive step often takes the form of a direct proof that "for all $n \in \mathbb{N}$, $P(n)$ implies $P(n+1)$." This gives us the preferred form for the outline of a proof using the PMI:

PROOF OF $(\forall n \in \mathbb{N})P(n)$ **USING THE PMI**
Proof.
   (i)   (Basis step) Show that $P(1)$ is true.
   (ii)   (Inductive step) Suppose $P(n)$ for some $n \in \mathbb{N}$.

   . . .

   Therefore $P(n+1)$.
   (iii)   Therefore, by the PMI, $(\forall n \in \mathbb{N})P(n)$ is true.   ∎

Proofs by induction may be used to establish inequalities and divisibility properties. Notice in the following examples that it is not enough just to figure out what the correct statement is for $n + 1$. To construct a valid inductive step, look for a connection between what we know about some number n and what we want to know about the next number $n + 1$.

Example.   For all $n \in \mathbb{N}$, $n + 3 < 5n^2$.

Proof.

   (i) $1 + 3 < 5 \cdot 1^2$, so the statement is true for 1.
   (ii)   Assume that for some $n \in \mathbb{N}$, $n + 3 < 5n^2$. Then

$$(n + 1) + 3 = n + 3 + 1$$
$$< 5n^2 + 1$$
$$< 5n^2 + 10n + 5$$
$$= 5(n + 1)^2.$$

   Thus the statement is true for $n + 1$.
   (iii)   By the Principle of Mathematical Induction, $n + 3 < 5n^2$ for every $n \in \mathbb{N}$.   ∎

Example.   The polynomial $x - y$ divides the polynomials $x^2 - y^2$ and $x^3 - y^3$ because $x^2 - y^2 = (x - y)(x + y)$ and $x^3 - y^3 = (x - y)(x^2 + xy + y^2)$. This suggests the possibility that for every natural number n, $x - y$ divides $x^n - y^n$. We prove this by induction.

Proof.

   (i) $x - y$ divides $x^1 - y^1 = x - y$ because $x - y = 1(x - y)$. Thus the statement holds for $n = 1$.

(ii)   Assume that $x - y$ divides $x^n - y^n$ for some n ⟨*this is the hypothesis of induction*⟩. We must show that $x - y$ divides $x^{n+1} - y^{n+1}$. We write

$$x^{n+1} - y^{n+1} = xx^n - yy^n$$
$$= xx^n - yx^n + yx^n - yy^n$$
$$= (x - y)x^n + y(x^n - y^n).$$

Now $x - y$ divides the first term because that term contains the factor $(x - y)$. Also $x - y$ divides the second term because it divides $x^n - y^n$ ⟨*by the hypothesis of induction*⟩. Therefore, $x - y$ divides the sum. That is, $x - y$ divides $x^{n+1} - y^{n+1}$.

(iii)   By the PMI, $x - y$ divides $x^n - y^n$ for every natural number n.   ∎

Recall that *sigma notation* is a compact way to write sums. We may write the sum $1 + 3 + 5 + \cdots + (2n - 1)$ as $\sum_{i=1}^{n} (2i - 1)$. Our first result proved by induction was that $\sum_{i=1}^{n} (2i - 1) = n^2$. The notation for products uses the capital Greek letter Π. For example

$$\prod_{i=1}^{4} (i^2 + 1) = 2 \cdot 5 \cdot 10 \cdot 17 = 1700.$$

Note that products with $n + 1$ factors may be rewritten, as in

$$\prod_{i=1}^{n+1} (2i) = (2n + 2) \cdot \prod_{i=1}^{n} (2i).$$

Our next example involves both factorial and product notation.

Example.   Prove that for all $n \in \mathbb{N}$, $\prod_{i=1}^{n} (4i - 2) = \dfrac{(2n)!}{n!}$.

Proof.

(i)   The statement is true for $n = 1$ because $\prod_{i=1}^{1} (4i - 2) = 2$ and $\dfrac{(2 \cdot 1)!}{1!} = 2$.

(ii)   Assume that $\prod_{i=1}^{n} (4i - 2) = \dfrac{(2n)!}{n!}$ for some $n \in \mathbb{N}$. ⟨*We now use the hypothesis of induction to prove that* $\prod_{i=1}^{n+1} (4i - 2) = \dfrac{(2(n+1))!}{(n+1)!}$.⟩ Then

$$\prod_{i=1}^{n+1} (4i - 2) = \left[\prod_{i=1}^{n} (4i - 2) \right] (4(n + 1) - 2)$$

$$= \frac{(2n)!}{n!} (4n + 2).$$

We also compute

$$\frac{(2(n+1))!}{(n+1)!} = \frac{(2n+2))!}{(n+1)!}$$
$$= \frac{(2n+2)(2n+1)(2n)!}{(n+1)n!}$$
$$= \frac{2(2n+1)(2n)!}{n!}$$
$$= \frac{(4n+2)(2n)!}{n!}.$$

Since these expressions are equal, the statement is true for $n+1$.

(iii) By the Principle of Mathematical Induction, $\displaystyle\prod_{i=1}^{n}(4i-2) = \frac{(2n)!}{n!}$ for all $n \in \mathbb{N}$. ∎

The following examples show some other situations where induction is used.

Example. Consider any "map" formed by drawing straight lines in a plane to represent boundaries. Figure 2.4.1 shows ten countries, labeled A through J, formed by drawing four lines in the plane. The problem is to color the countries so that adjoining countries (those with a line segment as a common border) have different colors. This has been done in Figure 2.4.2 using only two colors—blue and white. We will use induction to prove that *every* map formed by drawing n straight lines can be colored using exactly two colors.

Proof.

(i) If a map is made by drawing one straight line, then there are only two countries. Thus every map formed with one line can be colored with two colors.

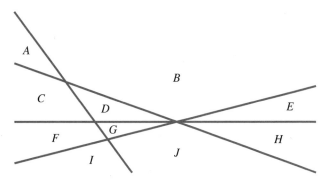

Figure 2.4.1

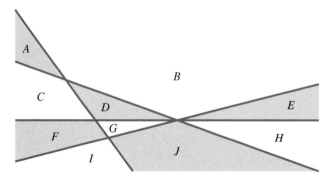

Figure 2.4.2

(ii)   Assume that for some n, *every* map formed by drawing n lines can be colored with exactly two colors. Now consider a map with $n + 1$ lines. Before coloring this map, choose any one of the lines and label it L. Now color the map as though the line L were not there, using exactly two colors. This can be done, initially, by the hypothesis of induction. ⟨*Such a coloring is shown in Figure 2.4.3, with the line L shown as a dashed line. Of course, only part of the plane can be shown.*⟩ To color the map *with* line L, proceed as follows. Call one half-plane determined by L side 1, and the other half-plane side 2. Leave all colors on side 1 exactly as they were but change every color on side 2 to the other color. This gives a coloring to every country in the map with line L. (See Figure 2.4.4.) It remains to verify that adjacent countries in this map with $n + 1$ lines have different colors.

Suppose we have two adjacent countries. There are two cases to consider:

Case 1.   Suppose L is the border between the two countries, which means that one country is on side 1 and the other on side 2. Initially, the two countries had the same color because they were *parts of the same country* in the map with n lines. When L was added to the map, the color of the country on side 2 switched to a different color from the country on side 1.

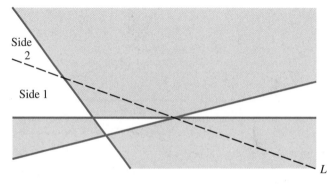

Figure 2.4.3

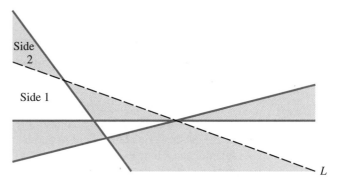

Figure 2.4.4

Case 2. Suppose L is not the border between the two countries. Then both countries are either on side 1 or side 2. If both countries are on side 1, they were initially colored differently and remain so when L is added. If both countries are on side 2, their colors were initially different, but are now switched, and still different.

In both cases, the two adjoining countries have different colors.

(iii) By the PMI, every map can be colored using only two colors. ∎

The next example involves computations using trigonometry and complex numbers.

Theorem 2.4.1

De Moivre's Theorem
Let θ be a real number. For all $n \in \mathbb{N}$, $(\cos \theta + i \sin \theta)^n = \cos n\theta + i \sin n\theta$.

Proof. ⟨*In this proof we use addition and multiplication of complex numbers and the following "sum of angles" formulas from trigonometry:*

$$\cos (\alpha + \beta) = (\cos \alpha)(\cos \beta) - (\sin \alpha)(\sin \beta)$$
$$\sin (\alpha + \beta) = (\sin \alpha)(\cos \beta) + (\cos \alpha)(\sin \beta).⟩$$

(i) For $n = 1$, the equation is $(\cos \theta + i \sin \theta)^1 = \cos \theta + i \sin \theta$, which is certainly true.

(ii) Assume that $(\cos \theta + i \sin \theta)^k = \cos k\theta + i \sin k\theta$, for some natural number k. Then, using the sum of angles formulas,

$(\cos \theta + i \sin \theta)^{k+1}$
$= (\cos \theta + i \sin \theta)^k (\cos \theta + i \sin \theta)$
$= (\cos k\theta + i \sin k\theta)(\cos \theta + i \sin \theta)$ ⟨*by the hypothesis of induction*⟩
$= \cos k\theta \cos \theta + i \sin k\theta \cos \theta + i \cos k\theta \sin \theta + i^2 \sin k\theta \sin \theta$
$= (\cos k\theta \cos \theta - \sin k\theta \sin \theta) + i(\sin k\theta \cos \theta + \cos k\theta \sin \theta)$

$$= \cos(k\theta + \theta) + i \sin(k\theta + \theta)$$

$$= \cos(k + 1)\theta + i \sin(k + 1)\theta.$$

(iii)   By steps (i) and (ii) and the PMI, $(\cos\theta + i \sin\theta)^n = \cos n\theta + i \sin n\theta$ is true for all $n \in \mathbb{N}$.   ∎

As one more example of the use of the Principle of Mathematical Induction, we prove a seemingly simple but useful result, known as the Archimedean Principle, about the comparative sizes of natural numbers. Archimedes* said that given a fulcrum and a long enough lever, he could move the world. See Figure 2.4.5. This statement illustrates the principle of physics that relates the forces at the ends of a lever to their distances from the fulcrum point. Even though it would take a very large force to move the Earth, and a person could exert only a small force, the force is multiplied when applied to a long lever.

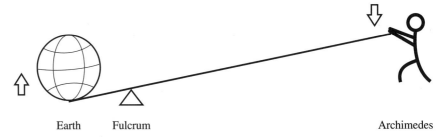

Earth   Fulcrum   Archimedes

Figure 2.4.5

To understand the next theorem, think of a and b as any two natural numbers, with a being much larger than b. The Archimedean Principle says that a can eventually be surpassed by taking natural number multiples of b. We give a proof by induction.

Theorem 2.4.2

Archimedean Principle for $\mathbb{N}$

For all natural numbers a and b, there exists a natural number s such that $sb > a$.

Proof.   Let b be a fixed natural number. The proof proceeds by induction on a.

(i)   If $a = 1$, choose s to be 2. Then $sb = 2b > a$.

(ii)   Suppose the statement is true when $a = n$, for some natural number n. Then there is a natural number t such that $tb > n$. Choose s to be $t + 1$. Then we have $sb = (t + 1)b = tb + b > n + 1$, so the statement is true when $a = n + 1$.

(iii)   By parts (i) and (ii) and the PMI, the statement is true for all natural numbers a and b.   ∎

* Archimedes (c. 287 B.C.E.–c. 212 B.C.E.) is considered the greatest scientist of his era, having made fundamental discoveries in mathematics, astronomy, physics, and engineering. Many of his drawings of proposed machines proved to be very effective devices. His "method of exhaustion" to calculate areas under curves is similar to the methods of integral calculus used today.

Some statements are not true for all natural numbers, but are true for numbers in some inductive subset of $\mathbb{N}$. To prove such statements in such cases, we need a slightly generalized form of the PMI, where the basis step starts at some number other than 1.

Generalized Principle of Mathematical Induction

Let k be a natural number. Let S be a subset of $\mathbb{N}$ with these two properties:
  (i) $k \in S$,
  (ii) for all $n \in \mathbb{N}$ with $n \geq k$, if $n \in S$, then $n + 1 \in S$.

Then S contains all natural numbers greater than or equal to k.

Example.   Prove by induction that $n^2 - n - 20 > 0$ for all $n > 5$.

Proof.   ⟨*We will use the Generalized PMI starting at $n = 6$.*⟩

  (i)   For $n = 6$, $6^2 - 6 - 20 = 10$, which is greater than zero.
  (ii)   Assume for some natural number $k > 5$ that $k^2 - k - 20 > 0$. Then

$$(k + 1)^2 - (k + 1) - 20 = k^2 + 2k + 1 - k - 1 - 20$$
$$= k^2 - k - 20 + 2k.$$

Since $k^2 - k - 20 > 0$ ⟨*by the induction hypothesis*⟩ and $2k > 0$ ⟨*since k is a natural number*⟩, $k^2 - k - 20 + 2k > 0$. ⟨*The sum of two positive integers is positive.*⟩ Thus $(k + 1)^2 - (k + 1) - 20 > 0$.
  (iii) By the Generalized PMI, $n^2 - n - 20 > 0$ is true for all $n > 5$. ∎

We note that an algebraic proof of the last example is possible: Since $n^2 - n - 20 = (n + 4)(n - 5)$ and $n + 4 > 0$ for all natural numbers n, $n^2 - n - 20 > 0$ for $n > 5$. Neither proof is "more correct" than the other. We chose the first proof to demonstrate the Generalized PMI.

Exercises 2.4

1. Which of these sets have the inductive property?
  (a) $\{20, 21, 22, 23, \dots\}$ ⋆ (b) $\{2, 4, 6, 8, 10, \dots\}$
  (c) $\{1, 2, 4, 5, 6, 7, \dots\}$ ⋆ (d) $\{17\}$
  (e) $\{x \in \mathbb{N}: x^2 \leq 1000\}$ ⋆ (f) $\{1, 2, 3, 4, 5, 6, 7, 8\}$

2. Suppose S is inductive. Which of the following must be true?
   (a) If $n + 1 \in S$, then $n \in S$.   ⋆ (b) If $n \in S$, then $n + 2 \in S$.
   (c) If $n + 1 \notin S$, then $n \notin S$.   (d) If $6 \in S$, then $11 \in S$.
   ⋆ (e) $6 \in S$ and $11 \in S$.

3. (a) Prove that $\mathbb{N}$ is inductive.
   (b) Prove that $\varnothing$ is inductive.

4. Evaluate or simplify each.
   (a) $4!$ (b) $7!$ (c) $\dfrac{97!}{96!}$

   (d) $\dfrac{8!}{3! \cdot 5!}$ ⋆ (e) $\dfrac{(n + 2)!}{n!}$ (f) $\dfrac{(n + 3)!}{(n + 1)!}$

   (g) $(n^2 + 3n + 2)n!$

5. Give an inductive definition for each:
   (a) $\{n: n = 5k \text{ for some } k \in \mathbb{N}\}$.
   (b) $\{n: n \in \mathbb{N} \text{ and } n > 10\}$.
   ⋆ (c) $\{n: n = 2^k \text{ for some } k \in \mathbb{N}\}$.
   (d) $\{a, a + d, a + 2d, a + 3d, \ldots\}$, where a and d are real numbers. (The elements of the set form an arithmetic progression.)
   (e) $\{a, ar, ar^2, ar^3, \ldots\}$, where a and r are real numbers. (The elements of the set form a geometric progression.)
   (f) $\displaystyle\bigcup_{i=1}^{n} A_i$, for some indexed family $\{A_i : i \in \mathbb{N}\}$.
   (g) The product $\displaystyle\prod_{i=1}^{n} x_i = x_1 \cdot x_2 \cdot x_3 \cdot \cdots \cdot x_n$ of n real numbers.

6. Use the PMI to prove the following for all natural numbers n.
   ⋆ (a) $\displaystyle\sum_{i=1}^{n} (3i - 2) = \frac{n}{2}(3n - 1)$.
   (b) $3 + 11 + 19 + \cdots + (8n - 5) = 4n^2 - n$.
   (c) $\displaystyle\sum_{i=1}^{n} 2^i = 2^{n+1} - 2$.
   (d) $1 \cdot 1! + 2 \cdot 2! + 3 \cdot 3! + \cdots + n \cdot n! = (n + 1)! - 1$.
   (e) $1^3 + 2^3 + \cdots + n^3 = \left[\dfrac{n(n + 1)}{2}\right]^2$
   (f) $\displaystyle\sum_{i=1}^{n} (2i - 1)^3 = n^2(2n^2 - 1)$.
   (g) $\dfrac{1}{1 \cdot 2} + \dfrac{1}{2 \cdot 3} + \dfrac{1}{3 \cdot 4} + \cdots + \dfrac{1}{n(n + 1)} = \dfrac{n}{n + 1}$.
   (h) $\dfrac{1}{2!} + \dfrac{2}{3!} + \cdots + \dfrac{n}{(n + 1)!} = 1 - \dfrac{1}{(n + 1)!}$.
   (i) $\displaystyle\sum_{i=1}^{n} \dfrac{1}{(2i - 1)(2i + 1)} = \dfrac{n}{2n + 1}$.
   (j) $\displaystyle\prod_{i=1}^{n} \left(1 - \dfrac{1}{i + 1}\right) = \dfrac{1}{n + 1}$.

(k) $\displaystyle\prod_{i=1}^{n}(2i-1) = \frac{(2n)!}{n!\,2^n}$.

(l) (Sum of a finite geometric series)
$$\sum_{i=0}^{n-1} ar^i = \frac{a(r^n - 1)}{r - 1} \text{ for } r \in \mathbb{R}, r \neq 1, \text{ and } n \geq 1.$$

7. Use the PMI to prove the following for all natural numbers:

  (a) $n^3 + 5n + 6$ is divisible by 3.

  (b) $4^n - 1$ is divisible by 3.

  (c) $n^3 - n$ is divisible by 6.

  (d) $(n^3 - n)(n + 2)$ is divisible by 12.

  (e) 8 divides $5^{2n} - 1$.

  (f) $10^{n+1} + 3 \cdot 4^{n-1} + 5$ is divisible by 9.

  (g) 8 divides $9^n - 1$.

  (h) $3^n \geq 1 + 2^n$.

  (i) $3^{n+3} > (n + 3)^3$.

  (j) $4^{n+4} > (n + 4)^4$.

  (k) $\displaystyle\sum_{i=1}^{n} \frac{1}{i^2} \leq 2 - \frac{1}{n}$.

  (l) For every positive real number x, $(1 + x)^n \geq 1 + nx$.

  (m) $\dfrac{n^3}{3} + \dfrac{n^5}{5} + \dfrac{7n}{15}$ is an integer.

  (n) Using the differentiation formulas $\frac{d}{dx}(x) = 1$ and $\frac{d}{dx}(fg) = f\frac{dg}{dx} + g\frac{df}{dx}$, prove that for all $n \in \mathbb{N}$, $\frac{d}{dx}(x^n) = nx^{n-1}$.

  (o) If a set A has n elements, then $\mathcal{P}(A)$ has 2^n elements.

8. Use the Generalized PMI to prove the following.

  (a) $n^3 < n!$ for all $n \geq 6$.

  (b) $2^n > n^2$ for all $n > 4$.

  (c) $(n + 1)! > 2^{n+3}$ for $n \geq 5$.

  (d) $n! > 3n$ for $n \geq 4$.

  (e) $\displaystyle\prod_{i=2}^{n} \frac{i^2 - 1}{i^2} = \frac{n + 1}{2n}$ for all $n \geq 2$.

  (f) $\displaystyle\prod_{i=1}^{n} \frac{1}{i} \leq 2^{-n}$ for $n \geq 4$.

  (g) For all $n > 2$, the sum of the angle measures of the interior angles of a convex polygon of n sides is $(n - 2) \cdot 180°$.

  (h) $\sqrt{n} < \dfrac{1}{\sqrt{1}} + \dfrac{1}{\sqrt{2}} + \dfrac{1}{\sqrt{3}} + \cdots + \dfrac{1}{\sqrt{n}}$ for $n \geq 2$.

9. Use the PMI to prove DeMorgan's Laws for an indexed family $\{A_i : i \in \mathbb{N}\}$. You may use De Morgan's Laws for two sets.

  (a) $\left(\displaystyle\bigcap_{i=1}^{n} A_i\right)^c = \displaystyle\bigcup_{i=1}^{n} A_i^c$ for all $n \in \mathbb{N}$.

  (b) $\left(\displaystyle\bigcup_{i=1}^{n} A_i\right)^c = \displaystyle\bigcap_{i=1}^{n} A_i^c$ for all $n \in \mathbb{N}$.

☆ **10.** Let $P_1, P_2, \ldots, P_n$ be n points in a plane with no three points collinear. Show that the number of line segments joining all pairs of points is $\dfrac{n^2 - n}{2}$. See the figure for $n = 5$.

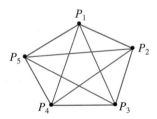

☆ **11.** A puzzle called the Towers of Hanoi consists of a board with 3 pegs and several disks of differing diameters that fit over the pegs. In the starting position all the disks are placed on one peg, with the largest at the bottom, and the others with smaller and smaller diameters up to the top disk (see the figure). A move is made by lifting the top disk off a peg and placing it on another peg so that there is no smaller disk beneath it. The object of the puzzle is to transfer all the disks from one peg to another.

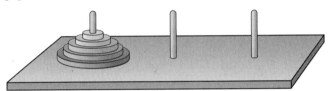

With a little practice, perhaps using coins of various sizes, you should convince yourself that if there are 3 disks, the puzzle can be solved in 7 moves. With 4 disks, 15 moves are required. Use the PMI to prove that with n disks, the puzzle can be solved in $2^n - 1$ moves. (*Hint:* In the inductive step you must *describe the moves* with $n + 1$ disks, and use the hypothesis of induction to count them.)

12. In a certain kind of tournament, every player plays every other player exactly once and either wins or loses. There are no ties. Define a *top* player to be a player who, for every other player x, either beats x or beats a player y who beats x.
(a) Show that there can be more than one top player.
(b) Use the PMI to show that every n-player tournament has a top player.

Proofs to Grade

13. Assign a grade of A (correct), C (partially correct), or F (failure) to each. Justify assignments of grades other than A.

★ (a) **Claim.** All horses have the same color.
"Proof." We must show that for all $n \in \mathbb{N}$, in every set of n horses, all horses in the set have the same color. Clearly in every set containing exactly 1 horse, all horses have the same color.

Now suppose all horses in every set of n horses have the same color. Consider a set of $n + 1$ horses. If we remove one horse, the horses in the remaining set of n horses all have the same color. Now consider a set of n horses obtained by removing some other horse. All horses in this set have the same color. Therefore all horses in the set of $n + 1$ horses have the same color. By the PMI, the statement is true for every $n \in \mathbb{N}$. ∎

★ **(b) Claim.** For all $n \in \mathbb{N}$, $n^3 + 44n$ is divisible by 3.

"**Proof.**"

(i) $1^3 + 44(1) = 45$, which is divisible by 3, so the statement is true for $n = 1$.

(ii) Assume the statement is true for some $n \in \mathbb{N}$. Then $n^3 + 44n$ is divisible by 3. Therefore $(n + 1)^3 + 44(n + 1)$ is divisible by 3.

(iii) By the PMI, the statement is true for all $n \in \mathbb{N}$. ∎

(c) Claim. For every natural number n, $n^2 + n$ is odd.

"**Proof.**" The number $n = 1$ is odd. Suppose $n \in \mathbb{N}$ and $n^2 + n$ is odd. Then

$$(n + 1)^2 + (n + 1) = n^2 + 2n + 1 + n + 1$$
$$= (n^2 + n) + (2n + 2)$$

is the sum of an odd and an even number. Therefore, $(n + 1)^2 + (n + 1)$ is odd. By the PMI, the property that $n^2 + n$ is odd is true for all natural numbers n. ∎

(d) Claim. For every natural number n, the matrix

$$\begin{bmatrix} 1 & 1 \\ 0 & 1 \end{bmatrix}^n = \begin{bmatrix} 1 & n \\ 0 & 1 \end{bmatrix}.$$

"**Proof.**" Let

$$S = \left\{ n \in \mathbb{N} : \begin{bmatrix} 1 & 1 \\ 0 & 1 \end{bmatrix}^n = \begin{bmatrix} 1 & n \\ 0 & 1 \end{bmatrix} \right\}.$$

Clearly,

$$\begin{bmatrix} 1 & 1 \\ 0 & 1 \end{bmatrix}^1 = \begin{bmatrix} 1 & 1 \\ 0 & 1 \end{bmatrix}.$$

so $1 \in S$. Assume that

$$\begin{bmatrix} 1 & 1 \\ 0 & 1 \end{bmatrix}^{n+1} = \begin{bmatrix} 1 & n + 1 \\ 0 & 1 \end{bmatrix}.$$

Then $n + 1 \in S$, so $n \in S$ implies $n + 1 \in S$. By the PMI, $S = \mathbb{N}$. ∎

★ **(e) Claim.** Every natural number greater than 1 has a prime factor.

"**Proof.**"

(i) Let $n = 2$. Then n is prime.

(ii) Suppose k has a prime factor x. Then $k = xy$ for some y. Thus $k + 1 = xy + 1 = (x + 1)(y + 1)$, which is a prime factorization.

(iii) By the PMI, the theorem is proved. ∎

(f) Claim. For all natural numbers $n \geq 4$, $2^n < n!$.

"**Proof.**" $2^4 = 16$ and $4! = 24$, so the statement is true for $n = 4$. Assume that $2^n < n!$ for some $n \in \mathbb{N}$. Then $2^{n+1} = 2(2^n) < 2(n!) \leq (n + 1)(n!) = (n + 1)!$, so $2^{n+1} < (n + 1)!$. By the PMI, the statement is true for all $n \geq 4$. ∎

2.5   Equivalent Forms of Induction

In the previous section, we used the Principle of Mathematical Induction to prove a variety of statements about natural numbers. The goal of this section is to learn how to use two other versions of induction. The value of these new forms of induction is that they may be used in situations where it would be difficult to apply the PMI. Both new forms are equivalent to the PMI, which means that either of them could replace the PMI in the list of axioms for $\mathbb{N}$.

To prove that a statement is true for all natural numbers using the PMI, the key step is to assume that a statement is true for an arbitrary natural number n and then show that the statement is true for $n + 1$. When there might be no apparent connection between the statement for n and the statement for $n + 1$, there may be a connection between the statement for the $n + 1$ case and the statement for some value or values less than n. There is a variation of the PMI to handle this situation. A much stronger assumption is made in this alternate form of induction, called **complete** (or **strong**) induction.*

Principle of Complete Induction (PCI)

Suppose S is a subset of $\mathbb{N}$ with this property:

For all natural numbers n, if $\{1, 2, 3, \ldots, n - 1\} \subseteq S$, then $n \in S$.

Then $S = \mathbb{N}$.

This form of induction begins with the assumption that a statement is true for *every* natural number less than n and shows that the statement is also true for n. Thus, we are allowed to assume the statement is true for each of $k = 1, 2, 3$, all the way through $n - 1$, rather than just for $n - 1$.

Notice that the statement of the PCI does not require a basis step in which we show that $1 \in S$. Nevertheless, for $n = 1$, the PCI property has the form

$$\varnothing \subseteq S \Rightarrow 1 \in S,$$

which is equivalent to $1 \in S$. Practically speaking then, it is almost always best to begin a PCI proof by verifying that 1 is in S. Sometimes special consideration is also needed for $n = 2$ or larger integers. We saw in Section 2.4 that this caution may be necessary for the PMI as well (see Exercise 13(a) of Section 2.4).

Our first example of a proof using the PCI revisits the first example from Section 2.4.

Example.   Prove for all natural numbers n that $1 + 3 + 5 + \cdots + (2n - 1) = n^2$.

Proof.   Let $S = \{n \in \mathbb{N}: 1 + 3 + 5 + \cdots + (2n - 1) = n^2\}$. ⟨*We must show that* $\{1, 2, 3, \ldots, m - 1\} \subseteq S \Rightarrow m \in S$ *and then conclude* $S = \mathbb{N}$ *by the PCI.*⟩

---

* Because this form of induction employs such a strong assumption, the Principle of Mathematical Induction is sometimes referred to as "weak induction."

Suppose m is a natural number and also that $\{1, 2, 3, \ldots, m - 1\} \subseteq S$. In the special case that $m = 1$, we note that $1 = 1^2$ and so $m \in S$. Otherwise, $m - 1 \in S$ ⟨*since* $m - 1 \in \{1, 2, 3, \ldots, m - 1\} \subseteq S$⟩ so $1 + 3 + 5 + \cdots + (2(m - 1) - 1 = (m - 1)^2$. Adding $2m - 1$ to both sides of this equality yields

$$1 + 3 + 5 + \cdots + (2(m - 1) - 1) + (2m - 1)$$
$$= (m - 1)^2 + (2m - 1)$$
$$= m^2 - 2m + 1 + 2m - 1$$
$$= m^2.$$

Thus $m \in S$ and we conclude $S = \mathbb{N}$ by the PCI. ∎

Like the PMI, the PCI has variations where the induction starts at some number greater than one. To prove, for example, that some property holds for all numbers greater than 6 we would verify that for all natural numbers $n > 6$, if $\{7, 8, \ldots, n - 1\} \subseteq S$, then $n \in S$. Our next example is a proof that every natural number greater than 1 has a prime factor. This fact was used without proof in Chapter 1 because we did not yet have induction available as a method of proof. It is a good example of a proof where the PCI is much more natural to use than the PMI.

Example. Prove that every natural number greater than 1 has a prime factor.

Proof. Let S be $\{n \in \mathbb{N}: n > 1$ and n has a prime factor$\}$. Notice that 1 is not in S, but 2 is in S. Let m be a natural number greater than 1. Assume that for all $k \in \{2, \ldots, m - 1\}$, $k \in S$. We must show that $m \in S$. If m has no factors other than 1 and m, then m is prime, and so m has a prime factor—itself. If m has a factor x other than 1 and m, then $1 < x < m$ so $x \in S$. Therefore x has a prime factor ⟨*by the induction hypothesis*⟩, which must also be a prime factor of m. In either case, $m \in S$. Therefore, $S = \{n \in \mathbb{N}: n > 1\}$, and every natural number greater than 1 has a prime factor. ∎

Like the PMI, the PCI can be used to create inductive definitions, one of which is the definition of the sequence $1, 1, 2, 3, 5, 8, 13, \ldots$ examined by Leonardo Fibonacci[*] in the 13th century (see Exercise 4). These numbers have played important roles in applications as diverse as population growth, flower petal patterns, and highly efficient file sorting algorithms in computer science. For each natural number n, the nth Fibonacci number f_n is defined inductively by

$$f_1 = 1, f_2 = 1, \text{ and } f_{n+2} = f_{n+1} + f_n \quad \text{for } n \geq 1.$$

We see that $f_3 = 2, f_4 = 3, f_5 = 5, f_6 = 8$, and so on.

---

[*] Leonardo of Pisa, also called Leonardo Fibonacci (c. 1170–c. 1250) was a prominent mathematician in the Middle Ages. His text, *Liber Abaci* (Book of Calculation) was influential in the European adoption of Hindu-Arabic numerals. He did not invent the sequence named for him, but used it as an example.

Inductive proofs of properties of the Fibonacci numbers usually involve the PCI because we need to "reach back" to both f_{n-1} and f_{n-2} to prove the property for f_n. Here is a typical example.

Example. Let α be the positive solution to the equation $x^2 = x + 1$. (The value of α is $\dfrac{(1 + \sqrt{5})}{2}$, approximately 1.618.) Prove that $f_n \leq \alpha^{n-1}$ for all $n \geq 1$.

Proof. In the special cases of $m = 1$ and $m = 2$, the inequality $f_m \leq \alpha^{m-1}$ is true since $f_1 = 1 \leq \alpha^0 = 1$ and $f_2 = 1 \leq \alpha^1 = (1 + \sqrt{5})/2$. Let m be a natural number, $m \geq 3$, and assume that $f_k \leq \alpha^{k-1}$ for all $k \in \{1, 2, 3, \ldots, m - 1\}$. For $m \geq 3$, we have

$$
\begin{aligned}
f_m &= f_{m-1} + f_{m-2} \\
&\leq \alpha^{m-2} + \alpha^{m-3} \quad \langle \textit{by the induction hypothesis for } m - 1 \textit{ and } m - 2 \rangle \\
&= \alpha^{m-3}(\alpha + 1) \\
&= \alpha^{m-3}(\alpha^2) \quad \langle \textit{since } \alpha \textit{ is a solution to } x^2 = x + 1, \alpha + 1 = \alpha^2 \rangle \\
&= \alpha^{m-1}.
\end{aligned}
$$

Therefore, $f_m \leq \alpha^{m-1}$.

By the PCI, we conclude that $f_n \leq \alpha^{n-1}$ for all $n \geq 1$. ∎

Theorem 2.5.4, at the end of this section, shows that the PMI and PCI are equivalent. Thus both properties are true for $\mathbb{N}$.

A third property that characterizes the set $\mathbb{N}$ is the Well-Ordering Principle. Although it is quite simple to state, the WOP turns out to be a powerful tool for constructing proofs. The WOP, like the PCI, may be derived from the Peano Axioms and is equivalent to the PMI. (See Theorem 2.5.5.)

Well-Ordering Principle (WOP)*

Every nonempty subset of $\mathbb{N}$ has a smallest element.

Proofs using the WOP frequently take the form of assuming that some desired property does not hold for all natural numbers. This produces a nonempty set of natural numbers that do not have the property. By working with the *smallest* such number, one can often find a contradiction.

In the next example, we prove again that every natural number greater than 1 has a prime factor. Compare the structure of this proof using the WOP with that of the PCI proof given earlier.

Example. Every natural number $n > 1$ has a prime factor.

---

* Some mathematicians refer to the Well-Ordering Principle as the Well-Ordering Property. They reserve the use of the term Well-Ordering Principle for the statement that for *every* set there is an order that makes the set a well-ordered set. Orderings are discussed in Section 3.4.

Proof.   If n is prime, then n is a prime factor of n. If n is composite, then n has factors other than 1 and n. Therefore the set

$$T = \{m \in \mathbb{N} : m \text{ divides } n, m \neq n, \text{ and } m \neq 1\}$$

is a nonempty subset of $\mathbb{N}$. By the WOP, T has a smallest element, which we call p. ⟨*We will show by contradiction that p is prime.*⟩

Suppose p is composite. Then p has a divisor d, with $d \neq p$, and $d \neq 1$. Since d divides p and p divides n, d divides n. Therefore, $d \in T$. But this is a contradiction since $d < p$ and p is the smallest element of T. Therefore p is a prime factor of n. ∎

As another example, we will prove that for every $n \in \mathbb{N}$, $n + 1 = 1 + n$. The purpose of this example is not to establish the fact that $n + 1 = 1 + n$, but to see how a proof using the WOP is done. So imagine for a moment that we did not know that addition is commutative, and we will show how the statement can be proved (from the associative property) by using the WOP.

Example.   For every natural number n, $n + 1 = 1 + n$.

Proof.   Suppose there exists a natural number n such that $n + 1 \neq 1 + n$. Let b be the smallest such number. Obviously, $1 + 1 = 1 + 1$, so $b \neq 1$. Thus b must be of the form $b = c + 1$ for some $c \in \mathbb{N}$. ⟨*See the successor properties for* $\mathbb{N}$.⟩ Then $(c + 1) + 1 \neq 1 + (c + 1)$. Therefore, by the associative property for $\mathbb{N}$, $(c + 1) + 1 \neq (1 + c) + 1$. Subtracting 1 ⟨*from the right side*⟩ from each expression, we have $c + 1 \neq 1 + c$. But this is a contradiction because $c < b$ and b is the smallest natural number with the property. We conclude that $n + 1 = 1 + n$ for all natural numbers n. ∎

The next three theorems were stated without proof earlier in the text. Each may be proved using the WOP. The Division Algorithm was the primary result that we used in Section 1.7 to develop interesting results about the greatest common divisor (gcd) of two integers and linear combinations of the integers.

Theorem 2.5.1

The Division Algorithm
For all integers a and b, with $a \neq 0$, there exist unique integers q and r such that $b = aq + r$ and $0 \leq r < |a|$.

Proof.   Let a and b be integers and $a \neq 0$. Assume that $a > 0$. ⟨*The proof in the case $a < 0$ is similar, and is left as an exercise.*⟩ We must first show the existence of q and r.

Let $S = \{b - ak : k \text{ is an integer and } b - ak \geq 0\}$. If 0 is in S, then a divides b, and we may take q to be the integer $\frac{b}{a}$ and $r = 0$. Now assume that $0 \notin S$.

It follows from the assumption $0 \notin S$ that $b \neq 0$. ⟨*Otherwise $b - a0 = 0 \in S$.*⟩ Now if $b > 0$ then $b - a0 \in S$, and if $b < 0$ then $b - a(2b) = b(1 - 2a) \in S$. In either case, the set S is nonempty.

By the Well-Ordering Principle, S has a smallest element, which we will call r. Then $r = b - aq$ for some integer q, so $b = aq + r$, and $r \geq 0$. We must also show that $r < |a| = a$. Suppose $r \geq a$. Then $b - a(q + 1) = b - aq - a = r - a \geq 0$, so $b - a(q + 1) \in S$. But $b - a(q + 1) < b - aq$ and $b - aq$ is the smallest member of S. We conclude that $r < |a|$.

To complete the proof, we must show that q and r are unique. Suppose there exist integers q, q', r, and r' such that

$$b = aq + r \text{ with } 0 \leq r < |a| \text{ and}$$
$$b = aq' + r' \text{ with } 0 \leq r' < |a|.$$

Without loss of generality, we may assume that $r' \geq r$. ⟨*Otherwise, we could relabel r and r'.*⟩ Then $aq + r = aq' + r'$, which implies $a(q - q') = r' - r$. Then a divides $r' - r$, and $0 \leq r' - r \leq r' < |a|$. Then $r' - r$ must be 0, so $r' = r$. Since $a(q - q') = 0$, $q' = q$. ∎

Section 1.7 also contained the following result about linear combinations of integers. The short proof using the WOP is Exercise 8.

Theorem 2.5.2   Let a and b be nonzero integers. Then there is a smallest positive linear combination of a and b.

The Fundamental Theorem of Arithmetic, stated in the *Preface to the Student*, may also be proved using the WOP. See Exercise 9.

Theorem 2.5.3   ## The Fundamental Theorem of Arithmetic
Every natural number greater than 1 is prime or can be expressed uniquely as a product of primes.

The final two theorems of this section show that the PMI, the PCI, and the WOP are all equivalent.

Theorem 2.5.4   The Principle of Mathematical Induction and the Principle of Complete Induction are equivalent.

Proof.   ⟨*The proof proceeds in two parts: The first part shows that PMI implies the PCI, and the second part shows the converse.*⟩

Part 1.   Assume that the PMI holds for $\mathbb{N}$. Suppose S is a subset of $\mathbb{N}$ with this property:

   * for all natural numbers m, if $\{1, 2, 3, \ldots, m - 1\} \subseteq S$, then $m \in S$.

As a step toward proving $S = \mathbb{N}$, we first use the PMI to show that for every natural number n, $\{1, 2, 3, \ldots, n\} \subseteq S$.

   (i)   For $n = 1$, the set $\{1, 2, 3, \ldots, n - 1\}$ is the empty set. Thus, $\{1, 2, 3, \ldots, n - 1\} \subseteq S$. Therefore, by the property * for S, $1 \in S$. Thus, $\{1\} \subseteq S$. Hence, for $n = 1$, we have $\{1, 2, 3, \ldots, n\} \subseteq S$.

   (ii)   Assume that $\{1, 2, 3, \ldots, n\} \subseteq S$. ⟨*We must show $\{1, 2, 3, \ldots, n + 1\} \subseteq S$.*⟩ Since $\{1, 2, 3, \ldots, n\} \subseteq S$, by the property *, we have $n + 1 \in S$. Therefore, $\{1, 2, 3, \ldots, n, n + 1\} \subseteq S$.

(iii)   By steps (i) and (ii), and the PMI, $\{1, 2, 3, \ldots, n\} \subseteq S$ for every natural number n. Now let n be a natural number. Then $\{1, 2, 3, \ldots, n\} \subseteq S$, so $n \in \mathbb{N}$. This shows $\mathbb{N} \subseteq S$. Since S is a subset of $\mathbb{N}$, we conclude that $S = \mathbb{N}$.

Part 2.   Now assume that the PCI holds for $\mathbb{N}$. ⟨*To show that the PMI is true, we assume its hypothesis and use the PCI to show that the conclusion of the PMI ($S = \mathbb{N}$) must also be true.*⟩ Suppose S is a subset of $\mathbb{N}$ with two properties:

(i) $1 \in S$.
(ii)   For all natural numbers n, if $n \in S$, then $n + 1 \in S$.

We will prove $S = \mathbb{N}$. ⟨*To prove that $S = \mathbb{N}$, we show that S satisfies the hypothesis for the PCI; namely, that for all $m \in \mathbb{N}$, if $\{1, 2, 3, \ldots, m - 1\} \subseteq S$, then $m \in S$.*⟩ Let m be a natural number such that $\{1, 2, 3, \ldots, m - 1\} \subseteq S$. There are two cases:

Case 1   If $m = 1$, then $1 \in S$ by the first property of S. Thus the statement $\{1, 2, 3, \ldots, m - 1\} \subseteq S$ implies $m \in S$ is true when $m = 1$.
Case 2   If $m > 1$, then from $\{1, 2, 3, \ldots, m - 1\} \subseteq S$, we have $m - 1 \in S$. But then by the second property for S, we have $m \in S$. In this case, too, we have $\{1, 2, 3, \ldots, m - 1\} \subseteq S$ implies $m \in S$.

We conclude that the statement $\{1, 2, 3, \ldots, m - 1\} \subseteq S$, implies $m \in S$ is true for all natural numbers m. Therefore, by the PCI, $S = \mathbb{N}$.   ∎

Theorem 2.5.5   The Well-Ordering Principle is equivalent to the Principle of Mathematical Induction.

Proof.   ⟨*This proof, like Theorem 2.5.4, proceeds in two parts: The first part shows that the PMI implies the WOP, and the second part shows the converse.*⟩

Part 1.   Assume that the PMI holds for $\mathbb{N}$. ⟨*To show that the WOP is true, we show that every nonempty subset of $\mathbb{N}$ has a smallest element.*⟩ Suppose T is a nonempty subset of $\mathbb{N}$. Let $S = \mathbb{N} - T$. Since $T \neq \varnothing$, $S \neq \mathbb{N}$. ⟨*The proof now proceeds by contradiction. We suppose T has no smallest element and use the PMI to show that $S = \mathbb{N}$.*⟩ Suppose that T has no smallest element.

(i)   Since 1 is the smallest element of $\mathbb{N}$ and T has no smallest element, $1 \notin T$. Therefore, $1 \in S$.
(ii)   Suppose that $n \in S$. No number less than n belongs to T, because, if any of the numbers $1, 2, 3, \ldots, n - 1$ were in T, then one of those numbers would be the smallest element of T. We know $n \notin T$ because $n \in S$. Therefore, $n + 1$ cannot be in T, or else it would be the smallest element of T. Thus, $n + 1 \in S$.
(iii)   By parts (i) and (ii) and the PMI, $S = \mathbb{N}$, which is a contradiction to $S \neq \mathbb{N}$. Therefore, T has a smallest element.

Part 2.   The proof that the WOP implies the PMI is Exercise 12(b).   ∎

Exercises 2.5

1. Use the PCI to prove that

    ☆ **(a)** every natural number greater than 22 can be written in the form $3s + 4t$, where s and t are integers, $s \geq 3$ and $t \geq 2$.

    (b) every natural number greater than 33 can be written in the form $4s + 5t$, where s and t are integers, $s \geq 3$ and $t \geq 2$.

2. Let $a_1 = 2$, $a_2 = 4$, and $a_{n+2} = 5a_{n+1} - 6a_n$ for all $n \geq 1$. Prove that $a_n = 2^n$ for all natural numbers n.

3. In this exercise you are to prove some well-known facts about numbers as a way of demonstrating use of the WOP. Use the WOP to prove the following:

    (a) If $a > 0$, then for every natural number n, $a^n > 0$.

    (b) For all positive integers a and b, $b \neq a + b$. (*Hint*: Suppose for some a there is b such that $b = a + b$. By the WOP, there is a smallest a_0 such that, for some b, $b = a_0 + b$. Apply the WOP again.)

    ☆ **(c)** $\sqrt{2}$ is irrational.

4. In 1202, Leonardo Fibonacci posed the following problem: Suppose a particular breed of rabbit breeds one new pair of rabbits each month, except that a 1-month-old pair is too young to breed. Suppose further that no rabbit breeds with any other except its paired mate and that rabbits live forever. At 1 month we have our original pair of rabbits. At 2 months we still have the single pair. At 3 months, we have two pairs (the original and their one pair of offspring). At 4 months we have three pairs (the original pair, one older pair of offspring, and one new pair of offspring).

    (a) Show that at n months, there are f_n pairs of rabbits.

    (b) Calculate the first ten Fibonacci numbers $f_1, f_2, f_3, \ldots, f_{10}$.

    (c) Find a formula for $f_{n+3} - f_{n+1}$.

5. Use the PMI to show that each of the following statements about Fibonacci numbers is true:

    ☆ **(a)** f_{3n} is even and both f_{3n+1} and f_{3n+2} are odd for all natural numbers n.

    (b) $\gcd(f_n, f_{n+1}) = 1$ for all natural numbers n.

    (c) $\gcd(f_n, f_{n+2}) = 1$ for all natural numbers n.

    ★ **(d)** $f_1 + f_2 + f_3 + \cdots + f_n = f_{n+2} - 1$ for all natural numbers n.

6. Use the PCI to prove the following properties of Fibonacci numbers:

    (a) f_n is a natural number for all natural numbers n.

    (b) $f_{n+6} = 4f_{n+3} + f_n$ for all natural numbers n.

    (c) For any natural number a, $f_a f_n + f_{a+1} f_{n+1} = f_{a+n+1}$ for all natural numbers n.

    ☆ **(d)** (Binet's formula) Let α be the positive solution and β the negative solution to the equation $x^2 = x + 1$. (The values are $\alpha = \dfrac{1 + \sqrt{5}}{2}$ and $\beta = \dfrac{1 - \sqrt{5}}{2}$.) Show for all natural numbers n that

    $$f_n = \frac{\alpha^n - \beta^n}{\alpha - \beta}.$$

7. Complete the proof of the Division Algorithm (Theorem 2.5.1) for the case $a < 0$. That is, show that for all integers a and b, with $a < 0$, there exist unique integers q and r such that $b = aq + r$ and $0 \leq r < |a| = -a$.

8. Let a and b be nonzero integers. Prove that there is a smallest positive linear combination of a and b. (Theorem 2.5.2)

9. Prove the Fundamental Theorem of Arithmetic: Every natural number greater than 1 is prime or can be expressed uniquely as a product of primes.

10. In the tournament described in Exercise 12 of Section 2.4, a top player is defined to be one who either beats every other player or beats someone who beats the other player. Use the WOP to show that in every such tournament with n players ($n \in \mathbb{N}$), there is at least one top player.

11. Let the "Fibonacci-2" numbers g_n be defined as follows:

$$g_1 = 2, \ g_2 = 2, \text{ and } g_{n+2} = g_{n+1} g_n \quad \text{for all } n \geq 1.$$

  (a) Calculate the first five "Fibonacci-2" numbers.
  (b) Show that $g_n = 2^{f_n}$.

12. Complete the proof of the equivalences of the PMI, PCI, and WOP by
  (a) using the PCI to prove the WOP.
  ★ (b) using the WOP to prove the PMI.
  (c) using the WOP to prove the PCI.

Proofs to Grade 13. Assign a grade of A (correct), C (partially correct), or F (failure) to each. Justify assignments of grades other than A.
  (a) **Claim.** For all natural numbers n, 5 divides $8^n - 3^n$.
  "Proof." Suppose there is $n \in \mathbb{N}$ such that 5 does not divide $8^n - 3^n$. Then by the WOP there is a smallest such natural number t. Now $t \neq 1$ since 5 does divide $8^1 - 3^1$. Therefore $t - 1$ is a natural number smaller than t, so 5 divides $8^{t-1} - 3^{t-1}$. But then 5 divides $8(8^{t-1} - 3^{t-1})$ and 5 divides $5(3^{t-1})$, so 5 divides their sum, which is $8^t - 3^t$. This is a contradiction. Therefore, 5 divides $8^n - 3^n$ for all $n \in \mathbb{N}$. ∎
  ★ (b) **Claim.** For every natural number n, 3 divides $n^3 + 2n + 1$.
  "Proof." Suppose there is a natural number n such that 3 does not divide $n^3 + 2n + 1$. By the WOP, there is a smallest such number. Call this number m. Then $m - 1$ is a natural number and 3 does divide

$$(m - 1)^3 + 2(m - 1) + 1 = m^3 - 3m^2 + 3m - 1 + 2m - 2 + 1$$
$$= m^3 - 3m^2 + 5m - 2.$$

  But 3 also divides $3m^2 - 3m + 3$, so 3 divides the sum of these two expressions, which is $m^3 + 2m + 1$. This contradicts what we know about m. Therefore, the set $\{n \in \mathbb{N} : 3 \text{ does not divide } n^3 + 2n + 1\}$ must be empty. Therefore, 3 divides $n^3 + 2n + 1$, for every natural number n. ∎
  (c) **Claim.** The PCI implies the WOP.
  "Proof." Assume the PCI. Let T be a nonempty subset of $\mathbb{N}$. Then T has some element x. Then $\{1, 2, \ldots, x - 1\} \subseteq \mathbb{N} - T$. By the PCI,

$x \in \mathbb{N} - T$. This is a contradiction, because $x \in T$ and $x \in \mathbb{N} - T$. Therefore T has a smallest element. ∎

(d)   Claim.   The WOP implies the PCI.

"Proof."   Assume the WOP. To prove the PCI, let S be a subset of $\mathbb{N}$ such that for all natural numbers m, $\{1, 2, 3, \ldots, m - 1\} \subseteq S$. Let $k \in \mathbb{N}$. Then $k + 1$ is an integer, so $\{1, 2, \ldots, (k + 1) - 1\} \subseteq S$. However, $(k + 1) - 1 = k$, so $k \in S$. Thus every natural number is in S, so $S = \mathbb{N}$. ∎

2.6   Principles of Counting

Recall from the *Preface to the Student* that a set A is **finite** iff $A = \varnothing$ or A has n elements for some $n \in \mathbb{N}$. For a finite set A the number of elements in A is denoted $\overline{\overline{A}}$. For example, if $A = \{p, q\}$ and $B = \{3, 2, 1, 5, 9\}$ then $\overline{\overline{A}} = 2$ and $\overline{\overline{B}} = 5$. This section describes some of the fundamental techniques for counting the number of elements in finite sets.

A more precise development of the concepts of the cardinality (number of elements) and finiteness of a set appears in Chapter 5. For this reason, proofs of the basic counting rules in this section appear in Section 5.1.

Theorem 2.6.1   **The Sum Rule**

Let A and B be finite sets with m and n elements, respectively. If A and B are disjoint then $\overline{\overline{A \cup B}} = m + n$.

We use the Sum Rule so often, we don't have to think about it: If a basket has 5 oranges and 6 apples, then there are 11 pieces of fruit in the basket. The rule can be extended to any finite number of sets. We prove the Generalized Sum Rule by using the Principle of Mathematical Induction.

Theorem 2.6.2   **The Generalized Sum Rule**

For all $n \in \mathbb{N}$ and for every family $\mathscr{A} = \{A_i : i = 1, 2, 3, \ldots, n\}$ of n distinct pairwise disjoint sets, if $\overline{\overline{A_i}} = a_i$ for $1 < i < n$, then

$$\overline{\overline{\bigcup_{i=1}^{n} A_i}} = \sum_{i=1}^{n} a_i.$$

Proof.   The proof is by induction on the number n of sets in the family $\mathscr{A}$.

(i)   If $n = 1$, then $\overline{\overline{\bigcup_{i=1}^{1} A_i}} = \overline{\overline{A_1}} = a_1 = \sum_{i=1}^{1} a_i$.

(ii).   Suppose for some $n \in \mathbb{N}$ that $\overline{\overline{A_i}} = a_i$ for $i = 1, 2, \ldots, n$ and $\overline{\overline{\bigcup_{i=1}^{n} A_i}} = \sum_{i=1}^{n} a_i$

for every family $\mathscr{A} = \{A_i : i = 1, 2, 3, \ldots, n\}$ of n distinct pairwise disjoint sets. Let $\mathscr{A} = \{A_1, A_2, A_3, \ldots, A_{n+1}\}$ be a family of $n + 1$ distinct pairwise disjoint sets with $\overline{\overline{A_i}} = a_i$ for $i = 1, 2, \ldots, n + 1$. Then

$$\overline{\bigcup_{i=1}^{n+1} A_i} = \overline{\left(\bigcup_{i=1}^{n} A_i\right) \cup A_{n+1}}$$

$$= \overline{\bigcup_{i=1}^{n} A_i} + \overline{A_{n+1}} \quad \langle \text{by the Sum Rule since } \bigcup_{i=1}^{n} A_i \text{ and } A_{n+1} \text{ are disjoint}\rangle$$

$$= \sum_{i=1}^{n} a_i + a_{n+1} \quad \langle \text{by the hypothesis of induction since}$$

$$\{A_1, A_2, A_3, \ldots A_n\} \text{ is a collection of}$$

$$n \text{ distinct pairwise disjoint sets}\rangle$$

$$= \sum_{i=1}^{n+1} a_i.$$

(iii)   By the PMI, the Generalized Sum Rule is true for every family of n distinct pairwise disjoint sets, for all $n \in \mathbb{N}$.   ∎

The Generalized Sum Rule is useful in situations where it would be practically impossible for any individual to make an acceptably accurate count. For example, a good estimate of the total population of a country on a fixed date (a census) may be accomplished by summing the results of the combined work of thousands of individuals, each of whom does a count for a designated small geographic area.

If sets A and B are not disjoint (see Figure 2.6.1), then determining the number of elements in $A \cup B$ by simply adding $\overline{\overline{A}}$ and $\overline{\overline{B}}$ overcounts $\overline{\overline{A \cup B}}$ by counting twice each element of $A \cap B$. Theorem 2.6.3 corrects this error by subtracting $\overline{\overline{A \cap B}}$ from $\overline{\overline{A}} + \overline{\overline{B}}$.

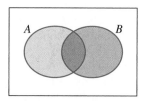

Figure 2.6.1

Theorem 2.6.3   For finite sets A and B, $\overline{\overline{A \cup B}} = \overline{\overline{A}} + \overline{\overline{B}} - \overline{\overline{A \cap B}}$.

Proof.   By the Sum Rule, $\overline{\overline{A}} = \overline{\overline{A - B}} + \overline{\overline{A \cap B}}$ and $\overline{\overline{B}} = \overline{\overline{B - A}} + \overline{\overline{A \cap B}}$. Applying the Generalized Sum Rule to the distinct and pairwise disjoint sets $A - B$, $A \cap B$, and $B - A$, we have $\overline{\overline{A \cup B}} = \overline{\overline{A - B}} + \overline{\overline{A \cap B}} + \overline{\overline{B - A}} = \overline{\overline{A}} + \overline{\overline{B}} - \overline{\overline{A \cap B}}$.   ∎

If we know the number of elements in A, B, and $A \cup B$, we can use Theorem 2.6.3 to determine $\overline{\overline{A \cap B}}$:

$$\overline{\overline{A \cap B}} = \overline{\overline{A}} + \overline{\overline{B}} - \overline{\overline{A \cup B}}.$$

Example. During one week a total of 46 patients were treated by Dr. Medical for either a broken leg or a sore throat. Of these, 32 had a broken leg and 20 had a sore throat. How many did she treat for both ailments?

Letting B be the set of patients with broken legs and S the set of patients with sore throats, the solution is

$$\overline{\overline{B \cap S}} = \overline{\overline{B}} + \overline{\overline{S}} - \overline{\overline{B \cup S}} = 32 + 20 - 46 = 6.$$

Applying the Sum Rule to the disjoint sets $B - S$ and $B \cap S$, we could also find that $\overline{\overline{B - S}} = \overline{\overline{B}} - \overline{\overline{B \cap S}} = 32 - 6$, so there were 26 patients with a broken leg but no sore throat. Similarly, we could determine that 14 patients had just a sore throat. See Figure 2.6.2.

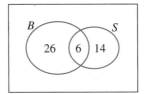

Figure 2.6.2

Example. French and German are two of the four national languages of Switzerland. Suppose 80% of Swiss residents speak German fluently, 66% speak French fluently, and 52% are fluent in both languages. What percentage of Swiss residents are not fluent in either French or German?

To solve this problem, we first find the percentage of residents fluent in at least one of the two languages $(80 + 66 - 52 = 94)$ and subtract this result from 100%. Based on the given estimates, 6% of residents are not fluent in either language.

Theorem 2.6.3 can be extended to three or more sets by the **Principle of Inclusion and Exclusion.** The idea is that, in counting the number of elements in the union of several sets by counting the number of elements in each set, we have included too many elements more than once; so some need to be excluded, or subtracted, from the total. When more than two sets are involved, this first attempt at exclusion will subtract too many elements, so that some need to be added back or included again, and so forth. For three sets A, B, C, the Principle of Inclusion and Exclusion states that

$$\overline{\overline{A \cup B \cup C}} = [\overline{\overline{A}} + \overline{\overline{B}} + \overline{\overline{C}}] - [\overline{\overline{A \cap B}} + \overline{\overline{A \cap C}} + \overline{\overline{B \cap C}}] + \overline{\overline{A \cap B \cap C}}.$$

The inclusion and exclusion formulas for more than 3 sets are lengthier. (See Exercise 5.) The Principle is often applied to determine the number of elements not in any of several sets, as in Exercise 6.

The next basic counting rule is as simple to state as the Sum Rule.

Theorem 2.6.4

The Product Rule

If A and B be finite sets with m and n elements, respectively, then $\overline{\overline{A \times B}} = mn$.

The Product Rule may be applied to counting the total number of ways to perform two *independent* tasks (jobs or activities). By independent we mean that the occurrence of one task has no influence on the occurrences of other tasks. For example, if we want to select one prime number (task #1) and one composite number (task #2) from the set $\{10, 11, 12, \ldots 20\}$, there are $4 \cdot 7 = 28$ possible ways to perform the two tasks.

Like the Sum Rule, the Product Rule can be extended to more than 2 sets.

Theorem 2.6.5

The Generalized Product Rule

For all $n \in \mathbb{N}$ and for every family $\mathscr{A} = \{A_i : i = 1, 2, 3, \ldots, n\}$ of n sets, if $\overline{\overline{A_i}} = a_i$ for $1 < i < n$, then

$$\overline{\overline{A_1 \times A_2 \times \cdots \times A_n}} = \prod_{i=1}^{n} a_i.$$

Proof.   See Exercise 12.   ■

Example.   To find the number of three-digit positive integers, we must perform three tasks: Choose each of the three digits. There are 10 possibilities for the units digit and 10 for the tens digit, but only 9 possibilities for the hundreds digit, because it can't be zero. By the product rule there are $10 \cdot 10 \cdot 9 = 900$ ways to form a three-digit positive integer. We can check this result easily: Of the 999 numbers from 1 to 999, the first 99 have only one or two digits, so 900 have three digits.

Example.   To find the number of three-digit positive integers *with no repeated digits*, we might begin by observing that there are 10 possibilities for the units digit and nine remaining possibilities for the tens digit. At this point we see that the task of choosing the hundreds digit is not independent of the other tasks: The number of possibilities depends on whether 0 is chosen for either the units or the tens digit. To use the Product Rule we must think of a different sequence of tasks, or perhaps of carrying out these tasks in a different order. Beginning with the hundreds digit there are 9 possibilities (everything but 0), then 9 possibilities for the tens digit, and 8 for the units digit. Thus there are $9 \cdot 9 \cdot 8 = 648$ three-digit positive integers with no repeated digits.

Example.   To find the number of *odd* three-digit positive integers with no repeated digits, one method is to choose the units digit (there are 5 possibilities), then the hundreds digit (8 possibilities, to avoid 0 and the chosen units digit), and finally the tens digit (again, 8 possibilities). Thus, there are 320 odd three-digit positive integers with no repeated digits.

It is instructive to use another method for this problem. A situation similar to that of the previous example arises if we begin with the units digit, then the tens digit, and finally the hundreds digit. To resolve this problem, we consider two disjoint sets of three-digit integers: those with 0 as the tens digit and those with a nonzero tens digit. For the first set, there are 5 possible units digits, only one possible tens digit, and 8 possible hundreds digits. There are 40 such integers. For the second set, there are 5 possible units digits, 8 possible tens digits, and 7 possible hundreds digits, making 280 integers in this set. By the Sum Rule, there are a total of 320 odd three-digit positive integers with no repeated digits.

If the set A has n elements, then forming a subset of A amounts to carrying out n independent tasks, where each task is to decide whether to place the element in the subset. Since each task has two outcomes, there are 2^n ways this process can be carried out, so $\mathscr{P}(A)$ has 2^n elements. This argument is a restatement of the proof of Theorem 2.1.4.

DEFINITION   A **permutation** of a set with n elements is an arrangement of the elements of the set in a specific order.

Example.   To find all permutations of the set $A = \{a, b, c, d\}$, we might begin by writing down all the arrangements of elements of A that begin with the element a. These are:

$$abcd \quad abdc \quad acbd \quad acdb \quad adbc \quad adcb$$

The other permutations of A are:

$$
\begin{array}{cccccc}
bacd & badc & bcad & bcda & bdac & bdca \\
cabd & cadb & cbad & cbda & cdab & cdba \\
dabc & dacb & dbac & dbca & dcab & dcba
\end{array}
$$

There are 24 permutations of the 4-element set A.

Recall that $n!$ (n-factorial) is defined inductively by

$$1! = 1$$

$$n! = n(n-1)! \quad \text{for } n > 1$$

or explicitly as $n! = n(n-1)(n-2) \cdots \cdot 3 \cdot 2 \cdot 1$. We also define $0! = 1$.

Theorem 2.6.6   The number of permutations of a set of n elements is $n!$.

Proof.   See Exercise 13.   ∎

Example.   A shuffle of the playlist on a portable music device is simply a permutation of the set of song titles. For a playlist of 10 songs there are $10! = 3,628,800$ possible different playlists.

Example.   Find the number of possible user passwords with 7 characters that consist of digits or letters of the alphabet, without repetition.
    Ignoring the case of the letters, we can think of the problem as having to select 7 different symbols without repetition from a set of 36, and then arranging them in

some order. For the first symbol there are 36 choices; for the second symbol there are 35 choices, and so on. The number of arrangements is $36 \cdot 35 \cdot 34 \cdot 33 \cdot 32 \cdot 31 \cdot 30 = \dfrac{36!}{29!}$, according to the next theorem.

Theorem 2.6.7

The Permutation Rule

If n is a natural number and r is an integer such that $0 \leq r \leq n$, then the number of permutations of any r distinct objects from a set of n objects is

$$\frac{n!}{(n-r)!}$$

Proof. See Exercise 14. ∎

Example. An entertainment agent has five celebrity clients and wants to list three of them on an Internet pop-up ad. Celebrities want to be listed first, not last, so the order is important. The agent has $\dfrac{5!}{(5-2)!} = \dfrac{5!}{2!} = 60$ permutations from which to choose.

If the ad were animated with the three pictures rotating around a circle, the order of selection would not be important—we simply select a group of three clients from the five. This is a combination of 5 people taken 3 at a time.

> **DEFINITIONS** For a natural number n and an integer r with $0 \leq r \leq n$, a **combination of n elements taken r at a time** is a subset with r elements from a set with n elements.
>
> The number of combinations of n elements taken r at a time is called the **binomial coefficient** $\dbinom{n}{r}$ read "n choose r" or "n binomial r."

Example. Choosing three people from a set of five people is the same as forming a 3-element subset. There are $\dbinom{5}{3}$ different possible combinations. If we identify the five people as R, C, W, H, and P, the 3-element subsets are $\{R, C, W\}$, $\{R, C, H\}$, $\{R, C, P\}$, $\{R, W, H\}$, $\{R, W, P\}$, $\{R, H, P\}$, $\{C, W, H\}$, $\{C, W, P\}$, $\{C, H, P\}$, and $\{W, H, P\}$. Thus, $\dbinom{5}{3} = 10$.

For any set with n elements, there is only one way to select a subset of n elements. Therefore,

$$\binom{n}{n} = 1 \text{ for all } n \in \mathbb{N}.$$

Also, there is only one way to construct a subset with zero elements. Therefore,

$$\binom{n}{0} = 1 \text{ for all } n \in \mathbb{N}.$$

In particular, there is only one 0-element subset of $\varnothing$—the empty set itself. Thus

$$\binom{0}{0} = 1.$$

Example.   The set $A = \{a, b, c, d\}$ has four subsets with one element and four subsets with three elements. A has six subsets with two elements. They are $\{a, b\}$, $\{a, c\}$, $\{a, d\}$, $\{b, c\}$, $\{b, d\}$, and $\{c, d\}$. Thus, $\binom{4}{1} = 4$, $\binom{4}{2} = 6$, and $\binom{4}{3} = 4$.

The next theorem develops a simple calculation for binomial coefficients. The proof is a good example of a technique called *two-way counting,* in which expressions for a given quantity are determined using two different counting approaches, thereby creating an equality.

Theorem 2.6.8

The Combination Rule
Let n be a positive integer and r be an integer such that $0 \le r \le n$. Then

$$\binom{n}{r} = \frac{n!}{r!(n - r)!}.$$

Proof.   ⟨*The quantity we count in two different ways is the number of ways to arrange the n objects in an n-element set.*⟩ Let A be a set with n elements. By Theorem 2.6.6, the number of permutations of all n objects in A is $n!$

The n elements of the set A may also be arranged as follows: Select r objects, order them, and then order the remaining $n - r$ objects. Selecting r objects can be done in $\binom{n}{r}$ ways; ordering the r objects can be done in $r!$ ways; and ordering the remaining $n - r$ objects can be done in $(n - r)!$ ways. Thus, the number of permutations of all n objects in A is $\binom{n}{r} \cdot r! \cdot (n - r)!$.

Comparing the two methods for counting the number of permutations of the elements of A, we have $\binom{n}{r} \cdot r! \cdot (n - r)! = n!$. Therefore $\binom{n}{r} = \frac{n!}{r!(n - r)!}$.   ■

Example.   In a company with 15 employees, suppose 5 are selected for bonus pay. The number of ways the 5 employees could be selected is

$$\binom{15}{5} = \frac{15!}{5!\,10!} = \frac{15 \cdot 14 \cdot 13 \cdot 12 \cdot 11}{5 \cdot 4 \cdot 3 \cdot 2 \cdot 1} = 3003.$$

For this calculation, we assumed that all 5 employees will receive the same bonus amount, so that there is no need to think of the 5 employees as being selected in any order. They may be selected simultaneously as a subset of the 15.

If the 5 selected employees are to get different bonus amounts, we need to arrange these employees in order. There are $5!$ ways to order 5 employees. Thus the number of ways to give 5 different bonuses is the number of combinations times the number of permutations within each combination, or $3003(5!) = 360{,}360$.

If we know from the start that there will be five different bonus amounts, we use the Permutation Rule to conclude that the five employees can be selected, in order, in $15 \cdot 14 \cdot 13 \cdot 12 \cdot 11 = 360,360$ ways.

Example. Let $A = \{2, 3, 6, 18, 38, 81, 442, 469, 574, 608\}$. In how many ways can four elements of A be selected so that their sum is **(a)** less than 400? **(b)** odd? **(c)** even and less than 400?

(a) For the sum to be less than 400, we can choose any four of the six elements of A that are less than 100. This can be done in $\binom{6}{4} = 15$ ways.

(b) There are three odd numbers in A, so for the sum to be odd we must select either all three of them or exactly one. There is only $\binom{3}{3} = 1$ way to choose all three of them, and then $\binom{7}{1} = 7$ ways to choose the fourth summand from the seven even numbers. By the Product Rule there are $1 \cdot 7 = 7$ combinations using all three odd numbers. To form an odd sum with only one odd summand, there are $\binom{3}{1} = 3$ ways to choose the odd number and $\binom{7}{3} = 35$ ways to choose three even numbers from A. By the Product Rule there are $3 \cdot 35 = 105$ combinations involving one odd number. Thus there are $7 + 105 = 112$ combinations of four elements of A whose sum is odd.

(c) A contains two odd numbers less than 400; for an even sum we must use both of them or neither. There are $\binom{2}{2}\binom{4}{2} = 1 \cdot 6 = 6$ ways to choose two odd and two even numbers and $\binom{2}{0}\binom{4}{4} = 1 \cdot 1 = 1$ way to choose four even numbers less than 400. Thus, there are seven combinations whose sum is even and less than 400. As an alternative, we could compute that an odd sum less than 400 requires one of two odd elements of A and three of the four even elements. There are $\binom{2}{1} \cdot \binom{4}{3} = 2 \cdot 4 = 8$ such combinations, which leaves 7 of the 15 sums that are even.

The next theorem describes some relationships among binomial coefficients. First, part (a) explains why $\binom{n}{r}$ is called a binomial coefficient: The coefficient of $a^r b^{n-r}$ in the expansion of $(a + b)^r$ is $\binom{n}{r}$. For example, $(a + b)^5 = a^5 + 5a^4b + 10a^3b^2 + 10a^2b^3 + 5ab^4 + b^5$. Thus, the coefficient of a^3b^2 is $\binom{5}{3} = 10$, and the coefficient of a^1b^4 is $\binom{5}{1} = 5$.

Part (b) tells us that there are as many ways to choose r objects out of a set with n elements as there are ways to choose $n - r$ objects from the set. This must be true because choosing r elements to take out is the same as choosing $n - r$ objects to leave behind. We will consider an interpretation of part (c) at the end of this section.

Relationships among binomial coefficients can often be proved either algebraically or combinatorially. An algebraic proof is one that applies formulas such as those in Theorem 2.6.8. Combinatorial proofs are based on the meaning of the binomial coefficients. In Theorem 2.6.9, we give a combinatorial proof for part (a) and an algebraic proof for part (b).

Theorem 2.6.9   Let n be a positive integer and r be an integer such that $0 \leq r \leq n$.

(a)   For $a, b \in \mathbb{R}$, $(a + b)^n = \sum_{r=0}^{n} \binom{n}{r} a^r b^{n-r}$.

(b) $\binom{n}{r} = \binom{n}{n - r}$.

(c) $\binom{n}{r} = \binom{n - 1}{r} + \binom{n - 1}{r - 1}$ for $r \geq 1$.

Proof.

(a)   Since $(a + b)^n = \underbrace{(a + b)(a + b) \ldots (a + b)}_{n \text{ factors}}$, each term of the expansion

of $(a + b)^n$ contains one term from each of the n factors $(a + b)$. Thus, each term of $(a + b)^n$ contains a total of n a's and b's and, therefore, each term includes a factor of the form $a^r b^{n-r}$ for some $0 \leq r < n$. For a given r, the coefficient of $a^r b^{n-r}$ is the number of times $a^r b^{n-r}$ is obtained in the expansion of $(a + b)^n$. Since the term with $a^r b^{n-r}$ is obtained by choosing a from exactly r of the factors $(a + b)$, the coefficient for $a^r b^{n-r}$ is $\binom{n}{r}$.

(b) $\binom{n}{r} = \dfrac{n!}{r!(n - r)!} = \dfrac{n!}{(n - r)!(n - (n - r))!} = \binom{n}{n - r}$.

(c)   See Exercise 21.   ∎

Part (a) of Theorem 2.6.8 provides another way to count the number of subsets of a finite set. If a set A has n elements, we start with $\binom{n}{0}$, the number of 0-element subsets of A, plus $\binom{n}{1}$, the number of 1-element subsets of A, and so on, up through $\binom{n}{n}$, the number of n-element subsets of A. The sum, $\sum_{r=0}^{n} \binom{n}{r}$ is the number of subsets of A. By part (a), $2^n = (1 + 1)^n = \sum_{r=0}^{n} \binom{n}{r} 1^r 1^{n-r} = \sum_{r=0}^{n} \binom{n}{r}$. Therefore, A has 2^n subsets.

To explain the relationship among coefficients in part (c) of Theorem 2.6.9, we refer to Pascal's* triangle, shown in Figure 2.6.3. The triangle provides a simple

---

* Blaise Pascal (1623–1662) was a French mathematician, physicist, and philosopher. He made profound contributions to projective geometry. He used the triangle, which was known centuries earlier by Chinese, Indian, and Arabian mathematicians, to advance the study of probability.

means for computing binomial coefficients. For example, we read on row $n = 4$ (rows are labeled on the left) the coefficients for $a^r b^{n-r}$, for r increasing from 0 to 4. Thus

$$(a + b)^4 = 1a^4 + 4a^3b + 6a^2b^2 + 4ab^3 + 1b^4.$$

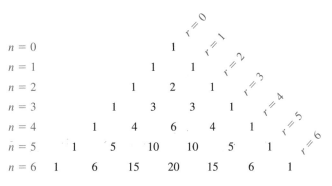

Figure 2.6.3

Pascal's triangle illustrates part (c) of Theorem 2.6.9. The triangle is constructed by beginning with the first two rows

$$1$$
$$1 \quad 1$$

and constructing the next row by putting 1 on the far left and far right. All other entries in a row are found by adding the two entries immediately to the left and right in the preceding row. Thus, the first 10 in the fifth row is the sum of 4 and 6 from the fourth row. Part (c) of Theorem 2.6.9 tells precisely how each entry in one row of the triangle is formed from the two entries in the row above.

Exercises 2.6

1. Find the number of elements in each set.
   (a) $\{n \in \mathbb{Z}: n^2 < 41\}$.
   (b) $\{2, 6, 2, 6, 2\}$.
   (c) $\{x \in \mathbb{R}: x^2 = -1\}$.
   (d) $\{n \in \mathbb{N}: n + 1 = 4n - 10\}$.

2. Suppose $\overline{\overline{A}} = 24$, $\overline{\overline{B}} = 21$, $\overline{\overline{A \cup B}} = 37$, $\overline{\overline{A \cap C}} = 11$, $\overline{\overline{B - C}} = 10$, and $\overline{\overline{C - B}} = 12$. Find

   $\overline{\overline{A}} = \#A$

   (a) $\overline{\overline{A \cap B}}$.
   ★ (b) $\overline{\overline{A - B}}$.
   (c) $\overline{\overline{B - A}}$.
   (d) $\overline{\overline{B \cup C}}$.
   (e) $\overline{\overline{C}}$.
   (f) $\overline{\overline{A \cup C}}$.

3. How many natural numbers less than or equal to 1 million are either squares or cubes of natural numbers?

4. Of the four teams in a softball league, one team has four pitchers and the other teams have three each. Give the counting rules that apply to determine each of the following.
   (a) The number of possible selections of pitchers for an all-star team, if exactly four pitchers are to be chosen.
   (b) The number of possible selections if one pitcher is to be chosen from each team.
   (c) The number of possible selections of four pitchers, if exactly two of the five left-handed pitchers in the league must be selected.
   (d) The number of possible orders in which the four pitchers, once they are selected, can appear (one at a time) in the all-star game.

☆ 5. State the Principle of Inclusion and Exclusion for four sets, A, B, C, and D.

6. Among the 40 campers at Camp Forlorn one week, 14 fell into the lake during the week, 13 suffered from poison ivy, and 16 got lost trying to find the dining hall. Three of these campers had poison ivy rash and fell into the lake, 5 fell into the lake and got lost, 8 had poison ivy and got lost, and 2 experienced all three misfortunes. How many campers got through the week without any of these mishaps?

7. (a) If you have 10 left shoes and 9 right shoes and do not care whether they match, how many "pairs" of shoes can you select?
   (b) A cafeteria has 3 entrée selections, 2 side dishes, and 4 dessert selections for a given meal. If a meal consists of one entrée, one side dish, and one dessert, how many different meals could be constructed?
   (c) There are 3 roads from Abbottville to Bakerstown, 4 roads from Bakerstown to Cadez, and 5 roads from Cadez to Detour Village. How many different routes are there from Abbottville through Bakerstown and then Cadez to Detour Village?

8. Calculate the number of even three-digit positive integers with no repeated digits by finding the number of such integers that have (a) units digit 0 and (b) nonzero units digit. Verify your answer by comparing the number of odd three-digit positive integers with no repeated digits with the total number of three-digit positive integers with no repeated digits.

9. (a) Find the number of four-digit positive integers with no repeated digits.
   (b) Find the number of odd four-digit positive integers with no repeated digits.
   (c) Without using your results from (a) and (b), find the number of even four-digit positive integers with no repeated digits.

★ 10. A square is bisected vertically and horizontally into 4 smaller squares, and each of the 4 smaller squares is to be painted so that adjacent squares have different colors. If there are 20 paints available, in how many ways can the 4 smaller squares be painted?

11. Prove that if A and B are disjoint and C is any other set, then $\overline{A \cup B \cup C} = \overline{\overline{A}} + \overline{B} + \overline{C} - \overline{A \cap C} - \overline{B \cap C}$.

12. Prove Theorem 2.6.5 by induction on the number of sets.

13. Prove Theorem 2.6.6 by induction on the number of elements.

14. Prove Theorem 2.6.7.
    (a) by using the Product Rule.
    (b) by induction on n. Use part (c) of Theorem 2.6.9.

15. Find the number of passwords that use each of the digits 3, 4, 5, 6, 7, 8, 9 exactly once.

16. In how many of the passwords of Exercise 15
    (a) are the first three digits even?
    (b) are the three even digits consecutive?
    (c) are the four odd digits consecutive?
    (d) are no two odd digits consecutive?

17. The number of four-digit numbers that can be formed using exactly the digits 1, 3, 3, 7 is less than 4!, because the two 3's are indistinguishable. Prove that the number of permutations of n objects, m of which are alike, is $n!/m!$. Generalize to the case when m_1 are alike and m_2 others are alike.

18. Among ten lottery finalists, four will be selected to win individual amounts of $1000, $2000, $5000, and $10,000. In how many ways can the money be distributed?

19. A vacationer is selecting 3 out of 19 recommended books to take along for reading at the beach. Eleven are fiction books.
    (a) How many selections are possible?
    (b) How many of these selections have exactly 2 of the 11 fiction books?
    (c) How many of these selections have exactly one fiction book?

20. Among 14 astronauts training for a Mars landing, 5 have advanced training in exobiology. If 4 astronauts are to be selected for a mission, how many selections can be made in which 2 astronauts have expertise in exobiology?

21. Prove these parts of Theorem 2.6.9 as follows:
    ☆ (a) Prove part (a) by induction on n.
    (b) Prove part (c) algebraically.
    ★ (c) Prove part (c) using a combinatorial argument.

22. Find
    (a) $(a + b)^6$.
    (b) $(a + 2b)^4$.
    (c) the coefficient of $a^3 b^{10}$ in the expansion of $(a + b)^{13}$.
    (d) the coefficient of $a^2 b^{10}$ in the expansion of $(a + 2b)^{12}$.

23. (a) Give a combinatorial proof that if n is an odd integer, then the number of ways to select an even number of objects from a set of n objects is equal to the number of ways to select an odd number of objects.
    ☆ (b) Give a combinatorial proof of Vandermonde's identity: For positive integers m and n, and r an integer such that $0 \le r \le n + m$,

$$\binom{n+m}{r} = \binom{n}{0}\binom{m}{r} + \binom{n}{1}\binom{m}{r-1} + \binom{n}{2}\binom{m}{r-2}$$
$$+ \cdots + \binom{n}{r}\binom{m}{0}.$$

(c) Prove that

$$\binom{2n}{n}+\binom{2n}{n+1}=\frac{1}{2}\binom{2n+2}{n+1}$$

Proofs to Grade   **24.** Assign a grade of A (correct), C (partially correct), or F (failure) to each. Justify assignments of grades other than A.

(a) **Claim.** For all $n \geq 0$, $\dfrac{n^2 + n}{2} = n + \dfrac{n^2 - n}{2}$.

"*Proof.*"   Consider a set of $n + 1$ elements, and let one of these elements be x. There are $\binom{n+1}{n-1} = \dfrac{n^2+n}{2}$ ways to choose $n - 1$ of these elements. Of these, there are $\binom{n}{n-1} = n$ ways to choose the $n - 1$ elements without choosing x, and $\binom{n}{n-2} = \dfrac{n^2-n}{2}$ ways to choose $n - 1$ elements including x. Therefore, $\dfrac{n^2+n}{2} = n + \dfrac{n^2-n}{2}$. ∎

(b) **Claim.** For $n \geq 1$,

$$\binom{n}{0}-\binom{n}{1}+\binom{n}{2}-\cdots+(-1)^k\binom{n}{k}+\cdots+(-1)^n\binom{n}{n}=0.$$

"*Proof.*"

$$0 = (-1+1)^n = \sum_{k=0}^{n}\binom{n}{k}(-1)^k(1)^{n-k}$$

$$= \binom{n}{0}-\binom{n}{1}+\binom{n}{2}-\cdots+(-1)^k\binom{n}{k}+\cdots+(-1)^n\binom{n}{n}.\ \blacksquare$$

(c) **Claim.** For $n \geq 1$, the number of ways to select an even number of objects from n is equal to the number of ways to select an odd number. "*Proof.*" From part (b) of this exercise ⟨*The claim made there is correct.*⟩, we have that

$$\binom{n}{0}+\binom{n}{2}+\cdots=\binom{n}{1}+\binom{n}{3}+\cdots$$

The left side of this equality gives the number of ways to select an even number of objects from n and the right side is the number of ways to select an odd number. ∎

Relations and Partitions

Given a set of objects, we may want to say that certain pairs of objects are related in some way. For example, we may say that two people are related if they have the same citizenship or the same blood type, or if they like the same kinds of food. If a and b are integers, we might say that a is related to b when a divides b. In this chapter we will study the idea of "is related to" by making precise the notion of a relation and then concentrating on certain relations called equivalence relations. The last two sections of the chapter introduce order relations and the theory of graphs.

3.1   Cartesian Products and Relations

When we speak of a relation on a set, we identify the notion of "a is related to b" with the ordered pair (a, b). For the set of all people, if Phoebe and Monica were born on the same day of the year, then the pair (Phoebe, Monica) is in the relation "has the same birthday as." Thus a relation may be defined simply as a set of ordered pairs.

> **DEFINITIONS**   Let A and B be sets. R is a **relation from A to B** iff R is a subset of $A \times B$. A relation from A to A is called a **relation on A.**
>
>   If $(a, b) \in R$, we write $a \, R \, b$ and say a is R-**related** (or simply **related**) to b. If $(a, b) \notin R$, we write $a \, \not{R} \, b$.

Examples.   If $A = \{-1, 2, 3, 4\}$ and $B = \{1, 2, 4, 5, 6\}$, let

$$R = \{(-1, 5), (2, 4), (2, 1), (4, 2)\},$$
$$S = \{(5, 2), (4, 3), (1, 3)\}, \text{ and}$$
$$T = \{(-1, 3), (2, 3), (4, 4)\}.$$

Then R is a relation from A to B, S is a relation from B to A and the set T is a relation on A.

We could describe the relation R by writing $-1\,R\,5$, $2\,R\,4$, $2\,R\,1$, and $4\,R\,2$. Since $(3, 5) \notin R$, we write $3\,\cancel{R}\,5$. We can also describe R by listing the pairs of R in a two-column table, by displaying the relation with an arrow diagram, or by drawing the graph of R as in Figure 3.1.1.

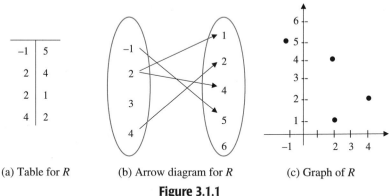

| (a) Table for R | (b) Arrow diagram for R | (c) Graph of R |

Figure 3.1.1

An equation, inequality, expression, or graph is often used to describe a relation, especially when listing all pairs is impractical or impossible. For example, the relation $LT = \{(x, y) \in \mathbb{R} \times \mathbb{R} : x < y\}$ is the familiar "less than" relation on $\mathbb{R}$, since $x\,LT\,y$ iff $x < y$. The graph of LT is shown (shaded) in Figure 3.1.2.

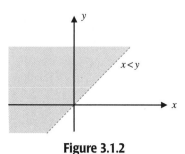

Figure 3.1.2

You have worked with the graphs of relations in previous courses, because, as we will see in Chapter 4, functions are relations that satisfy an additional condition.

Example.   The phone faceplate pictured on the next page may be used to define a relation from the set of digits $\Delta = \{0, 1, 2, \ldots, 9\}$ to the set of 26 letters $\Gamma = \{A, B, C, \ldots\}$. The relation R defined by "appear on the same phone button" is a subset of $\Delta \times \Gamma$ containing 24 pairs. The pair $(4, G) \in R$ since 4 and G appear on the same button. Likewise, $9\,R\,Y$ and $6\,R\,M$ are true. $(3, T) \notin R$ since 3 and T do not appear together. Also $1\,\cancel{R}\,E$ and $4\,\cancel{R}\,P$ are true.

Consider the relation S on the set $\mathbb{N} \times \mathbb{N}$ given by $(m, n) \, S \, (k, j)$ iff $m + n = k + j$. Then $(3, 17) \, S \, (12, 8)$, but $(5, 4)$ is not S-related to $(6, 15)$. Notice that S is a relation from $\mathbb{N} \times \mathbb{N}$ to $\mathbb{N} \times \mathbb{N}$ and consists of ordered pairs whose entries are themselves ordered pairs. For this reason, the description above is somewhat simpler than defining S with set notation:

$$S = \{((m, n), (k, j)) \in (\mathbb{N} \times \mathbb{N}) \times (\mathbb{N} \times \mathbb{N}): m + n = k + j\}.$$

The empty set $\varnothing$ and the set $A \times B$ are relations from A to B. In general, there are many different relations from a set A to a set B because *every* subset of $A \times B$ is a relation from A to B. In Exercise 12 you are asked to prove that if A has m elements and B has n elements, then there are 2^{mn} different relations from A to B.

DEFINITIONS   The **domain** of the relation R from A to B is the set

$$\text{Dom} \, (R) = \{x \in A: \text{there exists } y \in B \text{ such that } x \, R \, y\}.$$

The **range** of the relation R is the set

$$\text{Rng} \, (R) = \{y \in B: \text{there exists } x \in A \text{ such that } x \, R \, y\}.$$

Thus the domain of R is the *set of all first coordinates of ordered pairs in R*, and the range of R is the *set of all second coordinates*. By definition, Dom $(R) \subseteq A$ and Rng $(R) \subseteq B$.

For the relation $R = \{(-1, 5), (2, 4), (2, 1), (4, 2)\}$, Dom $(R) = \{-1, 2, 4\}$ and Rng $(R) = \{1, 2, 4, 5\}$. For the relation LT on $\mathbb{R}$, where $x \, LT \, y$ iff $x < y$, both the domain and range are $\mathbb{R}$. For the relation defined by "appears on same phone button," the domain is $\{2, 3, 4, 5, 6, 7, 8, 9\}$ and the range is the set of all capital letters except Q and Z.

Every set of ordered pairs is a relation. If M is any set of ordered pairs, then M is a relation from A to B, where A and B are any sets for which Dom $(M) \subseteq A$ and Rng $(M) \subseteq B$.

Example. Let $S = \left\{(x, y) \in \mathbb{R} \times \mathbb{R}: \dfrac{x^2}{324} + \dfrac{y^2}{64} \leq 1\right\}$. The graph of S is the shaded area in Figure 3.1.3. The domain is $[-18, 18]$ and the range is $[-8, 8]$.

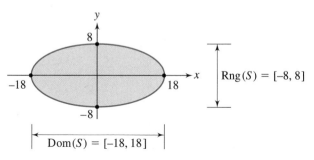

Figure 3.1.3

We can use a **directed graph** or **digraph** to represent a relation R on a small finite set A. We think of the objects in A as points (called **vertices**) and the relation R as telling us which vertices are connected by **arcs**. Arcs are drawn as arrows: There is an arc from vertex a to vertex b iff $(a, b) \in R$. An arc from a vertex to itself is called a **loop**. For example, let $A = \{2, 5, 6, 12\}$ and $R = \{(6, 12), (2, 6), (2, 12), (6, 6), (12, 2)\}$. The digraph for R is given in Figure 3.1.4.

The digraph of the relation "divides" on the set $\{3, 6, 9, 12\}$ has a loop at each vertex, as shown in Figure 3.1.5.

DEFINITION   For any set A, the relation $I_A = \{(x, x): x \in A\}$ is called the **identity relation on** A.

For $A = \{1, 2, a, b\}$, $I_A = \{(1, 1), (2, 2), (a, a), (b, b)\}$. Clearly, for any set A, Dom $(I_A) = A$ and Rng $(I_A) = A$. The graph of the identity relation on $[-2, \infty)$ is shown in Figure 3.1.6.

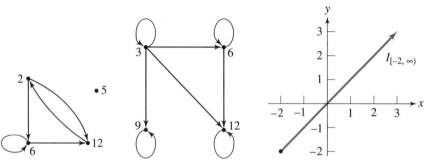

Figure 3.1.4   **Figure 3.1.5**   **Figure 3.1.6**

The remainder of this section is devoted to methods of constructing new relations from given relations. These ideas are important in the study of relations, and will be used again when we study functions.

Since relations from set A to set B are subsets of $A \times B$, the union and intersection of two relations from A to B are again relations from A to B.

Example. Let $X = [2, 4]$ and $Y = (1, 3) \cup \{4\}$. Let S be the relation on $\mathbb{R}$ defined by $x \, S \, y$ iff $x \in X$, and let T be the relation on $\mathbb{R}$ defined by $x \, T \, y$ iff $y \in Y$. The graphs of S and T are given in Figures 3.1.7(a) and (b). Figure 3.1.7(c) shows the graph of $S \cap T$. Note that $S = X \times \mathbb{R}$, $T = \mathbb{R} \times Y$, and $S \cap T = X \times Y$. Figure 3.1.7(d) shows the graph of $S \cup T$.

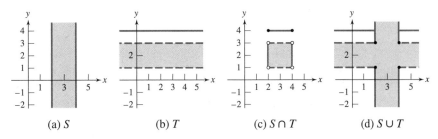

(a) S (b) T (c) $S \cap T$ (d) $S \cup T$

Figure 3.1.7

DEFINITION If R is a relation from A to B, then the **inverse** of R is the relation

$$R^{-1} = \{(y, x): (x, y) \in R\}.$$

Since inversion is a matter of switching the order of each pair in a relation, if R is a relation from A to B, then R^{-1} is a relation from B to A.

Examples. The inverse of the relation $R = \{(1, b), (1, c), (2, c)\}$ is the relation $R^{-1} = \{(b, 1), (c, 1), (c, 2)\}$. For any set A, the inverse of I_A is I_A itself. For the real numbers, the inverse of the "less than" relation $LT = \{(x, y) \in \mathbb{R} \times \mathbb{R}: x < y\}$ is the "greater than" relation on $\mathbb{R}$ because

$$(x, y) \in LT^{-1} \text{ iff } (y, x) \in LT$$
$$\text{iff } y < x$$
$$\text{iff } x > y.$$

In case R is a relation on A, the digraph of R^{-1} is obtained from the digraph of R by copying all the loops and arcs, but reversing the direction of the arrows for arcs. Figure 3.1.8 shows the digraphs of R and R^{-1}, where R is the relation $\subseteq$ on the set $\{\emptyset, \{1\}, \{3\}, \{1, 2\}\}$.

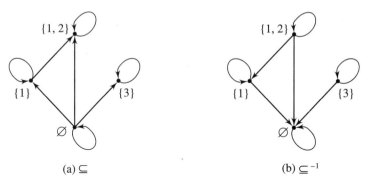

(a) $\subseteq$ (b) $\subseteq^{-1}$

Figure 3.1.8

Example.   Let EXP be the relation on $\mathbb{R}$ given by x EXP y iff $y = e^x$. The inverse of EXP is given by x EXP^{-1} y iff $x = e^y$. We know that $x = e^y$ iff $y = \ln x$ iff $x \ln y$, where ln is the natural logarithm. Thus, the inverse of EXP is the relation ln. The familiar graphs of EXP and ln are given in Figure 3.1.9.

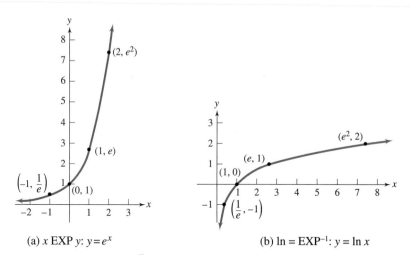

(a) x EXP y: $y = e^x$ (b) $\ln = $ EXP^{-1}: $y = \ln x$

Figure 3.1.9

In the previous example, Dom (EXP) $= \mathbb{R}$ and Rng (EXP) $= (0, \infty)$, while Dom (ln) $= (0, \infty)$ and Rng(ln) $= \mathbb{R}$. The next theorem says that this switch of the domain and range of a relation to the range and domain of inverse relation always happens.

Theorem 3.1.2   Let R be a relation from A to B.

(a)   Dom $(R^{-1}) = $ Rng (R).
(b)   Rng $(R^{-1}) = $ Dom (R).

Proof.

(a) $b \in \text{Dom}\,(R^{-1})$ iff there exists $a \in A$ such that $(b, a) \in R^{-1}$ iff there exists $a \in A$ such that $(a, b) \in R$ iff $b \in \text{Rng}\,(R)$.

(b) The proof is similar to the proof for part (a). ■

Given a relation from A to B and another from B to C, composition is a method of constructing a relation from A to C.

DEFINITION Let R be a relation from A to B, and let S be a relation from B to C. The **composite** of R and S is

$$S \circ R = \{(a, c): \text{there exists } b \in B \text{ such that } (a, b) \in R \text{ and } (b, c) \in S\}.$$

The relation $S \circ R$ is a relation from A to C since $S \circ R \subseteq A \times C$. It is always true that $\text{Dom}\,(S \circ R) \subseteq \text{Dom}\,(R)$ but it is not always true that $\text{Dom}\,(S \circ R) = \text{Dom}\,(R)$. (See Exercise 9.)

We have adopted the right-to-left notation for $S \circ R$ that is commonly used in analysis courses. To determine $S \circ R$, you need to remember that R is the relation from the first set to the second and S is the relation from the second set to the third. Thus, to determine $S \circ R$, we apply the relation R first and then S.

Example. Let $A = \{1, 2, 3, 4, 5\}$, and $B = \{p, q, r, s, t\}$, and $C = \{x, y, z, w\}$. Let R be the relation from A to B:

$$R = \{(1, p), (1, q), (2, q), (3, r), (4, s)\}$$

and S the relation from B to C:

$$S = \{(p, x), (q, x), (q, y), (s, z), (t, z)\}.$$

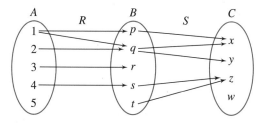

Figure 3.1.10

These relations are illustrated in Figure 3.1.10 by arrows from one set to another. An element a from A is related to an element c from C under $S \circ R$ if there is at least one "intermediate" element b of B such that $(a, b) \in R$ *and* $(b, c) \in S$.

For example, since $(1, p) \in R$ and $(p, x) \in S$, then $(1, x) \in S \circ R$. By following all possible paths along the arrows from A to B and B to C in Figure 3.1.10, we have

$$S \circ R = \{(1, x), (1, y), (2, x), (2, y), (4, z)\}.$$

If R is a relation from A to B, and S is a relation from B to A, then $R \circ S$ and $S \circ R$ are both defined, but you should not expect that $R \circ S = S \circ R$. Even when R and S are relations on the same set, it may happen that $R \circ S \neq S \circ R$.

Example.   Let $R = \{(x, y) \in \mathbb{R} \times \mathbb{R}: y = x + 1\}$ and $S = \{(x, y) \in \mathbb{R} \times \mathbb{R}: y = x^2\}$. Then

$$\begin{aligned} R \circ S &= \{(x, y): (x, z) \in S \text{ and } (z, y) \in R \text{ for some } z \in \mathbb{R}\} \\ &= \{(x, y): z = x^2 \text{ and } y = z + 1 \text{ for some } z \in \mathbb{R}\} \\ &= \{(x, y): y = x^2 + 1\}. \\ S \circ R &= \{(x, y): (x, z) \in R \text{ and } (z, y) \in S \text{ for some } z \in \mathbb{R}\} \\ &= \{(x, y): z = x + 1 \text{ and } y = z^2 \text{ for some } z \in \mathbb{R}\} \\ &= \{(x, y): y = (x + 1)^2\}. \end{aligned}$$

Clearly, $S \circ R \neq R \circ S$, since $x^2 + 1$ is seldom equal to $(x + 1)^2$.

The last theorem of this section presents several results about inversion, composition, and the identity relation. We prove part (b) and the first part of (c), leaving the rest as Exercise 10.

Theorem 3.1.3   Suppose A, B, C, and D are sets. Let R be a relation from A to B, S be a relation from B to C, and T be a relation from C to D.

(a) $(R^{-1})^{-1} = R$.
(b) $T \circ (S \circ R) = (T \circ S) \circ R$, so composition is associative.
(c) $I_B \circ R = R$ and $R \circ I_A = R$.
(d) $(S \circ R)^{-1} = R^{-1} \circ S^{-1}$.

Proof.

(b)   The pair $(x, w) \in T \circ (S \circ R)$ for some $x \in A$ and $w \in D$
          iff $(\exists z \in C)[(x, z) \in S \circ R$ and $(z, w) \in T]$
          iff $(\exists z \in C)[(\exists y \in B)((x, y) \in R$ and $(y, z) \in S)$ and $(z, w) \in T]$
          iff $(\exists z \in C)(\exists y \in B)[(x, y) \in R$ and $(y, z) \in S$ and $(z, w) \in T]$
          iff $(\exists y \in B)(\exists z \in C)[(x, y) \in R$ and $(y, z) \in S$ and $(z, w) \in T]$

iff $(\exists y \in B)[(x, y) \in R$ and $(\exists z \in C)((y, z) \in S$ and $(z, w) \in T)]$
iff $(\exists y \in B)[(x, y) \in R$ and $(y, w) \in T \circ S]$
iff $(x, w) \in (T \circ S) \circ R$.
Therefore, $T \circ (S \circ R) = (T \circ S) \circ R$.

(c) ⟨*We first show that* $I_B \circ R \subseteq R$.⟩ Suppose $(x, y) \in I_B \circ R$. Then there exists $z \in B$ such that $(x, z) \in R$ and $(z, y) \in I_B$. Since $(z, y) \in I_B, z = y$. Thus $(x, y) \in R$ ⟨*since* $(x, y) = (x, z) \in R$⟩.
Conversely, suppose $(p, q) \in R$. Then $(q, q) \in I_B$ and thus $(p, q) \in I_B \circ R$. Thus $I_B \circ R = R$. ∎

The storage and manipulation of data in tables (n-tuple relations) is an important field of computer science called *relational databases*. Operations such as union and composition for ordered pairs may be extended to operations on n-tuples. One generalization of composition in relational databases is the "join" of two tables.

Example. Suppose the student information at a small university includes both directory information and billing information. We let A be the set of first names, B be last names, C be 4-digit student ID numbers, D be names of campus residence halls, E be residence hall room numbers, F be tuition amounts due, and G be room charges due.

The student records in the directory may be described in a table R:

R (directory)

First Name	Last Name	Student ID	Residence Hall	Room Number
Krista	Maire	1234	Orlando	77
Harold	Dorman	2490	Mountain	455
Ferlin	Husky	5555	Dove	213A
Martha	Reeves	3215	Vandella	238
Kim	Anen	6920	Bowie	1979

The directory relation R is a subset of $A \times B \times C \times D \times E$ consisting of five 5-tuples. The 5-tuple (Krista, Maire, 1234, Orlando, 77) is one student record in the directory R.

The financial information relation S is a subset of $C \times F \times G$:

S (financial)

Student ID	Tuition	Room Charges
1234	$80	$40
2490	$150	$20
5555	$75	$25
3215	$0	$0
6920	$0	$60

The **join** of these two tables, denoted $R \otimes S$, is a table with 7 columns. The rows of the table are obtained by merging 5-tuples from R and 3-tuples from S that share a common ID number:

$R \otimes S$

First Name	Last Name	Student ID	Residence Hall	Room Number	Tuition	Room Charges
Krista	Maire	1234	Orlando	77	$80	$40
Harold	Dorman	2490	Mountain	455	$150	$20
Ferlin	Husky	5555	Dove	213A	$75	$25
Martha	Reeves	3215	Vandella	238	$0	$0
Kim	Anen	6920	Bowie	1979	$0	$60

The join operation is one of several database operations that allow a manager to create tables in response to requests for information (queries). There are many advantages to storing data in simple tables like R and S, but requests such as "What is the room charge for Harold Dorman?" cannot be answered using either of the tables by itself.

Exercises 3.1

1. Let T be the relation $\{(3, 1), (2, 3), (3, 5), (2, 2), (1, 6), (2, 6), (1, 2)\}$. Find
   (a) Dom (T).   (b) Rng (T).
   (c) T^{-1}.   (d) $(T^{-1})^{-1}$.

2. Find the domain and range for the relation W on $\mathbb{R}$ given by $x \, W \, y$ iff
   ★ (a) $y = 2x + 1$.   (b) $y = x^2 + 3$.

   ★ (c) $y = \sqrt{x - 1}$.   (d) $y = \dfrac{1}{x^2}$.

   ★ (e) $y \leq x^2$.   (f) $|x| < 2$ and $y = 3$.
   (g) $|x| < 2$ or $y = 3$.   (h) $y \neq x$.

   ☆ 3. Sketch the graph of each relation in Exercise 2.

4. The inverse of $R = \{(x, y) \in \mathbb{R} \times \mathbb{R} : y = 2x + 1\}$ may be expressed in the form $R^{-1} = \left\{ (x, y) \in \mathbb{R} \times \mathbb{R} : y = \dfrac{x - 1}{2} \right\}$, the set of all pairs (x, y) subject to some condition. Use this form to give the inverses of the following relations. In (i), (j), and (k), P is the set of all people.
   ★ (a) $R_1 = \{(x, y) \in \mathbb{R} \times \mathbb{R} : y = x\}$
   (b) $R_2 = \{(x, y) \in \mathbb{R} \times \mathbb{R} : y = -5x + 2\}$
   ★ (c) $R_3 = \{(x, y) \in \mathbb{R} \times \mathbb{R} : y = 7x - 10\}$
   (d) $R_4 = \{(x, y) \in \mathbb{R} \times \mathbb{R} : y = x^2 + 2\}$
   ★ (e) $R_5 = \{(x, y) \in \mathbb{R} \times \mathbb{R} : y = -4x^2 + 5\}$

(f) $R_6 = \{(x, y) \in \mathbb{R} \times \mathbb{R}: y < x + 1\}$

★ **(g)** $R_7 = \{(x, y) \in \mathbb{R} \times \mathbb{R}: y > 3x - 4\}$

(h) $R_8 = \left\{(x, y) \in \mathbb{R} \times \mathbb{R}: y = \dfrac{2x}{x - 2}\right\}$

★ **(i)** $R_9 = \{(x, y) \in P \times P: y \text{ is the father of } x\}$

(j) $R_{10} = \{(x, y) \in P \times P: y \text{ is a sibling of } x\}$

(k) $R_{11} = \{(x, y) \in P \times P: y \text{ loves } x\}$

5. Let $R = \{(1, 5), (2, 2), (3, 4), (5, 2)\}, S = \{(2, 4), (3, 4), (3, 1), (5, 5)\}$, and $T = \{(1, 4), (3, 5), (4, 1)\}$. Find

(a) $R \circ S.$ ★ **(b)** $R \circ T.$

(c) $T \circ S.$ ★ **(d)** $R \circ R.$

(e) $S \circ R.$ **(f)** $T \circ T.$

(g) $R \circ (S \circ T).$ **(h)** $(R \circ S) \circ T.$

6. Find these composites for the relations defined in Exercise 4.

★ **(a)** $R_1 \circ R_1$ **(b)** $R_1 \circ R_2$

(c) $R_2 \circ R_2$ ★ **(d)** $R_2 \circ R_3$

(e) $R_2 \circ R_4$ **(f)** $R_4 \circ R_2$

★ **(g)** $R_4 \circ R_5$ **(h)** $R_6 \circ R_2$

(i) $R_6 \circ R_4$ ★ **(j)** $R_6 \circ R_6$

(k) $R_7 \circ R_7$ **(l)** $R_5 \circ R_5$

(m) $R_8 \circ R_8$ ★ **(n)** $R_3 \circ R_8$

(o) $R_8 \circ R_3$ ☆ **(p)** $R_9 \circ R_9$

7. Give the digraphs for these relations on the set $\{1, 2, 3\}$.

(a) $=$ **(b)** $S = \{(1, 3), (2, 1)\}$

(c) $\leq$ **(d)** S^{-1}, where $S = \{(1, 3), (2, 1)\}$

(e) $\neq$ **(f)** $S \circ S$, where $S = \{(1, 3), (2, 1)\}$

8. Let $A = \{a, b, c, d\}$. Give an example of relations R, S, and T on A such that

(a) $R \circ S \neq S \circ R.$ **(b)** $(S \circ R)^{-1} \neq S^{-1} \circ R^{-1}.$

(c) $S \circ R = T \circ R$ but $S \neq T.$

(d) R and S are nonempty, and $R \circ S$ and $S \circ R$ are empty.

9. Let R be a relation from A to B and S be a relation from B to C.

(a) Prove that Dom $(S \circ R) \subseteq$ Dom $(R).$

(b) Show by example that Dom $(S \circ R) =$ Dom (R) may be false.

(c) Which of these two statements must be true:

$$\text{Rng}(S) \subseteq \text{Rng}(S \circ R) \quad \text{or} \quad \text{Rng}(S \circ R) \subseteq \text{Rng}(S)?$$

Give an example to show that the other statement may be false.

10. Complete the proof of Theorem 3.1.3.

11. Show by example that $(A \times B) \times C = A \times (B \times C)$ may be false.

12. Prove that if A has m elements and B has n elements, then there are 2^{mn} different relations from A to B.

13. **(a)** Let R be a relation from A to B. For $a \in A$, define the **vertical section of**
R at a to be $R_a = \{y \in B : (a, y) \in R\}$. Prove that $\bigcup_{a \in A} R_a = \text{Rng } (R)$.

(b) Let R be a relation from A to B. For $a \in A$, define the **horizontal section**
of R at b to be $_bR = \{x \in A : (x, b) \in R\}$. Prove that $\bigcup_{b \in B} {_bR} = \text{Dom } (R)$.

14. We may define ordered triples in terms of ordered pairs by saying that
$(a, b, c) = ((a, b), c)$. Use this definition to prove that $(a, b, c) = (x, y, z)$ iff
$a = x$ and $b = y$ and $c = z$.

Proofs to Grade 15. Assign a grade of A (correct), C (partially correct), or F (failure) to each. Justify
assignments of grades other than A.

★ **(a)** **Claim.** $(A \times B) \cup C = (A \times C) \cup (B \times C)$.
   "Proof." $x \in (A \times B) \cup C$
   iff $x \in A \times B$ or $x \in C$
   iff $x \in A$ and $x \in B$ or $x \in C$
   iff $x \in A \times C$ or $x \in B \times C$
   iff $x \in (A \times C) \cup (B \times C)$. ■

★ **(b)** **Claim.** If $A \subseteq B$ and $C \subseteq D$, then $A \times C \subseteq B \times D$.
   "Proof." Suppose $A \times C \not\subseteq B \times D$. Then there exists $(a, c) \in A \times C$
   with $(a, c) \notin B \times D$. But $(a, c) \in A \times C$ implies that $a \in A$ and $c \in C$,
   whereas $(a, c) \notin B \times D$ implies that $a \notin B$ and $c \notin D$. However, $A \subseteq B$
   and $C \subseteq D$, so $a \in B$ and $c \in D$. This is a contradiction. Therefore,
   $A \times C \subseteq B \times D$. ■

(c) **Claim.** If $A \times B = A \times C$ and $A \neq \emptyset$, then $B = C$.
   "Proof." Suppose $A \times B = A \times C$. Then
   $$\frac{A \times B}{A} = \frac{A \times C}{A}.$$
   Therefore $B = C$. ■

★ **(d)** **Claim.** If $A \times B = A \times C$ and $A \neq \emptyset$, then $B = C$.
   "Proof." To show $B = C$, suppose $b \in B$. Choose any $a \in A$.
   Then $(a, b) \in A \times B$. But since $A \times B = A \times C$, $(a, b) \in A \times C$. Thus
   $b \in C$. This proves $B \subseteq C$. A proof of $C \subseteq B$ is similar. Therefore,
   $B = C$. ■

(e) **Claim.** Let R and S be relations from A to B and from B to C, respec-
   tively. Then $S \circ R = (R \circ S)^{-1}$.
   "Proof." The pair $(x, y) \in S \circ R$ iff $(y, x) \in R \circ S$ iff $(x, y) \in (R \circ S)^{-1}$
   Therefore, $S \circ R = (R \circ S)^{-1}$. ■

(f) **Claim.** Let R be a relation from A to B. Then $I_A \subseteq R^{-1} \circ R$.
   "Proof." Suppose $(x, x) \in I_A$. Choose any $y \in B$ such that $(x, y) \in R$.
   Then, $(y, x) \in R^{-1}$. Thus $(x, x) \in R^{-1} \circ R$. Therefore, $I_A \subseteq R^{-1} \circ R$. ■

(g) **Claim.** Suppose R is a relation from A to B. Then $R^{-1} \circ R \subseteq I_A$.
   "Proof." Let $(x, y) \in R^{-1} \circ R$. Then for some $z \in B$, $(x, z) \in R$ and
   $(z, y) \in R^{-1}$. Thus $(y, z) \in R$. Since $(x, z) \in R$ and $(y, z) \in R$, $x = y$. Thus
   $(x, y) = (x, x)$ and $x \in A$, so $(x, y) \in I_A$. ■

3.2	**Equivalence Relations**

The goal of this section is to describe a way to equate objects in a set according to some value, property, or meaning. We might say that among all students who completed a certain math class, students are equivalent if they had the same numeric score on the final exam. With this meaning of equivalence, a student with a score of 87 on the final exam is related to every other student with a score of 87 and not related to any other student. We could also have said that two students are equivalent if they have the same favorite movie, or if they have the same blood type.

The three properties we define next, when taken together, comprise what we mean by objects being equivalent.

DEFINITIONS   Let A be a set and R be a relation on A.

R is **reflexive on** A iff for all $x \in A$, $x\,R\,x$.
R is **symmetric** iff for all x and $y \in A$, if $x\,R\,y$, then $y\,R\,x$.
R is **transitive** iff for all x, y, and $z \in A$, if $x\,R\,y$ and $y\,R\,z$, then $x\,R\,z$.

The relation R, defined as "had the same final exam score," on the set C of all students in a given class has all three of these properties. R is symmetric because if student x had the same score as student y, then student y must have had the same score as student x. R is transitive because if student x had the same score as student y and student y had the same score as student z, then x had the same score as z. Finally, for every student x in C, x must have had the same score as x. Thus R is reflexive on C.

To prove that a relation R is symmetric or transitive, we usually give a direct proof, because these properties are defined by conditional sentences. A proof that R is reflexive on A is different. What we must do is show that for all $x \in A$, x is R-related to x.

For a relation R on a nonempty set A, only the reflexive property actually asserts that some ordered pairs belong to R. The empty relation $\varnothing$ is not reflexive on a set A except in the special case when A is the empty set. The empty relation $\varnothing$ is, however, symmetric and transitive for any set A. See Exercise 4. For each of the three properties there is an alternate condition (involving the identity relation or the operations of inversion or composition) that may be used to prove that a relation has or does not have that property. See Exercise 13.

To prove that a relation R on a set A is not reflexive on A, we must show that there exists some $x \in A$ such that $x\,\not\!R\,x$. Since the denial of "If $x\,R\,y$ then $y\,R\,x$" is "$x\,R\,y$ and not $y\,R\,x$," a relation R is not symmetric iff there are elements x and y in A such that $x\,R\,y$ and $y\,\not\!R\,x$. Likewise, R is not transitive iff there exist elements x, y, and z in A such that $x\,R\,y$ and $y\,R\,z$ but $x\,\not\!R\,z$.

Examples. For $B = \{2, 5, 6, 7\}$, let $S = \{(2, 5), (5, 6), (2, 6), (7, 7)\}$ and $T = \{(2, 6), (5, 6)\}$. Since $6 \not{S} 6$ and $2 \not{T} 2$, neither S nor T is reflexive on B. The relation S is not symmetric because $2\, S\, 5$, but $5 \not{S} 2$. Likewise, T is not symmetric because $5\, T\, 6$ but $6 \not{T} 5$.

Both S and T are transitive relations. To verify that S is transitive we check all pairs (x, y) in S with all pairs of the form (y, z). We have $(2, 5)$ and $(5, 6)$ in S, so we must have $(2, 6)$; we have $(7, 7)$ and $(7, 7)$ in S so we must have $(7, 7)$. The relation T is transitive for a different reason: there do not exist x, y, z in B such that $x\, T\, y$ and $y\, T\, z$. Because its antecedent is false, the conditional sentence "If $x\, T\, y$ and $y\, T\, z$, then $x\, T\, z$" is true.

Example. Let R be the relation "is a subset of" on $\mathscr{P}(\mathbb{Z})$, the power set of $\mathbb{Z}$. R is reflexive on $\mathscr{P}(\mathbb{Z})$ since every set is a subset of itself. R is transitive by Theorem 2.1.1(c). Notice that $\{1, 2\} \subseteq \{1, 2, 3\}$ but $\{1, 2, 3\} \not\subseteq \{1, 2\}$. Therefore, R is not symmetric.

Example. Let STNR designate the relation $\{(x, y) \in \mathbb{Z} \times \mathbb{Z}: xy > 0\}$ on $\mathbb{Z}$. In this example, x STNR x for all x in $\mathbb{Z}$ except the integer 0; hence the relation STNR is not reflexive on $\mathbb{Z}$. STNR is symmetric since, if x and y are integers and $xy > 0$, then $yx > 0$. STNR is also transitive. To verify this, we assume that x STNR y and y STNR z. Then $xy > 0$ and $yz > 0$. If y is positive, then both x and z are positive; so $xz > 0$. If y is negative, then both x and z are negative; so $xz > 0$. Thus in either case, x STNR z. This relation gets its name from the fact that it is symmetric, transitive, and not reflexive on $\mathbb{Z}$.

For a relation R on a set A, the properties of reflexivity on A, symmetry, and transitivity can also be characterized by properties in the digraph of R:

  R is reflexive on A iff every vertex of the digraph has a loop.
  R is symmetric iff between any two vertices there are either no edges or an edge in both directions.
  R is transitive iff whenever there is an edge from vertex x to y and an edge from vertex y to z, there is an edge (a direct route) from x to z.

Examples. Figure 3.2.1 shows the digraphs of three relations on $A = \{2, 3, 6\}$. Figure 3.2.1(a) is the digraph of the relation "divides" and Figure 3.2.1(b) is the digraph of ">." Figure 3.2.1(c) is the digraph of the relation S, where $x\, S\, y$ iff $x + y > 7$.

There is a loop at every vertex in Figure 3.2.1(a) because the relation "divides" is reflexive: Every integer divides itself. The relations ">" and S are not reflexive; there is no loop at 2 in Figure 3.2.1(b) or (c).

S is a symmetric relation, but the others are not. In Figure 3.2.1(a) there is an arc from 2 to 6, but not in the reverse direction; in Figure 3.2.1(b) there is an arc from 6 to 2, but not from 2 to 6.

The relation S is not transitive—there is an arc from 2 to 6 and one from 6 to 3, but no arc from 2 to 3. The other two relations are transitive. Note that for the digraph in Figure 3.2.1(a), every pair of arcs to be checked for transitivity involves a loop. For example, there is an arc from 3 to 3 and an arc from 3 to 6; the shortcut is to go directly from 3 to 6.

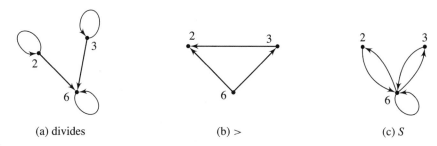

(a) divides   (b) >   (c) S

Figure 3.2.1

For every set A, the identity relation I_A is reflexive on A, symmetric, and transitive. The identity relation is, in fact, the relation "equals," because $x\,I_A\,y$ iff $x = y$. Equality is a way of comparing objects according to whether they are the same. Equivalence relations, defined next, are a means for relating objects according to whether they are, if not identical, at least alike in the sense that they share a common trait. For example, if T is the set of all triangles, we might say two triangles are "the same" (equivalent) when they are congruent. This generates the relation $R = \{(x, y) \in T \times T : x \text{ is congruent to } y\}$ on T, which is reflexive on T, symmetric, and transitive. The notion of equivalence, then, is embodied in these three properties.

DEFINITION   A relation R on a set A is an **equivalence relation on A** iff R is reflexive on A, symmetric, and transitive.

Suppose we say two integers are related iff they have the same parity. For this relation, $R = \{(x, y) \in \mathbb{Z} \times \mathbb{Z} : x + y \text{ is even}\}$, we see that all the odd integers are related to one another (since the sum of two odd numbers is even) and all the evens are related to each other. The relation R is reflexive on $\mathbb{Z}$, symmetric, and transitive and is, therefore, an equivalence relation.

For the set P of all people, let L be the relation on P given by $x\,L\,y$ iff x and y have the same family name. We have Lucy Brown L Charlie Brown, James Madison L Dolly Madison, and so on. If we make the assumption that everyone has exactly one family name, then L is an equivalence relation on P.

The subset of P consisting of all people who are L-related to Charlie Brown is the set of all people whose family name is Brown. This set contains Charlie by reflexivity. It also contains Sally Brown, James Brown, Buster Brown, Leroy Brown, and all other people who are like Charlie Brown in the sense that they have Brown as a family name. The same is true for the Madisons: The set of people L-related to Dolly Madison is the set of all people with the family name Madison.

DEFINITIONS   Let R be an equivalence relation on a set A. For $x \in A$, the **equivalence class** of x determined by R is the set

$$x/R = \{y \in A : x\, R\, y\}.$$

When R is fixed throughout a discussion or clear from the context, the notations $[x]$ and $\bar{x}$ are commonly used instead of x/R.

We read x/R as "the class of x modulo R," or simply "x mod R."

The set $A/R = \{x/R : x \in A\}$ of all equivalence classes is called **A modulo R**.

The equivalence class of Charlie Brown modulo L is the set of all people whose family name is Brown. Furthermore, Buster Brown/L is the same set as Charlie Brown/L.

Example.   The relation $H = \{(1, 1), (2, 2), (3, 3), (1, 2), (2, 1)\}$ is an equivalence relation on the set $A = \{1, 2, 3\}$. Here $1/H = 2/H = \{1, 2\}$ and $3/H = \{3\}$. Thus $A/H = \{\{1, 2\}, \{3\}\}$.

Example.   Let $S = \{(x, y) \in \mathbb{R} \times \mathbb{R} : x^2 = y^2\}$. S is an equivalence relation on $\mathbb{R}$. We have $\bar{2} = \{2, -2\}$, $\bar{\pi} = \{-\pi, \pi\}$, etc. Also, $\bar{0} = \{0\}$. In this example, for every $x \in \mathbb{R}$ the equivalence class of x and the equivalence class of $-x$ are the same. $\mathbb{R}$ modulo S is $\mathbb{R}/S = \{\{x, -x\} : x \in \mathbb{R}\}$.

Example.   For the equivalence relation $R = \{(x, y) \in \mathbb{Z} \times \mathbb{Z} : x + y \text{ is even}\}$ on $\mathbb{Z}$, there are only two equivalence classes: D, the set of all odd integers and E, the set of even integers. Thus $\mathbb{Z}/R = \{D, E\}$.

Note that in the examples above—A/H, $\mathbb{R}/S$, and $\mathbb{Z}/R$—any two equivalence classes are either equal or disjoint. The next theorem tells us for all equivalence relations, distinct equivalence classes never "overlap."

Theorem 3.2.1   Let R be an equivalence relation on a nonempty set A. For all x, y in A,

(a) $x/R \subseteq A$ and $x \in x/R$. Thus every equivalence class is a nonempty subset of A.

(b) $x\, R\, y$ iff $x/R = y/R$. Thus elements of A are related iff their equivalence classes are identical.

(c) $x\, \not{R}\, y$ iff $x/R \cap y/R = \varnothing$. Thus elements of A are unrelated iff their equivalence classes are disjoint.

Proof.

(a)   By the definition of x/R, $x/R \subseteq A$. Since R is reflexive on A, $x\,R\,x$. Thus $x \in x/R$.

(b)   **(i)**   Suppose $x\,R\,y$. To show $x/R = y/R$, we first show $x/R \subseteq y/R$. Let $z \in x/R$. Then $x\,R\,z$. From $x\,R\,y$, by symmetry, $y\,R\,x$. Then, by transitivity, $y\,R\,z$. Thus $z \in y/R$. The proof that $y/R \subseteq x/R$ is similar.

   (ii)   Suppose $x/R = y/R$. Since $y \in y/R$, $y \in x/R$. Thus $x\,R\,y$.

(c)   **(i)**   If $x/R \cap y/R = \varnothing$, then, since $y \in y/R$, $y \notin x/R$. Thus $x\,\not\!R\,y$.

   (ii)   Finally, we show $x\,\not\!R\,y$ implies $x/R \cap y/R = \varnothing$. ⟨*We prove the contrapositive.*⟩ Suppose $x/R \cap y/R \neq \varnothing$. Let $k \in x/R \cap y/R$. Then $x\,R\,k$ and $y\,R\,k$. Therefore, $x\,R\,k$ and $k\,R\,y$. Thus $x\,R\,y$.   ∎

For the rest of this section, we explore the properties of an equivalence relation that has a multitude of important applications. This relation, called congruence, provides a valuable way to deal with questions associated with divisibility in the integers. The notion of congruence, first introduced by Carl Friedrich Gauss,* leads to modular arithmetic, which is an abstraction of our usual arithmetic, and this leads in turn to methods for converting computational problems with large integers into more manageable problems.

DEFINITIONS   Let m be a fixed positive integer. For $x, y \in \mathbb{Z}$, we say **x is congruent to y modulo m** iff m divides $(x - y)$. We write $\pmb{x \equiv_m y}$, or simply $\pmb{x = y \pmod{m}}$. The number m is called the **modulus** of the congruence.

Examples.   Using 3 as the modulus, $4 = 1 \pmod 3$ because 3 divides $4 - 1$. Likewise, $10 = 16 \pmod 3$ because 3 divides $10 - 16 = -6$. Since 3 does not divide $5 - (-9) = 14$, we have $5 \neq -9 \pmod 3$. It is easy to see that 0 is congruent to $0, 3, -3, 6,$ and -6 and, in fact, 0 is congruent modulo 3 to every multiple of 3.

Theorem 3.2.2   For every fixed positive integer m, $\equiv_m$ is an equivalence relation on $\mathbb{Z}$.

Proof.   We note that $\equiv_m$ is a set of ordered pairs of integers and, hence, is a relation on $\mathbb{Z}$. ⟨*Now we show that $\equiv_m$ is reflexive on $\mathbb{Z}$, symmetric, and transitive.*⟩

(i)   To show reflexivity on $\mathbb{Z}$, let x be an integer. We show that $x = x \pmod m$. Since $m \cdot 0 = 0 = x - x$, m divides $x - x$. Thus $\equiv_m$ is reflexive on $\mathbb{Z}$.

(ii)   For symmetry, suppose $x = y \pmod m$. Then m divides $x - y$. Thus there is an integer k so that $x - y = km$. But this means that $-(x - y) = -(km)$, or that $y - x = (-k)m$. Therefore, m divides $y - x$, so that $y = x \pmod m$.

---

* The German Carl Friedrich Gauss (1777–1855), one of the greatest mathematicians of all time, also made major contributions to astronomy and physics. Congruence and modular arithmetic (and much more) appeared in his masterwork *Disquisitiones Arithmeticae*, which he completed at the age of 21. He proved the Fundamental Theorem of Algebra and the Prime Number Theorem, among many other results in number theory, statistics, analysis, and differential geometry.

(iii)   Suppose $x = y \pmod{m}$ and $y = z \pmod{m}$. Thus m divides both $x - y$ and $y - z$. Therefore, there exist integers h and k such that $x - y = hm$ and $y - z = km$. But then $h + k$ is an integer, and

$$x - z = (x - y) + (y - z) = hm + km = (h + k)m.$$

Thus m divides $x - z$, so $x = z \pmod{m}$. Therefore, $\equiv_m$ is transitive.   ∎

> **DEFINITION**   The set of equivalence classes for the relation $\equiv_m$ is denoted $\mathbb{Z}_m$.

We can now determine the set $\mathbb{Z}_3$ of all equivalence classes modulo 3. For $x \in \mathbb{Z}$, the equivalence class of x is $\{y \in \mathbb{Z} : x \equiv_3 y\}$, which we now denote by $\bar{x}$. Since the integers congruent to 0 (mod 3) are exactly the multiples of 3, we have

$$\bar{0} = \{\dots, -6, -3, 0, 3, 6, \dots\}.$$

To form the equivalence class of 1, denoted $\bar{1}$, we begin with 1 (because $1 \equiv_3 1$) and repeatedly add or subtract 3. This produces the positive integers 4, 7, 10, 13, ... and the negative integers $-2, -5, -8, \dots$ that are congruent to 1 modulo 3, so

$$\bar{1} = \{\dots, -8, -5, -2, 1, 4, 7, 10, 13, \dots\}.$$

In the same way we form

$$\bar{2} = \{\dots, -4, -1, 2, 5, 8, \dots\}.$$

If we compute $\bar{3} = \{\dots, -6, -3, 0, 3, 6, \dots\}$ we find that $\bar{3} = \bar{0}$ and in fact $\bar{4} = \bar{1}, \bar{5} = \bar{2}, \bar{6} = \bar{0}$, etc., so there are really only three different equivalence classes. We have found that $\mathbb{Z}_3 = \{\bar{0}, \bar{1}, \bar{2}\}$.

Notice that the class of 0 modulo 3 above is not the same as the congruence class of 0 modulo 4. The class of 0 modulo 4 contains 0, ± 4, ± 8, ± 12, and all the other multiples of 4. See Exercise 9.

Using the notation $\bar{x}$ for the equivalence class of x modulo m works well as long as the modulus remains unchanged, but suppose we want to compare computations with two different moduli. To work with elements of, say, $\mathbb{Z}_6$ as well as elements of $\mathbb{Z}_3$, we will write elements of $\mathbb{Z}_6$ as [0], [1], [2], [3], [4], and [5], to distinguish them from the elements $\bar{0}$, $\bar{1}$, and $\bar{2}$, of $\mathbb{Z}_3$.

The 12 hours on the clock correspond to the 12 classes in $\mathbb{Z}_{12}$. Rather than talking about hours beyond 12 o'clock, we start over again with 1 o'clock instead of 13 o'clock because $13 = 1 \pmod{12}$, and 2 o'clock instead of 14 o'clock because $14 = 2 \pmod{12}$, etc. The hours on a clock face show only the hours since the previous midnight or noon. We are so accustomed to working with equivalence classes modulo 12 that we routinely do arithmetic with them: 9 hours after 8 o'clock is 5 o'clock, because $8 + 9 = 17$ and $17 = 5$ (modulo 12) and 4 hours before 3 o'clock is 11 o'clock, because $3 - 4 = -1 = 11$ (modulo 12).

Our next theorem will show that there are always m different equivalence classes for the relation $\equiv_m$ and the set $\mathbb{Z}_m$ is always $\{\bar{0}, \bar{1}, \bar{2}, \ldots, \overline{m-1}\}$. It is helpful to observe that $0, 1, 2, \ldots,$ and $m - 1$ are exactly all the possible remainders when integers are divided by m. For this reason the elements of $\mathbb{Z}_m$ are sometimes called the *residue* (or remainder) classes modulo m.

Theorem 3.2.3   Let m be a fixed positive integer. Then

(a)   For integers x and y, $x = y \pmod{m}$ iff the remainder when x is divided by m equals the remainder when y is divided by m.

(b) $\mathbb{Z}_m$ consists of m distinct equivalence classes: $\mathbb{Z}_m = \{\bar{0}, \bar{1}, \bar{2}, \ldots, \overline{m-1}\}$.

Proof.

(a)   Let x and y be integers. By the Division Algorithm, there exist integers q, r, t, and s such that $x = mq + r$, with $0 \le r < m$ and $y = mt + s$, with $0 \le s < m$. ⟨*We must show that $x = y \pmod{m}$ iff $r = s$.*⟩ Then

$$x = y \pmod{m} \text{ iff } m \text{ divides } x - y$$
$$\text{iff } m \text{ divides } (mq + r) - (mt + s)$$
$$\text{iff } m \text{ divides } m(q - t) + (r - s)$$
$$\text{iff } m \text{ divides } r - s$$
$$\text{iff } r = s. \langle \textit{This is because } 0 \le r < m \textit{ and } 0 \le s < m. \rangle$$

(b)   ⟨*We first show that $\mathbb{Z}_m = \{\bar{0}, \bar{1}, \bar{2}, \ldots, \overline{m-1}\}$.*⟩ For each k, where $0 \le k \le m - 1$, the set $\bar{k}$ is an equivalence class, so $\{\bar{0}, \bar{1}, \bar{2}, \ldots, \overline{m-1}\}$ is a subset of $\mathbb{Z}_m$. Now suppose $\bar{x} \in \mathbb{Z}_m$ for some integer x. By the Division Algorithm, there exist integers q and r such that $x = mq + r$, with $0 \le r < m$. Then $x - r = mq$, so m divides $x - r$. Thus $x = r \pmod{m}$. By Theorem 3.2.1(b) $\bar{x} = \bar{r}$. Therefore $\mathbb{Z}_m \subseteq \{\bar{0}, \bar{1}, \bar{2}, \ldots, \overline{m-1}\}$.

Finally we will know that $\mathbb{Z}_m$ has exactly m elements when we show that the equivalence classes $\bar{0}, \bar{1}, \bar{2}, \ldots, \overline{m-1}$ are all distinct. Suppose $\bar{k} = \bar{r}$ where $0 \le r \le k \le m - 1$. Then $k = r \pmod{m}$, and thus m divides $k - r$. But $0 \le k - r \le m - 1$, so $k - r = 0$. Then $k = r$. Therefore the m equivalence classes are distinct. ∎

Exercises 3.2

1.   Indicate which of the following relations on the given sets are reflexive on a given set, which are symmetric, and which are transitive.
   ⋆   (a) $\{(1, 2)\}$ on $\{1, 2\}$ (b) $\le$ on $\mathbb{N}$
         (c) $=$ on $\mathbb{N}$ (d) $<$ on $\mathbb{N}$

⋆ **(e)** $\geq$ on $\mathbb{N}$ **(f)** $\neq$ on $\mathbb{N}$

(g) "divides" on $\mathbb{N}$ **(h)** $\{(x, y) \in \mathbb{Z} \times \mathbb{Z} : x + y = 10\}$

(i) $\{(1, 5), (5, 1), (1, 1)\}$ on the set $A = \{1, 2, 3, 4, 5\}$

(j) $\perp = \{(l, m) : l$ and m are lines and l is perpendicular to $m\}$ on the set of all lines in a plane

(k) R, where $(x, y) R (z, w)$ iff $x + z \leq y + w$, on the set $\mathbb{R} \times \mathbb{R}$

⋆ **(l)** S, where $x S y$ iff x is a sibling of y, on the set P of all people

(m) T, where $(x, y) T (z, w)$ iff $x + y \leq z + w$, on the set $\mathbb{R} \times \mathbb{R}$

2. Let $A = \{1, 2, 3\}$. List the ordered pairs and draw the digraph of a relation on A with the given properties.

⋆ **(a)** not reflexive, not symmetric, and not transitive

(b) reflexive, not symmetric, and not transitive

(c) not reflexive, symmetric, and not transitive

⋆ **(d)** reflexive, symmetric, and not transitive

(e) not reflexive, not symmetric, and transitive

(f) reflexive, not symmetric, and transitive

(g) not reflexive, symmetric, and transitive

(h) reflexive, symmetric, and transitive

☆ **3.** For each part of Exercise 2, give an example of a relation on $\mathbb{R}$ with the desired properties.

4. Let R be a relation on a set A. Prove that

(a) if A is nonempty, the empty relation $\varnothing$ is not reflexive on A.

(b) the empty relation $\varnothing$ is symmetric and transitive for every set A.

5. For each of the following, prove that the relation is an equivalence relation. Then give information about the equivalence classes as specified.

(a) The relation R on $\mathbb{R}$ given by $x R y$ iff $x - y \in \mathbb{Q}$. Give the equivalence class of 0; of 1, of $\sqrt{2}$.

(b) The relation R on $\mathbb{N}$ given by $m R n$ iff m and n have the same digit in the tens places. Find an element of $106/R$ that is less than 50; between 150 and 300; greater than 1,000. Find three such elements in the equivalence class $635/R$.

(c) The relation V on $\mathbb{R}$ given by $x V y$ iff $x = y$ or $xy = 1$. Give the equivalence class of 3; of $-\frac{2}{3}$; of 0.

☆ **(d)** On $\mathbb{N}$, the relation R given by $a R b$ iff the prime factorizations of a and b have the same number of 2's. For example, $16 R 80$ because $16 = 2^4$ and $80 = 2^4 \cdot 5$. Name three elements in each of these classes: $1/R$, $4/R$, $72/R$.

(e) The relation T on $\mathbb{R} \times \mathbb{R}$ given by $(x, y) T (a, b)$ iff $x^2 + y^2 = a^2 + b^2$. Sketch the equivalence class of $(1, 2)$; of $(4, 0)$.

(f) For the set $X = \{m, n, p, q, r, s\}$, let R be the relation on $\mathcal{P}(X)$ given by $A R B$ iff A and B have the same number of elements. List all the elements in $\{m\}/R$; in $\{m, n, p, q, r, s\}/R$. How many elements are in X/R? How many elements are in $\mathcal{P}(X)/R$?

(g) The relation P on $\mathbb{R} \times \mathbb{R}$ defined by $(x, y)\, P\, (z, w)$ iff $|x - y| = |z - w|$. Name at least one ordered pair in each quadrant that is related to $(3, 0)$. Describe all ordered pairs in the equivalence class of $(0, 0)$; in the class of $(1, 0)$.

(h) Let R be the relation on the set of all differentiable functions defined by $f\, R\, g$ iff f and g have the same first derivative, that is, $f' = g'$. Name three elements in each of these classes: x^2/R, $(4x^3 + 10x)/R$. Describe x^3/R and $7/R$.

(i) The relation T on $\mathbb{R}$ given by $x\, T\, y$ iff $\sin x = \sin y$. Describe the equivalence class of 0; of $\pi/2$; of $\pi/4$.

6. Let R be the relation on $\mathbb{Q}$ defined by $\frac{p}{q}\, R\, \frac{s}{t}$ iff $pt = qs$. Show that R is an equivalence relation. Describe all ordered pairs in the equivalence class of $\frac{2}{3}$.

7. Which of these digraphs represent relations that are (i) reflexive? (ii) symmetric? (iii) transitive?

★ (a)   (b)

★ (c)

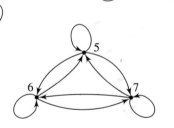

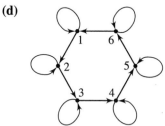

(d)

8. Determine the equivalence classes for the relation of
★ (a) congruence modulo 5.   (b) congruence modulo 8.
   (c) congruence modulo 1.   (d) congruence modulo 7.

9. Name a positive integer and a negative integer that are
   (a) congruent to 0 (mod 5) and not congruent to 0 (mod 6).
   (b) congruent to 0 (mod 5) and congruent to 0 (mod 6).
   (c) congruent to 2 (mod 4) and congruent to 8 (mod 6).
   (d) congruent to 3 (mod 4) and congruent to 3 (mod 5).
   (e) congruent to 1 (mod 3) and congruent to 1 (mod 7).

10. Using the fact that $\equiv_m$ is an equivalence relation on $\mathbb{Z}$ and without reference to Theorems 3.2.1 and 3.2.3, prove that for all x and y in $\mathbb{Z}$:
    (a) $x \in \bar{x}$.   ☆ (b) $\bar{x} \neq \varnothing$.
    ☆ (c) if $x \equiv_m y$, then $\bar{x} = \bar{y}$.   (d) if $\bar{x} = \bar{y}$, then $x \equiv_m y$.
    (e) if $\bar{x} \cap \bar{y} \neq \varnothing$, then $\bar{x} = \bar{y}$.   (f) if $\bar{x} \cap \bar{y} = \varnothing$, then $\bar{x} \neq \bar{y}$.

11. Consider the relation S on $\mathbb{N}$ defined by $x \, S \, y$ iff 3 divides $x + y$. Prove that S is not an equivalence relation.

12. Suppose that R and S are equivalence relations on a set A. Prove that $R \cap S$ is an equivalence relation on A.

13. The properties of reflexivity, symmetry, and transitivity are related to the identity relation and the operations of inversion and composition. Prove that
    (a) R is a reflexive relation on A iff $I_A \subseteq R$.
    ★ (b) R is symmetric iff $R = R^{-1}$.
    (c) R is transitive iff $R \circ R \subseteq R$.

14. Prove that if R is a symmetric, transitive relation on A and the domain of R is A, then R is reflexive on A.

15. Let R be a relation on the set A.
    ★ (a) Prove that $R \cup R^{-1}$ is symmetric. ($R \cup R^{-1}$ is the **symmetric closure** of R.)
    (b) Prove that if S is a symmetric relation on A and $R \subseteq S$, then $R^{-1} \subseteq S$.

16. Let R be a relation on the set A. Define $T_R = \{(x, y) \in A \times A:$ for some $n \in \mathbb{N}$ there exists $a_0 = x, a_1, a_2, \ldots, a_n = y \in A$ such that $(a_0, a_1), (a_1, a_2), (a_2, a_3), \ldots, (a_{n-1}, a_n) \in R\}$.
    (a) Prove that T_R is transitive. (T_R is the **transitive closure** of R.)
    (b) Prove that if S is a transitive relation on A and $R \subseteq S$, then $T_R \subseteq S$.

17. The **complement** of a digraph has the same vertex set as the original digraph, and an arc from x to y exactly when the original digraph does not have an arc from x to y. The two digraphs shown below are complementary. Call a digraph symmetric (transitive) iff its relation is symmetric (transitive).

    (a) Show that the complement of a symmetric digraph is symmetric.
    (b) Show by example that the complement of a transitive digraph need not be transitive.

☆ 18. Let L be a relation on a set A that is reflexive on A and transitive but not necessarily symmetric. Let R be the relation defined on A by $x \, R \, y$ iff $x \, L \, y$ and $y \, L \, x$. Prove that R is an equivalence relation.

Proofs to Grade   19. Assign a grade of A (correct), C (partially correct), or F (failure) to each. Justify assignments of grades other than A.

(a) **Claim.** If the relation R is symmetric and transitive, it is also reflexive.
"*Proof.*" Since R is symmetric, if $(x, y) \in R$, then $(y, x) \in R$. Thus $(x, y) \in R$ and $(y, x) \in R$, and since R is transitive, $(x, x) \in R$. Therefore, R is reflexive. ∎

(b) **Claim.** The relation T on $\mathbb{R} \times \mathbb{R}$ given by $(x, y) \, T \, (r, s)$ iff $x + y = r + s$ is symmetric.
"*Proof.*" Suppose $(x, y) \in \mathbb{R} \times \mathbb{R}$. Then $(x, y) \, T \, (y, x)$ because $x + y = y + x$. Therefore, T is symmetric. ∎

(c) **Claim.** The relation W on $\mathbb{R} \times \mathbb{R}$ given by $(x, y) \, W \, (r, s)$ iff $x - r = y - s$ is symmetric.
"*Proof.*" Suppose (x, y) and (r, s) are in $\mathbb{R} \times \mathbb{R}$ and $(x, y) \, W \, (r, s)$. Then $x - r = y - s$. Therefore, $r - x = s - y$, so $(r, s) \, W \, (x, y)$. Thus W is symmetric. ∎

(d) **Claim.** If the relations R and S are symmetric, then $R \cap S$ is symmetric.
"*Proof.*" Let R be the relation of congruence modulo 10 and S the relation of congruence modulo 6 on the integers. Both R and S are symmetric. If $(x, y) \in R \cap S$, then 6 and 10 divide $x - y$. Therefore, 2, 3, and 5 all divide $x - y$, so 30 divides $x - y$. Also if 30 divides $x - y$, then 6 and 10 divide $x - y$, so $R \cap S$ is the relation of congruence modulo 30. Therefore, $R \cap S$ is symmetric. ∎

(e) **Claim.** If the relations R and S are symmetric, then $R \cap S$ is symmetric.
"*Proof.*" Suppose $(x, y) \in R \cap S$. Then $(x, y) \in R$ and $(x, y) \in S$. Since R and S are symmetric, $(y, x) \in R$ and $(y, x) \in S$. Therefore, $(y, x) \in R \cap S$. ∎

⋆ (f) **Claim.** If the relations R and S are transitive, then $R \cap S$ is transitive.
"*Proof.*" Suppose $(x, y) \in R \cap S$ and $(y, z) \in R \cap S$. Then $(x, y) \in R$ and $(y, z) \in S$. Therefore, $(x, z) \in R \cap S$. ∎

3.3 Partitions

Partitioning is frequently used to organize the world around us. The United States, for example, is partitioned in several ways—by postal zip codes, state boundaries, time zones, etc. In each case nonempty subsets of the United States are defined that do not overlap and that together comprise the entire country. This section introduces this concept of partitioning of a set and describes the close relationship between partitions and equivalence relations.

.DEFINITION   Let A be a nonempty set. $\mathscr{P}$ is a **partition of A** iff $\mathscr{P}$ is a set of subsets of A such that

(i) If $X \in \mathscr{P}$, then $X \neq \varnothing$.
(ii) If $X \in \mathscr{P}$ and $Y \in \mathscr{P}$, then $X = Y$ or $X \cap Y = \varnothing$.
(iii) $\displaystyle\bigcup_{X \in \mathscr{P}} X = A$.

The set W of all employees in a large work area can be partitioned into work groups by putting up physical partitions (walls) to form cubicles. If we are careful so that (i) every cubicle contains at least one worker, (ii) no worker is assigned to two different cubicles, and (iii) every worker must be in some cubicle, then we have formed a partition of W. Notice that the workers are not elements of the partition; each element of the partition is a set of workers within a common cubicle. In Figure 3.3.1, W is a set of 6 workers and the partition of W consists of four sets—two sets each with two workers and two sets each with a single worker.

Figure 3.3.1

Examples.   The 2-element family $\mathscr{P} = \{E, D\}$, where E is the even integers and D is the odd integers, is a partition of $\mathbb{Z}$. The 3-element collection $\mathscr{K} = \{\mathbb{N}, \{0\}, \mathbb{Z}^-\}$, where $\mathbb{Z}^-$ is the set of negative integers is also a partition of $\mathbb{Z}$. For each $k \in \mathbb{Z}$, let $A_k = \{3k, 3k + 1, 3k + 2\}$. The family $\mathscr{T} = \{A_k : k \in \mathbb{Z}\}$ is an infinite family that is a partition of $\mathbb{Z}$. Some elements of $\mathscr{T}$ are $A_0 = \{0, 1, 2\}$, $A_1 = \{3, 4, 5\}$, and $A_{-1} = \{-3, -2, -1\}$.

Two other partitions of $\mathbb{Z}$ are $\{\ldots, \{-3\}, \{-2\}, \{-1\}, \{0\}, \{1\}, \{2\}, \{3\}, \ldots\}$ and $\{\mathbb{Z}\}$. In fact, for any nonempty set A, the families $\{\{x\}: x \in A\}$ and $\{A\}$ are partitions of A.

Example.   For each $n \in \mathbb{Z}$, let $G_n = [n, n + 1)$. The collection $\{G_n: n \in \mathbb{Z}\}$ of half open intervals is a partition of $\mathbb{R}$.

By definition, a partition of A is a pairwise disjoint collection of nonempty subsets of A whose union is A. Recall from Section 2.3 that the definition of "pairwise disjoint" allows for the possibility that sets in a pairwise disjoint family may be equal.

Example. For the set $A = \{a, b, c, d, e\}$, the family $C = \{C_1, C_2, C_3\}$, where

$$C_1 = \{b, e\}, C_2 = \{a, c, d\}, \text{ and } C_3 = \{b, e\},$$

is a partition of A even though the sets C_1 and C_3 are not disjoint. The family $\{C_1, C_2, C_3\}$, is the same as the family $\{C_2, C_3\}$.

Let W be a set of six people and $C = \{\text{blue, green, red, white}\}$. For each $c \in C$, let

$$B_c = \{x \in W : x \text{ is wearing clothing with color } c\}.$$

and let $\mathcal{B} = \{B_{\text{blue}}, B_{\text{green}}, B_{\text{red}}, B_{\text{white}}\}$. The family $\mathcal{B}$ may not be a partition of W because any of the three parts of the definition might be violated. If no one is wearing red, then B_{red} is empty, so condition (i) fails. If someone is wearing green only, while a second person is wearing green and blue, then the different sets B_{blue} and B_{green} overlap, in violation of condition (ii). If someone is wearing only yellow clothing, then that person does not belong to any set in $\mathcal{B}$, in violation of condition (iii).

The first half of the connection between partitions and equivalence relations is: Every equivalence relation on a set determines a partition of that set.

Theorem 3.3.1   If R is an equivalence relation on a nonempty set A, then A/R, the set of equivalence classes for R, is a partition of A.

Proof.   By Theorem 3.2.1 every equivalence class x/R is a subset of A and is nonempty because it contains x, and any two equivalence classes are either equal or disjoint. All that remains is to show that the union over A/R is equal to A.

First $\bigcup_{x \in A} x/R \subseteq A$ because each $x/R \subseteq A$. To prove $A \subseteq \bigcup_{x \in A} x/R$, suppose $t \in A$. Since $t \in t/R$, $t \in \bigcup_{x \in A} x/R$. Thus $A = \bigcup_{x \in A} x/R$.   ∎

Example.   Let $A = \{4, 5, 6, 7\}$ and T be the equivalence relation

$$\{(4, 4), (5, 5), (6, 6), (7, 7), (5, 7), (7, 5), (7, 6), (6, 7), (5, 6), (6, 5)\}.$$

By Theorem 3.3.1, we can form a partition of A by finding the equivalence classes of T. These are $4/T = \{4\}$ and $5/T = 6/T = 7/T = \{5, 6, 7\}$. The partition produced by T is $A/R = \{\{4\}, \{5, 6, 7\}\}$.

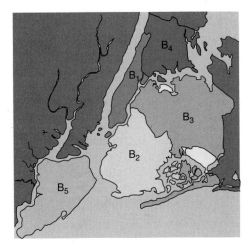

The Five Boroughs of New York City
  B_1: Manhattan
  B_2: Brooklyn
  B_3: Queens
  B_4: The Bronx
  B_5: Staten Island

Figure 3.3.2

New York City is divided into 5 boroughs (counties). The boroughs are labeled B_1 through B_5 in Figure 3.3.2. If A is the set of all residents of New York City, then A is partitioned into 5 subsets: the set of residents living in B_1, the residents living in B_2, and so on. How can we use this fact to define an equivalence relation on A? We say that two residents of New York City are equivalent iff they are in the same partition element; that is, they reside in the same borough.

The method we will use to produce an equivalence relation from a partition is based on this idea that two objects will be said to be related iff they belong to the same member of the partition. The next theorem proves that this method for defining a relation always produces an equivalence relation and, furthermore, the set of equivalence classes of the relation is the same as the original partition.

Theorem 3.3.2   Let $\mathcal{P}$ be a partition of the nonempty set A. For x and $y \in A$, define $x \, Q \, y$ iff there exists $C \in \mathcal{P}$ such that $x \in C$ and $y \in C$. Then

(a) Q is an equivalence relation on A.
(b) $A/Q = \mathcal{P}$.

Proof.

(a)   We prove Q is transitive and leave the proofs of symmetry and reflexivity on A for Exercise 10. Let $x, y, z \in A$. Assume $x \, Q \, y$ and $y \, Q \, z$. Then there are sets C and D in $\mathcal{P}$ such that $x, y \in C$ and $y, z \in D$. Since $\mathcal{P}$ is a partition of A, the sets C and D are either identical or disjoint; but since y is an element of both sets, they cannot be disjoint. Hence, there is a set $C \, (= D)$ that contains both x and z, so that $x \, Q \, z$. Therefore, Q is transitive.

(b)   We first show $A/Q \subseteq \mathcal{P}$. Let $x/Q \in A/Q$. Then choose $B \in \mathcal{P}$ such that $x \in B$. We claim $x/Q = B$. If $y \in x/Q$, then $x \, Q \, y$. Then there is some $C \in \mathcal{P}$ such that $x \in C$ and $y \in C$. Since $x \in C \cap B$, $C = B$, so $y \in B$. On the other hand, if $y \in B$, then $x \, Q \, y$, and so $y \in x/Q$. Therefore, $x/Q = B$.

To show $\mathcal{P} \subseteq A/Q$, let $B \in \mathcal{P}$. As an element of a partition, $B \neq \varnothing$. Choose any $t \in B$; then we claim $B = t/Q$. If $s \in B$, then $t \, Q \, s$, so $s \in t/Q$. On the other hand, if $s \in t/Q$, then $t \, Q \, s$; so s and t are elements of the same member of $\mathcal{P}$, which must be B. ∎

Example. Let $A = \{1, 2, 3, 4\}$ and $\mathcal{P} = \{\{1\}, \{2, 3\}, \{4\}\}$ be a partition of A with three sets. The equivalence relation Q associated with $\mathcal{P}$ is $\{(1, 1), (2, 2), (3, 3), (4, 4), (2. 3), (3, 2)\}$. The three equivalence classes for Q are $1/Q = \{1\}$, $2/Q = 3/Q = \{2, 3\}$, and $4/Q = \{4\}$. The set of all equivalence classes is precisely $\mathcal{P}$.

Example. The set $\mathcal{A} = \{A_0, A_1, A_2, A_3\}$ is a partition of $\mathbb{Z}$, where

$$A_0 = \{4k : k \in \mathbb{Z}\}.$$
$$A_1 = \{4k + 1 : k \in \mathbb{Z}\}.$$
$$A_2 = \{4k + 2 : k \in \mathbb{Z}\}.$$
$$A_3 = \{4k + 3 : k \in \mathbb{Z}\}.$$

Then integers x and y are in the same set A_i iff $x = 4n + i$ and $y = 4m + i$ for some integers n and m or, in other words, iff $x - y$ is a multiple of 4. Thus, the equivalence relation associated with the partition $\mathcal{A}$ is the relation of congruence modulo 4 and each A_i is the residue class of i modulo 4, for $i = 0, 1, 2, 3$.

We have seen that every equivalence relation on a set determines a partition for the set and every partition of a set determines a corresponding equivalence relation on that set. Furthermore, if we start with an equivalence relation, the partition we make is the set of equivalence classes, and if we use that partition to form an equivalence relation, the relation formed is the relation we started with. Thus, each concept may be used to describe the other. This is to our advantage, for we may use partitions and equivalence relations interchangeably, choosing the one that lends itself more readily to the situation at hand.

Exercises 3.3

1. Describe four different partitions of the set of all students enrolled at a university.
2. For the given set A, determine whether $\mathcal{P}$ is a partition of A.
   (a) $A = \{1, 2, 3, 4\}$, $\mathcal{P} = \{\{1, 2\}, \{2, 3\}, \{3, 4\}\}$
   (b) $A = \{1, 2, 3, 4, 5, 6, 7\}$, $\mathcal{P} = \{\{1, 2\}, \{3\}, \{4, 5\}\}$
   (c) $A = \{1, 2, 3, 4, 5, 6, 7\}$, $\mathcal{P} = \{\{1, 3\}, \{5, 6\}, \{2, 4\}, \{7\}\}$
   ★ (d) $A = \mathbb{N}$, $\mathcal{P} = \{1, 2, 3, 4, 5\} \cup \{n \in \mathbb{N} : n > 5\}$
   (e) $A = \mathbb{R}$, $\mathcal{P} = (-\infty, -1) \cup [-1, 1] \cup (1, \infty)$
   (f) $A = \mathbb{R}$, $\mathcal{P} = \{S_y : y \in \mathbb{R} \text{ and } y > 0\}$, where $S_y = \{x \in \mathbb{R} : x < y\}$

3. Describe the partition for each of the following equivalence relations.
   (a) For $x, y \in \mathbb{R}$, $x \, R \, y$ iff $x - y \in \mathbb{Z}$.
   (b) For $n, m \in \mathbb{Z}$, $n \, R \, m$ iff n and m have the same tens digit.
   (c) For $x, y \in \mathbb{R}$, $x \, R \, y$ iff $\sin x = \sin y$.
   (d) For $x, y \in \mathbb{R}$, $x \, R \, y$ iff $x^2 = y^2$.
   (e) For (x, y) and $(u, v) \in \mathbb{R} \times \mathbb{R}$, $(x, y) \, S \, (u, v)$ iff $xy = uv = 0$ or $xyuv > 0$.
   (f) $(x, y) \, R \, (u, v)$ iff $x + v = y + u$.

4. Let $C = \{i, -1, -i, 1\}$, where $i^2 = -1$. The relation R on C given by $x \, R \, y$ iff $xy = \pm 1$ is an equivalence relation on C. Give the partition of C associated with R.

5. Let C be as in Exercise 4. The relation S on $C \times C$ given by $(x, y) \, S \, (u, v)$ iff $xy = uv$ is an equivalence relation. Give the partition of $C \times C$ associated with S.

6. Describe the equivalence relation on each of the following sets with the given partition.
   (a) $\mathbb{N}, \{\{1, 2, \ldots 9\}, \{10, 11, \ldots 99\}, \{100, 101, \ldots 999\}, \ldots\}$
   (b) $\mathbb{Z}, \{\ldots, \{-2\}, \{-1\}, \{0\}, \{1\}, \{2\}, \{3, 4, 5, \ldots\}\}$
   (c) $\mathbb{R}, \{(-\infty, 0), \{0\}, (0, \infty)\}$
   (d) $\mathbb{R}, \{\ldots, (-3, -2), \{-2\}, (-2, -1), \{-1\}, (-1, 0), \{0\}, (0, 1), \{1\},$
       $(1, 2), \{2\}, (2, 3), \ldots\}$
   (e) $\mathbb{Z}, \{A, B\}$, where $A = \{x \in \mathbb{Z}: x < 3\}$ and $B = \mathbb{Z} - A$

7. For each $a \in \mathbb{R}$, let $A_a = \{(x, y) \in \mathbb{R} \times \mathbb{R}: y = a - x^2\}$.
   (a) Sketch a graph of the set A_a for $a = -2, -1, 0, 1$, and 2.
   (b) Prove that $\{A_a: a \in \mathbb{R}\}$ is a partition of $\mathbb{R} \times \mathbb{R}$.
   (c) Describe the equivalence relation associated with this partition.

8. List the ordered pairs in the equivalence relation on $A = \{1, 2, 3, 4, 5\}$ associated with these partitions:
   (a) $\{\{1, 2\}, \{3, 4, 5\}\}$ (b) $\{\{1\}, \{2\}, \{3, 4\}, \{5\}\}$
   (c) $\{\{2, 3, 4, 5\}, \{1\}\}$

9. Partition the set $D = \{1, 2, 3, 4, 5, 6, 7\}$ into two subsets: those symbols made from straight line segments only (like 4), and those that are drawn with at least one curved segment (like 2). Describe or draw the digraph of the corresponding equivalence relation on D.

10. Complete the proof of Theorem 3.3.2 by proving that if $\mathcal{P}$ is a partition of A, and $x \, Q \, y$ iff there exists $C \in \mathcal{P}$ such that $x \in C$ and $y \in C$, then
    (a) Q is symmetric.
    (b) Q is reflexive on A.

11. Let R be a relation on a set A that is reflexive and symmetric but not transitive. Let $R(x) = \{y: x \, R \, y\}$. [Note that $R(x)$ is the same as x/R except that R is not an equivalence relation in this exercise.] Does the set $\mathcal{A} = \{R(x): x \in A\}$ always form a partition of A? Prove that your answer is correct.

12. Repeat Exercise 11, assuming R is reflexive and transitive but not symmetric.

13. Repeat Exercise 11, assuming R is symmetric and transitive but not reflexive.

14. Let A be a set with at least three elements.
   ★ (a) If $\mathscr{P} = \{B_1, B_2\}$ is a partition of A with $B_1 \neq B_2$, is $\{B_1^c, B_2^c\}$ a partition of A? Explain. What if $B_1 = B_2$?
   (b) If $\mathscr{P} = \{B_1, B_2, B_3\}$ is a partition of A, is $\{B_1^c, B_2^c, B_3^c\}$ a partition of A? Explain. Consider the possibility that two or more of the elements of $\mathscr{P}$ may be equal.
   (c) If $\mathscr{P} = \{B_1, B_2\}$ is a partition of A, $\mathscr{C}_1$ is a partition of B_1, and $\mathscr{C}_2$ is a partition of B_2, and $B_1 \neq B_2$, prove that $\mathscr{C}_1 \cup \mathscr{C}_2$ is a partition of A.

Proofs to Grade 15. Assign a grade of A (correct), C (partially correct), or F (failure) to each. Justify assignments of grades other than A.
   (a) **Claim.** Let R be an equivalence relation on the set A, and let x, y, and z be elements of A. If $x \in y/R$ and $z \notin x/R$, then $z \notin y/R$.
   "Proof." Assume that $x \in y/R$ and $z \in x/R$. Then $y\,R\,x$ and $x\,R\,z$. By transitivity, $y\,R\,z$, so $z \in y/R$. Therefore, if $x \in y/R$ and $z \notin x/R$, then $z \notin y/R$. ∎

   (b) **Claim.** Let R be an equivalence relation on the set A, and let x, y, and z be elements of A. If $x \in y/R$ and $z \notin x/R$, then $z \notin y/R$.
   "Proof." Assume that $x \in y/R$ and assume that $z \in y/R$. Then $y\,R\,x$ and $y\,R\,z$. By symmetry, $x\,R\,y$, and by transitivity, $x\,R\,z$. Therefore, $z \in x/R$. We conclude that if $x \in y/R$ and $z \notin x/R$, then $z \notin y/R$. ∎

   (c) **Claim.** If $\mathscr{A}$ is a partition of a set A and $\mathscr{B}$ is a partition of a set B, then $\mathscr{A} \cup \mathscr{B}$ is a partition of $A \cup B$.
   "Proof."
   (i) If $X \in \mathscr{A} \cup \mathscr{B}$, then $X \in \mathscr{A}$, or $X \in \mathscr{B}$. In either case $X \neq \varnothing$.
   (ii) If $X \in \mathscr{A} \cup \mathscr{B}$ and $Y \in \mathscr{A} \cup \mathscr{B}$, then $X \in \mathscr{A}$ and $Y \in \mathscr{A}$, or $X \in \mathscr{A}$ and $Y \in \mathscr{B}$, or $X \in \mathscr{B}$ and $Y \in \mathscr{A}$, or $X \in \mathscr{B}$ and $Y \in \mathscr{B}$. Since both $\mathscr{A}$ and $\mathscr{B}$ are partitions, in each case either $X = Y$ or $X \cap Y = \varnothing$.
   (iii) Since $\bigcup_{X \in \mathscr{A}} X = A$ and $\bigcup_{x \in \mathscr{B}} X = B$, $\bigcup_{X \in \mathscr{A} \cup \mathscr{B}} X = A \cup B$. ∎

   ★ (d) **Claim.** If $\mathscr{B}$ is a partition of A, and if $x\,Q\,y$ iff there exists $C \in \mathscr{B}$ such that $x \in C$ and $y \in C$, then the relation Q is symmetric.
   "Proof." First, $x\,Q\,y$ iff there exists $C \in \mathscr{B}$ such that $x \in C$ and $y \in C$. Also, $y\,Q\,x$ iff there exists $C \in \mathscr{B}$ such that $y \in C$ and $x \in C$. Therefore, $x\,Q\,y$ iff $y\,Q\,x$. ∎

3.4   Ordering Relations

Familiar ordering relations for $\mathbb{N}$, $\mathbb{Z}$, and $\mathbb{R}$ such as "less than," "greater than," and "less than or equal to" are basic to our understanding of number systems but they are not equivalence relations. For instance, $<$ is not reflexive on $\mathbb{R}$ because $3 < 3$ is false, and is not symmetric because $2 < \pi$ is true but $\pi < 2$ is false. The relation $<$ is transitive, because the conjunction $x < y$ and $y < z$ implies $x < z$. This section describes those properties of relations that characterize orderings like $<$ and $\leq$. We begin with some examples.

Example. In addition to transitivity and reflexivity on $\mathbb{R}$, the relation $\leq$ on $\mathbb{R}$ has two properties we have not previously considered. The first of these properties is **comparability**: every two elements of $\mathbb{R}$ are comparable. This means that for all $x, y \in \mathbb{R}$, either $x \leq y$ or $y \leq x$. The other property is that for all $x, y \in \mathbb{R}$, if $x \leq y$ and $y \leq x$, then $x = y$.

Example. We saw earlier that the relation "divides" is reflexive on $\mathbb{N}$. While we did not use the term "transitive" in Section 1.4, in effect we proved in that section that "divides" is transitive. Two other properties of this relation are notable. If a divides b and b divides a, then $a = b$. Also, there are elements of $\mathbb{N}$ that are not comparable. That is, there are natural numbers x and y (for example, 10 and 21) such that both "x divides y" and "y divides x" are false.

Example. Let X be a set. The set inclusion relation $\subseteq$ on the power set of X is reflexive on $\mathscr{P}(X)$ and transitive. Also, if A and B are subsets of X with $A \subseteq B$ and $B \subseteq A$ then $A = B$. In this relation some pairs of elements are not comparable. For example, if $X = \{1, 2, 3, 4\}$, then $\{1, 3\}$ and $\{1, 4\}$ are elements of $\mathscr{P}(X)$ but both $\{1, 3\} \subseteq \{1, 4\}$ and $\{1, 4\} \subseteq \{1, 3\}$ are false.

Example. Let Y be the relation "is the same age in years or younger than" on a fixed set P of people. Then Y is reflexive on P and transitive. This relation also has the property that any two elements of P are comparable. However, the relation Y has a property that is undesirable for an ordering. If a and b are two different people in P, and both a and b are 20 years old, then $a\,Y\,b$ and $b\,Y\,a$, but $a \neq b$.

Although we find it acceptable in an ordering for two elements to not be comparable, we wish to avoid the situation in the previous example where two different objects are both related to each other. The property we want is called antisymmetry.

> **DEFINITION** A relation R on a set A is **antisymmetric** iff for all $x, y \in A$, if $x\,R\,y$ and $y\,R\,x$, then $x = y$.

Examples. We have already noted that the relations "divides" on $\mathbb{N}$, $\leq$ on $\mathbb{R}$, and $\subseteq$ on $\mathscr{P}(A)$ are antisymmetric. The relation $<$ differs from the relation $\leq$ on $\mathbb{R}$ because $<$ is not reflexive on $\mathbb{R}$. Like $\leq$, the relation $<$ is antisymmetric but for a different reason: the statement "For all x, y in $\mathbb{R}$, if $x < y$ and $y < x$ then $x = y$" is true because the antecedent is false.

The relation "divides" is an antisymmetric relation on $\mathbb{N}$. However, "divides" is not an antisymmetric relation on $\mathbb{Z}$. For example, 6 divides -6 and -6 divides 6, but $6 \neq -6$.

Antisymmetry is an important concept for maintaining the chain of command in the military where the relation "can give orders to" must be explicit. It would be chaotic if two different officers could give orders to each other.

A relation may be antisymmetric and not symmetric, symmetric and not antisymmetric, both, or neither. See Exercise 2. In Exercise 3, you are asked to show that if R is an antisymmetric relation, then $x \mathrel{R} y$ and $x \neq y$ implies $y \mathrel{\not R} x$. That is, the only possible symmetry that an antisymmetric relation may exhibit is that an object may be related to itself.

DEFINITION A relation R on a set A is a **partial order** (or **partial ordering**) for A if R is reflexive on A, antisymmetric, and transitive. A set A with partial order R is called a **partially ordered set**, or **poset**.

Three relations discussed above: "divides" on $\mathbb{N}$, $\leq$ on $\mathbb{R}$, and $\subseteq$ on $\mathcal{P}(X)$ for any set X, are examples of partial orderings.

Example. Let W be the relation on $\mathbb{N}$ given by $x \mathrel{W} y$ iff $x + y$ is even and $x \leq y$. Then W is a partial order. For example, $2 \mathrel{W} 4$, $4 \mathrel{W} 6$, $6 \mathrel{W} 8, \ldots$, and $1 \mathrel{W} 3$, $3 \mathrel{W} 5$, $5 \mathrel{W} 7, \ldots$, but we never have $m \mathrel{W} n$ where m and n have opposite parity. We verify that W is a partial order:

Proof.

(i) $\langle$*Show W is reflexive on $\mathbb{N}$.*$\rangle$ Let $x \in \mathbb{N}$. Then $x + x = 2x$ is even and $x \leq x$, so $x \mathrel{W} x$.

(ii) $\langle$*Show W is antisymmetric.*$\rangle$ Suppose $x \mathrel{W} y$ and $y \mathrel{W} x$. Then $x + y$ is even, $x \leq y$, and $y \leq x$. By antisymmetry of $\leq$ on $\mathbb{N}$, $x = y$.

(iii) $\langle$*Show W is transitive.*$\rangle$ Suppose $x \mathrel{W} y$ and $y \mathrel{W} z$. Then $x \leq y$, $x + y$ is even, $y \leq z$, and $y + z$ is even. By transitivity of $\leq$ on $\mathbb{N}$, $x \leq z$. Also, $x + z$ is even because $x + z = (x + y) + (y + z) + (-2y)$ is the sum of three even numbers. Therefore, $x \mathrel{W} z$. ∎

Suppose R is a partial order on the set A and a, b, c are three distinct elements of A. Further suppose that $a \mathrel{R} b$, $b \mathrel{R} c$, and $c \mathrel{R} a$. A portion of the digraph of R is shown in Figure 3.4.1. The chain of relationships $a \mathrel{R} b$, $b \mathrel{R} c$, $c \mathrel{R} a$ is called a closed path (of length 3) in the digraph. (See the next section for more about paths in graphs.) The path is closed because as we move from vertex to vertex along the path, we can start and end at the same vertex. From $a \mathrel{R} b$ and $b \mathrel{R} c$, by transitivity we must have $a \, R \, c$. (The arc from a to c is not shown in the portion of the digraph in Figure 3.4.1.) But $c \, R \, a$ is also true, and this contradicts the antisymmetry property of R. Using this reasoning, we conclude that the digraph of a partial order can never contain a closed path except for loops at individual vertices.

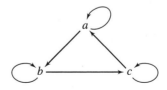

Figure 3.4.1

Theorem 3.4.1
If R is a partial order for a set A and $x \, R \, x_1$, $x_1 \, R \, x_2$, $x_2 \, R \, x_3$, ..., $x_n \, R \, x$, then $x = x_1 = x_2 = x_3 = \cdots = x_n$.

Proof. ⟨*We prove this by induction on n.*⟩ For $n = 1$, suppose we have $x \, R \, x_1$ and $x_1 \, R \, x$. By antisymmetry, we conclude that $x = x_1$.

Now suppose that for some natural number k, whenever $x \, R \, x_1$, $x_1 \, R \, x_2$, $x_2 \, R \, x_3$, ..., $x_k \, R \, x$, then $x = x_1 = x_2 = x_3 = \cdots = x_k$ and suppose that $x \, R \, x_1$, $x_1 \, R \, x_2$, $x_2 \, R \, x_3$, ..., $x_k \, R \, x_{k+1}$, $x_{k+1} \, R \, x$. By transitivity ⟨*applied to $x_k \, R \, x_{k+1}$ and $x_{k+1} \, R \, x$*⟩ we have $x_k \, R \, x$. From $x \, R \, x_1$, $x_1 \, R \, x_2$, ..., $x_k \, R \, x$ and the hypothesis of induction, we have $x = x_1 = x_2 = \cdots = x_k$. Since $x_k = x$ we have $x \, R \, x_{k+1}$ and $x_{k+1} \, R \, x$, so $x = x_{k+1}$. Therefore, $x = x_1 = x_2 = \cdots = x_{k+1}$. ∎

DEFINITION Let R be a partial ordering on a set A and let $a, b \in A$ with $a \neq b$. Then a is an **immediate predecessor** of b iff $a \, R \, b$ and there does not exist $c \in A$ such that $a \neq c$, $b \neq c$, $a \, R \, c$ and $c \, R \, b$.

In other words, a is an immediate predecessor of b when $a \, R \, b$ and no other element lies "between" a and b.

Example. For $A = \{1, 2, 3, 4, 5\}$, $\mathcal{P}(A)$ is partially ordered by the set inclusion relation $\subseteq$. For the set $\{2, 3, 5\}$, there are three immediate predecessors in $\mathcal{P}(A)$: $\{2, 3\}$, $\{2, 5\}$, and $\{3, 5\}$. The empty set has no immediate predecessor. Also, $\varnothing$ is the only immediate predecessor for $\{3\}$. We have $\{4\} \subseteq \{2, 4, 5\}$, but $\{4\}$ is not an immediate predecessor of $\{2, 4, 5\}$ because $\{4\} \neq \{4, 5\}$, $\{4, 5\} \neq \{2, 4, 5\}$, $\{4\} \subseteq \{4, 5\}$, and $\{4, 5\} \subseteq \{2, 4, 5\}$.

Let $M = \{1, 2, 3, 5, 6, 10, 15, 30\}$ be the set of all positive divisors of 30. The relation "divides" is a partial order for M whose digraph is given in Figure 3.4.2(a). We can simplify the digraph significantly. First, since we know that every vertex must have a loop, we need not include them in the digraph. Also, since there are no closed paths, we can orient the digraph so that all edges point upward; thus we may eliminate the arrowheads, assuming that each edge has the arrowhead on the upper end. We can also remove edges that can be recovered by transitivity. For example, since there is an edge from 2 to 10 and another from 10 to 30, we do not need to include the edge from 2 to 30. In other words, we need only include those edges that relate immediate predecessors. The resulting simplified digraph, Figure 3.4.2(b), is called a **Hasse diagram** of the partial order "divides."

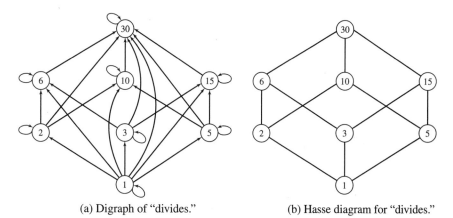

(a) Digraph of "divides."   (b) Hasse diagram for "divides."

Figure 3.4.2

Example. Let $A = \{1, 2, 3\}$. The Hasse diagram for $\mathcal{P}(A)$ partially ordered by $\subseteq$ is given in Figure 3.4.3. It bears a striking resemblance to Figure 3.4.2(b) for good reason. Except for the naming of the elements in the sets, the orderings are the same. In fact, it can be shown that every partial order is "the same" as the set inclusion relation on subsets of some set. Although we need the concepts of Chapter 4 to

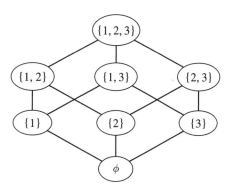

Hasse diagram for $\subseteq$.

Figure 3.4.3

make precise what we mean by "same," Exercise 19 outlines how one might start to show this.

DEFINITIONS   Let R be a partial order for a set A. Let B be any subset of A and $a \in A$. Then

a is an **upper bound** for B iff $b \, R \, a$ for every $b \in B$.
a is a **lower bound** for B iff $a \, R \, b$ for every $b \in B$.
a is a **least upper bound** for B (or **supremum** for B) iff

    (i) a is an upper bound for B, and
    (ii) $a \, R \, x$ for every upper bound x for B.

a is a **greatest lower bound** for B (or **infimum** for B) iff

    (i) a is a lower bound for B, and
    (ii) $x \, R \, a$ for every lower bound x for B.

We write **sup (B)** to denote a supremum of B and **inf (B)** for an infimum of B.

We shall soon see (Theorem 3.4.2) that there is at most one supremum and one infimum for a set.

Examples.   For $A = \{1, 2, 3, 4, 5, 6, 7, 8, 9, 10\}$, let $B = \{\{1, 4, 5, 7\}, \{1, 4, 7, 8\}, \{2, 4, 7\}\}$. B is a subset of $\mathcal{P}(A)$. Using the partial order $\subseteq$ for $\mathcal{P}(A)$, we see that $\{1, 2, 3, 4, 5, 6, 7, 8\}$ is an upper bound for B because

$$\{1, 4, 5, 7\} \subseteq \{1, 2, 3, 4, 5, 6, 7, 8\},$$
$$\{1, 4, 7, 8\} \subseteq \{1, 2, 3, 4, 5, 6, 7, 8\}, \text{ and}$$
$$\{2, 4, 7\} \subseteq \{1, 2, 3, 4, 5, 6, 7, 8\}.$$

Another upper bound for B is $\{2, 4, 5, 7, 8, 9, 10\}$. The least upper bound for B is $\sup(B) = \{1, 2, 4, 5, 7, 8\}$.

Elements of $\mathcal{P}(X)$ that are lower bounds for B are $\varnothing$, $\{4\}$, $\{7\}$, and $\{4, 7\}$. The greatest lower bound for B is $\inf(B) = \{4, 7\}$.

You should notice in the example above that $\sup(B)$ is the union of the sets in B and $\inf(B)$ is the intersection of the sets in B. This is true in general: for any nonempty set A with $\mathcal{P}(A)$ partially ordered by $\subseteq$, if B is a set of subsets of A, then $\sup(B) = \bigcup_{X \in B} X$ and $\inf(B) = \bigcap_{X \in B} X$. See Exercise 14.

Example.   Here are least upper bounds and greatest lower bounds for some subsets of $\mathbb{R}$ with the usual ordering $\leq$:

for $A = [0, 4)$, $\sup(A) = 4$ and $\inf(A) = 0$.
for $B = \{1, 6, 3, 9, 12, -4, 10\}$, $\sup(B) = 12$ and $\inf(B) = -4$.

for $C = \{2^k : k \in \mathbb{N}\}$, sup(C) does not exist and inf(C) = 2.

for $D = \{2^{-k} : k \in \mathbb{N}\}$, sup(D) = $\frac{1}{2}$ and inf(D) = 0.

Example. Let A be the set of all positive divisors of 1000 with the ordering relation "divides" on A. Let $B = \{10, 20, 25, 100\}$. Both 500 and 1000 are upper bounds for B; the least upper bound is 100. The greatest lower bound for B is 5. Note that for "divides," the least upper bound is the lcm (least common multiple) and the greatest lower bound is the gcd (greatest common divisor).

Theorem 3.4.2   Let R be a partial order for a set A and $B \subseteq A$. Then if sup(B) exists, it is unique. Also, if inf(B) exists, it is unique.

Proof.   Suppose that x and y are both least upper bounds for B. ⟨*We prove that* $x = y$.⟩ Since x and y are least upper bounds, then x and y are upper bounds. Since x is an upper bound and y is a least upper bound, we must have $y\,R\,x$. Likewise, since y is an upper bound and x is a least upper bound, we must have $x\,R\,y$. From $x\,R\,y$ and $y\,R\,x$, we conclude that $x = y$ by antisymmetry. Thus, if it exists, sup(B) is unique.

The proof for inf(B) is left as an exercise.   ∎

We have seen examples of sets B where, when they exist, the least upper and greatest lower bounds for B are in B and other examples where they are not in B.

DEFINITION   Let R be a partial order for a set A. Let $B \subseteq A$. If the greatest lower bound for B exists and is an element of B, it is called the **smallest element** (or **least element**) of B. If the least upper bound for B is in B, it is called the **largest element** (or **greatest element**) of B.

The usual ordering of the number systems has the comparability property: for any x and y, either $x \leq y$ or $y \leq x$. A partial ordering with this property is called linear.

DEFINITION   A partial ordering R on A is called a **linear order** (or **total order**) on A if for any two elements x and y of A, either $x\,R\,y$ or $y\,R\,x$.

Examples.   Each of $\mathbb{N}$, $\mathbb{Z}$ and $\mathbb{R}$ with the ordering $\leq$ is linearly ordered. $\mathscr{P}(A)$ with set inclusion, where $A = \{1, 2, 3\}$, is not a linearly ordered set because the two elements $\{1, 2\}$ and $\{1, 3\}$ cannot be compared. Likewise, the relation "divides" is not a linear order for $\mathbb{N}$ because 3 and 5 are not related (neither divides the other).

If R is a linear order on A, then by antisymmetry, if x and y are distinct elements of A, $x\,R\,y$ or $y\,R\,x$ (but not both). The Hasse diagram for a linear ordered set is a set of points on a vertical line.

For a given linear order on a set it is not always true that every subset has a smallest or largest element. The set of integers with $\leq$ is linearly ordered but the set $B = \{1, 3, 5, 7, \dots\}$ has neither upper bounds nor a least upper bound. Likewise, $\{-2, -4, -8, -16, -32, \dots\}$ has no greatest lower bound (and hence no smallest element).

> **DEFINITION**   Let L be a linear ordering on a set A. L is a **well ordering** on A if every nonempty subset B of A contains a smallest element.

In Chapter 2 we proved the Well-Ordering Principle from the Principle of Mathematical Induction. Using the terminology of this section, the Well-Ordering Principle says that the natural numbers are well ordered by $\leq$. The integers, $\mathbb{Z}$, on the other hand, are not well ordered by $\leq$ because we have seen that $\{-2, -4, -8, -16, -32, \dots\}$ is a nonempty subset that has no smallest element.

Finally, we state without proof a remarkable result.

Theorem 3.4.3

Well-Ordering Theorem
Every set can be well ordered.

The Well-Ordering Theorem should not be confused with the Well-Ordering Principle of Section 2.5, which is a property of the natural numbers. The theorem says for any nonempty set A there is always a way to define a linear ordering on the set so that every nonempty subset of A has a least element. Even the set of real numbers, which we know is not well ordered by the usual linear order $\leq$, has some other linear ordering so that $\mathbb{R}$ is well ordered by that ordering. The proof of the Well-Ordering Theorem requires a new property of sets, the Axiom of Choice. (See Section 5.5.)

Exercises 3.4

1. Which of these relations on the given set are antisymmetric?
   ⋆ **(a)** $A = \{1, 2, 3, 4, 5\}$, $R = \{(1, 3), (1, 1), (2, 4), (3, 2), (5, 4), (4, 2)\}$.
     (b) $A = \{1, 2, 3, 4, 5\}$, $R = \{(1, 4), (1, 2), (2, 3), (3, 4), (5, 2), (4, 2), (1, 3)\}$.
   ⋆ **(c)** $\mathbb{Z}$, $x \, R \, y$ iff $x^2 = y^2$.
     (d) $\mathbb{R}$, $x \, R \, y$ iff $x \leq 2^y$.
     (e) $\mathbb{R} \times \mathbb{R}$, $x \, S \, y$ iff $y = x - 1$.
   ⋆ **(f)** $A = \{1, 2, 3, 4\}$, R as given in the digraph:

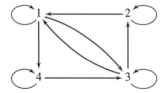

(g) $A = \{1, 2, 3, 4\}$, R as given in the digraph:

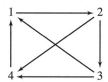

2. Let $A = \{a, b, c\}$. Give an example of a relation on A that is
    (a)   antisymmetric and symmetric.
    (b)   antisymmetric, reflexive on A, and not symmetric.
    (c)   antisymmetric, not reflexive on A, and not symmetric.
    (d)   symmetric and not antisymmetric.
    (e)   not symmetric and not antisymmetric.

— 3. Let R be an antisymmetric relation on the set A and $x, y \in A$.
    (a)   Prove that if $x R y$ and $x \neq y$, then $y \not{R} x$.
    (b)   Prove that if R is symmetric and $\text{Dom}(R) = A$, then $R = I_A$.

— 4. **(a)**   Give an example of a relation R on a set A that is antisymmetric and such that $x R x$ for some, but not all, x in A.
    (b)   Give an example of a relation S on the set $A = \{a, b, c, d\}$ such that S is transitive, antisymmetric, and **irreflexive** (that is, $x R x$ is false for all x in A).

5. Show that the relation R on $\mathbb{N}$ given by $a R b$ iff $b = 2^k a$ for some integer $k \geq 0$ is a partial ordering.

6. Define the relation R on $\mathbb{R} \times \mathbb{R}$ by $(a, b) R (x, y)$ iff $a \leq x$ and $b \leq y$. Prove that R is a partial ordering for $\mathbb{R} \times \mathbb{R}$.

7. Define the relation R on $\mathbb{C}$ by $(a + bi) R (c + di)$ iff $a^2 + b^2 \leq c^2 + d^2$. Is R a partial order for $\mathbb{C}$? Justify your answer.

— 8. Let A be a partially ordered set, called "the alphabet." Let W be the set of all "words" of length two—that is, all permutations of two letters of the alphabet. Define the relation $\leq$ on W as follows: for $x_1 x_2 \in W$ and $y_1 y_2 \in W$, $x_1 x_2 \leq y_1 y_2$ iff (i) $x_1 < y_1$ or (ii) $x_1 = y_1$ and $x_2 \leq y_2$. Prove that $\leq$ is a partial ordering for W (called the **lexicographic ordering**, as in a dictionary).

9. Draw the Hasse diagram for the poset $\mathscr{P}(A)$ with the set inclusion relation, where $A = \{a, b, c, d\}$.

10. For each Hasse diagram, list all pairs of elements in the relation on the indicated set.
    ★   **(a)** $A = \{a, b, c\}$ **(b)** $A = \{a, b, c, d\}$ **(c)** $A = \{a, b, c, d\}$

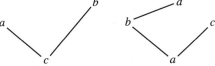

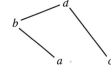

11. Use your own judgment about which tasks should precede others to draw a Hasse diagram for the partial order among the tasks for each of the following projects.

⋆ (a) To make his special stew, Fubini must perform 9 tasks:

t_1: wash the vegetables
t_2: cut up the vegetables
t_3: put vegetables in cooking pot
t_4: cut up the meat
t_5: brown the meat in a skillet
t_6: add seasoning to the skillet
t_7: add flour to the skillet
t_8: put the skillet ingredients in the pot
t_9: cook the stew for 30 minutes

(b) To back a car out of the garage, Kim must perform 11 tasks:

t_1: put the key in the ignition
t_2: step on the gas
t_3: check to see if the driveway is clear
t_4: start the car
t_5: adjust the mirror
t_6: open the garage door
t_7: fasten the seat belt
t_8: adjust the position of the driver's seat
t_9: get in the car
t_{10}: put the car in reverse gear
t_{11}: step on the brake

— 12. Let A be a nonempty set and let $\mathcal{P}(A)$ be partially ordered by set inclusion. Show that

⋆ (a) if $B \in \mathcal{P}(A)$ and $x \in B$, then $B - \{x\}$ is an immediate predecessor of B.

(b) if $B \in \mathcal{P}(A)$ and $x \notin B$, then B is an immediate predecessor of $B \cup \{x\}$.

13. Let R be the rectangle shown here, including the edges. Let H be the set of all rectangles whose sides have positive length, are parallel to the sides of R, and lie within R. H is partially ordered by set inclusion.

(a) Does every subset of H have an upper bound? a least upper bound?

⋆ (b) Does every subset of H have a largest element?

⋆ (c) Does every subset of H have a lower bound?

(d) Does every subset of H have a smallest element?

— 14. Let A be a set and $\subseteq$ be the ordering for $\mathcal{P}(A)$.

☆ (a) Let C and D be subsets of A. Prove that the least upper bound of $\{C, D\}$ is $C \cup D$ and the greatest lower bound of $\{C, D\}$ is $C \cap D$.

☆ (b) Let $\mathcal{P}$ be a family of subsets of A. Prove that the least upper bound of $\mathcal{P}$ is $\bigcup_{B \in \mathcal{P}} B$ and the greatest lower bound of $\mathcal{P}$ is $\bigcap_{B \in \mathcal{P}} B$.

15. Which are linear orders on $\mathbb{N}$? Prove your answers.

(a) T, where $m \, T \, n$ iff $m < 2n$

(b) V, where $m \, V \, n$ iff

m is odd and n is even, or

m and n are even and $m \leq n$, or

m and n are odd and $m \leq n$

(c) $S = \{(m, n): m, n \in \mathbb{N}, m \leq n$ and $m \neq 5\} \cup \{(m, 5): m \in \mathbb{N}\}$

(d) $T = \{(m, n): m, n \in \mathbb{N}, m \leq n$ and $n \neq 5\} \cup \{(5, m): m \in \mathbb{N}\}$

16. Prove that the relation V in Exercise 15(b) is a well ordering.

17. In determining whether a given relation is a well ordering, it is not necessary to verify all the conditions for a linear order as well as the additional condition for a well ordering:

− **(a)** Prove that a partial order R on a set A is a well ordering iff every nonempty subset of A has a smallest element.

(b) Prove that a relation R on a set A is a well ordering iff every nonempty subset B of A contains a unique element that is R-related to every element of B.

18. Prove that every subset of a well-ordered set is well ordered.

− **19.** *This exercise provides the steps necessary to prove that every partial ordering is in a sense the same as the set inclusion relation on a collection of subsets of a set.* Let A be a set with a partial order R. For each $a \in A$, let $S_a = \{x \in A: x \, R \, a\}$. Let $\mathscr{F} = \{S_a: a \in A\}$. Then $\mathscr{F}$ is a subset of $\mathscr{P}(A)$ and thus may be partially ordered by $\subseteq$.

(a) Show that if $a \, R \, b$, then $S_a \subseteq S_b$.

(b) Show that if $S_a \subseteq S_b$, then $a \, R \, b$.

(c) Show that for every $b \in A$, an immediate predecessor of b in A corresponds to an immediate predecessor of S_b in $\mathscr{F}$.

(d) Show that if $B \subseteq A$ and x is the least upper bound for B, then S_x is the least upper bound for $\{S_b: b \in B\}$.

Proofs to Grade − **20.** Assign a grade of A (correct), C (partially correct), or F (failure) to each. Justify assignments of grades other than A.

(a) **Claim.** Let A be a set with a partial order R. If $C \subseteq B \subseteq A$ and $\sup(C)$ and $\sup(B)$ exist, then $\sup(C) \leq \sup(B)$.

"***Proof.***" $\sup(B)$ is an upper bound for B. Therefore, $\sup(B)$ is an upper bound for C. Thus $\sup(C) \leq \sup(B)$. ∎

⋆ **(b)** **Claim.** Let A be a set with a partial order R. If $B \subseteq A$, u is an upper bound for B, and $u \in B$, then $\sup(B)$ exists and $u = \sup(B)$.

"***Proof.***" Since $u \in B$, $u \leq \sup(B)$. Since u is an upper bound, $\sup(B) \leq u$. Thus $u = \sup(B)$. ∎

(c) **Claim.** For A, $B \subseteq \mathbb{R}$ with the usual $\leq$ ordering, $\sup(A \cup B) = \sup(A) + \sup(B)$.

"***Proof.***" If $x \in A \cup B$, then $x \in A$ or $x \in B$. Therefore $x < \sup(A)$ or $x < \sup(B)$. Thus $x \leq \sup(A) + \sup(B)$, for all x in $A \cup B$. Therefore $\sup(A \cup B) \leq \sup(A) + \sup(B)$. Also $A \subseteq A \cup B$ and $B \subseteq A \cup B$, so by part (a), $\sup(A) \leq \sup(A \cup B)$ and $\sup(B) \leq \sup(A \cup B)$. Therefore $\sup(A) + \sup(B) \leq \sup(A \cup B)$. Thus $\sup(A) + \sup(B) = \sup(A \cup B)$. ∎

3.5 Graphs

In Section 3.1 we used a digraph—a collection of vertices and directed edges—to represent a relation on a set. In this section we present a similar, but different, method to represent some relations. The representation is called simply a *graph*.

Graph theory is a significant branch of mathematics with applications to many fields, such as computer science, linguistics, and chemistry. There is great variation in the terminology about graphs and types of graphs, in part because of all the diverse areas of applications.

Like a digraph of a relation R on a set A, a graph will have a vertex for each element of A. Vertices may be connected by edges, but unlike digraphs, the edges are not directed.

DEFINITIONS A **graph** G is a pair (V, E), where V is a nonempty set and E is a set of unordered pairs of distinct elements of V.

An element of V is called a **vertex** and an element of E is called an **edge**. An edge between vertices u and v is written uv (or vu) rather than as the set $\{u, v\}$.

A graph as defined above is also called a simple graph because the definition allows at most one edge between two vertices and does not allow loops at vertices. A more general definition allows multiple edges and loops.

We begin with an example representing conversations at a party. The five people (vertices) at the party are Doc, Grumpy, Sneezy, Dopey, and Happy. Rather than listing ordered pairs in the relation S on this set of people, where $x \, S \, y$ iff x had a conversation at the party with y, we describe the relation with the graph in Figure 3.5.1.

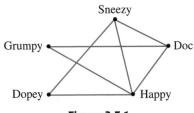

Figure 3.5.1

This graph has 7 edges; an edge connects vertices x and y exactly when person x has had a conversation with person y. It can be seen from the graph that Doc spoke with each of the others except Dopey and that Grumpy had a conversation only with Doc and Happy. The graph does not show anything about where the party-goers stood, how long they talked, or whether they had more than one conversation.

Other graphs could be used to represent the relation S of having had a conversation at the party. For example, both of the two graphs in Figure 3.5.2 convey exactly the same information as the graph in Figure 3.5.1, because these graphs have the same vertices and the same edges.

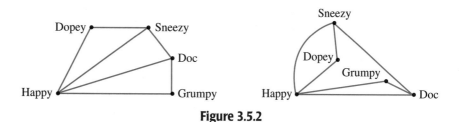

Figure 3.5.2

We call two graphs (V_1, E_1) and (V_2, E_2) **isomorphic** iff their vertices are the same except for renaming and whenever there is an edge in E_1 joining two vertices in V_1 there is a corresponding edge in E_2 that joins the corresponding vertices in V_2.

Example.   Each of the three graphs in Figure 3.5.3 is isomorphic to the other two. For the first two graphs, the vertices A, B, C, and D correspond, respectively to α, β, γ, and δ. The edge AB corresponds to the edge $\alpha\beta$, the edge BD corresponds to the edge $\beta\delta$, and so on.

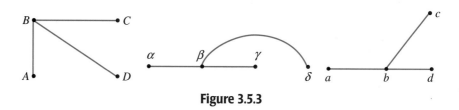

Figure 3.5.3

DEFINITIONS   Let $G = (V, E)$ be a graph. The **order** of the graph G is the number of vertices. The **size** of the graph G is the number of edges.

Vertices u and v are **adjacent** iff $uv \in E$; the edge uv is said to be **incident** with u and with v.

The **degree** of a vertex u is the number of edges incident with u.

A vertex is **isolated** iff it has degree zero.

The order of our conversation graph is 5 and the size is 7. Doc has degree 3, meaning he held conversations with three other people. Dopey and Sneezy are adjacent, whereas Dopey and Doc are not.

The definition of a graph $G = (V, E)$ allows for E to be empty. Such a graph is a **null graph**. Figure 3.5.4 is a null graph of order 5. Every vertex of a null graph is isolated.

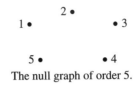

The null graph of order 5.

Figure 3.5.4

A graph in which every pair of distinct vertices are adjacent is called a **complete graph**. If $G = (V, E)$ has order n and is complete, then every vertex has degree $n - 1$. The complete graph of order n is denoted K_n. Figure 3.5.5 shows the complete graphs of order 5 and less.

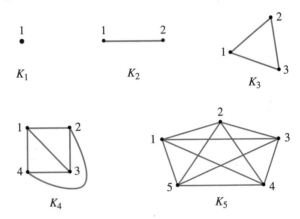

Figure 3.5.5

If you look again at Figure 3.5.1, you will see that the degrees of the vertices are 3, 3, 4, 2, and 2. The sum of the degrees is 14, which is even. The explanation for the fact that the sum of the degrees of the vertices of a graph is even is the same as the explanation for the fact that if a group of people shake hands, the total number of hands shaken must be even.

Theorem 3.5.1   (a)   **The Handshaking Lemma.** For every graph G, the sum of the degrees of the vertices is twice the number of edges. Thus the sum of the degrees is even.

(b)   For every graph G, the number of vertices of G having odd degree is even.

Proof.

(a)   Each edge is incident with two vertices. Thus the sum of the degrees of the vertices is exactly twice the number of edges. Therefore, the sum is even.

(b)   Obviously, the sum of the degrees of the vertices that have even degree is an even number. If there were an odd number of vertices with odd degree, then the sum of all the degrees would be odd. By the Handshaking Lemma, this is impossible.   ∎

For the complete graph K_n, each vertex has degree $n - 1$. Therefore, the sum of the degrees of the vertices is $n(n - 1)$. Since this number is twice the number of edges, the number of edges in K_n is $\dfrac{n(n - 1)}{2}$.

The graph (V', E') is a **subgraph** of the graph (V, E) if and only if $V' \subseteq V$ and $E' \subseteq E$. Thus we can form a subgraph of (V, E) by selecting some of the vertices from V and some of the edges from E, but it is understood that an edge cannot be in E' unless both its vertices are in V'. Figure 3.5.6 shows three subgraphs of our graph of conversations.

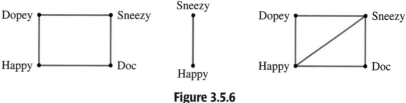

Figure 3.5.6

DEFINITIONS A **walk*** in a graph G is a finite sequence of vertices $v_0, v_1, v_2, v_3, \ldots, v_m$, where each $v_i v_{i+1}$ is an edge in G. The walk is said to **traverse** the vertices in the sequence, starting with the **initial vertex** v_0 and ending with the **terminal vertex** v_m. The **length** of the walk is m, the number of edges. If $v_0 = v_m$, the walk is **closed**.

A **path** in G is a walk where all the vertices, except for possibly the initial and terminal vertices, are distinct.

Some sequences in the graph of order 6 of Figure 3.5.7 are

$$p_1: 6, 3, 5$$
$$p_2: 1, 2, 5, 6, 4, 3$$
$$p_3: 3, 2, 5, 2, 3$$
$$p_4: 2, 5, 3, 4$$
$$p_5: 1, 2, 5, 4, 3, 6, 1.$$

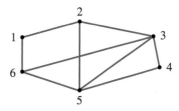

Figure 3.5.7

* What we call a *walk* is called by some a *path*, an *edge-sequence*, a *route*, a *trail*, or a *chain*.

The sequence p_1 is a walk because there is an edge from 6 to 3 and from 3 to 5; the initial vertex of p_1 is 6 and the terminal vertex 5. The sequence p_2 is not a walk because the graph has no edge from 6 to 4. The sequence p_3 is a walk but not a path because of the repeated vertex 2. The walk p_3 is closed. Walk p_4 is a path of length 3 and is not closed. Walk p_5 is a closed path of length 6 with initial and terminal vertices 1.

The graph in Figure 3.5.7 might represent airline routes among 6 cities. It's natural to think of planning a trip that would traverse certain vertices (cities) by taking a sequence of edges (flights). If we think of the graph in this way, then a salesperson's trip would be a walk through certain cities. Because the salesperson does not want to visit cities twice, the route should be a path. In addition, the path should be closed so that the trip starts and ends in the city of the salesperson's home office.

Since the vertices of a path p in a graph are distinct, the length of p is limited by the order of the graph.

Theorem 3.5.2   Let G be a graph of order n.

(a)   If there is a walk originating at v and terminating at u in a graph G, then there is a path from v to u.

(b)   The length of a path in G that is not closed is at most $n - 1$. The length of a closed path is at most n.

Proof.

(a)   Suppose $v, v_1, v_2, \ldots, u$ is a walk from v to u, with $v \neq u$. If the walk is not a path, then some vertex appears twice in the sequence. Let x be the first such vertex. Then the walk contains at least one closed walk of the form $x, v_j, v_{j+1}, \ldots, v_m, x$.

Delete the vertices $v_j, v_{j+1}, \ldots, v_m, x$ from the sequence $v, v_1, v_2, \ldots, u$. If the result is a path, we are done. Otherwise another such repeated vertex can be found and the deletion process repeated. Since we delete at least one vertex each time and there are finitely many vertices in $v, v_1, v_2, \ldots, u$, eventually no more vertices can be deleted, so this process must result in a path from v to u.

In the case $v = u$, the same process is applied to delete all repetitions of vertices except the initial and terminal vertex. The result is a closed path.

(b)   Consider a path in the graph G, where G has n vertices. If the path has length t, then there are $t + 1$ vertices traversed by the path. In a path that is not closed, all vertices are distinct so there are at most n vertices traversed. Thus $t + 1 \leq n$ and the path has length at most $n - 1$. In a closed path the initial and terminal vertices are the same and there is no other repeated vertex. Thus if the closed path has length t, there are t distinct vertices. Therefore, if G has n vertices, the length of a closed path is at most n. ∎

DEFINITIONS   Let G be a graph and u be a vertex of G. The vertex v is **reachable** (or **accessible**) from u if and only if there is a path from u to v. The number of edges in a path of minimum length from u to v is called the **distance from u to v**, denoted $d(u, v)$. For any vertex u, we say u is reachable from itself and $d(u, u) = 0$.

Example.   Let G be the graph with vertex set $V = \{a, b, c, q, e, f, g, h, i, j, k\}$ shown in Figure 3.5.8. The vertex q is reachable from the vertices q, c, g, h, j, and k and from no other vertices in G. The distances to vertex q are $d(q, q) = 0$, $d(q, c) = 1$, $d(q, g) = 2$, $d(q, h) = 1$, $d(q, j) = 3$, and $d(q, k) = 2$. Likewise, $d(a, f) = 1$, $d(e, i) = 1$, and $d(q, b)$ is not defined.

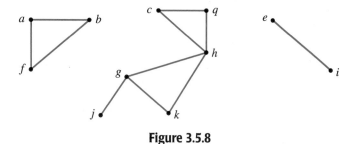

Figure 3.5.8

DEFINITION   Let G be a graph. If u is a vertex of G, the **component containing u** is the subgraph $C(u)$ whose vertex set consists of all vertices reachable from u and whose edge set is all the edges of G that are incident with those vertices.

The graph in Figure 3.5.8 has three components. The component $C(a) = C(b) = C(f)$ has vertex set $\{a, b, f\}$ and 3 edges. The vertices c, g, h, j, k, and q all have the same component that has 7 edges, and $C(e) = C(i)$ has vertex set $\{e, i\}$ and one edge.

If we think of the vertices of the graph G in Figure 3.5.8 as representing cities and the edges representing roads, the figure might represent cities and roads on three islands. The vertices of G are partitioned into three components ("islands"). Since every partition of a set is associated with an equivalence relation, it is not surprising that reachability determines an equivalence relation.

Theorem 3.5.3   Let G be a graph with vertex set V and let R be the relation on V defined by $u\,R\,v$ iff v is reachable from u. Then R is an equivalence relation on V and the equivalence classes for R are the vertex sets of the components.

Proof.   By our definition, every vertex is reachable from itself. Thus, R is reflexive on V.

If v is reachable from u, then there is a path from u to v. By reversing the order of the edges of this path, we have a path from v to u. Thus, u is reachable from v. Thus, R is symmetric.

Let v be reachable from u and w reachable from v. By following the path from u to v then the path from v to w, we have a *walk* from u to w. By Theorem 3.5.2(a) there is a path from u to w. Thus, the relation R is transitive.

Therefore, R is an equivalence relation on V. For each $u \in V$, the equivalence class determined by u is all the vertices reachable from u, which is precisely the vertex set of $C(u)$.   ■

DEFINITIONS   A graph G is **connected** if and only if every vertex is reachable from every other vertex. G is **disconnected** iff G is not connected.

When a graph is pictured as in Figure 3.5.8 it is easy to determine the components of G and that G is disconnected. The null graph is disconnected as long as the graph has at least two vertices. The complete graph K_n is connected for every $n \in \mathbb{N}$.

Choose any of the three components of the graph G shown in Figure 3.5.8, say the component $C(a)$ with vertices a, b, and f. Notice that if we were to take any subgraph of G that included the vertices of the component and at least one more vertex, that subgraph would not be connected. We say the component is a maximally connected subgraph of G.

DEFINITION   Let $G = (V, E)$ be a graph and $G' = (V', E')$ be a subgraph of G. Then G' is a **maximally connected subgraph** of G iff

(i) G' is connected and
(ii)   for every subgraph G'' of G whose vertex set properly includes V', G'' is disconnected.

The three components of the graph G in Figure 3.5.8 are the only maximally connected subgraphs. For example, the subgraph $H = (\{c, g, h\}, \{ch, gh\})$ is connected but not maximally connected, because there exist subgraphs, such as the component $C(h)$, that are connected and have a vertex set that properly includes $\{c, g, h\}$.

The properties of components are collected in the next theorem.

Theorem 3.5.4   For each vertex v in a graph G, let $C(v)$ be the component of v in G. Then

(a) $C(v) = C(w)$ iff w is reachable from v.
(b) $C(v) \neq C(w)$ iff no vertex is in both $C(v)$ and $C(w)$.

(c) for each v, $C(v)$ is connected.

(d) for each v, $C(v)$ is a maximally connected subgraph of G.

Proof. Parts (a) and (b) follow from the fact that the vertex set of each component $C(v)$ is the equivalence class of v under the reachability relation. See Exercise 14.

(c) Let x and y be any two vertices in $C(v)$. Then both x and y are reachable in G from v, so y is reachable from v, and v is reachable from x. Since all the edges needed to reach x and y from v are also in $C(v)$, y is reachable from x in $C(v)$. Therefore, $C(v)$ is connected.

(d) We must show that if $G' = (V', E')$ is a subgraph of G, and V' properly contains the vertices of any component $C(v)$, then G' is disconnected. Suppose u is a vertex in V' that is not in $C(v)$. Then u is not reachable from v in G, so there can be no path from v to u in G'. Thus G' is disconnected. ∎

Theorem 3.5.4 tells us that every vertex belongs to exactly one component, and that the collection of components is pairwise disjoint. Further, components are maximally connected subgraphs. It follows that every isolated point forms a component, and that a graph is connected iff it has exactly one component.

Exercises 3.5

1. List the degrees of the vertices of each of these graphs. Verify both parts of Theorem 3.5.1 in each case.

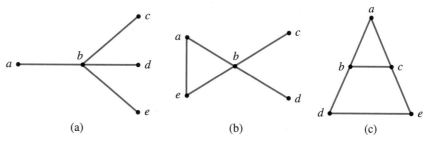

(a) (b) (c)

2. If possible, give an example of a graph with order 6 such that
   (a) the vertices have degrees 1, 1, 1, 1, 1, 5.
   (b) the vertices have degrees 1, 1, 1, 1, 1, 1.
   (c) the vertices have degrees 2, 2, 2, 2, 2, 2.
   ★ **(d)** the vertices have degrees 1, 2, 2, 2, 3, 3.
   (e) exactly two vertices have even degree.
   (f) exactly two vertices have odd degree.

3. If possible, give an example of a graph
   (a) with order 6 and size 6.
   (b) with order 4 and size 6.
   ★ **(c)** with order 3 and size 4.
   (d) with order 6 and size 3.

4. For the graph at the right, find all subgraphs
    (a) with two vertices.
    (b) with three vertices.

5. The **complement** of a graph $G = (V, E)$ is the graph with vertex set V in which two vertices are adjacent iff they are not adjacent in G. Give the complements of the graphs below.

      (a)   (b)   (c)   (d)

6. For the graph shown, give
    (a) all paths of length 4 and initial vertex g. (There are eight.)
    (b) all cycles of length 6 and initial vertex c. (There are eight.)
    (c) a path of length 7.
    (d) a walk of length 4 that is not a path.

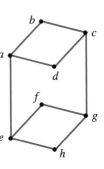

7. Give an example of a graph with 6 vertices having degrees 1, 1, 2, 2, 2, 2 that is
    (a) connected.
    (b) disconnected.

8. Give an example of a graph with 6 vertices having
    (a) one component.
    (b) two components.
    (c) three components.
    (d) six components.

☆ 9. Prove that in every graph of order $n \geq 2$ there are two vertices with the same degree.

10. Give an example of a graph with order 6 such that
    (a) two vertices u and v have distance 5.
    (b) for any two vertices u and v, $d(u, v) \leq 2$.

11. Verify these properties for the distance between vertices in a connected graph:
    (a) $d(u, v) \geq 0$.
    (b) $d(u, v) = 0$ iff $u = v$.
    (c) $d(u, w) \leq d(u, v) + d(v, w)$.

12. Let u and r be vertices in a graph such that $d(u, r) \geq 2$. Show that there exists a vertex w such that $d(u, w) + d(w, r) = d(u, r)$.

13. An edge e of a connected graph is called a **bridge** iff, when e is removed from the edge set, the resulting subgraph is disconnected. For example, the edges in the graph below that are bridges are ab, bc, and fg. Give an example of a connected graph of order 7
    (a) with no bridges. (b) with one bridge. (c) with 6 bridges.

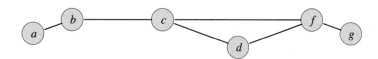

14. Prove parts (a) and (b) of Theorem 3.5.4. Keep in mind that $C(v)$ and $C(w)$ are graphs; to be the same they must have the same vertices and the same edges.

Proofs to Grade

15. Assign a grade of A (correct), C (partially correct), or F (failure) to each. Justify assignments of grades other than A.
    (a) **Claim.** If v and w are vertices in a graph such that $d(u, v) = 3$ and $d(u, w) = 4$, then $d(v, w) \leq 7$.
    "Proof." Suppose $d(u, v) = 3$ and $d(u, w) = 4$. Then there is a path u, x_1, x_2, v from u to v with length 3 and a path u, y_1, y_2, y_3, w from u to w with length 4. Then $v, x_2, x_1, u, y_1, y_2, y_3, w$ is a path from v to w with length 7. The distance from v to w is the length of the shortest path from v to w and there is a path of length 7, so $d(v, w) \leq 7$. ∎
    (b) **Claim.** Every connected graph G of order n has a closed path of order n.
    "Proof." Let G be connected graph of order n with vertices $x_1, x_2, x_3, \ldots, x_n$. Since G is connected, each of the vertices x_i is reachable from x_{i-1} and x_1 is reachable from x_n. Thus connecting these paths there is a path from x_1 to x_2 to $x_3, \ldots,$ to x_n and back to x_1. By Theorem 3.5.2(b) the length of any closed path, including this one, is at most n. ∎

CHAPTER 4

Functions

The notion of a function is familiar to you from previous study in algebra, trigonometry, and calculus. The *Preface to the Student* reviews the concept of a function as a rule of correspondence and the basic properties of and notations for functions. In this chapter, where we view functions as single-valued relations, our goals are (1) to develop a deeper understanding of methods of constructing functions and the properties of being one-to-one and onto, and (2) to write proofs establishing that a relation is a function, or has (or does not have) these properties. The techniques and results developed here are used throughout the remainder of the text.

4.1   Functions as Relations

The concept of a function is very old, but the word *function* was not explicitly used until 1694 by G. W. Leibnitz.* It is only relatively recently that it has become standard practice to treat a function as we define it below—as a relation with special properties. This is possible because the rule that makes an element in one set correspond to an element from a second set may be viewed as forming a collection of ordered pairs.

---

* Gottfried Wilhelm Leibnitz (1646–1716) was a versatile German scholar, lawyer, and diplomat who made major contributions to mathematics, philosophy, logic, technology, and physics. Although they worked independently, both he and Isaac Newton developed calculus. Leibnitz devised the now standard $\frac{dy}{dx}$ and $\int f(x)dx$ notations, referring to dy and dx as "infinitesimals." His development of the binary number system is the basis of all modern computing devices.

> **DEFINITIONS**   A **function** (or **mapping**) **from A to B** is a relation f from A to B such that
>
> **(i)**   the domain of f is A, and
> **(ii)**   if $(x, y) \in f$ and $(x, z) \in f$, then $y = z$.
>
> We write $f: A \to B$ and this is read "f is a function from A to B," or "f maps A to B." The set B is called the **codomain of f**. In the case where $B = A$, we say f is a **function on A**.

No restriction is placed on the sets A and B. They may be sets of numbers, sets of ordered pairs, or even sets of functions.

The conditions for f to be a function from A to B have a lot to say about the first coordinates of the ordered pairs in f: Condition (i) ensures that every element of A is a first coordinate in f, and condition (ii) says that each first coordinate appears in just one ordered pair in f. There are no corresponding requirements for second coordinates: It may happen that some elements of B are not used as second coordinates, or that some elements of B are used as second coordinates in two or more different ordered pairs.

Examples.   Let $A = \{1, 2, 3\}$ and $B = \{4, 5, 6\}$. All of the sets

$$R_1 = \{(1, 4), (2, 5), (3, 6), (2, 6)\}$$
$$R_2 = \{(1, 4), (2, 6), (3, 5)\}$$
$$R_3 = \{(1, 5), (2, 5), (3, 4)\}$$
$$R_4 = \{(1, 4), (3, 6)\}$$

are relations from A to B. Since $(2, 5)$ and $(2, 6)$ are distinct ordered pairs with the same first coordinate, R_1 is not a function from A to B. Both R_2 and R_3 satisfy conditions (i) and (ii) and are functions from A to B. The domain of R_4 is the set $\{1, 3\}$, which is not equal to A, so R_4 is not a function from A to B. However, R_4 satisfies condition (ii), so it is correct to say that R_4 is a function from $\{1, 3\}$ to B.

The codomain B for a function $f: A \to B$ is the set of all objects available for use as second coordinates (images). As with any relation, the **range** of f is

$$\text{Rng}(f) = \{v \in B : \text{there is } u \in A \text{ such that } (u, v) \in f\},$$

which is the set of objects that are actually used as second coordinates. The range of f is always a subset of the codomain. In the examples above, the range of R_2 is the same as its codomain, but the range of R_3 is $\{4, 5\} \neq B$. We say that R_3 is a function from A to B, but we could also say that R_3 is also a function from A to $\{4, 5\}$, and R_3 is a function from A to $\{\sqrt{3}, \pi, 4, 5, 8\}$ or to any other set that contains both 4 and 5. A function has only one domain and one range, but many possible codomains, because any set that includes the range may be considered to be a codomain.

> **DEFINITIONS** Let $f: A \to B$. We write $y = f(x)$ when $(x, y) \in f$. We say that y is the **value of f at x** (or the **image of f at x**) and that x is a **pre-image of y under f**.

Suppose $f: A \to B$. It is condition (ii) of the definition that makes f a *single-valued correspondence*, which means that for every $x \in A$ there corresponds a unique (single) value in B. This condition allows us to refer to *the* image of x, rather than *an* image and to write the familiar $f(x)$ for the image of x.

Example. Let $F = \{(x, y) \in \mathbb{Z} \times \mathbb{Z}: y = x^2\}$. Then F is a mapping with domain $\mathbb{Z}$ and $F(x) = x^2$. The image of 4 is 16, $F(-3) = 9$, $F(t + 2) = (t + 2)^2$, and the value of F at 10 is 100. Both 5 and -5 are pre-images of 25. Since 7 has no pre-image in $\mathbb{Z}$, 7 is not in the range of F. The range is $\text{Rng}(F) = \{0, 1, 4, 9, 16 \ldots\}$.

Example. Prove that $g = \{(x, y) \in \mathbb{R} \times \mathbb{R}: x = y^3\}$ is a function from $\mathbb{R}$ to $\mathbb{R}$.

Proof. First, observe that g is a relation from $\mathbb{R}$ to $\mathbb{R}$.

(i) $\langle$*Show that* $\text{Dom}(g) = \mathbb{R}.\rangle$ Suppose $x \in \text{Dom}(g)$. By definition of g, the first coordinates of elements of g are real numbers, so $x \in \mathbb{R}$.
   Let $x \in \mathbb{R}$. Then $\sqrt[3]{x}$ is a real number and $x = (\sqrt[3]{x})^3$. Thus there exists $y \in \mathbb{R}$ (namely $y = \sqrt[3]{x}$) for which $(x, y) \in g$. Thus $x \in \text{Dom}(g)$.

(ii) $\langle$*Show that f is single-valued.*$\rangle$ Suppose $(x, u) \in g$ and $(x, v) \in g$. Then $x = u^3$ and $x = v^3$. Therefore $u^3 = v^3$, from which we conclude that $u = v$.

By parts (i) and (ii), g is a function from $\mathbb{R}$ to $\mathbb{R}$. ∎

To prove that a given relation r from A to B is *not* a function from A to B, we may either show that (i) some element of A is not a first coordinate (that is, some element of A is not in $\text{Dom}(r)$), or (ii) find some element x of A that is a first coordinate with two different second coordinates—thus showing the existence of some $(x, y) \in r$ and $(x, z) \in r$ with $y \neq z$.

The relation $H = \{(x, y) \in \mathbb{R} \times \mathbb{R}: x^2 + y^2 = 25\}$ with domain $[-5, 5]$ is not a function from $[-5, 5]$ to $\mathbb{R}$ because, for example, $(3, 4) \in H$ and $(3, -4) \in H$. (See the graph in Figure 4.1.1 on the next page). Since we have the graph to view, an easy way to tell that H is not a function is to apply the *Vertical Line Test* for a relation r on $\mathbb{R}$:

r is a function iff no vertical line intersects the graph of r more than once.

Visualizing the vertical line $x = 3$ helps us discover that H is not a function because $(3, 4)$ and $(3, -4)$ are both in H.

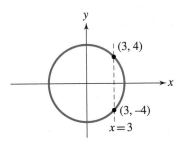

Figure 4.1.1

The graph of $g = \{(x, y) \in \mathbb{R} \times \mathbb{R} \colon x = y^3\}$ in the example above appears in Figure 4.1.2. If we apply the Vertical Line Test to the graph of g, the relation g appears to be a function. However, this observation does *not* constitute a proof because the graph represents only a portion of the relation g and our representation might not reveal small vertical segments of the graph.

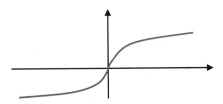

Figure 4.1.2

Note that arrow diagrams for relations with small finite domains may be used to determine whether the relation is a function. The diagram in Figure 4.1.3(a) represents a function with domain $\{a, b, c\}$ but the diagram in Figure 4.1.3(b) does not.

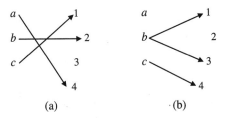

(a)   (b)

Figure 4.1.3

Functions whose domains and codomains are subsets of $\mathbb{R}$ are often referred to as "real functions." The words "f is defined on the interval I," mean that $I \subseteq \mathrm{Dom}(f)$. The domain of a real function is usually understood to be the largest possible subset of $\mathbb{R}$, so that, for example, the domain of the function $f(x) = \dfrac{1}{\sqrt{x}}$ is $(0, \infty)$.

Because functions are sets of ordered pairs, we may say that functions f and g are **equal** if and only $f \subseteq g$ and $g \subseteq f$. For example, let $f: \{-2, 3\} \rightarrow \{4, 9\}$ be given by

$$f(x) = x^2$$

and $g: \{-2, 3\} \rightarrow [0, \infty)$ be given by

$$g(x) = x + 6.$$

Although f and g have different rules and different codomains, the two functions have the same domain $\{-2, 3\}$ and both are equal to $\{(-2, 4), (3, 9)\}$. Therefore $f = g$. A very natural and useful way to express the idea that two functions are equal is to assert that they have the same domain (so they act on the same objects) and that for each object in the common domain the function images are the same.

Theorem 4.1.1

Two functions f and g are equal iff

(i) $\mathrm{Dom}(f) = \mathrm{Dom}(g)$ and
(ii) for all $x \in \mathrm{Dom}(f)$, $f(x) = g(x)$.

Proof. ⟨*We prove that conditions* (i) *and* (ii) *hold when* $f = g$. *The converse is left as Exercise 14.*⟩ Assume $f = g$.

(i) Suppose $x \in \mathrm{Dom}(f)$. Then $(x, y) \in f$ for some y and, since $f = g$, we have $(x, y) \in g$. Therefore $x \in \mathrm{Dom}(g)$. This shows $\mathrm{Dom}(f) \subseteq \mathrm{Dom}(g)$. Similar reasoning shows that $\mathrm{Dom}(g) \subseteq \mathrm{Dom}(f)$. Therefore, $\mathrm{Dom}(g) = \mathrm{Dom}(f)$.
(ii) Suppose $x \in \mathrm{Dom}(f)$. Then for some y, $(x, y) \in f$. Since $f = g$, $(x, y) \in g$. Therefore, $f(x) = y = g(x)$. ■

Examples. Suppose f, g, and h are real functions given by $f(x) = \dfrac{x}{x}$, $g(x) = 1$, and $h(x) = \dfrac{|x|}{x}$. Then $f \neq g$ because they have different domains: The number 0 is in $\mathrm{Dom}(g)$ but not in $\mathrm{Dom}(f)$. The functions f and h are different because they have different function values. For instance, $f(-2) = 1$ and $h(-2) = -1$.

The remainder of this section describes several types of functions, some of which will be familiar to you.

Let A be any set. The identity relation I_A is the **identity function** $I_A: A \rightarrow A$ given by $I_A(x) = x$. If A is a subset of B, we define the **inclusion function** $i: A \rightarrow A$ by $i(x) = x$ for all $x \in A$. Since they both have domain A and $I_A(x) = i(x) = x$ for all $x \in A$, $I_A = i$ by Theorem 4.1.1. There is no difference between these functions, but it is customary to write I_A when we think of the function from A to A and i when we think of the function from A to B.

Assume that a universe U has been specified, and that $A \subseteq U$. Define $X_A: U \rightarrow \{0, 1\}$ by

$$X_A(x) = \begin{cases} 1 & \text{if } x \in A \\ 0 & \text{if } x \in U - A \end{cases}.$$

Then $\chi_A(x)$ is called the **characteristic function of A.** For example, if $A = [1, 4)$, with the universe being the real numbers, then $\chi_A(x) = 1$ iff $1 \le x < 4$. Figure 4.1.4 is a graph of $\chi_{[1, 4)}$.

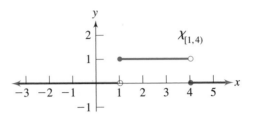

Figure 4.1.4

One variation of the characteristic function is a **step function.** Suppose $A \subseteq \mathbb{R}$, $\mathscr{C} = \{C_\delta \colon \delta \in \Delta\}$ is a partition of a set A, and for each $\delta \in \Delta$, b_δ is in the set B. Define $f \colon A \to B$ by $f(x) = b_\delta$ if $x \in C_\delta$.

As an example, let $A = [1, 5]$ with $C_1 = [1, 2]$, $C_2 = (2, 4)$, and $C_3 = [4, 5]$, and let $B = \mathbb{R}$ with $b_1 = 3$, $b_2 = 4$, $b_3 = 2$. The graph of the corresponding step function is given in Figure 4.1.5.

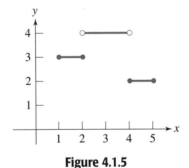

Figure 4.1.5

The **greatest integer function** is an example of a step function with domain $\mathbb{R}$ and range $\mathbb{Z}$. It assigns to each real number x the integer part of x, by which we mean the largest integer n such that $n \le x$. On graphing calculators this function is usually denoted as "int." For instance, $\mathrm{int}(5.9) = 5$, $\mathrm{int}(\sqrt{2}) = 1$, and $\mathrm{int}(-\pi) = -4$.

DEFINITION A function x with domain $\mathbb{N}$ is called an **infinite sequence,** or simply a **sequence.** The image of n is usually written as x_n instead of $x(n)$ and is called the ***n*th term of the sequence.**

For sequence x given by $x_n = \dfrac{1}{n+1}$, the 63rd term is $x_{63} = \dfrac{1}{64}$. The range of x is $\left\{\dfrac{1}{2}, \dfrac{1}{3}, \dfrac{1}{4}, \dfrac{1}{5}, \cdots\right\}$.

The terms of a sequence need not be distinct. The first ten terms of the sequence x given by $x_n = ((-1)^n + 1)n^2$ are 0, 8, 0, 32, 0, 72, 0, 128, 0, 200.

If R is an equivalence relation on the set X, then the function from X to X/R that maps each $a \in X$ to the equivalence class of a is called the **canonical map** for the relation R. Recall that in Chapter 3 we used $\bar{a}$ to denote the equivalence class of a. If f is the canonical map for the relation of congruence modulo 5 on $\mathbb{Z}$, the images of -3 and 9 are

$$f(-3) = \overline{-3} = \overline{2} = \{\ldots, -8, -3, 2, 7, 12, \ldots\},$$
$$f(9) = \overline{9} = \overline{4} = \{\ldots, -6, -1, 4, 9, 14, \ldots\}.$$

The equivalence classes under the family name relation L on the set P of all people with family names (see Section 3.1) are sets of people all having the same family name. Under the canonical map f from P to P/L, every person corresponds to his or her equivalence class. Thus f(Charlie Brown) is the set of all people with family name Brown and f(Charlie Brown) $= f$(Buster Brown). The canonical map is a natural function to consider, and it plays an essential role in the development of many mathematical structures.

Rules of correspondence between equivalence classes have interesting properties. Consider for example the classes $\overline{0}, \overline{1}, \overline{2}$, and $\overline{3}$ of $\mathbb{Z}_4$ and the rule that $\bar{x}$ in $\mathbb{Z}_4$ corresponds to the equivalence class $[2x]$ in $\mathbb{Z}_{10}$. (Note, for clarity, we use here the bar notation for the equivalence classes of $\mathbb{Z}_4$ and the bracket notation for equivalence classes in $\mathbb{Z}_{10}$.) Under this rule,

$$f(\overline{0}) = [0], \quad f(\overline{1}) = [2], \quad f(\overline{2}) = [4], \quad \text{and} \quad f(\overline{3}) = [6].$$

However, 0 and 4 are in the same class in $\mathbb{Z}_4$, so by the rule $f(\overline{4}) = [8]$. In $\mathbb{Z}_{10}$, however, $[0] \neq [8]$, so the rule assigns two different values to the same element, $\overline{0} = \overline{4}$, of $\mathbb{Z}_4$. Thus, f is not a function. In cases where an object in the domain has more than one representative (for instance, the object $\overline{0}$ can be represented by $\overline{0}, \overline{4}, \overline{-4}, \overline{8}, \overline{-8}, \ldots$) and a supposed function assigns different values that depend on the representative, we say "the function is not well defined," meaning that it is not really a function.

Exercises 4.1

1. Which of the following relations are functions? For those relations that are functions, give the domain and two sets that could be a codomain.
   ⋆ **(a)** $\{(0, \triangle), (\triangle, \square), (\square, \cap), (\cap, \cup), (\cup, 0)\}$
   (b) $\{(1, 2), (1, 3), (1, 4), (1, 5), (1, 6)\}$
   (c) $\{(1, 2), (2, 1)\}$
   (d) $\{(x, y) \in \mathbb{R} \times \mathbb{R} : x = \sin y\}$
   (e) $\{(x, y) \in \mathbb{N} \times \mathbb{N} : x \leq y\}$
   (f) $\{(x, y) \in \mathbb{Z} \times \mathbb{Z} : y^2 = x\}$
   (g) $\{(\varnothing, \{\varnothing\}), (\{\varnothing\}, \varnothing), (\varnothing, \varnothing), (\{\varnothing\}, \{\varnothing\})\}$

(h)

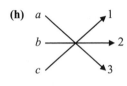

(i)

2. Give a relation r from $A = \{5, 6, 7\}$ to $B = \{3, 4, 5\}$ such that
   (a) r is not a function.
   (b) r is a function, but not a function from A to B.
   (c) r is a function from A to B, with $\text{Rng}(r) = B$.
   (d) r is a function from A to B, with $\text{Rng}(r) \neq B$.

3. Identify the domain, range, and another possible codomain for each of the following mappings.
   ★ **(a)** $\left\{ (x, y) \in \mathbb{R} \times \mathbb{R}: y = \dfrac{1}{x + 1} \right\}$
   (b) $\{(x, y) \in \mathbb{R} \times \mathbb{R}: y = x^2 + 5\}$
   (c) $\{(x, y) \in \mathbb{N} \times \mathbb{N}: y = x + 5\}$
   ★ **(d)** $\{(x, y) \in \mathbb{R} \times \mathbb{R}: y = \tan x\}$
   (e) $\{(x, y) \in \mathbb{R} \times \mathbb{R}: y = \mathcal{X}_{\mathbb{N}}(x)\}$
   (f) $\left\{ (x, y) \in \mathbb{R} \times \mathbb{R}: y = \dfrac{e^x + e^{-x}}{2} \right\}$
   (g) $\left\{ (x, y) \in \mathbb{R} \times \mathbb{R}: y = \dfrac{x^2 - 4}{x - 2} \right\}$
   (h) $\left\{ (x, y) \in \mathbb{Z} \times \mathbb{Z}: y = \dfrac{x^2 - 4}{x - 2} \right\}$

   $(x - 3)(x - 4)$

4. Assuming that the domain of each of the following functions is the largest possible subset of $\mathbb{R}$, find the domain and range of
   ★ **(a)** $f(x) = \dfrac{x^2 - 7x + 12}{x - 3}$.   **(b)** $f(x) = 2x + 5$.
   (c) $f(x) = \dfrac{1}{\sqrt{x + \pi}}$.   **(d)** $f(x) = \sqrt{5 - x}$.
   (e) $f(x) = \sqrt{5 - x} + \sqrt{x - 3}$.   **(f)** $f(x) = \sqrt{x + 2} + \sqrt{-2 - x}$.

★ **5. (a)** Let A be the set $\{1, 2, 3, 4\}$ and let R be the relation on A given by $\{(x, y): 2x + y$ is prime and not equal to $5\}$. Prove that R is a function with domain A.

(b) Let A be the set $\{1, 2, 3\}$ and let R be the relation on A given by $\{(x, y): 3x + y$ is prime$\}$. Prove that R is a function with domain A.

(c) Let R be the relation on $\mathbb{Z}$ given by $\{(x, y): x^2 + y = 2\}$. Prove that R is a function with domain $\mathbb{Z}$.

6. Show that the following relations are not functions on $\mathbb{R}$.

(a) $\{(x, y) \in \mathbb{R} \times \mathbb{R}: x^2 = y^2\}$

(b) $\{(x, y) \in \mathbb{R} \times \mathbb{R}: x^2 + y^2 = 1\}$

(c) $\{(x, y) \in \mathbb{R} \times \mathbb{R}: x = \cos y\}$

(d) $\{(x, y) \in \mathbb{R} \times \mathbb{R}: y^2 = \sqrt{x}\}$

7. Let the universe be $\mathbb{R}$ and $A = [1, 3)$. Sketch the graph of

(a) X_A. **(b)** X_{A^c}. **(c)** $X_{\{1/2\}}$. **(d)** $X_{\mathbb{N}}$.

8. Let U be the universe and $A \subseteq U$ with $A \neq \varnothing$, $A \neq U$. Let X_A be the characteristic function of A.

★ **(a)** What is $\{x \in U: X_A(x) = 1\}$? **(b)** What is $\{x \in U: X_A(x) = 0\}$?

(c) What is $\{x \in U: X_A(x) = 2\}$?

9. Give an example of a sequence x such that

★ **(a)** the range of x is the negative integers.

(b) the terms of x are alternately positive and negative.

☆ **(c)** all terms of x are distinct and between 3 and 4.

(d) the range of x has exactly 3 elements.

10. For the canonical map $f: \mathbb{Z} \to \mathbb{Z}_6$, find

★ **(a)** $f(3)$. **(b)** the image of 6.

(c) a pre-image of $\overline{3}$. **(d)** all pre-images of $\overline{1}$.

11. Which of the following are functions from the indicated domain to the indicated codomain? In each case, we represent an element of the domain as an equivalence class $\bar{x}$, and use the notation $[x]$ for equivalence classes in the codomain. For those relations that are not functions, show that the function is not well defined by naming an equivalence class in the domain that is assigned two different values.

(a) $f: \mathbb{Z}_3 \to \mathbb{Z}_6$ given by $f(\bar{x}) = [x]$

★ **(b)** $f: \mathbb{Z}_6 \to \mathbb{Z}_6$ given by $f(\bar{x}) = [x + 1]$

(c) $f: \mathbb{Z}_3 \to \mathbb{Z}_6$ given by $f(\bar{x}) = [2x]$

(d) $f: \mathbb{Z}_4 \to \mathbb{Z}_6$ given by $f(\bar{x}) = [2x + 1]$

★ **(e)** $f: \mathbb{Z}_3 \to \mathbb{Z}_4$ given by $f(\bar{x}) = [x]$

(f) $f: \mathbb{Z}_4 \to \mathbb{Z}_2$ given by $f(\bar{x}) = [3x]$

12. Explain why the functions $f(x) = \dfrac{9 - x^2}{x + 3}$ and $g(x) = 3 - x$ are not equal.

13. (a) Prove that the empty set $\varnothing$ is a function with domain $\varnothing$.

(b) Prove that if $f: A \to B$ and any one of f, A, or $\operatorname{Rng}(f)$ is empty, then all three are empty.

14. Complete the proof of Theorem 4.1.1. That is, prove that if (i) $\operatorname{Dom}(f) = \operatorname{Dom}(g)$ and (ii) for all $x \in \operatorname{Dom}(f)$, $f(x) = g(x)$, then $f = g$.

15. Let S be a relation from A to B. We define two **projection functions** $\pi_1: S \to A$ and $\pi_2: S \to B$ as follows: For all (a, b) in S, $\pi_1(a, b) = a$ and $\pi_2(a, b) = b$. In terms of S, find
    ★ **(a)** $\operatorname{Rng}(\pi_1)$.   **(b)** $\operatorname{Rng}(\pi_2)$.

16. A **metric** on a set X is a function $d: X \times X \to \mathbb{R}$ such that for all $a, b, c \in X$,
    (i) $d(a, b) \geq 0$.
    (ii) $d(a, b) = 0$ iff $a = b$.
    (iii) $d(a, b) = d(b, a)$.
    (iv) $d(a, b) + d(b, c) \geq d(a, c)$.
    Prove that each of the following is a metric for the indicated set.
    ★ **(a)** $X = \mathbb{N}$, $d(x, y) = |x - y|$
    (b) $X = \mathbb{R}$, $d(x, y) = \begin{cases} 0 & \text{if } x = y \\ 1 & \text{if } x \neq y \end{cases}$
    (c) $X = \mathbb{R} \times \mathbb{R}$, $d((x, y), (z, w)) = \sqrt{(x - z)^2 + (y - w)^2}$
    (d) $X = \mathbb{R} \times \mathbb{R}$, $d((x, y), (z, w)) = |x - z| + |y - w|$

17. Suppose that set A has m elements and set B has n elements. We have seen that $A \times B$ has mn elements and that there are 2^{mn} relations from A to B. Find the number of relations from A to B that are
    (a) functions from A to B.
    (b) functions with one element in the domain.
    ★ **(c)** functions with two elements in the domain.
    (d) functions whose domain is a subset of A.

18. **(a)** Let f be a function from A to B. Define the relation T on A by $x \, T \, y$ iff $f(x) = f(y)$. Prove that T is an equivalence relation on A.
    (b) In the case when $f: \mathbb{R} \to \mathbb{R}$ is given by $f(x) = x^2$, describe the equivalence class of 0; of 2; of 4.
    (c) In the case when $f: \mathbb{R} \to \mathbb{R}$ is the cosine function, describe the equivalence class of 0; of $\pi/2$; of $\pi/4$.

Proofs to Grade 19. Assign a grade of A (correct), C (partially correct), or F (failure) to each. Justify assignments of grades other than A.
    (a) **Claim.** The functions f and g are equal, where f and g are given by

$$f(x) = \frac{x}{|x|} \text{ and } g(x) = \begin{cases} 1 & \text{if } x \geq 0 \\ -1 & \text{if } x < 0 \end{cases}.$$

"Proof." Let x be a real number. If x is positive, then $\dfrac{x}{|x|} = \dfrac{x}{x} = 1$, so $f(x) = g(x)$. If x is negative, then $\dfrac{x}{|x|} = \dfrac{x}{-x} = -1$, so $f(x) = g(x)$. In every case, $f(x) = g(x)$, so $f = g$. ∎

(b) **Claim.** The functions $f(x) = 1 + \dfrac{1}{x}$ and $g(x) = \dfrac{x+1}{x}$ are equal.

"Proof." The domain of each function is assumed to be the largest possible subset of $\mathbb{R}$. Thus $\mathrm{Dom}(f) = \mathrm{Dom}(g) = \mathbb{R} - \{0\}$. For every $x \in \mathbb{R} - \{0\}$ we have

$$f(x) = 1 + \frac{1}{x} = \frac{x}{x} + \frac{1}{x} = \frac{x+1}{x} = g(x).$$

Therefore, by Theorem 4.1.1, $f = g$. ∎

(c) **Claim.** The relation $x^2 = y^3$ defines a function from $\mathbb{R}$ to $\mathbb{R}$.

"Proof." The graph of the relation $x^2 = y^3$ is given here:

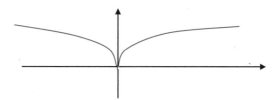

Since no vertical line crosses the graph more than once, $x^2 = y^3$ defines a function. ∎

(d) **Claim.** If $h: A \to B$ and $g: C \to D$, then $h \cup g: A \cup C \to B \cup D$.

"Proof." Suppose $(x, y) \in h \cup g$ and $(x, z) \in h \cup g$. Then $(x, y) \in h$ or $(x, y) \in g$, and $(x, z) \in h$ or $(x, z) \in g$. If $(x, y) \in h$ and $(x, z) \in h$, then $y = z$. Otherwise, $(x, y) \in g$ and $(x, z) \in g$; so again $y = z$. Therefore, $h \cup g$ is a function. Thus, we have that $\mathrm{Dom}(h \cup g) = \mathrm{Dom}(h) \cup \mathrm{Dom}(g) = A \cup C$, so $h \cup g: A \cup C \to B \cup D$. ∎

(e) **Claim.** The rule that assigns to each equivalence class $\bar{x}$ in $\mathbb{Z}_4$ the class $[x + 1]$ in $\mathbb{Z}_2$ is a function.

"Proof." Suppose $(\bar{x}, [y])$ and $(\bar{x}, [z])$ are two ordered pairs in the relation determined by the rule. We must show that $[y] = [z]$. According to the rule, $[y] = [x_1 + 1]$ and $[z] = [x_2 + 1]$ for some x_1, x_2 in the class $\bar{x}$. Since x_1 and x_2 are in the same equivalence class (mod 4), $x_1 - x_2 = 4k$ for some integer k. Then $(x_1 + 1) - (x_2 + 1) = 4k = 2(2k)$, so $x_1 + 1$ and $x_2 + 1$ are in the same equivalence class (mod 2). Therefore $[y] = [z]$. ∎

4.2 Constructions of Functions

This section discusses several methods for constructing new functions from given ones. You have already seen the operations of composition and inversion of relations in Chapter 3. Since every function is a relation, these operations are performed on functions in the same way.

For a function $F: A \rightarrow B$, the **inverse of F** is the relation from B to A:

$$F^{-1} = \{(x, y) : (y, x) \in F\}.$$

We are careful to say F^{-1} is a *relation* because the inverse of a function is a relation, but might not be a function. Conditions on F to ensure that F^{-1} is a function will be given in Section 4.4.

For functions $F: A \rightarrow B$ and $G: B \rightarrow C$ the **composite of F and G** is the relation from A to C:

$$G \circ F = \{(x, z) \in A \times C : (x, y) \in F \text{ and } (y, z) \in G, \text{ for some } y \in B\}.$$

Here, too, we say the *relation* $G \circ F$, but we will soon see in Theorem 4.2.1 that the composite of two functions is a function.

Examples. Let $F = \{(x, y): y = 2x + 1\}$ and $G = \{(x, y): y = x^2\}$. Then F and G are mappings with domain and codomain $\mathbb{R}$. The inverses of F and G are

$$\begin{aligned} F^{-1} &= \{(x, y) \in \mathbb{R} \times \mathbb{R}: (y, x) \in F\} \\ &= \{(x, y) \in \mathbb{R} \times \mathbb{R}: x = 2y + 1\} \\ &= \left\{(x, y) \in \mathbb{R} \times \mathbb{R}: y = \frac{x - 1}{2}\right\} \end{aligned}$$

and

$$\begin{aligned} G^{-1} &= \{(x, y) \in \mathbb{R} \times \mathbb{R}: (y, x) \in G\} \\ &= \{(x, y) \in \mathbb{R} \times \mathbb{R}: x = y^2\}. \end{aligned}$$

The inverse of F is a function. The inverse of G is not a function since, for instance, $(4, 2) \in G^{-1}$ and $(4, -2) \in G^{-1}$.

The composite of F and G is

$$\begin{aligned} G \circ F &= \{(x, z) \in \mathbb{R} \times \mathbb{R}: \exists y \in \mathbb{R} \text{ such that } (x, y) \in F \text{ and } (y, z) \in G\} \\ &= \{(x, z) \in \mathbb{R} \times \mathbb{R}: \exists y \in \mathbb{R} \text{ such that } y = 2x + 1 \text{ and } z = y^2\} \\ &= \{(x, z) \in \mathbb{R} \times \mathbb{R}: z = (2x + 1)^2\}. \end{aligned}$$

We can also compute other composites, such as

$$\begin{aligned} F \circ F &= \{(x, z) \in \mathbb{R} \times \mathbb{R}: \exists y \in \mathbb{R} \text{ such that } y = 2x + 1 \text{ and } z = 2y + 1\} \\ &= \{(x, z) \in \mathbb{R} \times \mathbb{R}: z = 2(2x + 1) + 1\} \\ &= \{(x, z) \in \mathbb{R} \times \mathbb{R}: z = 4x + 3\} \end{aligned}$$

and

$$\begin{aligned} F \circ G &= \{(x, z) \in \mathbb{R} \times \mathbb{R}: \exists y \in \mathbb{R} \text{ such that } y = x^2 \text{ and } z = 2y + 1\} \\ &= \{(x, z) \in \mathbb{R} \times \mathbb{R}: z = 2x^2 + 1\}. \end{aligned}$$

These examples show that $G \circ F$ and $F \circ G$ are not always equal. Thus, composition of functions is not commutative.

Theorem 4.2.1   Let A, B, and C be sets and $F: A \to B$ and $G: B \to C$. Then $G \circ F$ is a function from A to C and $\mathrm{Dom}(G \circ F) = A$.

Proof.   The relationships among mappings and their domains and codomains are given in Figure 4.2.1. In Section 3.1, we proved that $G \circ F$ is a relation from A to C. To show that $G \circ F$ is a function from A to C, let $(x, y) \in G \circ F$ and $(x, z) \in G \circ F$. ⟨We must show that $y = z$.⟩ Since $(x, y) \in G \circ F$, there exists $u \in B$ such that $(x, u) \in F$ and $(u, y) \in G$. Likewise, there exists $v \in B$ such that $(x, v) \in F$ and $(v, y) \in G$. Since F is a function, $(x, u) \in F$ and $(x, v) \in F$ imply that $u = v$. Since G is a function, $(u, y) \in G$, $(v, z) \in G$, and $u = v$ imply that $y = z$.

⟨We next show that $\mathrm{Dom}(G \circ F) = A$.⟩ In Section 3.1, we proved that $\mathrm{Dom}(G \circ F) \subseteq \mathrm{Dom}(F) = A$. ⟨We must now show that $A \subseteq \mathrm{Dom}(G \circ F)$.⟩ Suppose $a \in A$. Since $A = \mathrm{Dom}(F)$ there is $b \in B$ such that $(a, b) \in F$. Since $B = \mathrm{Dom}(G)$ there is $c \in C$ such that $(b, c) \in G$. Then $(a, c) \in G \circ F$. Therefore $a \in \mathrm{Dom}(G \circ F)$.   ∎

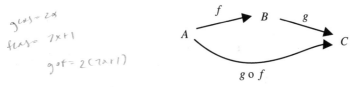

Figure 4.2.1

We can take advantage of the fact that each element of the domain of a function has a unique image to simplify the notation for composition. Let $H: A \to B$ and $K: B \to C$. Since $(x, y) \in H$ and $(y, z) \in K$ may be written in the form $y = H(x)$ and $z = K(y)$, we can write $z = K(H(x))$; that is,

$$(K \circ H)(x) = K(H(x)).$$

Notice that the first function applied in composition is the function on the right, which is closer to the variable x.

Examples.   For $H(x) = \sin x$, $K(x) = x^2 + 6$, and $L(x) = e^x$, the composites $H \circ K$, $K \circ L$, and $L \circ H$ are

$$(H \circ K)(x) = H(K(x)) = H(x^2 + 6) = \sin(x^2 + 6),$$
$$(K \circ L)(x) = K(L(x)) = K(e^x) = (e^x)^2 + 6 = e^{2x} + 6,$$

and

$$(L \circ H)(x) = L(H(x)) = L(\sin x) = e^{\sin x}.$$

Example.   In this example, we consider functions on the sets $\mathbb{Z}_6 = \{\overline{0}, \overline{1}, \overline{2}, \overline{3}, \overline{4}, \overline{5}\}$ and $\mathbb{Z}_3 = \{[0], [1], [2]\}$. (As usual, we use a different notation for an equivalence class when the modulus is different.) Let $H\colon \mathbb{Z}_6 \to \mathbb{Z}_3$ be given by $H(\overline{0}) = [0]$, $H(\overline{1}) = [1]$, $H(\overline{2}) = [2]$, $H(\overline{3}) = [0]$, $H(\overline{4}) = [1]$, and $H(\overline{5}) = [2]$. Let $K\colon \mathbb{Z}_3 \to \mathbb{Z}_3$ be given by $K([0]) = [0]$, $K([1]) = [2]$, and $K([2]) = [1]$. Then

$$(K \circ H)(\overline{0}) = K(H(\overline{0})) = K([0]) = [0],$$
$$(K \circ H)(\overline{4}) = K(H(\overline{4})) = K([1]) = [2],$$

and so on. In this example, K^{-1} is a function. In fact, $K^{-1} = K$. The composite $K^{-1} \circ K$ has images $(K^{-1} \circ K)([0]) = K^{-1}([0]) = [0]$, $(K^{-1} \circ K)([1]) = K^{-1}([2]) = [1]$, and $(K^{-1} \circ K)([2]) = K^{-1}([1]) = [2]$. Thus $K^{-1} \circ K$ is the identity function on $\mathbb{Z}_3$.

Generally, when we use composite functions, the domain of the composite is the domain of the first function applied. If it happens that $\text{Rng}(F)$ is not a subset of $\text{Dom}(G)$, we need to be aware that if $F(x)$ is not in the domain of G, then $(G \circ F)(x)$ is undefined.

For example, let F and G be the functions given by $F(x) = x^2$ and $G(x) = \dfrac{1}{x - 4}$. Then $2 \in \text{Dom}(F)$ but $F(2) = 4 \notin \text{Dom}(G)$. In this example, $\text{Dom}\,(G \circ F)$ is not the same as $\text{Dom}\,(F)$ because $(G \circ F)(2)$ is not defined.

In Chapter 3 we proved that composition of relations is associative. As a result, composition of functions is associative as well. Similarly, the result of forming the composite of a function f with the appropriate identity function yields the same function f. These properties are restated for functions here with proofs that take advantage of functional notation.

Theorem 4.2.2.   Let A, B, C, and D be sets and $f\colon A \to B$, $g\colon B \to C$, and $h\colon C \to D$. Then $(h \circ g) \circ f = h \circ (g \circ f)$. That is, composition of functions is associative.

Proof.   ⟨We must show that the domains of $(h \circ g) \circ f$ and $h \circ (g \circ f)$ are the same and that $((h \circ g) \circ f)(x) = (h \circ (g \circ f))(x)$.⟩ By Theorem 4.2.1 the domain of each function is A. Now let $x \in A$. Then $((h \circ g) \circ f)(x) = (h \circ g)(f(x)) = h(g(f(x))) = h((g \circ f)(x)) = (h \circ (g \circ f))(x)$.   ∎

The relationship $(h \circ g) \circ f = h \circ (g \circ f)$ in Theorem 4.2.2 is represented in the diagram in Figure 4.2.2. For any $x \in A$, by following the diagram from A to D along the upper route the image of x is $((h \circ g) \circ f)(x)$, while along the lower route the image of x is $(h \circ (g \circ f))(x)$. Theorem 4.2.2 says that these images are always the same, and consequently the figure is called a *commutative diagram*. This theorem allows us to avoid the use of parentheses for composition and to simply say $h \circ g \circ f$ is a function from A to D and the image of x is $(h \circ g \circ f)(x)$.

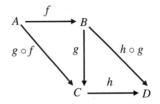

Figure 4.2.2

Theorem 4.2.3   Let $f: A \to B$. Then $f \circ I_A = f$ and $I_B \circ f = f$.

Proof. $\text{Dom}(f \circ I_A) = \text{Dom}(I_A) = A = \text{Dom}(f)$. If $x \in A$, then $(f \circ I_A)(x) = f(I_A(x)) = f(x)$. Therefore $f \circ I_A = f$. The proof that $I_B \circ f = f$ is left as Exercise 6. ∎

Theorem 4.2.4   Let $f: A \to B$ with $\text{Rng}(f) = C$. If f^{-1} is a function, then $f^{-1} \circ f = I_A$ and $f \circ f^{-1} = I_C$.

Proof. Suppose $f: A \to B$ and f^{-1} is a function. Then $\text{Dom}(f^{-1} \circ f) = \text{Dom}(f)$ ⟨by *Theorem 4.2.1*⟩. Thus $\text{Dom}(f^{-1} \circ f) = A = \text{Dom}(I_A)$. Suppose $x \in A$. From the fact that $(x, f(x)) \in f$, we have $(f(x), x) \in f^{-1}$. Therefore, $(f^{-1} \circ f)(x) = f^{-1}(f(x)) = x = I_A(x)$. This proves that $f^{-1} \circ f = I_A$.

The proof that $f \circ f^{-1} = I_C$ is left as Exercise 7. ∎

Every subset of a single-valued relation (i.e., a function) is single-valued, so a subset of a function is always single-valued. Thus, removing some of the ordered pairs from a given function $f: A \to B$ is yet another way to create a function. If $g \subseteq f$, we say the function g is a restriction of f. A restriction is usually defined by specifying what we want the new, smaller, domain to be.

DEFINITIONS Let $f: A \to B$ and let $D \subseteq A$. The **restriction of f to D** is the function

$$f|_D = \{(x, y): y = f(x) \text{ and } x \in D\}.$$

If g and h are functions and g is a restriction of h, we say h is an **extension** of g.

Examples. Let $A = \{1, 2, 3, 4\}$, $B = \{a, b, c, d\}$, and g be the function $\{(1, a), (2, a), (3, d), (4, c)\}$. Then $g|_{\{1, 4\}} = \{(1, a), (4, c)\}$, $g|_{\{3\}} = \{(3, d)\}$, and $g|_A = g$.

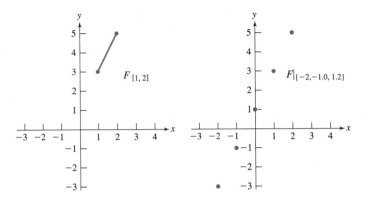

Figure 4.2.3

Let $F: \mathbb{R} \to \mathbb{R}$ be given by $F(x) = 2x + 1$. Figure 4.2.3 shows the graphs of $F|_{[1, 2]}$ and $F|_{\{-2, -1, 0, 1, 2\}}$.

Example. Recall that restricting the trigonometric function sin: $\mathbb{R} \to \mathbb{R}$ to a smaller domain is the first step in defining the inverse of sine. When the domain of sine is restricted to $\left[-\frac{\pi}{2}, \frac{\pi}{2}\right]$, the result is usually referred to as the Sine function (with a capital S), abbreviated Sin, and is the principal branch of the sine function. The graphs of sine and Sine are shown in Figures 4.2.4(a) and 4.2.4(b), respectively. $\sin \frac{\pi}{3} = \sin \frac{\pi}{3} = \frac{\sqrt{3}}{2}$ and $\sin 0 = \sin 0 = 0$, but $\sin \frac{2\pi}{3}$ is not defined because $\frac{2\pi}{3} \notin \left[-\frac{\pi}{2}, \frac{\pi}{2}\right]$.

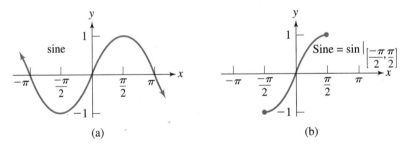

(a)   (b)

Figure 4.2.4

Because functions are sets (of ordered pairs), it is appropriate to ask about unions and intersections of functions. If h and g are functions, is $h \cap g$ a function? For the functions

$$H = \{(1, 2), (2, 6), (3, -9), (5, 7)\},$$
$$G = \{(1, 8), (2, 6), (4, 8), (5, 7), (8, 3)\},$$

$H \cap G$ is $\{(2, 6), (5, 7)\}$, which is a function. Notice that $\text{Dom}(H \cap G) = \{2, 5\}$ is a proper subset of $\text{Dom}(H) \cap \text{Dom}(G) = \{1, 2, 5\}$. This is because $1 \in \text{Dom}(H)$ and $1 \in \text{Dom}(G)$, but $H(1) \neq G(1)$.

It turns out that if h and g are functions, then $h \cap g$ is always a function (see Exercise 10), but $h \cap g$ can just as easily be expressed as a restriction of either h or g.

The situation regarding $h \cup g$ is much more interesting and useful. First, in general, $h \cup g$ need not be a function. In the case of the functions H and G above, the union is not a function because $(1, 2) \in H \cup G$ and $(1, 8) \in H \cup G$. However, if the domains of h and g are disjoint sets, then $h \cup g$ is a function. The next theorem states that we can put together functions with disjoint domains to define a function "piecewise."

Theorem 4.2.5 Let h and g be functions with $\text{Dom}(h) = A$ and $\text{Dom}(G) = B$. If $A \cap B = \varnothing$, then $h \cup g$ is a function with domain $A \cup B$. Furthermore,

$$(h \cup g)(x) = \begin{cases} h(x) & \text{if } x \in A \\ g(x) & \text{if } x \in B. \end{cases}$$

Proof. See Exercise 11. ∎

Example. Let $h(x) = x^2$ and $g(x) = 6 - x$. The restrictions $h|_{(-\infty, 2]}$ and $g|_{(2, \infty)}$ have disjoint domains. Their union f is an extension of each (but not an extension of h or g). See Figure 4.2.5. The function f may be described in two pieces:

$$f(x) = \begin{cases} x^2 & \text{if } x \leq 2 \\ 6 - x & \text{if } x > 2. \end{cases}$$

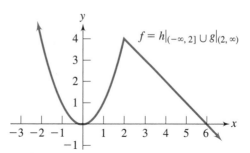

Figure 4.2.5

Functions can be constructed piecewise from three or more functions, and Theorem 4.2.5 may also be extended to the case where domains are not disjoint, provided that the functions agree on the intersection of the domains. See Exercise 13. The characteristic and step functions discussed in Section 4.1 are examples of piecewise defined functions.

We conclude this section with examples of proofs about increasing and decreasing functions. Recalling the definitions of these properties from previous study:

DEFINITIONS Let f be a function from a set of reals to $\mathbb{R}$ whose domain includes an interval I. We say f is **increasing on I** iff for all $x, y \in I$, if $x < y$, then $f(x) < f(y)$. Similarly, f is **decreasing on I** iff for all $x, y \in I$, if $x < y$, then $f(x) > f(y)$.

The graph of the function f in Figure 4.2.5 is increasing on the interval [0, 2] and decreasing on the intervals $(-\infty, 0]$ and $[2, \infty)$. This seems clear from the graph, but remember that looking at its graph does not constitute a proof because we cannot see the entire graph or all the details of its behavior. We give two examples of proofs of these properties.

Example.   Prove that the function f of Figure 4.2.5 is increasing on the interval [0, 2].

Proof.   Let $x, y \in [0, 2]$ and suppose $x < y$. Then $g(x) = x^2$ and $g(y) = y^2$. Since $x < y$ and $x \geq 0$, $x^2 \leq xy$. Since $x < y$ and $y > 0$, $xy < y^2$. Therefore, $x^2 < y^2$. Thus $g(x) < g(y)$, so g is increasing on [0, 2]. ∎

Example.   Let $f(x) = 2 + \dfrac{1}{x}$. Prove that f is decreasing for $x > 0$.

Proof.   Suppose $0 < x < y$. Then $f(x) = 2 + \dfrac{1}{x}$ and $f(y) = 2 + \dfrac{1}{y}$. Since x and y are positive and $x < y$ we have $\dfrac{1}{y} < \dfrac{1}{x}$. Thus $2 + \dfrac{1}{y} < 2 + \dfrac{1}{x}$; that is, $f(x) > f(y)$. Therefore f is decreasing on $(0, \infty)$. ∎

Exercises 4.2

1.   Find $f \circ g$ and $g \circ f$ for each pair of functions f and g. Use the understood domains for f and g.
   ★ **(a)** $f(x) = 2x + 5,$ $g(x) = 6 - 7x$
   (b) $f(x) = x^2 + 2x,$ $g(x) = 2x + 1$
   ★ **(c)** $f(x) = \sin x,$ $g(x) = 2x^2 + 1$
   (d) $f(x) = \tan x,$ $g(x) = \sin x$
   ★ **(e)** $f(x) = \{(t, r), (s, r), (k, l)\},$ $g(x) = \{(k, s), (t, s), (s, k)\}$
   (f) $f(x) = \{(1, 3), (2, 6), (3, 5), (4, 2), (5, 2)\},$
       $g(x) = \{(1, 5), (2, 3), (3, 7), (4, 3), (5, 4)\}$
   (g) $f(x) = \dfrac{x + 1}{x + 2},$ $g(x) = x^2 + 1$
   (h) $f(x) = 3x + 2,$ $g(x) = |x|$
   ★ **(i)** $f(x) = \begin{cases} x + 1 & \text{if } x \leq 0 \\ 2x & \text{if } x > 0 \end{cases},$ $g(x) = \begin{cases} 2x & \text{if } x \leq -1 \\ -x & \text{if } x > -1 \end{cases}$
   (j) $f(x) = \begin{cases} 2x + 3 & \text{if } x < 3 \\ x^2 & \text{if } x \geq 3 \end{cases},$ $g(x) = \begin{cases} 7 - 2x & \text{if } x \leq 2 \\ x + 1 & \text{if } x > 2 \end{cases}$

☆ 2.   Find the domain and range of each composite in Exercise 1.

3.   Give two different examples of
   ★ **(a)**   a pair of functions f and g such that $(f \circ g)(x) = (3x + 7)^2$.
   (b)   a pair of functions f and g such that $(f \circ g)(x) = \sqrt{2x^2 - 5}$.
   (c)   a pair of functions f and g such that $(f \circ g)(x) = \sin|2x + 4|$.

4. Let $\mathbb{Z}_8 = \{\overline{0}, \overline{1}, \overline{2}, \overline{3}, \overline{4}, \overline{5}, \overline{6}, \overline{7},\}$ and $\mathbb{Z}_4 = \{[0], [1], [2], [3]\}$. Define $f: \mathbb{Z}_8 \to \mathbb{Z}_4$, $g: \mathbb{Z}_4 \to \mathbb{Z}_8$, $h: \mathbb{Z}_8 \to \mathbb{Z}_8$, and $k: \mathbb{Z}_4 \to \mathbb{Z}_4$, as follows: $f(\overline{x}) = [x + 2]$, $g([x]) = \overline{2x}$, $h(\overline{x}) = \overline{2x + 4}$, and $k([x]) = [2x + 2])$. By comparing images, verify the following equalities.
   - **(a)** $(k \circ k)(x) = [2]$ for all x in $\mathbb{Z}_4$.
   - **(b)** $(g \circ f)(x) = h(x)$ for all x in $\mathbb{Z}_8$.
   - **(c)** $(f \circ g)(x) = k(x)$ for all x in $\mathbb{Z}_4$.
   - **(d)** $(h \circ (h \circ h))(x) = \overline{4}$ for all x in $\mathbb{Z}_8$.

5. For which of the following functions f is the relation f^{-1} a function? When f^{-1} is a function, write an explicit expression for $f^{-1}(x)$. Use the understood domain for each function.

   ⋆ **(a)** $f(x) = 5x + 2$ **(b)** $f(x) = 2x^2 + 1$ ⋆ **(c)** $f(x) = \dfrac{x + 1}{x + 2}$

   (d) $f(x) = \sin x$ ⋆ **(e)** $f(x) = e^{x+3}$ **(f)** $f(x) = \dfrac{1 - x}{-x}$

   (g) $f(x) = \dfrac{1}{1 - x}$ **(h)** $f(x) = -x + 3$ **(i)** $f(x) = \dfrac{-x}{3x - 4}$

6. Prove the remaining part of Theorem 4.2.3: if $f: A \to B$, then $I_B \circ f = f$.

7. Prove the remaining part of Theorem 4.2.4: if $f: A \to B$ with $\mathrm{Rng}(f) = C$, and if f^{-1} is a function, then $f \circ f^{-1} = I_C$.

8. Let $f(x) = 4 - 3x$ with domain $\mathbb{R}$ and $A = \{1, 2, 3, 4\}$. Sketch the graphs of the functions $f|_A$, $f|_{[-1, 3]}$, $f|_{(2, 4]}$, and $f|_{\{6\}}$. What is the range of $f|_\mathbb{N}$?

9. Describe two extensions of f with domain $\mathbb{R}$ for the function
   ⋆ **(a)** $f = \{(x, y) \in \mathbb{N} \times \mathbb{N}: y = x^2\}$.
   (b) $f = \{(x, y) \in \mathbb{N} \times \mathbb{N}: y = 3\}$.
   (c) $f = \{(x, y) \in [-1, 1] \times [-1, 1]: y = -x\}$.

10. Prove that, if f and g are functions, then $f \cap g$ is a function by showing that $f \cap g = g|_A$ where $A = \{x: g(x) = f(x)\}$.

11. Prove Theorem 4.2.5.

12. Let f be a function with domain D, and let g be an extension of f with domain A. Then by definition, $f = g|D$ and $D \subseteq A$. Let i be the inclusion mapping from D to A given by $i(x) = x$ for all $x \in D$. Prove that $f = g \circ i$.

☆ 13. Let $h: A \to B$, $g: C \to D$ and suppose $E = A \cap C$. Prove $h \cup g$ is a function from $A \cup C$ to $B \cup D$ if and only if $h|_E = g|_E$.

14. For each pair of functions h and g, determine whether $h \cup g$ is a function. In each case sketch a graph of $h \cup g$.
   ⋆ **(a)** $h: (-\infty, 0] \to \mathbb{R}$, $h(x) = 3x + 4$
      $g: (0, \infty) \to \mathbb{R}$, $g(x) = \dfrac{1}{x}$
   (b) $h: [-1, \infty) \to \mathbb{R}$, $h(x) = x^2 + 1$
      $g: (-\infty, -1] \to \mathbb{R}$, $g(x) = x + 3$
   (c) $h: (-\infty, 1] \to \mathbb{R}$, $h(x) = |x|$
      $g: [0, \infty) \to \mathbb{R}$, $g(x) = 3 - |x - 3|$

     (d) $h: (-\infty, 2] \to \mathbb{R}$, $h(x) = \cos x$
          $g: [2, \infty) \to \mathbb{R}$, $g(x) = x^2$

     (e) $h: (-\infty, 3) \to \mathbb{R}$, $h(x) = 3 - x$
          $g: (0, \infty) \to \mathbb{R}$, $g(x) = x + 1$

15. Let $f: A \to B$ and $g: C \to D$. Define $f \times g = \{((a, c), (b, d)): (a, b) \in f$ and $(c, d) \in g\}$.

     (a)   Prove that $f \times g$ is a function from $A \times C$ to $B \times D$.

     (b)   For $(a, c) \in A \times C$, write $(f \times g)(a, c)$ in terms of $f(a)$ and $g(c)$.

16. Prove each of these statements.

  ★ **(a)** f is increasing on $\mathbb{R}$, where $f(x) = 3x - 7$.

     (b) g is decreasing on $\mathbb{R}$, where $g(x) = 2 - 5x$.

     (c) h is increasing on $[0, \infty)$, where $h(x) = x^2$.

  ★ **(d)** f is increasing on $(-3, \infty)$, where $f(x) = \dfrac{x - 1}{x + 3}$.

17. Prove or give a counterexample:

     (a)   If f is a linear function with positive slope, f is increasing on $\mathbb{R}$.

     (b)   If f and g are decreasing functions on an interval I and $f \circ g$ is defined on I, then $f \circ g$ is decreasing on I.

     (c)   If f and g are decreasing functions on I and $f \circ g$ is defined on I, then $f \circ g$ is increasing on I.

  ★ **(d)**   If $\text{Dom}(f) = \mathbb{R}$ and f is increasing on the intervals $[-2, -1]$ and $[1, 2]$, then f is increasing on $[-2, 2]$.

     (e)   If f is decreasing on $(-\infty, 0)$ and decreasing on $[0, \infty)$, then f is decreasing on $\mathbb{R}$.

18. Let $f_1: \mathbb{R} \to \mathbb{R}$ and $f_2: \mathbb{R} \to \mathbb{R}$. Define the **pointwise sum** $f_1 + f_2$ and **pointwise product** $f_1 \cdot f_2$ as follows:

$$(f_1 + f_2)(x) = f_1(x) + f_2(x) \text{ for all } x \in \mathbb{R} \text{ and}$$
$$(f_1 \cdot f_2)(x) = f_1(x) \cdot f_2(x) \text{ for all } x \in \mathbb{R}.$$

  ☆ **(a)**   Prove that $f_1 + f_2$ and $f_1 \cdot f_2$ are functions with domain $\mathbb{R}$.

     (b)   Let $f(x) = 2x + 5$, $g(x) = 6 - 7x$, and $h(x) = 3x^2 - 7x + 2$. Compute $(f + g)(x)$, $(f \cdot g)(x)$, $(f + h)(x)$, and $(g \cdot h)(x)$.

     (c)   Prove or disprove: If f and g are increasing on $\mathbb{R}$, and h is decreasing on $\mathbb{R}$, then $(f + g) + h$ is increasing on $\mathbb{R}$.

19. Let $f: \mathbb{R} \to \mathbb{R}$ and $c \in \mathbb{R}$. Define the **scalar product** cf:

$$cf(x) = c \cdot f(x) \text{ for all } x \in \mathbb{R}.$$

Prove that cf is a function with domain $\mathbb{R}$.

Proofs to Grade   **20.** Assign a grade of A (correct), C (partially correct), or F (failure) to each. Justify assignments of grades other than A.

  ★ **(a)**   **Claim.**   Let $f: A \to B$. If f^{-1} is a function, then $f^{-1} \circ f = I_A$.
             "Proof."   Suppose $(x, y) \in f^{-1} \circ f$. Then there is z such that $(x, z) \in f$ and $(z, y) \in f^{-1}$. But this means that $(z, x) \in f^{-1}$ and $(z, y) \in f^{-1}$. Since f^{-1} is a function, $x = y$. Hence $(x, y) \in f^{-1} \circ f$

implies $(x, y) \in I_A$; that is, $f^{-1} \circ f \subseteq I_A$. Now suppose $(x, y) \in I_A$. Since $A = \text{Dom}(f)$, there is $w \in B$ such that $(x, w) \in f$. Hence $(w, x) \in f^{-1}$. But $(x, y) \in I_A$ implies $x = y$ and so $(w, y) \in f^{-1}$. But from $(x, w) \in f$ and $(w, y) \in f^{-1}$, we have $(x, y) \in f^{-1} \circ f$. This shows $I_A \subseteq f^{-1} \circ f$. Therefore, $I_A = f^{-1} \circ f$. ∎

(b) Claim. If f and f^{-1} are functions on A, and $f \circ f = f$, then $f = I_A$.

"Proof." Suppose $f: A \rightarrow A$ and $f^{-1}: A \rightarrow A$. Since $f = f \circ f$, $f^{-1} \circ f = f^{-1} \circ (f \circ f)$. By associativity, we have $f^{-1} \circ f = (f^{-1} \circ f) \circ f$. This gives $I_A = I_A \circ f$. Since $I_A \circ f = f$, by cancellation we have $I_A = f$. ∎

(c) Claim. If f, g, and f^{-1} are functions on A, then $g = f^{-1} \circ (g \circ f)$.

"Proof." Using associativity and Theorems 4.2.3 and 4.2.4,

$$f^{-1} \circ (g \circ f) = f^{-1} \circ (f \circ g) = (f^{-1} \circ f) \circ g = I_A \circ g = g. \qquad ∎$$

(d) Claim. If $f'(x) > 0$ on an open interval (a, b), then f is increasing on (a, b).

"Proof." Assume that $f'(x) > 0$ on the interval (a, b). Suppose x_1 and x_2 are in (a, b) and $x_1 < x_2$. We must show that $f(x_1) < f(x_2)$. We know from calculus that since f is differentiable on (a, b), it is continuous on $[x_1, x_2] \subseteq (a, b)$, and differentiable on (x_1, x_2). By the Mean Value Theorem, there exists c in (x_1, x_2) such that

$$\frac{f(x_2) - f(x_1)}{x_2 - x_1} = f'(c).$$

Therefore, $f(x_2) - f(x_1) = f'(c)(x_2 - x_1)$. By hypothesis $f'(c) > 0$, and $x_2 - x_1 > 0$ since $x_1 < x_2$. Therefore, $f(x_2) - f(x_1) > 0$. We conclude that $f(x_1) < f(x_2)$. ∎

4.3 Functions That Are Onto; One-to-One Functions

The definition of a function $f: A \rightarrow B$ is stated in terms of conditions on the first coordinates of f. Two important properties of functions are defined by requiring additional conditions on the second coordinates of f.

DEFINITION A function $f: A \rightarrow B$ is **onto B** (or is a **surjection**) iff $\text{Rng}(f) = B$. When f is a surjection, we write $f: A \xrightarrow{\text{onto}} B$

For $f: A \rightarrow B$, it is always the case that $\text{Rng}(f) \subseteq B$, so a function always maps to its codomain. We say f maps onto its codomain when the codomain is $\text{Rng}(f)$. From our discussion in Section 4.1, we know that the functions

$$f: \mathbb{N} \rightarrow \mathbb{R}, \text{ where } f(n) = 2n$$
$$\text{and } g: \mathbb{N} \rightarrow E^+, \text{ where } g(n) = 2n$$

are equal because they are the same sets of ordered pairs. The range of this function is the set E^+ of even natural numbers. This range is the same as the given codomain for g, so we say g maps onto E^+. It would be incorrect to say $f: \mathbb{N} \xrightarrow{\text{onto}} \mathbb{R}$.

Properly speaking, when we say a function f is onto, we should finish the sentence by saying what set it is that f maps onto. However, when the codomain of a surjection f is clear from the context, it is common practice, even if not perfectly correct, to say simply that "f is onto." If we are given a function f and we want to say that f is onto, what we must do first is determine the range of f. We can then say f is onto that range.

Since $\text{Rng}(f) \subseteq B$ is always true, $f: A \to B$ is a surjection if and only if $B \subseteq \text{Rng}(f)$; that is, iff every $b \in B$ has a pre-image. Therefore, to prove that $f: A \to B$ is onto its codomain B, one must show that for every $b \in B$, $b = f(a)$, for some $a \in A$.

Example.   Prove that $F: \mathbb{R} \to \mathbb{R}$, where $F(x) = x + 2$, is onto $\mathbb{R}$.

Proof.   ⟨*We must show that for every* $w \in \mathbb{R}$, *there exists* $t \in \mathbb{R}$ *such that* $F(t) = w$.⟩ Let $w \in \mathbb{R}$. We choose $t = w - 2$. ⟨*Since we want* $w = F(t) = t + 2$, *the pre-image we need is* $t = w - 2$.⟩ Then $F(t) = t + 2 = w$. Therefore, $F: \mathbb{R} \to \mathbb{R}$ is onto $\mathbb{R}$.   ∎

When $f: A \to B$ and the sets A and B are subsets of $\mathbb{R}$, it is often helpful to look at the graph of f and apply this *Horizontal Line Test for onto functions:*

> f maps onto B iff for every $b \in B$, the horizontal
> line $y = b$ intersects the graph of f.

A visual check is not the same as a proof, but can help us decide whether to attempt a proof that f is onto, or else to identify an element of B that is not in the range.

Examples.   Figure 4.3.1 shows the graphs of two functions, h and k, from $[1, 3]$ to $[1, 4]$. The function h can be shown to be onto $[1, 4]$ because every horizontal line with y-intercept b, where $1 \leq b \leq 4$, intersects the graph. The function k is not onto $[1, 4]$. The line $y = 1.5$ does not intersect the graph.

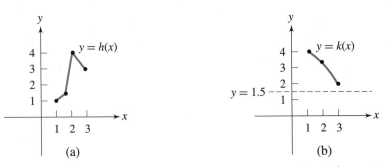

Figure 4.3.1

Example. Let $p: \mathbb{R} \to \mathbb{R}$ be the polynomial function $p(x) = x^3 + 3x^2 - 24x$ (Figure 4.3.2). Prove that p is onto $\mathbb{R}$.

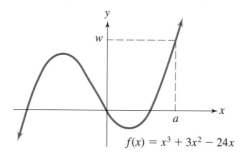

$f(x) = x^3 + 3x^2 - 24x$

Figure 4.3.2

Proof. Let $w \in \mathbb{R}$. *⟨We must show that $w = p(a)$ for some $a \in \mathbb{R}$.⟩* The equation $p(x) - w = 0$ is a third degree polynomial equation in the variable x. Since the degree of the equation is odd and since nonreal (complex) roots of all equations with coefficients in $\mathbb{R}$ occur in conjugate pairs, the equation $p(x) - w = 0$ has at least one real root $a \in \mathbb{R}$. Thus $p(a) - w = 0$. Therefore $p(a) = w$. Hence p is onto $\mathbb{R}$. ∎

Example. Let $G: \mathbb{R} \to \mathbb{R}$ be defined by $G(x) = x^2 + 1$. Then G is not onto $\mathbb{R}$. To show this, we find an element y in the codomain $\mathbb{R}$ that has no pre-image in the domain $\mathbb{R}$. Let y be -2. Since $x^2 + 1 \geq 1$ for every real number x, there is no $x \in \mathbb{R}$ such that $G(x) = -2$. Hence G is not onto $\mathbb{R}$.

Example. The function $M: \mathbb{Z} \times \mathbb{Z} \to \mathbb{Z}$, where $M(x, y) = xy$, is a surjection. For any $z \in \mathbb{Z}$, $(z, 1) \in \mathbb{Z} \times \mathbb{Z}$ and $M(z, 1) = z \cdot 1 = z$. Even though some integers have many pre-images (for example, $24 = M(3, 8) = M(12, 2) = M(4, 6)$), to prove that M maps onto $\mathbb{Z}$ we only need to show that there is at least one pre-image for each $z \in \mathbb{Z}$.

When you prove that a given function f maps A onto B by showing that every element of B has a pre-image, be sure to verify that each pre-image is in the domain A. This very important step is the statement "Then $x \in (-\infty, 0]$" in the following example.

Example. Let $s: (-\infty, 0] \to [-4, \infty)$ be defined by $s(x) = x^2 - 4$. Prove that s is onto $[-4, \infty)$.

Proof. Let $w \in [-4, \infty)$. Then $w \geq -4$, so $w + 4 \geq 0$. Choose $x = -\sqrt{w + 4}$. *⟨Note that we do not choose $x = \sqrt{w + 4}$.⟩* Then $x \in (-\infty, 0]$. It follows that

$$s(x) = (-\sqrt{w + 4})^2 - 4 = (w + 4) - 4 = w.$$

Therefore the function f maps onto $[-4, \infty)$. ∎

The next two theorems relate composition and the property of being a surjection.

Theorem 4.3.1   If $f: A \xrightarrow{\text{onto}} B$, and $g: B \xrightarrow{\text{onto}} C$, then $g \circ f: A \xrightarrow{\text{onto}} C$. That is, the composite of surjective functions is a surjection.

Proof.   Exercise 5.   ∎

Theorem 4.3.2   If $f: A \rightarrow B$, $g: B \rightarrow C$, and $g \circ f: A \xrightarrow{\text{onto}} C$, then g is onto C. That is, when the composite of two functions maps onto a set C, then the second function applied must map onto C.

Proof.   ⟨*We must show* $C \subseteq \text{Rng}(g)$.⟩ Suppose $c \in C$. Since $g \circ f$ maps onto C, there is $a \in A$ such that $(g \circ f)(a) = c$. Let $b = f(a)$, which is in B. Then $(g \circ f)(a) = g(f(a)) = g(b) = c$. Thus there is $b \in B$ such that $g(b) = c$, and g maps onto C.   ∎

For a function $f: A \rightarrow B$, we said that f is onto B if every element of B is used *at least once* as a second coordinate. For the property of being one-to-one, the condition is that every element of B is used as a second coordinate *at most once*.

DEFINITION   A function $f: A \rightarrow B$ is **one-to-one** (or is **an injection**) iff whenever $f(x) = f(y)$, then $x = y$. When f is an injection, we write $f: A \xrightarrow{1-1} B$.

A direct proof that $f: A \rightarrow B$ is one-to-one begins with the assumption that x and y are elements of A and that $f(x) = f(y)$; the rest of the proof shows that $x = y$. A proof by contraposition assumes that $x \neq y$ and shows that $f(x) \neq f(y)$. To show that f is not one-to-one, it suffices to exhibit two different elements of A with the same image.

Example.   Show that the function $F: \mathbb{R} \rightarrow \mathbb{R}$ defined by $F(x) = 2x + 1$ is one-to-one.

Proof.   Suppose x and z are real numbers and $F(x) = F(z)$. Then $2x + 1 = 2z + 1$. Therefore $2x = 2z$, so $x = z$.   ∎

Example.   The function $r(x) = |x|$ is not one-to-one because $2 \neq -2$ and $r(2) = r(-2)$.

Given the graph of a function $f: A \rightarrow B$ where the sets A and B are subsets of $\mathbb{R}$, we can apply this *Horizontal Line Test for one-to-one functions:*

f is one-to-one iff every horizontal line intersects the graph of f at most once.

From Figure 4.3.1(a) we see that the line $y = 3$ meets the graph of h twice. Therefore 3 has two pre-images and so h is not one-to-one. The graph in Figure 4.3.1(b) suggests that the function k is one-to-one.

The next example shows that we need to consider the domain when we determine whether a function is one-to-one.

Example.   Let $f: [0, \infty) \to [0, \infty)$ be given by $f(x) = x^2$. Show that f is one-to-one.

Proof.   Suppose $x, y \in [0, \infty)$ and $f(x) = f(y)$. Then $x^2 = y^2$, so either $x = y$ or $x = -y$. Since $x \geq 0$ and $y \geq 0$, we conclude that $x = y$. Therefore f is one-to-one.   ∎

Example.   Let $G: \mathbb{R} \to \mathbb{R}$ be given by $G(x) = \dfrac{1}{x^2 + 1}$. We attempt to show G is an injection by assuming that $G(x) = G(y)$. Then

$$\frac{1}{x^2 + 1} = \frac{1}{y^2 + 1}.$$

Therefore, $x^2 + 1 = y^2 + 1$, so $x^2 = y^2$. It does not follow from this that $x = y$. This unsuccessful proof suggests a way to find distinct real numbers with equal images. We note that $2^2 = (-2)^2$, and then compute $G(2) = G(-2) = \frac{1}{5}$. Thus G is not an injection.

The next two theorems relate composition with the property of being an injection. Compare these theorems with those above regarding surjections.

Theorem 4.3.3   If $f: A \xrightarrow{1-1} B$ and $g: B \xrightarrow{1-1} C$, then $g \circ f: A \xrightarrow{1-1} C$. That is, the composite of injective functions is an injection.

Proof.   Assume that $(g \circ f)(x) = (g \circ f)(z)$. Thus $g(f(x)) = g(f(z))$. Then $f(x) = f(z)$ since g is one-to-one. Then $x = z$ since f is one-to-one. Therefore, $g \circ f$ is one-to-one.   ∎

Theorem 4.3.4   If $f: A \to B$, $g: B \to C$, and $g \circ f: A \xrightarrow{1-1} C$, then $f: A \xrightarrow{1-1} B$. That is, if the composite of two functions is one-to-one, then the first function applied must be one-to-one.

Proof.   Exercise 6.   ∎

Mappings that are constructed by means of restrictions or unions may share injective or surjective properties. These results will be used in the study of cardinality in Chapter 5.

Theorem 4.3.5   **(a)**   A restriction of a one-to-one function is one-to-one.
(b)   If $h: A \xrightarrow{\text{onto}} C$, $g: B \xrightarrow{\text{onto}} D$, and $A \cap B = \varnothing$, then $h \cup g: A \cup B \xrightarrow{\text{onto}} C \cup D$.
(c)   If $h: A \xrightarrow{1-1} C$, $g: B \xrightarrow{1-1} D$, $A \cap B = \varnothing$, and $C \cap D = \varnothing$, then $h \cup g: A \cup B \xrightarrow{1-1} C \cup D$.

Proof.   Parts (a) and (b) are left as Exercise 7.

(c)   Suppose $h: A \xrightarrow{1-1} C$, $g: B \xrightarrow{1-1} D$, $A \cap B = \varnothing$, and $C \cap D = \varnothing$. Then by Theorem 4.2.5, $h \cup g$ is a function with domain $A \cup B$.

Suppose $x, y \in A \cup B$. Assume $(h \cup g)(x) = (h \cup g)(y)$.

(i)   If $x, y \in A$, then $h(x) = (h \cup g)(x) = (h \cup g)(y) = h(y)$. Since h is one-to-one, $x = y$.

(ii)   If $x, y \in B$, then by a similar argument, $g(x) = g(y)$, and g is one-to-one; so $x = y$.

(iii)   Suppose $x \in A$ and $y \in B$. Then $h(x) = g(y)$ and $h(x) \in C$ and $g(y) \in D$. But $C \cap D = \varnothing$. This case is impossible.

(iv)   Similarly, $x \in B$ and $y \in A$ is impossible.

In every possible case, $x = y$. Therefore, $h \cup g$ is one-to-one.   ∎

Examples.   Let H be the function on $\{r, s, t, u\}$ that sends r to 1, s to 2, t to 3, and u to 4. Then H is one-to-one and onto $\{1, 2, 3, 4\}$. The function $G = \{(x, 4), (y, 5)\}$ maps $\{x, y\}$ one-to-one and onto $\{4, 5\}$. Since the domains are disjoint, $H \cup G$ is a function. By Theorem 4.3.5 (b), $H \cup G$ is onto $\{1, 2, 3, 4\} \cup \{4, 5\} = \{1, 2, 3, 4, 5\}$. Notice that we cannot apply part (c) of Theorem 4.3.5, because the ranges of H and G are not disjoint. See Figure 4.3.3(a).

If we let $K = \{(w, 5), (z, 6)\}$, then the domains of H and K are disjoint and the ranges of H and K are disjoint. In this case, $H \cup K$ is a function that maps $\{r, s, t, u, w, z\}$ one-to-one and onto $\{1, 2, 3, 4, 5, 6\}$. See Figure 4.3.3(b).

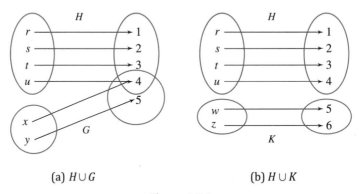

(a) $H \cup G$ (b) $H \cup K$

Figure 4.3.3

Exercises 4.3

1.   Which of the following functions map onto their indicated codomains? Prove each of your answers.

★   **(a)** $f: \mathbb{R} \to \mathbb{R}$, $f(x) = \frac{1}{2}x + 6$

(b) $f: \mathbb{Z} \to \mathbb{Z}$, $f(x) = -x + 1{,}000$

⋆ (c) $f: \mathbb{N} \to \mathbb{N} \times \mathbb{N}$, $f(x) = (x, x)$
   (d) $f: \mathbb{R} \to \mathbb{R}$, $f(x) = x^3$
   (e) $f: \mathbb{R} \to \mathbb{R}$, $f(x) = \sqrt{x^2 + 5}$
   (f) $f: \mathbb{R} \to \mathbb{R}$, $f(x) = 2^x$
   (g) $f: \mathbb{R} \to \mathbb{R}$, $f(x) = \sin x$
   (h) $f: \mathbb{R} \times \mathbb{R} \to \mathbb{R}$, $f(x, y) = x - y$
   (i) $f: \mathbb{R} \to [-1, 1]$, $f(x) = \cos x$
   (j) $f: \mathbb{R} \to [1, \infty)$, $f(x) = x^2 + 1$

⋆ (k) $f: [2, 3) \to [0, \infty)$, $f(x) = \dfrac{x - 2}{3 - x}$

   (l) $f: (1, \infty) \to (1, \infty)$, $f(x) = \dfrac{x}{x - 1}$

☆ **2.** Which of the functions in Exercise 1 are one-to-one? Prove each of your answers.

3. For each function, determine whether the function maps onto the given codomain and whether it is one-to-one. Prove your answers.
   (a) The identity function I_A from A to A.
   (b) The canonical map from $\mathbb{Z}$ to $\mathbb{Z}_5$.
   (c) The greatest integer function int: $\mathbb{R} \to \mathbb{Z}$.
   (d) The sequence $a: \mathbb{N} \to \mathbb{N}$ whose nth term is $a_n = 2^n$.

4. Let $A = \{1, 2, 3, 4\}$. Describe a codomain B and a function $f: A \to B$ such that f is
⋆ **(a)** onto B but not one-to-one.
   (b) one-to-one but not onto B.
   (c) both one-to-one and onto B.
   (d) neither one-to-one nor onto B.

5. Prove that if $f: A \xrightarrow{\text{onto}} B$, and $g: B \xrightarrow{\text{onto}} C$, then $g \circ f: A \xrightarrow{\text{onto}} C$ (Theorem 4.3.1).

6. Prove that if $f: A \to B$, $g: B \to C$, and $g \circ f: A \xrightarrow{1\text{-}1} C$, then $f: A \xrightarrow{1\text{-}1} B$ (Theorem 4.3.4).

7. Prove parts (a) and (b) of Theorem 4.3.5.

8. Find sets A, B, C, and functions $f: A \to B$ and $g: B \to C$ such that
⋆ **(a)** f is onto B, but $g \circ f$ is not onto C.
   (b) g is onto C, but $g \circ f$ is not onto C.
   (c) $g \circ f$ is onto C, but f is not onto B.
   (d) f is one-to-one, but $g \circ f$ is not one-to-one.
⋆ **(e)** g is one-to-one, but $g \circ f$ is not one-to-one.
   (f) $g \circ f$ is one-to-one, but g is not one-to-one.

9. Prove that

   (a) $f(x) = \begin{cases} 2 - x & \text{if } x \leq 1 \\ \dfrac{1}{x} & \text{if } x > 1 \end{cases}$ is one-to-one but not onto $\mathbb{R}$.

   (b) $f(x) = \begin{cases} x + 4 & \text{if } x \leq -2 \\ -x & \text{if } -2 < x < 2 \\ x - 4 & \text{if } x \geq 2 \end{cases}$ is onto $\mathbb{R}$ but not one-to-one.

⋆ **(c)** $f(x) = \begin{cases} \dfrac{x-2}{x+4} & \text{if } x \neq -4 \\ 1 & \text{if } x = -4 \end{cases}$ is one-to-one and onto $\mathbb{R}$.

(d) $f(x) = \begin{cases} |x| & \text{if } x \leq 2 \\ x-3 & \text{if } x > 2 \end{cases}$ is neither one-to-one nor onto $\mathbb{R}$.

10. Prove that if the real-valued function f is increasing (or decreasing), then f is one-to-one.

11. ⋆ **(a)** Let $f: \mathbb{Z}_4 \rightarrow \mathbb{Z}_8$ be given by $f(\bar{x}) = [2x]$, for each $\bar{x} \in \mathbb{Z}_4$. Prove that f is one-to-one but not onto $\mathbb{Z}_8$.

   (b) Let $f: \mathbb{Z}_4 \rightarrow \mathbb{Z}_2$ be given by $f(\bar{x}) = [3x]$, for each $x \in \mathbb{Z}_4$. Prove that f is onto $\mathbb{Z}_2$, but not one-to-one.

   (c) Let $f: \mathbb{Z}_6 \rightarrow \mathbb{Z}_6$ be given by $f(\bar{x}) = \overline{x+1}$, for each $\bar{x} \in \mathbb{Z}_6$. Prove that f is a bijection.

   (d) Let $f: \mathbb{Z}_4 \rightarrow \mathbb{Z}_4$ be given by $f(\bar{x}) = \overline{2x}$. Prove that f is not one-to-one and not onto $\mathbb{Z}_4$.

12. Give two examples of a sequence x of natural numbers (i.e., a function with domain $\mathbb{N}$ and range that is a subset of $\mathbb{N}$) such that

   (a) x is neither one-to-one nor onto $\mathbb{N}$.

   ⋆ **(b)** x is one-to-one and onto $\mathbb{N}$.

   (c) x is one-to-one and not onto $\mathbb{N}$.

   (d) x is onto $\mathbb{N}$ and not one-to-one.

13. Suppose the set A has m elements and the set B has n elements. By Exercise 17 in Section 4.1, there are 2^{mn} relations from A to B and n^m functions from A to B.

   (a) If $m < n$, find the number of one-to-one functions from A to B.

   (b) If $m = n$, find the number of one-to-one functions from A to B.

   ☆ **(c)** If $m > n$, find the number of one-to-one functions from A to B.

   (d) If $m < n$, find the number of functions from A onto B.

   (e) If $m = n$, find the number of functions from A onto B.

   ⋆ **(f)** If $m = n + 1$, find the number of functions from A onto B.

   (g) If $m = n$, find the number of one-to-one correspondences from A onto B.

Proofs to Grade   **14.** Assign a grade of A (correct), C (partially correct), or F (failure) to each. Justify assignments of grades other than A.

   ⋆ **(a)** **Claim.**   The function $f: \mathbb{R} \times \mathbb{R} \rightarrow \mathbb{R}$ given by $f(x, y) = 2x - 3y$ is a surjection.

   "***Proof.***"   Suppose $(x, y) \in \mathbb{R} \times \mathbb{R}$. Then $x \in \mathbb{R}$, so $2x \in \mathbb{R}$. Also, $y \in \mathbb{R}$, so $3y \in \mathbb{R}$. Therefore $2x - 3y \in \mathbb{R}$. Thus $f(x, y) \in \mathbb{R}$, so f is a surjection.   ∎

   ☆ **(b)** **Claim.**   The function $f: [1, \infty) \rightarrow (0, \infty)$ defined by $f(x) = \dfrac{1}{x}$ maps onto $(0, \infty)$.

   "***Proof.***"   Suppose $w \in (0, \infty)$. Choose $x = \dfrac{1}{w}$. Then $f(x) = \dfrac{1}{\frac{1}{w}} = w$.

   Therefore the function f is onto $(0, \infty)$.   ∎

(c) **Claim.** If $f\colon A \xrightarrow{\text{onto}} B$ and $g\colon B \xrightarrow{\text{onto}} C$, then $g \circ f\colon A \to C$ maps onto C. (Theorem 4.3.1)

"*Proof.*" Suppose $a \in A$. Then $f(a) \in B$. Since $f(a) \in B$, $g(f(a)) \in C$. Therefore, $(g \circ f)(a) = g(f(a)) \in C$, so $g \circ f$ is onto C. ■

★ **(d)** **Claim.** The function $f\colon \mathbb{R} \to \mathbb{R}$ given by $f(x) = 2x + 7$ is onto $\mathbb{R}$.

"*Proof.*" Suppose f is not onto $\mathbb{R}$. Then there exists $b \in \mathbb{R}$ with $b \notin \text{Rng}(f)$. Thus for all real numbers x, $b \neq 2x + 7$. But $a = \frac{1}{2}(b - 7)$ is a real number and $f(a) = b$. This is a contradiction. Thus f is onto $\mathbb{R}$. ■

(e) **Claim.** Let I be the interval $(0, 1)$. The function $f\colon I \times I \to I$ given by $f(x, y) = x^y$ is a surjection.

"*Proof.*" Let $t \in I$. Then $0 < t < 1$, so $0 < t^2 < t < 1$, so $t^2 \in I$. Choose $x = t^2$ and $y = \frac{1}{2} \in I$. Then $f(x, y) = x^y = (t^2)^{1/2} = t$. Therefore, $I \subseteq \text{Rng}(f)$, so the function f is onto I. ■

(f) **Claim.** If $f\colon A \xrightarrow{1-1} B$ and $g\colon B \xrightarrow{1-1} C$, then $g \circ f\colon A \xrightarrow{1-1} C$ (Theorem 4.3.3).

"*Proof.*" We must show that if (x, y) and (z, y) are elements of $g \circ f$, then $x = z$. If $(x, y) \in g \circ f$, then there is $u \in B$ such that $(x, u) \in f$ and $(u, y) \in g$. If $(z, y) \in g \circ f$, then there is $v \in B$ such that $(z, v) \in f$ and $(v, y) \in g$. However, $(u, y) \in g$ and $(v, y) \in g$ imply $u = v$ since g is one-to-one. Then $(x, u) \in f$ and $(z, v) \in f$ and $u = v$; therefore, $x = z$, since f is one-to-one. Hence (x, y) and (z, y) in $g \circ f$ imply $x = z$. Therefore, $g \circ f$ is one-to-one. ■

(g) **Claim.** The function $f\colon \mathbb{R} \to \mathbb{R}$ given by $f(x) = 2x + 7$ is one-to-one.

"*Proof.*" Suppose x_1 and x_2 are real numbers with $f(x_1) \neq f(x_2)$. Then $2x_1 + 7 \neq 2x_2 + 7$ and thus $2x_1 \neq 2x_2$. Hence $x_1 \neq x_2$, which shows that f is one-to-one. ■

(h) **Claim.** The function f in part (e) is an injection.

"*Proof.*" Suppose (x, y) and (x, z) are in $I \times I$ and $f(x, y) = f(x, z)$. Then $x^y = x^z$. Dividing by x^z, we have $x^{y-z} = x^0 = 1$. Since $x \neq 1$ and $x^{y-z} = 1$, $y - z$ must be 0. Therefore, $y = z$. This shows that $(x, y) = (x, z)$, so f is an injection. ■

(i) **Claim.** The function f in part (e) is not an injection.

"*Proof.*" Both $\left(\frac{1}{2}, \frac{1}{2}\right)$ and $\left(\frac{1}{4}, \frac{1}{4}\right)$ are in $I \times I$. But

$$f\left(\frac{1}{4}, \frac{1}{4}\right) = \left(\frac{1}{4}\right)^{\frac{1}{4}} = \left(\left(\frac{1}{2}\right)^2\right)^{\frac{1}{4}} = \left(\frac{1}{2}\right)^{\frac{1}{2}} = f\left(\frac{1}{2}, \frac{1}{2}\right).$$ ■

4.4 One-to-One Correspondences and Inverse Functions

In this section we consider functions that have both the desirable properties of being one-to-one and mapping onto their codomains. The key role played by these functions in succeeding chapters suggests their importance in advanced mathematics.

> **DEFINITION** A function $f: A \to B$ is a **one-to-one correspondence** (or a **bijection**) iff f is one-to-one and onto B.

Example. Let $A = \{a, b, c\}$ and $B = \{p, q, r\}$. The function $h: A \to B$, where $h(a) = p$, $h(b) = r$, and $h(c) = q$ is a bijection from A onto B.

Example. The function $d: \mathbb{N} \to E^+$ given by $d(n) = 2n$ is a one-to-one correspondence between the natural numbers and the set E^+ of positive even integers.

Example. Let $F: \mathbb{N} \times \mathbb{N} \to \mathbb{N}$ be defined by $F(m, n) = 2^{m-1}(2n - 1)$. For example, $F(1, 3) = 2^0(2 \cdot 3 - 1) = 5$, $F(3, 1) = 2^2(2 \cdot 1 - 1) = 4$, and $F(5, 5) = 2^4(2 \cdot 5 - 1) = 144$. The function F is a one-to-one correspondence.

Proof.

(i)   To show that F is onto $\mathbb{N}$, let $s \in \mathbb{N}$. We must show that $s = F(m, n)$ for some (m, n) in $\mathbb{N} \times \mathbb{N}$. If s is even, then s may be written as $2^k t$, where $k \geq 1$ and t is odd. Since t is odd, $t = 2n - 1$ for some $n \in \mathbb{N}$. Choosing $m = k + 1$, we have $F(m, n) = 2^{m-1}(2n - 1) = 2^k t = s$. If s is odd, then $s = 2n - 1$ for some $n \in \mathbb{N}$. For this n and $m = 1$, we find $F(m, n) = 2^0(2n - 1) = s$. Therefore, F is onto $\mathbb{N}$.

(ii)   To show that F is one-to-one, suppose (m, n) and (r, s) are in $\mathbb{N} \times \mathbb{N}$ and $F(m, n) = F(r, s)$. We first prove that $m = r$. Without loss of generality, we may assume that $m \geq r$. ⟨*If $m < r$, we could relabel the arguments.*⟩ From $F(m, n) = F(r, s)$, we have $2^{m-1}(2n - 1) = 2^{r-1}(2s - 1)$, which implies $2^{m-r}(2n - 1) = 2s - 1$. Since the right side of the equality is odd, the left side is odd. Thus $2^{m-r} = 1$. Therefore, $m - r = 0$, and we conclude that $m = r$.

Dividing both sides of the equation $2^{m-1}(2n - 1) = 2^{r-1}(2s - 1)$ by 2^{m-1} ⟨$2^{m-1} = 2^{r-1}$⟩, we have $2n - 1 = 2s - 1$, which implies $2n = 2s$, or $n = s$. Thus $m = r$ and $n = s$, which gives $(m, n) = (r, s)$. Hence the function F is one-to-one.   ■

The last theorem of the previous section is a useful tool for constructing bijections. Applying this theorem to the examples above, we can say that there exists a one-to-one correspondence between $(\mathbb{N} \times \mathbb{N}) \cup \{a, b, c\}$ and the set $\mathbb{N} \cup \{p, q, r\}$. Since the domains of F and h in the examples above are disjoint, the union $F \cup h$ is a function that maps onto $\mathbb{N} \cup \{p, q, r\}$. Since their ranges are disjoint, the function $F \cup h$ is one-to-one. This observation is easier than actually defining a correspondence.

Combining Theorems 4.3.1 and 4.3.3, we have the following theorem.

Theorem 4.4.1   If $f: A \xrightarrow[\text{onto}]{1-1} B$ and $g: B \xrightarrow[\text{onto}]{1-1} C$, then $g \circ f: A \xrightarrow[\text{onto}]{1-1} C$. That is, the composite of one-to-one correspondences is a one-to-one correspondence.

Example. Let d and F be the bijections defined earlier in this section. Then the function $d \circ F$ is a bijection from $\mathbb{N} \times \mathbb{N}$ to E^+. As a sample computation we find that $d \circ F(1, 1) = d(F(1, 1)) = d(2^0(1)) = 2$. In general, $d \circ F((m, n)) = 2(2^{m-1}(2n-1)) = 2^m(2n-1)$.

It was observed in Section 4.2 that the inverse of a function is not always a function. The situation is clarified when we understand the connection between inverses and one-to-one mappings.

Theorem 4.4.2 Let F be a function from set A to set B.

(a) F^{-1} is a function from $\text{Rng}(F)$ to A iff F is one-to-one.
(b)   If F^{-1} is a function, then F^{-1} is one-to-one.

Proof.

(a)   Assume that F^{-1} is a function from $\text{Rng}(F)$ to A. To show that F is one-to-one, assume that $F(x) = F(y) = z$. ⟨Now show that $x = y$⟩. Then $(x, z) \in F$ and $(y, z) \in F$. Therefore, $(z, x) \in F^{-1}$ and $(z, y) \in F^{-1}$. Since F^{-1} is a function, $x = y$.

Now assume that F is one-to-one. To show that F^{-1} is a function, let $(x, y) \in F^{-1}$ and $(x, z) \in F^{-1}$. Therefore, $(y, x) \in F$ and $(z, x) \in F$. Since F is one-to-one, $y = z$.

(b)   See Exercise 4.   ∎

Example. The trigonometric function $\sin: \mathbb{R} \to \mathbb{R}$ is clearly not one-to-one because, for example, $\sin(0) = \sin(2\pi) = 0$. In Section 4.2 we defined the restriction of sine, $\text{Sin}: \left[-\frac{\pi}{2}, \frac{\pi}{2}\right] \to \mathbb{R}$, to the domain $\left[-\frac{\pi}{2}, \frac{\pi}{2}\right]$. The Sine function has range $[-1, 1]$ and is one-to-one and onto $[-1, 1]$. By Theorem 4.4.2 Sine has an inverse function $\text{Sin}^{-1}: [-1, 1] \to \left[-\frac{\pi}{2}, \frac{\pi}{2}\right]$.

We must be careful not to conclude that if $F: A \xrightarrow{1-1} B$, then $F^{-1}: B \xrightarrow{1-1} A$, since F may not be onto B. Recall that the domain and range of a relation and its inverse are interchanged. Therefore if $F: A \xrightarrow{1-1} B$, then all we can say by Theorem 4.4.2(a) is $F^{-1}: \text{Rng}(F) \xrightarrow{1-1} A$.

Corollary 4.4.3 If $F: A \xrightarrow[\text{onto}]{1-1} B$, then $F^{-1}: B \xrightarrow[\text{onto}]{1-1} A$. That is, the inverse of a one-to-one correspondence is a one-to-one correspondence.

The next result relates the concepts of injection, surjection, composition, and inversion. It gives a simple, practical method using composition to determine whether a given function is the inverse of a function F and, thereby, indirectly proves that F is a bijection. Part (b) of the theorem is a useful shortcut when someone wants to verify the inverse of a one-to-one correspondence: It suffices to test only one of the composites.

Theorem 4.4.4 Let $F: A \to B$ and $G: B \to A$. Then

(a) $G = F^{-1}$ iff $G \circ F = I_A$ and $F \circ G = I_B$.
(b)   If F is one-to-one and onto B, then $G = F^{-1}$ iff $G \circ F = I_A$ or $F \circ G = I_B$.

Proof.

(a) If $G = F^{-1}$, then $G \circ F = I_A$ and $F \circ G = I_B$, by Theorem 4.2.3. ⟨*We use the fact that* $\text{Rng}(F) = \text{Dom}(F^{-1}) = B$.⟩

Assume now that $G \circ F = I_A$ and $F \circ G = I_B$. Then F is one-to-one by Theorem 4.3.4 and F maps onto B by Theorem 4.3.2. Thus F^{-1} is a function on B and $F^{-1} = F^{-1} \circ I_B = F^{-1} \circ (F \circ G) = (F^{-1} \circ F) \circ G = I_A \circ G = G$.

(b) See Exercise 6. ∎

Another way to read Theorem 4.4.4 is that it explains what inverse functions do to each other: Whatever a function F does to x, applying the inverse to $F(x)$ takes you right back to x. You know this idea already from previous study. For instance, as properties of the natural logarithm and exponential functions, Theorem 4.4.4(a) says that

$$\text{for all } x > 0, \quad e^{\ln x} = x \quad \text{and} \quad \text{for all } x \in \mathbb{R}, \quad \ln e^x = x.$$

Example. The function $H(x) = \begin{cases} x^2 + 2 & \text{if } x \leq 0 \\ 2 - x^2 & \text{if } x > 0 \end{cases}$ maps $\mathbb{R}$ one-to-one and onto $\mathbb{R}$. Prove that $H^{-1} = K$, where

$$K(x) = \begin{cases} -\sqrt{x - 2} & \text{if } x \geq 2 \\ \sqrt{2 - x} & \text{if } x < 2 \end{cases}.$$

Proof. There are two cases to consider. If $x \leq 0$, $(K \circ H)(x) = K(H(x)) = K(x^2 + 2)$. Since $x^2 + 2 \geq 2$, the value of K is $-\sqrt{(x^2 + 2) - 2} = -\sqrt{x^2} = -|x| = x$ when $x \leq 0$. If $x > 0$, $(K \circ H)(x) = K(H(x)) = K(2 - x^2)$. Since $2 - x^2 < 2$, the value of K is $\sqrt{2 - (2 - x^2)} = \sqrt{x^2} = x$ if $x > 0$. In either case, $(K \circ H)(x) = x$, so $K = H^{-1}$. ∎

Example. Let $F: \mathbb{R} \to \mathbb{R}$ and $G: \mathbb{R} \to \mathbb{R}$ be functions where $F(x) = 2x + 1$ and $G(x) = \dfrac{x - 1}{2}$. We can prove that F (and likewise G) is one-to-one and onto $\mathbb{R}$ by showing that G and F are inverse functions.

Proof. We calculate the two composites:

$$\text{For all } x \text{ in } \mathbb{R}, (G \circ F)(x) = G(F(x)) = G(2x + 1) = \frac{(2x + 1) - 1}{2}$$

$$= \frac{2x}{2} = x.$$

$$\text{For all } x \text{ in } \mathbb{R}, (F \circ G)(x) = F(G(x)) = F\left(\frac{x - 1}{2}\right) = 2\left(\frac{x - 1}{2}\right) + 1$$

$$= (x - 1) + 1 = x.$$

It follows that F and G are one-to-one and onto $\mathbb{R}$. ∎

We conclude this section with the special case of functions that are bijections from a set to itself. These functions will appear again in Chapter 6 because they are essential to the understanding of the algebraic structures called groups.

> **DEFINITION** Let A be a nonempty set. A **permutation of A** is one-to-one correspondence from A onto A.

If f is a permutation of the set A and $a \in A$, then $f(a)$ is another (perhaps different) element of A. Thus f has the effect of arranging (or permuting) the elements of A.

Examples. Let A be the set $\{1, 2, 3\}$. The identity function I_A on A, the function $t = \{(1, 2), (2, 1), (3, 3)\}$ and the function $s = \{(1, 2), (2, 3), (3, 1)\}$ are three different permutations of A. The resulting arrangements (or permutations) of the elements of A are obtained by listing the images of 1, 2, and 3 in order.

From I_A we get the arrangement 1 2 3.
From t we get the arrangement 2 1 3.
From s we get the arrangement 2 3 1.

The word permutation can also be used to describe the *result* of arranging the elements of A in some specified order; for example, the list *dbac* is a permutation of *abcd*. (This was the meaning we used in Section 2.6.) This use is the basis for a simplified notation for a permutation of a finite set. By *listing the images in order*, we write the function s in the example above as [2 3 1], to indicate that 2 is the image of 1, 3 is the image of 2, and 1 is the image of 3. The permutation t is written [2 1 3] and I_A is [1 2 3].

Examples. The function $h = [4\,2\,6\,5\,3\,1]$ is a permutation of the set $B = \{1, 2, 3, 4, 5, 6\}$. It maps 1 to 4, 2 to 2, 3 to 6, 4 to 5, 5 to 3, and 6 to 1. The identity permutation on B is [1 2 3 4 5 6].

Previous results about one-to-one correspondences and inverses can be combined to yield this important list of facts about permutations.

Theorem 4.4.5

Let A be a nonempty set. Then

(a) the identity mapping I_A is a permutation of A.
(b) the composite of permutations of A is a permutation of A.
(c) the inverse of a permutation of A is a permutation of A.
(d) if f is a permutation of A, then $f \circ I_A = I_A \circ f = f$.
(e) if f is a permutation of A, then $f \circ f^{-1} = f^{-1} \circ f = I_A$.
(f) if f and g are permutations of A, then $(g \circ f)^{-1} = f^{-1} \circ g^{-1}$.

Proof. These statements follow immediately from:

(a) Exercise 3 of Section 4.3.
(b) Theorem 4.4.1.
(c) Corollary 4.4.3.
(d) Theorem 4.2.4.
(e) Theorem 4.2.5.
(f) Theorem 3.1.3(d). ∎

Composites and inverses of permutations can be easily computed using the notation described above. Remember to evaluate a composite from right to left, because the function on the right is applied first.

Examples.   For $A = \{1, 2, 3, 4\}$ and the functions $t = [2\ 4\ 3\ 1]$ and $s = [2\ 3\ 1\ 4]$, $t \circ s$ is $[2\ 4\ 3\ 1] \circ [2\ 3\ 1\ 4] = [4\ 3\ 2\ 1]$. The thought process begins with $[2\ 3\ 1\ 4]$, where "1 goes to 2." Then we see that in $[2\ 4\ 1\ 3]$, "2 goes to 4." Thus in the composite "1 goes to 4." We find the other images similarly. For example, s sends 2 to 3 and t sends 3 to 3, so the composite sends 2 to 3.

To find the inverse of the permutation $t = [2\ 4\ 3\ 1]$, we think of reversing its action. The permutation t sends 1 to 2 so the inverse of t sends 2 to 1 Since t sends 2 to 4, t^{-1} sends 4 to 2, etc. In this way we find that $t^{-1} = [4\ 1\ 3\ 2]$. We can verify that this is correct by computing $[2\ 4\ 3\ 1] \circ [4\ 1\ 3\ 2] = [1\ 2\ 3\ 4]$, which is the identity permutation.

Exercises 4.4

1. Show that each of these functions is a one-to-one correspondence.

   (a) $f: (2, \infty) \to (-\infty, -1)$ given by $f(x) = \dfrac{-x}{x-2}$.

   (b) $g: (-\infty, -4) \to (-\infty, 0)$ given by $g(x) = -|x+4|$.

   (c) $h: 8\mathbb{N} \to 10\mathbb{N}$, where $8\mathbb{N} = \{8k: k \in \mathbb{N}\}$, $10\mathbb{N} = \{10k: k \in \mathbb{N}\}$, and $h(x) = 1.25x$.

   (d) $G: \mathbb{N} \times \mathbb{N} \to 8\mathbb{N}$ given by $G(m, n) = 2^{m+2}(2n - 1)$.

   (e) $k: \mathbb{R} \to (3, \infty)$ given by $k(x) = 2e^x + 3$. You may use the fact that an exponential function maps $\mathbb{R}$ one-to-one and onto $(0, \infty)$.

2. Find a one-to-one correspondence between each of these pairs of sets. Prove that your function is one-to-one and onto the given codomain.

   (a) $\{a, b, c, d, e, f\}$ and $\{2, 4, 8, 16, 32, 64\}$

   (b) $\mathbb{N}$ and $\mathbb{N} - \{1\}$

   (c) $(3, \infty)$ and $(5, \infty)$

   (d) $(-\infty, 1)$ and $(-1, \infty)$

   (e) $12\mathbb{N}$ and $20\mathbb{N}$, where $12\mathbb{N} = \{12k: k \in \mathbb{N}\}$ and $20\mathbb{N} = \{20k: k \in \mathbb{N}\}$

3. For each one-to-one-correspondence, find the inverse function. Verify your answer by computing the composite of the function and its inverse.

   (a) $f: (0, \infty) \to (0, \infty)$ given by $f(x) = \dfrac{1}{x}$.

   ★ (b) $g: (-2, \infty) \to (-\infty, 4)$ given by $g(x) = \dfrac{4x}{x+2}$.

   (c) $h: \mathbb{R} \to (0, \infty)$ given by $h(x) = e^{x+3}$.

   (d) $G: (3, \infty) \to (5, \infty)$ given by $G(x) = \dfrac{5(x-1)}{x-3}$.

4. Prove part (b) of Theorem 4.4.2: If $F: A \to B$ and F^{-1} is a function, then F is one-to-one.

5. (a) Assume that $F: A \xrightarrow[\text{onto}]{1-1} B$. Prove that $G = F^{-1}$ iff $G \circ F = I_A$ or $F \circ G = I_B$ (Theorem 4.4.4, part (b)).

   (b) Give an example of sets A and B and functions F and G such that $F: A \to B$, $G: B \to A$, $G \circ F = I_A$ and $G \neq F^{-1}$.

6. Let $F: A \to B$ and $G: B \to A$. Use the results of this section to prove that if $G \circ F = I_A$ and $F \circ G = I_B$, then $F: A \xrightarrow[\text{onto}]{1-1} B$ and $G: B \xrightarrow[\text{onto}]{1-1} A$.

7. Use the one-to-one correspondences $\ln: (0, \infty) \to \mathbb{R}$ and $f: (2, \infty) \to (0, \infty)$, where $f(x) = x - 2$, to describe a one-to-one correspondence

   (a) from $(0, \infty)$ onto $(2, \infty)$.

   (b) from $(2, \infty)$ onto $\mathbb{R}$.

   (c) from $\mathbb{R}$ onto $(2, \infty)$.

8. Prove that if $f: A \xrightarrow[\text{onto}]{1-1} B$, $g: B \xrightarrow[\text{onto}]{1-1} C$, and $h: C \xrightarrow[\text{onto}]{1-1} D$, then $f^{-1} \circ g^{-1} \circ h^{-1}$ is a one-to-one correspondence from D to A.

9. Use the notation of this section to write these permutations of the set $C = \{1, 2, 3, 4, 5, 6, 7\}$.

   (a) I_C (b) $u = \{(1, 5), (6, 7), (4, 4), (5, 1), (3, 2), (2, 6), (7, 3)\}$

   ★ (c) $v = \{(1, 2), (2, 5), (3, 4), (5, 1), (4, 6), (6, 7), (7, 3)\}$

   (d) $w = \{(1, 2), (2, 1), (3, 4), (4, 3), (5, 5), (6, 7), (7, 6)\}$

   ★ (e) $u \circ v$ (f) $v \circ u$ (g) $w \circ w$ (h) $u \circ v \circ w$ ★ (i) u^{-1}

   (j) v^{-1} (k) $(u \circ v)^{-1}$ (l) $v^{-1} \circ u^{-1}$ (m) $u^{-1} \circ v^{-1}$

Proofs to Grade

10. Assign a grade of A (correct), C (partially correct), or F (failure) to each. Justify assignments of grades other than A.

    (a) **Claim.** If f is a one-to-one correspondence from A to B and g is a one-to-one correspondence from B to A, then $g = f^{-1}$.

    "Proof." Suppose f is a one-to-one correspondence from A to B and g is a one-to-one correspondence from B to A. Then by Theorem 4.4.1, $g \circ f$ is a one-to-one correspondence from A to A. Likewise, $f \circ g$ is a one-to-one correspondence from A to A. Then $f^{-1} = f^{-1} \circ (g \circ f) = f^{-1} \circ (f \circ g) = (f^{-1} \circ f) \circ g = I_A \circ g = g$. ∎

    (b) **Claim.** If f is a permutation of A, then $I_A \circ f = f \circ I_A$.

    "Proof." Since I_A is the identity, $I_A \circ f = f$. Also I_A is the identity, so $f \circ I_A = f$. Therefore $I_A \circ f = f \circ I_A$. ∎

    (c) **Claim.** Let $r = [1\ 2\ 4\ 3\ 5]$ and $s = [4\ 2\ 3\ 1\ 5]$ be permutations of $\{1, 2, 3, 4, 5\}$. Then the inverse of $r \circ s$ is $[3\ 2\ 4\ 1\ 5]$.

    "Proof." Let $r = [1\ 2\ 4\ 3\ 5]$ and $s = [4\ 2\ 3\ 1\ 5]$. Then $r^{-1} = r$ and $s^{-1} = s$. Therefore $(r \circ s)^{-1} = r^{-1} \circ s^{-1} = r \circ s = [3\ 2\ 4\ 1\ 5]$. ∎

    (d) **Claim.** If f and g are permutations of A, then $(f \circ g)^{-1} = g^{-1} \circ f^{-1}$.

    "Proof." We know by Theorem 4.4.5 that $f \circ g$ is a permutation of A, so $(f \circ g)^{-1}$ is a permutation of A. Also by Theorem 4.4.5, g^{-1} and f^{-1} are permutations of A, so $g^{-1} \circ f^{-1}$ is a permutation of A. By Theorem 4.4.4, we can check whether $g^{-1} \circ f^{-1}$ is the inverse of $f \circ g$ by computing their composite. We find that $(f \circ g) \circ (g^{-1} \circ f^{-1}) = f \circ (g \circ (g^{-1} \circ f^{-1})) = f \circ (g \circ g^{-1}) \circ f^{-1} = f \circ I_A \circ f^{-1} = (f \circ I_A) \circ f^{-1} = f \circ f^{-1} = I_A$. Therefore $(f \circ g)^{-1} = g^{-1} \circ f^{-1}$. ∎

4.5   Images of Sets

Let f be a function from A to B. Up to this point we have considered the mapping of individual elements in A to their images in B or considered pre-images of individual elements in B. The next step is to ask about collections of points in A or in B and what corresponds to them in the other set.

DEFINITIONS   Let $f: A \to B$ and let $X \subseteq A$ and $Y \subseteq B$. The **image of** X or **image set of** X is

$$f(X) = \{y \in B: y = f(x) \text{ for some } x \in X\}$$

and the **inverse image of** Y is

$$f^{-1}(Y) = \{x \in A: f(x) \in Y\}.$$

Example.   Let $A = \{0, 1, 2, 3, -1, -2, -3\}$, $B = \{0, 1, 2, 4, 6, 9\}$, $X = \{-1, 3\}$, $Y = \{4, 6\}$, and $f: A \to B$ be given by $f(x) = x^2$. Figure 4.5.1 shows that

$$f(X) = f(\{-1, 3\}) = \{1, 9\}$$
$$f(\{-3, 3\}) = \{9\}$$
$$f(A) = \{0, 1, 4, 9\}$$

and

$$f^{-1}(Y) = f^{-1}(\{4, 6\}) = \{-2, 2\}$$
$$f^{-1}(\{6\}) = \varnothing$$
$$f^{-1}(B) = A.$$

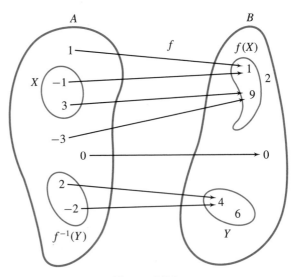

Figure 4.5.1

Note that f^{-1} is not a function from B to A, so it would not make sense to consider $f^{-1}(1)$. However, $f^{-1}(\{1\})$ is meaningful and equal to $\{1, -1\}$.

Examples. Let $f: \mathbb{N} \to \mathbb{N}$ be given by $f(x) = x + 4$. Then we have $f(1) = 5$ and $f(2) = 6$, so $f(\{1, 2\}) = \{5, 6\}$. Also, $f(\{10, 11, 12\}) = \{14, 15, 16\}$ and $f(\{x \in \mathbb{N}: x > 20\}) = \{x \in \mathbb{N}: x > 24\}$. The image set of $\mathbb{N}$ is $f(\mathbb{N}) = \{5, 6, 7, \ldots\}$, which is the range of f. There is no x in $\mathbb{N}$ such that $f(x) = 2$ or $f(x) = 3$, so $f^{-1}(\{2, 3\}) = \varnothing$. Also, $f^{-1}(\{5, 6\}) = \{1, 2\}$ is the same as $f^{-1}(\{1, 2, 3, 4, 5, 6\})$. The inverse image of $\mathbb{N}$ is $\text{Dom}(f) = \mathbb{N}$, which is the same as $f^{-1}(\{x \in \mathbb{N}: x \geq 5\})$.

Examples. Let $f: \mathbb{R} \to \mathbb{R}$ be given by $f(x) = x^2$. Then $f(\{-2, 2\}) = \{4\}$ since both $f(2) = 4$ and $f(-2) = 4$. From Figure 4.5.2 we see that $f([1, 2]) = [1, 4]$. Also, $f([-1, 0]) = [0, 1]$.

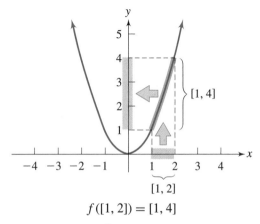

$$f([1, 2]) = [1, 4]$$

Figure 4.5.2

In this example it is tempting to guess that $f([-1, 2]) = [(-1)^2, 2^2] = [1, 4]$, but this is incorrect. By definition, $f([-1, 2])$ is the set of all images of elements of $[-1, 2]$. Since $-\frac{1}{2}$, 0, and 0.7 are in $[-1, 2]$, their images $\frac{1}{4}$, 0, and 0.49 must be in $f([-1, 2])$. Figure 4.5.3 shows that $f([-1, 2]) = [0, 4]$.

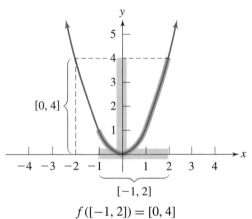

$$f([-1, 2]) = [0, 4]$$

Figure 4.5.3

We also have $f^{-1}(\{16\}) = \{-4, 4\}$, $f^{-1}(\{1, 4, 9\}) = \{-3, -2, -1, 1, 2, 3\}$, and $f^{-1}([-4, -3]) = \varnothing$ since all images of f are nonnegative. Even though $f([0, 2]) = [0, 4]$, $f^{-1}([0, 4]) = [-2, 2]$. Figure 4.5.4 illustrates why $f^{-1}([1, 4]) = [-2, -1] \cup [1, 2]$.

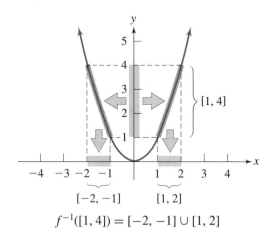

$$f^{-1}([1, 4]) = [-2, -1] \cup [1, 2]$$

Figure 4.5.4

Proofs involving set images take some special care because of the interplay between points, sets, and images of sets. Let $f: A \rightarrow B$, $D \subseteq A$, $E \subseteq B$, and $a \in A$. Here are some facts about images of sets that follow from the definitions:

important

(a)   If $a \in D$, then $f(a) \in f(D)$.
(b)   If $a \in f^{-1}(E)$, then $f(a) \in E$.
(c)   If $f(a) \in E$, then $a \in f^{-1}(E)$.
(d)   If $f(a) \in f(D)$, then $a \in D$, provided that f is one-to-one.

For part (d) we note that it is *not* correct to say that $f(a) \in f(D)$ always implies $a \in D$. For the function $f(x) = x^2$ and $D = [1, 2]$, we see that $f(-1) \in f(D)$, but $-1 \notin D$.

Examples.   Let $f: \mathbb{R} \rightarrow \mathbb{R}$ be the function given by $f(x) = x^2$. Let $A = [-3, 2]$ and $C = [1, 5]$. Then $f(A \cup C) = f([-3, 5]) = [0, 25]$ and $f(A) \cup f(C) = [0, 9] \cup [1, 25] = [0, 25]$, so $f(A \cup C) = f(A) \cup f(C)$.

On the other hand, $f(A \cap C) = f([1, 2]) = [1, 4]$ and $f(A) \cap f(C) = [1, 9]$, so $f(A \cap C) \subseteq f(A) \cap f(C)$, but it is not true that $f(A \cap C) = f(A) \cap f(C)$.

Theorem 4.5.1   Let $f: A \rightarrow B$, C and D be subsets of A, and E and F be subsets of B. Then

(a) $f(C \cap D) \subseteq f(C) \cap f(D)$.
(b) $f(C \cup D) = f(C) \cup f(D)$.
(c) $f^{-1}(E \cap F) = f^{-1}(E) \cap f^{-1}(F)$.
(d) $f^{-1}(E \cup F) = f^{-1}(E) \cup f^{-1}(F)$.

Proof.

(a)   Suppose $b \in f(C \cap D)$. Then $b = f(a)$ for some $a \in C \cap D$. Since $a \in C$
and $b = f(a)$, $b \in f(C)$. Also, $a \in D$ and $b = f(a)$, so $b \in f(D)$. Thus,
$b \in f(C) \cap f(D)$.

(c) $a \in f^{-1}(E \cap F)$
   iff $f(a) \in E \cap F$
   iff $f(a) \in E$ and $f(a) \in F$
   iff $a \in f^{-1}(E)$ and $a \in f^{-1}(F)$
   iff $a \in f^{-1}(E) \cap f^{-1}(F)$.

The proofs of parts (b) and (d) are left as exercises.   ∎

Finally, we note that if $f: A \rightarrow B$, then every subset X of A has a corresponding image set $f(X)$ that is a subset of B. This correspondence is a function from
$\mathcal{P}(A)$ to $\mathcal{P}(B)$ called the **induced function**, $f: \mathcal{P}(A) \rightarrow \mathcal{P}(B)$. Note that the same
letter, f, is used for the function and the induced function. The appropriate interpre-
tation is usually clear from the context.

Likewise, every subset $Y \subseteq B$ has a unique inverse image $f^{-1}(Y)$ that is a
subset of A. This correspondence is another induced function, $f^{-1}: \mathcal{P}(B) \rightarrow \mathcal{P}(A)$.

Exercises 4.5

1.   Let $A = \{1, 2, 3\}$, $B = \{4, 5, 6\}$, and $h = \{(1, 4), (2, 4), (3, 5)\}$.
★ **(a)**   Find the image of each of the 8 subsets of A.
   (b)   Find the inverse image of each of the 8 subsets of B.

2.   Let $f(x) = x^2 + 1$. Find
★ **(a)** $f([1, 3])$.   **(b)** $f([-1, 0] \cup [2, 4])$.
★ **(c)** $f^{-1}([-1, 1])$.   **(d)** $f^{-1}([-2, 3])$.
   (e) $f^{-1}([5, 10])$.   **(f)** $f^{-1}([-1, 5] \cup [17, 26])$.

3.   Let $f(x) = 1 - 2x$. Find
   (a) $f(A)$ where $A = \{-1, 0, 1, 2, 3\}$.   **(b)** $f(\mathbb{N})$.
   (c) $f^{-1}(\mathbb{R})$.   **(d)** $f^{-1}([2, 5])$.
   (e) $f((1, 4])$.   **(f)** $f(f^{-1}(f([3, 4]))$.

4.   Let $f: \mathbb{R} - \{0\} \rightarrow \mathbb{R}$ be given by $f(x) = x + \dfrac{1}{x}$. Find
   (a) $f((0, 2))$.   ★ **(b)** $f([1, 5])$.
★ **(c)** $f^{-1}((3, 4])$.   **(d)** $f^{-1}([0, 1))$.
   (e) $f(f^{-1}(\mathbb{R}))$.   **(f)** $f\left(f^{-1}\left(\left[\dfrac{-10}{3}, 10.1\right]\right)\right)$

5.   Let $f: \mathbb{N} \times \mathbb{N} \rightarrow \mathbb{N}$ be given by $f(m, n) = 2^m 3^n$. Find
   (a) $f(A \times B)$ where $A = \{1, 2, 3\}$, $B = \{3, 4\}$.
   (b) $f^{-1}(\{5, 6, 7, 8, 9, 10\})$.

6.   Let $f: \mathbb{R} \rightarrow \mathbb{R}$ be given by $f(x) = \begin{cases} 2x & \text{if } x \geq 1 \\ 2 - 2x & \text{if } x < 1 \end{cases}$. Give an example of

   (a) $a \in \mathbb{R}$ and $D \subseteq \mathbb{R}$ such that $f(a) \in f(D)$ and $a \notin D$.
   (b)   subsets A and C of $\mathbb{R}$ such that $f(A \cap C) \neq f(A) \cap f(C)$.
   (c)   a subset D of $\mathbb{R}$ such that $D \neq f^{-1}(f(D))$.
   (d)   a subset E of $\mathbb{R}$ such that $E \neq f(f^{-1}(E))$.

7.   Prove parts (b) and (d) of Theorem 4.5.1.

8.   Let $f: A \to B$, and let $\{D_\alpha: \alpha \in \Delta\}$ and $\{E_\beta: \beta \in \Gamma\}$ be families of subsets of A and B, respectively. Prove that

   ★   (a) $f\left(\bigcap_{\alpha \in \Delta} D_\alpha\right) \subseteq \bigcap_{\alpha \in \Delta} f(D_\alpha)$.

       (b) $f\left(\bigcup_{\alpha \in \Delta} D_\alpha\right) = \bigcup_{\alpha \in \Delta} f(D_\alpha)$.

       (c) $f^{-1}\left(\bigcap_{\beta \in \Gamma} E_\beta\right) = \bigcap_{\beta \in \Gamma} f^{-1}(E_\beta)$.

       (d) $f^{-1}\left(\bigcup_{\beta \in \Gamma} E_\beta\right) = \bigcup_{\beta \in \Gamma} f^{-1}(E_\beta)$.

☆   9.   Give an example of a function $f: \mathbb{R} \to \mathbb{R}$, and a family $\{D_\alpha: \alpha \in \Delta\}$ of subsets of $\mathbb{R}$ such that $f\left(\bigcap_{\alpha \in \Delta} D_\alpha\right) \neq \bigcap_{\alpha \in \Delta} f(D_\alpha)$.

10.   Let $f: A \to B$, $D \subseteq A$, and $E \subseteq B$. Prove that *use things on pg 222*
   ★   (a) $f(f^{-1}(E)) \subseteq E$.
       (b) $A - f^{-1}(E) \subseteq f^{-1}(B - E)$.
       (c) $f^{-1}(B - E) \subseteq A - f^{-1}(E)$.
   ★   (d) $E = f(f^{-1}(E))$ iff $E \subseteq \text{Rng}(f)$.
       (e) $D \subseteq f^{-1}(f(D))$.
       (f) $D = f^{-1}(f(D))$ iff $f(A - D) \subseteq B - f(D)$.

11.   Let $f: A \to B$ and let $X, Y \subseteq A$ and $U, V \subseteq B$. Prove that
       (a) $f(X) \subseteq U$ iff $X \subseteq f^{-1}(U)$.   ★   (b) $f(X) - f(Y) \subseteq f(X - Y)$.
       (c) $f^{-1}(U) - f^{-1}(V) = f^{-1}(U - V)$.

☆   12.   Let $f: A \to B$. Prove that if f is one-to-one, then $f(X) \cap f(Y) = f(X \cap Y)$ for all $X, Y \subseteq A$. Is the converse true? Explain.

13.   Let $f: A \to B$. Prove that if $X \subseteq A$ and f is one-to-one, then $f(A - X) = f(A) - f(X)$.

14.   Let $f: A \to B$. Prove that if $X \subseteq A$, $Y \subseteq B$, and f is a bijection, then $f(X) = Y$ iff $f^{-1}(Y) = X$.

15.   Let $f: A \to B$.
   ★   (a)   What condition on f will ensure that the induced function $f: \mathcal{P}(A) \to \mathcal{P}(B)$ is one-to-one?
       (b)   What condition on f will ensure that the induced function $f: \mathcal{P}(A) \to \mathcal{P}(B)$ is onto $\mathcal{P}(B)$?

16.   Let $f: A \to B$ and $K \subseteq B$. Prove that $f(f^{-1}(K)) = K \cap \text{Rng}(f)$.

17.   Let $f: A \to B$. Let T be the relation on A defined by $x \, T \, y$ iff $f(x) = f(y)$. By Exercise 18 (a) of Section 4.1, T is an equivalence relation on A. Describe the partition of A associated with T.

Proofs to Grade
18. Assign a grade of A (correct), C (partially correct), or F (failure) to each. Justify assignments of grades other than A.

★ **(a) Claim.** If $f: A \to B$ and $X \subseteq A$, then $f^{-1}(f(X)) \subseteq X$.
  "Proof." If $x \in f^{-1}(f(X))$, then by definition of f^{-1}, $f(x) \in f(X)$. Therefore $x \in X$. Thus $f^{-1}(f(X)) \subseteq X$. ∎

(b) Claim. If $f: A \to B$ and $X \subseteq A$, then $X \subseteq f^{-1}(f(X))$.
  "Proof." Suppose $z \in X$. Then $f(z) \in f(X)$. Therefore we conclude that $z \in f^{-1}(f(X))$, which proves the set inclusion. ∎

(c) Claim. If $f: A \to B$ and $\{D_\alpha : \alpha \in \Delta\}$ is a family of subsets of A, then
$$\bigcap_{\alpha \in \Delta} f(D_\alpha) \subseteq f\left(\bigcap_{\alpha \in \Delta} D_\alpha\right).$$
  "Proof." Suppose $y \in \bigcap_{\alpha \in \Delta} f(D_\alpha)$. Then $y \in f(D_\alpha)$ for all α. Thus there exists $x \in D_\alpha$ such that $f(x) = y$, for all α. Then $x \in \bigcap_{\alpha \in \Delta} D_\alpha$ and $f(x) = y$, so $y \in f\left(\bigcap_{\alpha \in \Delta} D_\alpha\right)$. Therefore, $\bigcap_{\alpha \in \Delta} f(D_\alpha) \subseteq f\left(\bigcap_{\alpha \in \Delta} D_\alpha\right)$. ∎

4.6   Sequences

In calculus sequences play a central role in the representation of functions using infinite series. Sequences are also important because of their usefulness in characterizing a number of important properties of the real numbers. This section, which is devoted to sequences of real numbers and the fundamentals of convergent and divergent sequences, is a prerequisite for Section 7.4.

As defined in Section 4.1, a **sequence** is a function with domain $\mathbb{N}$. If x is a sequence and $n \in \mathbb{N}$, the image of n, usually written x_n instead of $x(n)$, is called the **nth term of the sequence x.**

Examples. The sequence x of odd positive integers has nth term $x_n = 2n - 1$. The first few terms of x are $1, 3, 5, 7, \ldots$. The sequence y, where $y_n = (-1)^n$, has range $\{-1, 1\}$ since its terms are alternately -1 and 1.

The sequence x whose nth term is $x_n = \dfrac{(-1)^n}{n + 1}$ illustrates the convergence property of sequences. The first few terms of x are $-\dfrac{1}{2}, \dfrac{1}{3}, -\dfrac{1}{4}, \dfrac{1}{5}, -\dfrac{1}{6}, \ldots$ as shown in Figure 4.6.1. The 99th term is $-\dfrac{1}{100}$ and the 1000th term is $\dfrac{1}{1001}$. Evidently

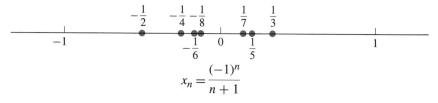

Figure 4.6.1

x_n is near 0 when n is large, and the farther out we go in the sequence, the closer the nth term is to 0. We say that the limit of this sequence is 0, and make this notion precise in the following definition.

DEFINITIONS   For a sequence x of real numbers and a real number L, we say x **has limit L** (or x **converges to L**) iff for every $\varepsilon > 0$ there exists a natural number N such that if $n > N$, then $|x_n - L| < \varepsilon$.
When x converges to the real number L, we write $\lim\limits_{n \to \infty} x_n = L$ or $x_n \to L$.
If no such number L exists we say x **diverges** or $\lim\limits_{n \to \infty} x_n$ **does not exist.**

In the definition of $x_n \to L$, we usually think of ε as being a small positive number, so the expression $|x_n - L| < \varepsilon$ means that the distance between x_n and L is small. What the definition guarantees is that no matter how small ε may be, all terms beyond a certain point in the sequence are within ε distance of L. The point where we can be sure this happens is the Nth term. When we work with a particular sequence, we need to be aware that if we were to make ε smaller, we would probably need to go farther out in the sequence (choose a larger number for N) to be sure the terms are close enough to L.

In symbols, we may write the definition of $\lim\limits_{n \to \infty} x_n = L$ as

$$(\forall \varepsilon > 0)(\exists N \in \mathbb{N})(\forall n \in \mathbb{N})(n > N \Rightarrow |x_n - L| < \varepsilon).$$

Based on this form, a proof of the statement $\lim\limits_{n \to \infty} x_n = L$ will usually have this structure:

Proof.   Let ε be a real number greater than 0.
Choose $N = $ ____ (Specify some value for N, typically in terms of ε.)
Let $n \in \mathbb{N}$ and suppose that $n > N$.
$\cdots$
Therefore $|x_n - L| < \varepsilon$.
We conclude that $\lim\limits_{n \to \infty} x_n = L$ ∎

The intermediate steps are often discovered through some preliminary scratch work, working backwards from $|x_n - L| < \varepsilon$ and continuing until we find a relationship between n and ε that will suggest a choice for N.

Example.   Earlier we claimed that the sequence x with nth term $x_n = \dfrac{(-1)^n}{n+1}$ converges and $\lim\limits_{n \to \infty} \dfrac{(-1)^n}{n+1} = 0$. Before proving this result, we first use the inequality

$|x_n - 0| < \varepsilon$ to derive a relationship between x_n and ε. For any $\varepsilon > 0$, since $|x_n - 0| = |x_n| = \left| \dfrac{(-1)^n}{n+1} \right| = \dfrac{1}{n+1}$, we need

$$\frac{1}{n+1} < \varepsilon,$$

$$n + 1 > \frac{1}{\varepsilon},$$

$$n > \frac{1}{\varepsilon} - 1.$$

This last inequality tells us how large n must be (and hence tells us a value for N) to ensure that $|x_n - 0| < \varepsilon$. Here is the proof:

Proof. Let ε be a positive real number. Let N be any integer greater than $\frac{1}{\varepsilon} - 1$. Suppose $n > N$. Then $n > \frac{1}{\varepsilon} - 1$, so $n + 1 > \frac{1}{\varepsilon}$ and $\dfrac{1}{n+1} < \varepsilon$. Since $|x_n - 0| = \left| \dfrac{(-1)^n}{n+1} \right| = \dfrac{1}{n+1}$, $|x_n - 0| < \varepsilon$. Therefore, $\displaystyle \lim_{n \to \infty} \frac{(-1)^n}{n+1} = 0$. ∎

The next theorem says that once we know that a sequence converges to a limit L, we know it cannot converge to any other number.

Theorem 4.6.1   If a sequence x converges, then its limit is unique.

Proof. Suppose $x_n \to L$ and $x_n \to M$ and $L \neq M$. Let $\varepsilon = \frac{1}{3}|L - M|$. ⟨*The idea of the proof is to suppose there are two different limits and select ε so small that the terms cannot simultaneously be within ε of each limit.*⟩ See Figure 4.6.2.

Since $x_n \to L$ and $x_n \to M$, there are natural numbers N_1 and N_2 such that $n > N_1$ implies $|x_n - L| < \varepsilon$ and $n > N_2$ implies $|x_n - M| < \varepsilon$. Let N be the larger of N_1, N_2. Suppose $n \in \mathbb{N}$ and $n > N$. Then $n > N_1$ and $n > N_2$, so

$$
\begin{aligned}
|L - M| &= |(L - x_n) + (x_n - M)| \\
&\leq |L - x_n| + |x_n - M| \\
&= |x_n - L| + |x_n - M| \\
&< \varepsilon + \varepsilon \\
&= \frac{2}{3}|L - M|.
\end{aligned}
$$

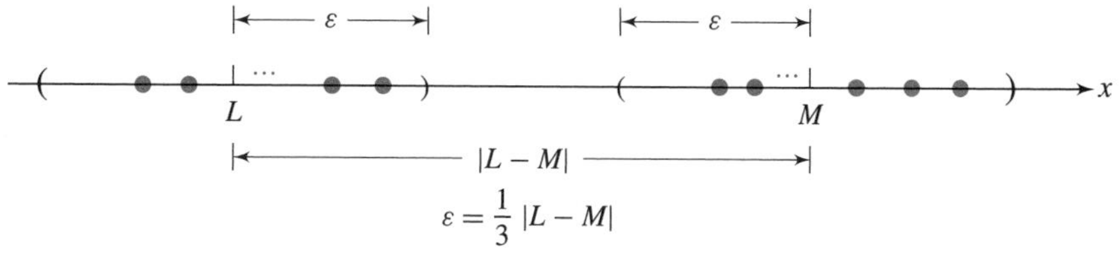

Figure 4.6.2

Thus the assumption $L \neq M$ leads to $|L - M| < \frac{2}{3}|L - M|$. Since $|L - M| > 0$, this is impossible. We conclude that the limit of x is unique. ∎

With some practice, you will be able to discern limits of some sequences. Sometimes it helps to calculate several terms to see the trend. For example, for the sequence x, where $x_n = \dfrac{\sin\left(\frac{1}{n}\right)}{\frac{1}{n}}$, you will be correct in guessing that $\lim\limits_{n \to \infty} x_n = 1$ by calculating several values such as $x_5 = 0.993346$, $x_{20} = 0.995833$, and $x_{60} = 0.999954$. Be careful with estimating limits, because appearances can be deceiving.

It should be clear that the terms of a constant sequence c given by $c_n = K$ for some real number K, are very close, in fact equal, to the number K. The proof that c converges to K is Exercise 4(a).

Sequences involving rational expressions can often be quickly evaluated. For example, let $x_n = \dfrac{12n^4 + 5n + 1}{70n^2 - 18n^4}$. For large values of n, the term $5n + 1$ is rather small compared to $12n^4$. Likewise, $-18n^4$ overpowers $70n^2$ for large values of n. Thus, for large values of n, the sequence x_n behaves much like the sequence $w_n = \dfrac{12n^4}{-18n^4} = -\dfrac{2}{3}$. We claim (correctly) that $\lim\limits_{n \to \infty} x_n = -\dfrac{2}{3}$.

Here is another example of a proof that a sequence converges. We need two preliminary steps in which we first estimate the limit and then find a relationship between n and ε.

Example. The sequence x given by $x_n = \dfrac{3n^2}{n^2 + 1}$ converges.

Scratchwork. We make a guess that x converges to 3. Next we must show that the limit is 3 by demonstrating that, for every $\varepsilon > 0$, there is a natural number N such that $n > N$ implies $\left|\dfrac{3n^2}{n^2 + 1} - 3\right| < \varepsilon$. Since $\left|\dfrac{3n^2}{n^2 + 1} - 3\right| = \left|\dfrac{-3}{n^2 + 1}\right| = \dfrac{3}{n^2 + 1}$, we require an integer N such that $n > N$ implies $\dfrac{3}{n^2 + 1} < \varepsilon$ or, equivalently, $n^2 + 1 > \dfrac{3}{\varepsilon}$. We know that $n^2 + 1 > n$, so by selecting N to be any natural number greater than $\dfrac{3}{\varepsilon}$, we have that $n > N$ implies $n^2 + 1 > N > \dfrac{3}{\varepsilon}$.

This scratchwork leads to the formal proof that follows.

Proof.   Let $\varepsilon > 0$. Let N be a natural number greater than $\frac{3}{\varepsilon}$. Suppose $n > N$. Then $n > \frac{3}{\varepsilon}$, and since $n^2 + 1 > n$, $n^2 + 1 > \frac{3}{\varepsilon}$. Therefore, if $n > N$, then $\dfrac{3}{n^2 + 1} < \varepsilon$.

Thus $\left| \dfrac{3n^2}{n^2 + 1} - 3 \right| < \varepsilon$. Therefore, $\displaystyle\lim_{n \to \infty} \frac{3n^2}{n^2 + 1} = 3$.   ∎

The sequence y given by $y_n = (-1)^n$ is an alternating sequence whose terms are $-1, 1, -1, 1, \ldots$. You would expect that y diverges because if the limit existed it would have to be close to both -1 and 1. To prove that y diverges, we prove a denial of the definition of convergence. For this purpose we choose ε to be 1, because no number can be less than 1 away from both -1 and 1.

Example.   Prove that the sequence y given by $y_n = (-1)^n$ diverges.

Proof.   Suppose that L is a real number. Let $\varepsilon = 1$. ⟨*We will show that for all $N \in \mathbb{N}$, there exists $n \in \mathbb{N}$ such that $n > N$ and $|y_n - L| \geq 1$.*⟩ Let $N \in \mathbb{N}$. If $L > 0$, let n be any odd integer greater than N; then $y_n = -1$ and $|y_n - L| = |-1 - L| = 1 + L > \varepsilon$. If $L \leq 0$, let n be any even integer greater than N; then $y_n = 1$ and $|y_n - L| = |1 - L| = 1 - L \geq 1 = \varepsilon$. In both cases we have shown that there is an $n \in \mathbb{N}$ such that $n > N$ and $|y_n - L| \geq 1$. Since for all real numbers L, y does not converge to L, the sequence y diverges.   ∎

Example.   Prove that the sequence x given by $x_n = 2n$ diverges.

Proof.   Assume that $\displaystyle\lim_{n \to \infty} x_n = L$ for some real number L. Let $\varepsilon = 1$. Then for some $N \in \mathbb{N}$, $|x_n - L| < 1$ for all $n > N$. Suppose $n > N$ and $n > \frac{1}{2}|L| + 1$. Then $2n > |L| + 1$, so $|x_n - L| = |2n - L| \geq 1 = \varepsilon$. This is a contradiction. We conclude that x diverges.   ∎

The next theorem is useful for determining and verifying limits without directly using to the definition. Sometimes referred to as the *Sandwich* or *Squeeze Theorem*, it states that if a sequence b has its nth term "sandwiched" below by a_n and above by c_n for all $n \in \mathbb{N}$, and both a and c converge to a number L, then b must also converge to L.

Theorem 4.6.2

Suppose a, b, and c are sequences of real numbers such that $a_n \leq b_n \leq c_n$ for all $n \in \mathbb{N}$. If $a_n \to L$ and $c_n \to L$, then $b_n \to L$.

Proof.   Suppose $a_n \to L$ and $c_n \to L$. Let $\varepsilon > 0$. There are natural numbers N_1 and N_2 such that $n > N_1$ implies $|a_n - L| < \varepsilon$ and $n > N_2$ implies $|c_n - L| < \varepsilon$.

Let N be the larger of N_1, N_2. Since $a_n \le b_n \le c_n$, $a_n - L \le b_n - L \le c_n - L$. Therefore, for $n > N$, $-\varepsilon < a_n - L \le b_n - L \le c_n - L < \varepsilon$. Thus $|b_n - L| < \varepsilon$ for all $n > N$, so $b_n \to L$. ∎

Example.   To illustrate Theorem 4.6.2, consider the sequence whose nth term is $x_n = \dfrac{\sin n}{n}$. Since sine is a function with range $[-1, 1]$, $-\dfrac{1}{n} \le \dfrac{\sin n}{n} \le \dfrac{1}{n}$, for all $n \in \mathbb{N}$. Because both $-\dfrac{1}{n} \to 0$ and $\dfrac{1}{n} \to 0$, we conclude that $\dfrac{\sin n}{n} \to 0$.

Exercises 4.6

1. List the first five terms of each sequence:

    (a) $a_n = \dfrac{n + 1}{2n + 3}$

    (b) $b_n = \sin\left(\dfrac{n\pi}{2}\right)$

    (c) $c_n = \dfrac{1}{n!}$

    (d) $d_n = 1 - 2^{-n}$

    (e) $e_n = \dfrac{n!}{2^n}$

2. Determine whether each sequence in Exercise 1 converges. If the sequence converges, identify or estimate the limit.

3. For each sequence x, estimate $\displaystyle\lim_{n \to \infty} x_n$ or determine that it does not exist.

    ★ (a) $x_n = 10n$

    (b) $x_n = \dfrac{10}{n}$

    ★ (c) $x_n = \dfrac{4n^2 + 7n + 12}{11 - n + 5n^2}$

    (d) $x_n = \dfrac{6n^3 + 5n^2 + 3n + 8}{10n^2 + 7n + 5}$

    ★ (e) $x_n = \dfrac{8n^2 + 4n + 1}{11n^3 - n + 5}$

    (f) $x_n = \dfrac{6n^3 + 5n^2 + 3n + 8}{10n^3 + 7n^2 + 5n - 8}$

    ★ (g) $x_n = \left(1 + \dfrac{1}{n}\right)^{2n}$

    (h) $x_n = \left(1 + \dfrac{1}{n}\right)^{-n}$

    ★ (i) $x_n = (-0.9)^n$

    (j) $x_n = (-1.1)^n$

    (k) $x_n = ((-1)^n + 1)\left(1 + \dfrac{1}{n}\right)^n$

    (l) $x_n = \dfrac{2^n + 3^n}{5^n}$

    (m) $x_n = \dfrac{n!}{n^n}$

    (n) $x_n = (n)^{\frac{1}{n}}$

4. (a) Let $K \in \mathbb{R}$, and let c be the constant sequence given by $c_n = K$. Prove that c converges.

    (b) Describe all possible sequences x of natural numbers that converge with limit 2.

5. For each sequence x, prove that x converges or diverges.

☆ (a) $x_n = 2^n$ (b) $x_n = \dfrac{n+1}{n}$

☆ (c) $x_n = n^2$ (d) $x_n = \dfrac{(-1)^n n}{2n+1}$

(e) $x_n = \dfrac{\cos n}{n}$ ☆ (f) $x_n = \sqrt{n+1} - \sqrt{n}$

(g) $x_n = \dfrac{7(1-n^2)}{n^2+2}$ (h) $x_n = \dfrac{6}{2^n}$

(i) $x_n = \dfrac{5{,}000}{n!}$ (j) $x_n = \sin\left(\dfrac{n\pi}{2}\right)$

(k) $x_n = \dfrac{n!}{n^n}$ (l) $x_n = \left(\dfrac{n}{2}\right)^n$

6. Prove that if $x_n \to L$ and $y_n \to M$ and $r \in \mathbb{R}$, then

★ (a) $x_n + y_n \to L + M$.   (b) $x_n - y_n \to L - M$.

   (c) $-x_n \to -L$.   (d) $r x_n \to rL$.

   (e) $x_n y_n \to LM$.   ☆ (f) $|x_n| \to |L|$.

☆ 7. (a) Prove that if $x_n \to L$ and $L \neq 0$, then there is a number N such that if $n \geq N$, then $|x_n| > \dfrac{|L|}{2}$.

   (b) Prove that if $x_n \to L$, $x_n \neq 0$ for all n, $L \neq 0$, and if $y_n \to M$, then $\dfrac{y_n}{x_n} \to \dfrac{M}{L}$.

8. A sequence y_n is a **subsequence** of x_n if and only if there is an increasing function $f: \mathbb{N} \to \mathbb{N}$ such that $y_n = x_{f(n)}$. For example, $y_n = x_{2n}$ is the sequence whose terms are just the even-numbered term of the sequence x_n.

   (a) Let $x_n = \dfrac{(-1)^n + 1}{n}$. Describe the subsequences x_{2n} and x_{2n-1}.

   (b) Prove that if a sequence x converges to L then for every real $\varepsilon > 0$, there exists a subsequence y of x such that $|y_n - L| < \varepsilon$ for all $n \in \mathbb{N}$.

   (c) Prove that if x_n converges to L and y_n is a subsequence of x, then y converges to L.

   (d) Prove that if x contains two convergent subsequences y and z, $y_n \to M$ and $z_n \to L$, and $M \neq L$, then x diverges.

9. A sequence x may be defined inductively by specifying a value for the first term and then specifying x_n for $n > 1$ in terms of earlier values in the sequence.

   (a) Let $x_1 = 10$ and, for $n > 1$, $x_n = \frac{1}{2} x_{n-1}$. Find the first six terms of x and determine whether x converges.

   (b) Let $x_1 = 1$ and, for $n > 1$, $x_n = 1 - x_{n-1}$. Find the first six terms of x and determine whether x converges.

   (c) Find the first ten terms of the Fibonnaci sequence f, where $f_1 = 1, f_2 = 1$, and, for $n > 2$, $f_n = f_{n-1} + f_{n-2}$.

(d)   If it exists, estimate the limit of the sequence x, where
$$x_n = \begin{cases} 3 & \text{if } n = 1 \\ \dfrac{2}{x_{n-1}} & \text{if } n \geq 2 \end{cases}.$$

(e)   If it exists, estimate the limit of the sequence x, where
$$x_n = \begin{cases} 1 & \text{if } n = 1 \\ 2\sqrt{x_{n-1}} & \text{if } n \geq 2 \end{cases}.$$

Proofs to Grade

10.   Assign a grade of A (correct), C (partially correct), or F (failure) to each. Justify assignments of grades other than A.

(a)   **Claim.**   If two sequences x and y both diverge, then $x + y$ diverges.

"Proof."   Suppose $\lim\limits_{n \to \infty} (x_n + y_n) = L$. Since x diverges, there exists $\varepsilon_1 > 0$ such that for all $N \in \mathbb{N}$ there exists $n > N$ such that $\left| x_n - \dfrac{L}{2} \right| \geq \varepsilon_1$. Since y diverges, there exists $\varepsilon_2 > 0$ such that, for all $N \in \mathbb{N}$, there exists $n > N$ such that $\left| y_n - \dfrac{L}{2} \right| \geq \varepsilon_2$. Let $\varepsilon = \dfrac{1}{2} \min\{\varepsilon_1, \varepsilon_2\}$. Then for all $N \in \mathbb{N}$, there exists $n > N$ such that

$$\begin{aligned} |(x_n + y_n) - L| &= \left| \left(x_n - \frac{L}{2} \right) + \left(y_n - \frac{L}{2} \right) \right| \\ &\geq \left| x_n - \frac{L}{2} \right| + \left| y_n - \frac{L}{2} \right| \\ &\geq \varepsilon_1 + \varepsilon_2 \\ &\geq \frac{1}{2}\varepsilon + \frac{1}{2}\varepsilon \\ &= \varepsilon. \end{aligned}$$

Therefore, $\lim\limits_{n \to \infty} (x_n + y_n) \neq L$.   ∎

⋆   **(b)**   **Claim.**   If the sequence x converges and the sequence y diverges, then $x + y$ diverges.

"Proof."   Suppose $x_n + y_n \to K$ for some real number K. Since $x_n \to L$ for some number L, $(x_n + y_n) - x_n \to K - L$; that is, $y_n \to K - L$. This is a contradiction. Thus $x_n + y_n$ diverges.   ∎

CHAPTER 5

Cardinality

How many elements are in the following set?

$$A = \left\{ \pi, 28, \sqrt{2}, \frac{1}{5}, -3, \star, X, \alpha \right\}.$$

After a short pause, you said "eight." You probably looked at π and thought "1,"
then looked at 28 and thought "2," and so on up through α and thought "8." What
you have done is match up the set A and the known set of eight elements $\{1, 2, 3, 4,$
$5, 6, 7, 8\}$ to conclude that A has eight elements. Counting the number of elements
in a set is essentially setting up a one-to-one correspondence between the set and a
known (standard) set of elements. Here is another example.

A shepherd has many dozens of sheep in his flock, but he cannot count beyond
ten. Each day he takes all his sheep out to graze, and each night he brings them back
into the fold. How can he be sure all his sheep have returned? The answer is that
he can count them with a one-to-one correspondence. He needs two containers, one
empty and one containing many pebbles, one pebble for each sheep. When the
sheep return in the evening, he transfers pebbles from one container to the other,
one at a time for each returning sheep. Whenever there are pebbles left over, he
knows that there are lost sheep. The solution to the shepherd's problem illustrates
the point that even though we may not have counted the sheep, we know that the
set of missing sheep and the set of leftover pebbles have the same number of
elements—because there is a one-to-one correspondence between them.

In this chapter we will make precise the idea informally introduced in Chapter 2
about the number of elements in a set. We will discuss finite and infinite sets and
discover that there are different sizes of infinite sets.

5.1 Equivalent Sets; Finite Sets

To determine whether two sets have the same number of elements, we see whether it
is possible to match the elements of the sets in a one-to-one fashion. This idea may

be conveniently described in terms of a one-to-one correspondence (a bijection) from one set to another.

DEFINITION   Two sets A and B are **equivalent** iff there exists a one-to-one function from A onto B. A and B are also said to be **in one-to-one correspondence**, and we write $A \approx B$.
    If A and B are not equivalent, we write $A \not\approx B$.

Example.   The sets $A = \{5, 8, \varphi\}$ and $B = \{r, p, m\}$ are equivalent. The function $f: A \rightarrow B$ given by $f(5) = r, f(8) = p$, and $f(\varphi) = m$ is one of six possible bijections that verify this.

Example.   The sets $C = \{x, y\}$ and $D = \{q, r, s\}$ are not equivalent. There are nine different functions from C to D. An examination of all nine will show that none of them is onto D. Since there is no one-to-one correspondence from C to D, the sets are not equivalent.

Example.   The set E of even integers is equivalent to D, the set of odd integers. To prove this, we let $f: E \rightarrow D$ be given by $f(x) = x + 1$. The function is one-to-one, because $f(x) = f(y)$ implies $x + 1 = y + 1$, which yields $x = y$. Also, f is onto D because if z is any odd integer, then $w = z - 1$ is even and $f(w) = w + 1 = (z - 1) + 1 = z$.

Example.   For $a, b, c, d \in \mathbb{R}$, with $a < b$ and $c < d$, the open intervals (a, b) and (c, d) are equivalent.

Proof.   ⟨*There are many bijections from* (a, b) *to* (c, d). *We choose the simplest: a linear function.*⟩ Let $f: (a, b) \rightarrow (c, d)$ be given by

$$f(x) = \frac{d - c}{b - a}(x - a) + c.$$

See Figure 5.1.1. Exercise 3 asks you to prove that f is a bijection.   ∎

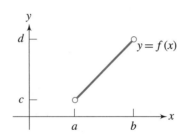

Figure 5.1.1

This last example shows that any two open intervals are equivalent, even when the intervals have different lengths. It also says, for example, that the open interval $(5, 6)$ is equivalent to the open interval $(1, 9)$, even though $(1, 9)$ is a longer interval and $(5, 6)$ is a proper subset of $(1, 9)$.

Example.   Let $\mathscr{F}$ be the set of all functions from $\mathbb{N}$ to $\{0, 1\}$. $\mathscr{F}$ is the set of all binary sequences. We will show $\mathscr{F} \approx \mathscr{P}(\mathbb{N})$, the power set of $\mathbb{N}$.

Proof.   ⟨*The key to this proof is to think of $\mathscr{F}$ as the set of all characteristic functions with domain $\mathbb{N}$. We associate each function with the appropriate subset of $\mathbb{N}$.*⟩ To show $\mathscr{F} \approx \mathscr{P}(\mathbb{N})$, we define $H: \mathscr{F} \to \mathscr{P}(\mathbb{N})$ as follows:

$$\text{for } g \in \mathscr{F}, \quad H(g) = \{x \in \mathbb{N}: g(x) = 1\}.$$

⟨*Note that under the function H, every function in $\mathscr{F}$ has an image in $\mathscr{P}(\mathbb{N})$.*⟩
     To show H is one-to-one, let $g_1, g_2 \in \mathscr{F}$ and $g_1 \neq g_2$. ⟨*We must show that $H(g_1) \neq H(g_2)$.*⟩ Since g_1 and g_2 are different functions with the same domain $\mathbb{N}$, there exists $n \in \mathbb{N}$ such that $g_1(n) \neq g_2(n)$. Without loss of generality, assume $g_1(n) = 1$ and $g_2(n) = 0$. ⟨*The case where $g_1(n) = 0$ and $g_2(n) = 1$ is similar.*⟩ Then $n \in \{x \in \mathbb{N}: g_1(x) = 1\} = H(g_1)$ and $n \notin \{x \in \mathbb{N}: g_2(x) = 1\} = H(g_2)$. Thus, $H(g_1) \neq H(g_2)$.
     To show that H is onto $\mathscr{P}(\mathbb{N})$, let $A \in \mathscr{P}(\mathbb{N})$. Then $A \subseteq \mathbb{N}$ and the characteristic function $\chi_A: \mathbb{N} \to \{0, 1\}$ is an element of $\mathscr{F}$. Furthermore, $H(\chi_A) = \{x \in \mathbb{N}: \chi_A(x) = 1\} = A$. Therefore, H is onto $\mathscr{P}(\mathbb{N})$.
     Because H is a bijection, $\mathscr{F} \approx \mathscr{P}(\mathbb{N})$.   ∎

Theorem 5.1.1   Equivalence of sets is an equivalence relation on the class of all sets.

Proof.   We must show that the relation $\approx$ is reflexive on the class of all sets, and is symmetric and transitive. (See Exercise 1.)   ∎

The next lemma will be particularly useful for showing equivalences of sets.

Lemma 5.1.2   Suppose A, B, C, and D are sets with $A \approx C$ and $B \approx D$.

(a)   If A and B are disjoint and C and D are disjoint, then $A \cup B \approx C \cup D$.
(b) $A \times B \approx C \times D$.

Proof.   Assume $A \approx C$ and $B \approx D$. Then there exist one-to-one correspondences $h: A \to C$ and $g: B \to D$.

(a)   By Theorem 4.3.5, $h \cup g: A \cup B \to C \cup D$ is a one-to-one correspondence. Therefore $A \cup B \approx C \cup D$.

(b)   Let $f: A \times B \to C \times D$ be given by $f(a, b) = (h(a), g(b))$. We leave it as Exercise 4 to show that f is a one-to-one correspondence. Therefore $A \times B \approx C \times D$.   ∎

Examples. The set $\{a, b\}$ is equivalent to $\{1, 2\}$ and the set $\{3, 4, x\}$ is equivalent to $\{5, 6, x\}$. To apply Lemma 5.1.2(a), we note that $\{a, b\}$ and $\{3, 4, x\}$ are disjoint and that $\{1, 2\}$ and $\{5, 6, x\}$ are disjoint. Therefore, $\{a, b, 3, 4, x\} \approx \{1, 2, 5, 6, x\}$.

By Lemma 5.1.2(b) the product of $\{a, b\}$ and $\{3, 4, x\}$ is equivalent to the product of $\{1, 2\}$ and $\{5, 6, x\}$:

$$\{(a, 3), (a, 4), (a, x), (b, 3), (b, 4), (b, x)\} \approx \{(1,5), (1,6), (1,x), (2,5), (2,6), (2,x)\}.$$

> **DEFINITIONS**   For each natural number k, let $\mathbb{N}_k = \{1, 2, 3, \ldots, k\}$.
> A set S is **finite** iff $S = \varnothing$ or $S \approx \mathbb{N}_k$ for some $k \in \mathbb{N}$.
> A set S is **infinite** iff S is not a finite set.

You should think of the set $\mathbb{N}_k$ as the standard finite set with k elements against which the sizes of other sets may be compared. For example, $\mathbb{N}_4 = \{1, 2, 3, 4\}$ is the standard set with 4 elements. The set $S = \{t, \frac{1}{2}, c, 99\}$ is finite because $S \approx \mathbb{N}_4$. The function $f: S \rightarrow \mathbb{N}_4$, where $f(t) = 1$, $f\left(\frac{1}{2}\right) = 2$, $f(c) = 3$, and $f(99) = 4$, is a bijection.

Sets such as $\mathbb{N}$, $\mathbb{R}$ and $\left\{\frac{1}{2}, \frac{1}{3}, \frac{1}{4}, \ldots\right\}$ are examples of infinite sets. These and other infinite sets will be discussed in the next section.

> **DEFINITIONS**   Let S be a finite set. If $S \approx \mathbb{N}_k$ for some natural number k, S has **cardinal number k** (or **cardinality k**) and we write $\overline{\overline{S}} = k$.
>
> If $S = \varnothing$ we say S has **cardinal number 0** (or **cardinality 0**) and write $\overline{\overline{\varnothing}} = 0$.

The set $S = \{t, \frac{1}{2}, c, 99\}$ has cardinality 4, and we write $\overline{\overline{S}} = 4$, because S is equivalent to $\mathbb{N}_4$. The set $A = \{\pi, 28, \sqrt{2}, \frac{1}{5}, -3, \star, X, \alpha\}$ has cardinality 8. The set $B = \{8, 7, 3, 7, 2, 7, 8\}$ is finite and $\overline{\overline{B}} = 4$ since $B = \{8, 7, 3, 2\} \approx \mathbb{N}_4$. Because the identity function $I_{\mathbb{N}_k}: \mathbb{N}_k \rightarrow \mathbb{N}_k$ is a one-to-one correspondence, $\mathbb{N}_k \approx \mathbb{N}_k$ and $\overline{\overline{\mathbb{N}_k}} = k$.

Our definition of $\overline{\overline{A}}$ for a finite set A agrees with our intuitive notion that $\overline{\overline{A}}$ is the number of elements in A. We use the same symbol $\overline{\overline{A}}$ for the cardinality of a finite set A as we used in Section 2.6 for the number of elements in A.

Because the definition of finite has two parts, proofs that a set is finite usually have two cases—the empty set case and the case in which the set is equivalent to $\mathbb{N}_k$ for some $k \in \mathbb{N}$.

Theorem 5.1.3   If A is finite and $B \approx A$, then B is finite.

Proof.   Suppose A is finite and $B \approx A$. If $A = \varnothing$, then $B = \varnothing$ (see Exercise 5). If $A \approx \mathbb{N}_k$ for some natural number k, then $B \approx \mathbb{N}_k$ by transitivity of $\approx$. In either case, B is finite. ∎

Our next goal is to show that every subset of a finite set is finite. The proof uses the results of the following two lemmas.

Lemma 5.1.4

If S is a finite set with cardinality k and x is any object not in S, then $S \cup \{x\}$ is finite and has cardinality $k + 1$.

Proof. If $S = \emptyset$, then S has cardinality 0. Then $S \cup \{x\} = \{x\}$ is finite because it is equivalent to $\mathbb{N}_1$. In this case $S \cup \{x\}$ has cardinality $0 + 1 = 1$.

If $S \neq \emptyset$, then $S \approx \mathbb{N}_k$ for some natural number k. Also, $\{x\} \approx \{k + 1\}$. Therefore, by Theorem 5.1.2(a), $S \cup \{x\} \approx \mathbb{N}_k \cup \{k + 1\} = \mathbb{N}_{k+1}$. Thus, $S \cup \{x\}$ is finite and has cardinality $k + 1$. ∎

Lemma 5.1.5

For every $k \in \mathbb{N}$, every subset of $\mathbb{N}_k$ is finite.

Proof. Let k be a natural number. ⟨*We prove by induction that every subset of $\mathbb{N}_k$ is finite.*⟩

(i)   If $k = 1$ and $A \subseteq \mathbb{N}_k$, then $A = \emptyset$ or $A = \mathbb{N}_1$. In both cases, A is finite.

(ii)   Suppose all subsets of $\mathbb{N}_k$ are finite for some number k. Let $A \subseteq \mathbb{N}_{k+1}$. Then $A - \{k + 1\}$ is a subset of $\mathbb{N}_k$ and, by the induction hypothesis, is finite. If $A = A - \{k + 1\}$ then A is finite. Otherwise, $A = (A - \{k + 1\}) \cup \{k + 1\}$, which is finite by Lemma 5.1.4. In both cases, A is finite.

(iii)   By the PMI, every subset of $\mathbb{N}_k$ is finite for every $k \in \mathbb{N}$. ∎

Theorem 5.1.6

Every subset of a finite set is finite.

Proof. Assume S is a finite and $T \subseteq S$. If $T = \emptyset$, then T is finite. Thus we may assume $T \neq \emptyset$ and hence $S \neq \emptyset$. Since $S \approx \mathbb{N}_k$ for some $k \in \mathbb{N}$, there is a one-to-one function f from S onto $\mathbb{N}_k$. Then the restriction of $f|_T$ of f to T is a one-to-one function from T onto $\mathrm{Rng}\,(f|_T)$. Therefore, T is equivalent to $\mathrm{Rng}\,(f|_T)$ (see Figure 5.1.2). But $\mathrm{Rng}\,(f|_T)$ is a subset of the finite set $\mathbb{N}_k$ and is finite by Lemma 5.1.5. Therefore, since T is equivalent to a finite set, T is finite. ∎

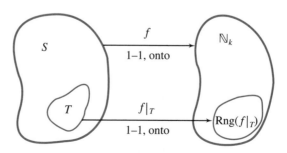

Figure 5.1.2

At this point you may think that Lemmas 5.1.4 and 5.1.5 and Theorem 5.1.6 accomplish nothing more than a proof of the obvious result that a subset of a finite set is finite. However, the real value of these results lies in the reasoning and in the use of functions to establish facts about cardinalities of finite sets. This experience will be helpful when we deal with infinite sets, because there our intuition may fail us.

The next main result is that the union of a finite number of finite sets is finite. We first prove the case for two disjoint sets. Notice that the proof is a rigorous development of the Sum Rule (Theorem 2.6.1) which says that if A has m elements, B has n elements, and A and B are disjoint, then $A \cup B$ has $m + n$ elements.

Theorem 5.1.7

(a)   If A and B are finite disjoint sets, then $A \cup B$ is finite and $\overline{\overline{A \cup B}} = \overline{\overline{A}} + \overline{\overline{B}}$.

(b)   If A and B are finite sets, then $A \cup B$ is finite and $\overline{\overline{A \cup B}} = \overline{\overline{A}} + \overline{\overline{B}} - \overline{\overline{A \cap B}}$.

(c)   If $A_1, A_2, \ldots, A_n$ are finite sets, then $\displaystyle\bigcup_{i=1}^{n} A_i$ is finite.

Proof.

(a)   Suppose A and B are finite sets and $A \cap B = \varnothing$. If $A = \varnothing$, then $A \cup B = B$; if $B = \varnothing$, then $A \cup B = A$. In either case $A \cup B$ is finite, and since $\overline{\overline{\varnothing}} = 0$, $\overline{\overline{A \cup B}} = \overline{\overline{A}} + \overline{\overline{B}}$.

Now suppose that $A \neq \varnothing$ and $B \neq \varnothing$. Let $A \approx \mathbb{N}_m$ and $B \approx \mathbb{N}_n$, and suppose that $f: A \to \mathbb{N}_m$ and $g: B \to \mathbb{N}_n$ are one-to-one correspondences. Let $H = \{m + 1, m + 2, \ldots, m + n\}$. Then $h: \mathbb{N}_n \to H$ given by $h(x) = m + x$ is a one-to-one correspondence, and thus $\mathbb{N}_n \approx H$. See Figure 5.1.3. Therefore, $B \approx H$ by transitivity. Finally, by Lemma 5.1.2, $A \cup B \approx \mathbb{N}_m \cup H = \mathbb{N}_{m+n}$, which proves that $A \cup B$ is finite and that $\overline{\overline{A \cup B}} = m + n$.

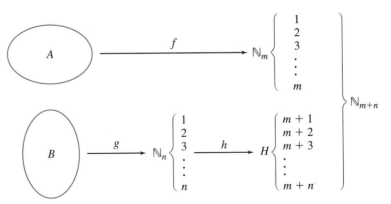

Figure 5.1.3

(b) Assume that A and B are finite sets. Since $B - A \subseteq B$, $B - A$ is finite. Therefore, by part (a), $A \cup B = A \cup (B - A)$ is finite. The proof that $\overline{\overline{A \cup B}} = \overline{\overline{A}} + \overline{\overline{B}} - \overline{\overline{A \cap B}}$ is Exercise 7.

(c) The proof of this part uses mathematical induction. See Exercise 8. ∎

Lemma 5.1.4 shows that adding one element to a finite set increases its cardinality by one. It is also true that removing one element from a finite set reduces the cardinality by one. The proof of Lemma 5.1.8 is left as Exercise 13.

Lemma 5.1.8   Let $r \in \mathbb{N}$ with $r > 1$. For all $x \in \mathbb{N}_r$, $\mathbb{N}_r - \{x\} \approx \mathbb{N}_{r-1}$.

There is a property of finite sets popularly known as the Pigeonhole Principle. In its informal version it says: "If a flock of n pigeons comes to roost in a house with r pigeonholes and $n > r$, then at least one hole contains more than one pigeon." If we think of the set of pigeons as $\mathbb{N}_n$ and the set of pigeonholes as $\mathbb{N}_r$, then the Pigeonhole Principle says any assignment of pigeons to pigeonholes (function from $\mathbb{N}_n$ to $\mathbb{N}_r$) is not one-to-one.

Theorem 5.1.9

The Pigeonhole Principle

Let $n, r \in \mathbb{N}$ and $f: \mathbb{N}_n \to \mathbb{N}_r$. If $n > r$, then f is not one-to-one.

Proof.   The proof proceeds by induction on the number n. Since $n > r$ and r is a natural number, we begin with $n = 2$.

(i) If $n = 2$, then $r = 1$. In this case f is a constant function with $f(1) = 1$ and $f(2) = 1$. Thus, f is not one-to-one.

(ii) Suppose the Pigeonhole Principle holds for some integer n; that is, suppose for all $r < n$, there is no one-to-one function from $\mathbb{N}_n$ to $\mathbb{N}_r$. Let $r < n + 1$. ⟨*The proof now proceeds by contradiction.*⟩ Suppose there is a one-to-one function $h: \mathbb{N}_{n+1} \to \mathbb{N}_r$. The restriction $h|_{\mathbb{N}_n}$ of h to $\mathbb{N}_n$ is one-to-one. The range of this function does not contain $h(n + 1)$. We may assume that $r > 1$, because otherwise h would be a constant function, which is not one-to-one. Now by Lemma 5.1.8, there is a one-to-one function $g: \mathbb{N}_r - \{h(n + 1)\} \to \mathbb{N}_{r-1}$. Let $f = g \circ (h|_{\mathbb{N}_n})$. Then $f: \mathbb{N}_n \to \mathbb{N}_r$ is one-to-one because the composite of one-to-one functions is one-to-one. This is a contradiction to the hypothesis of induction.

(iii) By the PMI, for every $n \in \mathbb{N}$ if $r < n$ there is no one-to-one function from $\mathbb{N}_{n+1}$ to $\mathbb{N}_r$. ∎

The Pigeonhole Principle is surprisingly powerful. See the discussion and references in Martin Gardner's *The Last Recreations* (Springer-Verlag, 1997) and the examples in Exercise 21. It also provides the following useful result about finite sets.

Corollary 5.1.10   If A is finite, then A is not equivalent to any of its proper subsets.

Proof.   We will show that $\mathbb{N}_k$ is not equivalent to any of its proper subsets and leave the general case as Exercise 14.

If $k = 1$, the only proper subset of $\mathbb{N}_1 = \{1\}$ is $\varnothing$ and $\{1\}$ is not equivalent to $\varnothing$. Thus, we may assume that $k > 1$. Suppose A is a proper subset of $\mathbb{N}_k$ and $A \approx \mathbb{N}_k$. Then there is a one-to-one function $f \colon \mathbb{N}_k \to A$ that is onto A.

Case 1. Suppose $k \notin A$. Then $A \subseteq \mathbb{N}_{k-1}$. In this case the function f maps $\mathbb{N}_k$ to $\mathbb{N}_{k-1}$, and f is one-to-one. This contradicts the Pigeonhole Principle.

Case 2. Suppose $k \in A$. Since A is a proper subset of $\mathbb{N}_k$, the set $\mathbb{N}_k - A$ is nonempty. Choose an element $y \in \mathbb{N}_k - A$. Let $A' = (A - \{k\}) \cup \{y\}$. Then $A \approx A'$ because the function $I_{A - \{k\}} \cup \{(k, y)\}$ is a one-to-one correspondence. Thus $A' \approx \mathbb{N}_k$, A' is a proper subset of $\mathbb{N}_k$, and $k \notin A'$. This is the situation of Case 1 with $\mathbb{N}_k$ and A' and again yields a contradiction. ∎

Corollary 5.1.10 tells us that our definition of cardinality for a finite set A corresponds to our informal understanding that $\overline{\overline{A}}$ is the number of elements in A: *The cardinality of a finite set is unique.* That is, if $A \approx \mathbb{N}_n$ and $A \approx \mathbb{N}_m$, then $n = m$. See Exercise 15.

Exercises 5.1

1. Prove Theorem 5.1.1. That is, show that the relation $\approx$ is reflexive, symmetric, and transitive on the class of all sets.

2. Which of the following sets are finite?
   ★ (a) the set of all grains of sand on Earth
     (b) the set of all positive integer powers of 2
   ★ (c) the set of all five-letter words in English
     (d) the set of rational numbers
   ★ (e) the set of rationals in $(0, 1)$ with denominator 2^k for some $k \in \mathbb{N}$
   ★ (f) $\{x \in \mathbb{R} \colon x^2 + 4x + 5 < 0\}$
     (g) the set of all stars within 100 light years of Earth.
     (h) $\{x \in \mathbb{R} \colon x^2 + 1 = 0\}$
     (i) $\{1, 3, 5\} \times \{2, 4, 6, 8\}$
     (j) $(1, 4) - (2, 3)$
     (k) $\{x \in \mathbb{N} \colon x \text{ is a prime}\}$
     (l) $\{x \in \mathbb{N} \colon x \text{ is composite}\}$
     (m) $\{x \in \mathbb{N} \colon x^2 + x \text{ is prime}\}$
     (n) $\{x \in \mathbb{R} \colon x \text{ is a solution to } 4x^8 - 5x^6 + 12x^4 - 18x^3 + x^2 - x = 0\}$
     (o) $\{x \in \mathbb{N} \colon x(10 - x) > 0\}$
     (p) the set of all complex numbers $a + bi$ such that $a^2 + b^2 = 1$

3. Complete the proof that any two open intervals (a, b) and (c, d) are equivalent by showing that $f(x) = \left(\dfrac{d - c}{b - a}\right)(x - a) + c$ maps one-to-one and onto (c, d).

4. Complete the proof of Lemma 5.1.2(b) by showing that if $h \colon A \to C$ and $g \colon B \to D$ are one-to-one correspondences, then $f \colon A \times B \to C \times D$ given by $f(a, b) = (h(a), g(b))$ is a one-to-one correspondence.

5. Show that if $A \approx \emptyset$, then $A = \emptyset$. (See also Exercise 13, Section 4.1.)

6. Let A and B be sets. Prove that
   ★ (a) if A is finite, then $A \cap B$ is finite.
   (b) if A is infinite and $A \subseteq B$, then B is infinite.

☆ 7. Using the methods of this section, prove that if A and B are finite sets, then $\overline{\overline{A \cup B}} = \overline{\overline{A}} + \overline{\overline{B}} - \overline{\overline{A \cap B}}$. This fact is a restatement of Theorem 2.6.1.

8. Prove part (c) of Theorem 5.1.7.

☆ 9. (a) Show that $A \approx A \times \{x\}$, for every set A and every object x.
   (b) Use Theorem 5.1.7(c) to prove that if A and B are finite, $A \times B$ is finite.

☆ 10. Define B^A to be the set of all functions from A to B. Show that if A and B are finite, then B^A is finite.

11. If possible, give an example of each of the following:
   (a) an infinite subset of a finite set.
   (b) a collection $\{A_i : i \in \mathbb{N}\}$ of finite sets whose union is finite.
   ★ (c) a finite collection of finite sets whose union is infinite.
   (d) finite sets A and B such that $\overline{\overline{A \cup B}} \neq \overline{\overline{A}} + \overline{\overline{B}}$.

12. Prove that if A is finite and B is infinite, then $B - A$ is infinite.

☆ 13. Prove that if $r > 1$ and $x \in \mathbb{N}_r$, then $\mathbb{N}_r - \{x\} \approx \mathbb{N}_{r-1}$ (Lemma 5.1.8).

14. Complete the proof of Corollary 5.1.10 by showing that if A is finite and B is a proper subset of A, then $B \not\approx A$.

☆ 15. Let A be a finite set. Prove that if $A \approx \mathbb{N}_n$ and $A \approx \mathbb{N}_m$, then $n = m$.

16. Prove or disprove:
   (a) If C is an infinite set and $C = A \cup B$, then at least one of the sets A or B is infinite.
   (b) Suppose A is a set and p is an object *not* in A. If $A \approx A \cup \{p\}$, then A is infinite.

17. Prove by induction on n that if $r < n$ and $f: \mathbb{N}_r \to \mathbb{N}_n$, then f is not onto $\mathbb{N}_n$.

18. Let A and B be finite sets with $A \approx B$. Suppose $f: A \to B$.
   ☆ (a) If f is one-to-one, show that f is onto B.
   ☆ (b) If f is onto B, prove that f is one-to-one.

☆ 19. Prove that if the domain of a function is finite, then the range is finite.

★ 20. Let A and B be finite sets with $\overline{\overline{A}} = m$ and $\overline{\overline{B}} = n$, and let f be a function from A to B. Prove that if $m > n$, then f is not one-to-one.

21. Give a proof using the Pigeonhole Principle:
   (a) The Italian village of Solomeo, near Perugia, has a population of 400. Prove that there are at least two village residents with the same birthday.
   ☆ (b) Let $S \subseteq \mathbb{N}_{99}$ such that S contains exactly 10 elements. Prove that S has two disjoint subsets with identical sums. For example, if S contains 4, 12, 18, 27, 36, 50, 61, 62, 70, and 98, then the elements of the sets $\{4, 12, 27, 36\}$ and $\{18, 61\}$ both add up to 79.

Proofs to Grade   22. Assign a grade of A (correct), C (partially correct), or F (failure) to each. Justify assignments of grades other than A.

(a) **Claim.**   If A and B are finite, then $A \cup B$ is finite.

"*Proof.*"   If A and B are finite, then there exist $m, n \in \mathbb{N}$ such that $A \approx \mathbb{N}_m$ and $B \approx \mathbb{N}_n$. Let $f \colon A \xrightarrow[\text{onto}]{1-1} \mathbb{N}_m$ and $h \colon B \xrightarrow[\text{onto}]{1-1} \mathbb{N}_n$. Then $f \cup h \colon A \cup B \xrightarrow[\text{onto}]{1-1} \mathbb{N}_{m+n}$, which shows that $A \cup B \approx \mathbb{N}_{m+n}$. Thus $A \cup B$ is finite.   ∎

★ (b) **Claim.**   If S is a finite, nonempty set, then $S \cup \{x\}$ is finite.

"*Proof.*"   Suppose S is finite and nonempty. Then $S \approx \mathbb{N}_k$ for some integer k.

Case 1. $x \in S$. Then $S \cup \{x\} = S$, so $S \cup \{x\}$ has k elements and is finite.

Case 2. $x \notin S$. Then $S \cup \{x\} \approx \mathbb{N}_k \cup \{x\} \approx \mathbb{N}_k \cup \mathbb{N}_1 \approx \mathbb{N}_{k+1}$. Thus $S \cup \{x\}$ is finite.   ∎

(c) **Claim.**   If $A \times B$ is finite, then A is finite.

"*Proof.*"   Choose any $b^* \in B$. Then $A \approx A \times \{b^*\}$. But $A \times \{b^*\} = \{(a, b^*) \colon a \in A\} \subseteq A \times B$. Since $A \times B$ is finite, $A \times \{b^*\}$ is finite. Since A is equivalent to a finite set, A is finite.   ∎

(d) **Claim.**   The set $\mathbb{N}$ is finite.

"*Proof.*"   For every n in $\mathbb{N}$, the set $\mathbb{N}_n$ is finite, because $\mathbb{N}_n \approx \mathbb{N}_n$. By Theorem 5.1.7, we know $\displaystyle\bigcup_{n=1}^{\infty} \mathbb{N}_n$ is finite. Since $\displaystyle\bigcup_{n=1}^{\infty} \mathbb{N}_n = \mathbb{N}$, we see that $\mathbb{N}$ is finite.   ∎

5.2   Infinite Sets

In this section we will verify the not-at-all-surprising result that some familiar sets, such as the sets of natural numbers, integers, and real numbers, are infinite. The result that many people find surprising is that there are different sizes of infinite sets. We will describe two infinite cardinal numbers and find that we can use them to "count" all of the elements of certain infinite sets.

Recall that an infinite set is defined as a nonempty set that cannot be put into a one-to-one correspondence with any of the sets $\mathbb{N}_k$. To prove that a set is infinite using this definition, we assume that the set is finite and that such a correspondence exists for some natural number k. We then find a contradiction. Another approach to proving that a set A is infinite is to make use of the contrapositive of Corollary 5.1.10:

If A is equivalent to one of its proper subsets, then A is infinite.

We can interpret this statement as a test for whether a set could be finite. To use this test we look for a suitable proper subset of A and a one-to-one correspondence between A and the subset. If we find such a set and correspondence, we conclude that A is not finite.

To demonstrate the use of these methods, we give two different proofs that $\mathbb{N}$ is infinite. Notice that the first proof resembles Euclid's proof that there are infinitely many primes (see Section 1.5).

Theorem 5.2.1   The set $\mathbb{N}$ of natural numbers is infinite.

First Proof.   Suppose $\mathbb{N}$ is finite. Since $\mathbb{N} \neq \varnothing$, there exists a natural number k such that $\mathbb{N} \approx \mathbb{N}_k$. Therefore, there exists a one-to-one function f from $\mathbb{N}_k$ onto $\mathbb{N}$. ⟨*We will show that f is not onto $\mathbb{N}$ by constructing a number that is not an image.*⟩ Let $n = \max\{f(1), f(2), \ldots, f(n)\} + 1$. Then $n \neq f(i)$ for any $i \in \mathbb{N}_k$. Therefore f is not onto $\mathbb{N}$, a contradiction. Hence $\mathbb{N}$ is an infinite set.   ∎

Second Proof.   Let E^+ be the set of even positive integers. The function $f: \mathbb{N} \to E^+$ defined by $f(x) = 2x$ is a one-to-one correspondence from $\mathbb{N}$ onto E^+. Thus, $\mathbb{N} \approx E^+$. Since E^+ is a proper subset of $\mathbb{N}$, we conclude that $\mathbb{N}$ is infinite.   ∎

The set of natural numbers is our first example of a set with infinite cardinality. The standard symbol for the cardinality of $\mathbb{N}$ uses the letter $\aleph$, aleph, which is the first letter of the Hebrew alphabet.

DEFINITIONS   Let S be a set. S is **denumerable** if and only if $S \approx \mathbb{N}$. For a denumerable set S, we say S has **cardinal number** $\aleph_0$ (or **cardinality** $\aleph_0$) and write $\overline{\overline{S}} = \aleph_0$.

Because $\mathbb{N}$ is equivalent to itself, $\mathbb{N}$ is denumerable and the cardinality of $\mathbb{N}$ is $\aleph_0$. The subscript 0, variously read "naught" or "null," indicates that $\aleph_0$ is the smallest infinite cardinal number, just as the integer 0 is the smallest finite cardinal number. The set $\mathbb{N}$ is our "standard" set for the cardinal number $\aleph_0$.

We showed earlier that the set E^+ of even positive integers is equivalent to $\mathbb{N}$. Therefore, E^+ is denumerable. Even though E^+ is a proper subset of $\mathbb{N}$, E^+ has the same number ($\aleph_0$) of elements as $\mathbb{N}$. Thus, although our intuition might tell us that only half of the natural numbers are even, it would be misleading to say that $\mathbb{N}$ has twice as many elements as E^+, or even to say that $\mathbb{N}$ has more elements than E^+.

Results like this may be surprising if you rely only on your knowledge of finite cardinal numbers to guide your insight into infinite cardinals. We have seen that if A and B are finite disjoint sets, where A has m elements and B has n elements, then $A \cup B$ has $m + n$ elements. The situation is more complicated when either A or B is infinite. We must rely on one-to-one correspondences to determine cardinality.

In Section 5.5 we will see that *every* infinite set is equivalent to one of its proper subsets. Together with Corollary 5.1.10, this will characterize infinite sets:

A set is infinite iff it is equivalent to one of its proper subsets.

The next theorem will show that the set of all integers is denumerable. Our proof constructs a bijection between $\mathbb{N}$ and $\mathbb{Z}$.

Theorem 5.2.2   The set $\mathbb{Z}$ is denumerable.

Proof.   Define the function $f: \mathbb{N} \to \mathbb{Z}$ by

$$f(x) = \begin{cases} \dfrac{x}{2} & \text{if } x \text{ is even} \\[2ex] \dfrac{1-x}{2} & \text{if } x \text{ is odd} \end{cases}.$$

We see that $f(1) = 0$, $f(2) = 1$, $f(3) = -1$, $f(4) = 2$, $f(5) = -2$, $f(6) = 3$, $f(7) = -3$, and so on.

To show that f is one-to-one, assume $f(x) = f(y)$ for some $x, y \in \mathbb{N}$. We first observe that if one of x or y is even and the other is odd, then only one of $f(x)$ or $f(y)$ is positive, so $f(x) \neq f(y)$. Therefore, x and y must have the same parity. If x and y are both even, $f(x) = f(y)$ implies $\frac{x}{2} = \frac{y}{2}$, and therefore, $x = y$. If x and y are both odd, then $\dfrac{1-x}{2} = \dfrac{1-y}{2}$, and again, $x = y$.

To show that f maps onto $\mathbb{Z}$, suppose $w \in \mathbb{Z}$. If $w > 0$, then $2w$ is even and $f(2w) = \dfrac{2w}{2} = w$. If $w \leq 0$, then $1 - 2w$ is an odd natural number and $f(1 - 2w) = \dfrac{1-(1-2w)}{2} = \dfrac{2w}{2} = w$. In both cases, $w \in \text{Rng}(f)$. Thus, f maps onto $\mathbb{Z}$.

Therefore $\mathbb{Z}$ is equivalent to $\mathbb{N}$.   ■

Example.   Let P be the set of reciprocals of positive integer powers of 2. By writing the set P as

$$P = \left\{ \frac{1}{2^k} : k \in \mathbb{N} \right\},$$

we see that there is a natural one-to-one correspondence between $\mathbb{N}$ and P. Since the function $h: \mathbb{N} \to P$ given by $h(n) = \dfrac{1}{2^n}$ is a bijection, P is denumerable.

Example.   Suppose we want to prove that the set $K = \{p, q, r\} \cup \{n \in \mathbb{N}: n \neq 5\}$ is denumerable. Then we need a bijection from $\mathbb{N}$ to K. Although there are many such functions, we'd like to construct one that can easily be seen to be one-to-one and onto K. Let g be the piecewise function from $\mathbb{N}$ to K given by

$$g(n) = \begin{cases} p & \text{if } n = 1 \\ q & \text{if } n = 2 \\ r & \text{if } n = 3 \\ n - 3 & \text{if } 4 \leq n \leq 7 \\ n - 2 & \text{if } n \geq 8 \end{cases}.$$

See Figure 5.2.1, which illustrates how we constructed this function. Then g is a bijection. Thus $\mathbb{N} \approx K$ and K is denumerable.

$$\mathbb{N}:\quad 1 \quad 2 \quad 3 \quad 4 \quad 5 \quad 6 \quad 7 \quad 8 \quad 9 \quad \dots \ n \ \dots$$
$$\downarrow \ \ \downarrow \ \ \downarrow \ \ \downarrow \ \ \downarrow \ \ \downarrow \ \ \downarrow \ \ \downarrow \ \ \downarrow \qquad\qquad \downarrow$$
$$K:\quad p \quad q \quad r \quad 1 \quad 2 \quad 3 \quad 4 \quad 6 \quad 7 \quad \dots \ n{-}2 \ \dots$$

Figure 5.2.1

Theorem 5.2.3

(a)   The set $\mathbb{N} \times \mathbb{N}$ is denumerable.

(b)   If A and B are denumerable sets, then $A \times B$ is denumerable.

Proof.

(a)   ⟨*We need to show that there is a bijection* $F\colon \mathbb{N} \times \mathbb{N} \to \mathbb{N}$.⟩ The function F in Section 4.4 given by $F(m, n) = 2^{m-1}(2n - 1)$ is one such function. Therefore, $\mathbb{N} \times \mathbb{N}$ is denumerable.

(b)   Since $A \approx \mathbb{N}$ and $B \approx \mathbb{N}$, $A \times B \approx \mathbb{N} \times \mathbb{N}$ by Theorem 5.1.2(a). By part (a), $\mathbb{N} \times \mathbb{N} \approx \mathbb{N}$. Therefore, $A \times B \approx \mathbb{N}$. Hence, $A \times B$ is denumerable.   ∎

DEFINITIONS   A set S is **countable** if and only if S is finite or denumerable. S is **uncountable** if and only if it is not countable.

Sets that are finite or denumerable are called countable because their elements can be "counted" using some or all of the natural numbers. "Counting" elements in a nonempty countable set S means setting up a one-to-one correspondence between S and $\mathbb{N}_k$ (when S is finite) or between S and the entire set $\mathbb{N}$ (when S is denumerable).

Figure 5.2.2 shows the relationship between finite, infinite, denumerable, countable, and uncountable sets. We see that every finite set is countable and every uncountable set is infinite. Since denumerable sets are those sets that are both infinite and countable, denumerable sets are sometimes referred to as *countably infinite* sets.

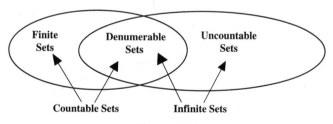

Figure 5.2.2

Examples.   Some sets that are both infinite and countable (that is, denumerable) are $\mathbb{N}$, $\mathbb{Z}$, and the set of even positive integers. Some countable finite sets include $\mathbb{N}_k, \varnothing, \{11, 7, 77, 3, 15, 79\}$ and $\{x \in \mathbb{R}: x^5 + 12x^3 - 21x^2 + 3x + 11 = 0\}$. There are infinite sets that are uncountable, such as $\mathbb{R}$ and $(0, 1)$, as we will now see.

Before showing that the open interval $(0, 1)$ is uncountable, we need to review decimal expressions for real numbers. In its decimal form, any real number in $(0, 1)$ may be written as $0.a_1a_2a_3a_4 \ldots$, where each a_i is an integer, $0 \le a_i \le 9$. In this form, $\frac{7}{12} = 0.583333\ldots$, which is abbreviated to $0.58\overline{3}$ to indicate that the 3 is repeated. The number $x = 0.a_1a_2a_3 \ldots$ is said to be in **normalized form** iff there is no k such that for all $n > k$, $a_n = 9$. For example, $0.82\overline{142857}$ and $\frac{2}{5} = 0.4\overline{0}$ are in normalized form, but $0.4\overline{9}$ is not. *Every real number can be expressed uniquely in normalized form.* Both $0.4\overline{9}$ and $0.5\overline{0}$ represent the same real number $\frac{1}{2}$, but only $0.5\overline{0}$ is normalized. The importance of normalizing decimals is that *two decimal numbers in normalized form are equal iff they have identical digits in each decimal position.*

Theorem 5.2.4   The open interval $(0, 1)$ is uncountable.

Proof.   ⟨*We must show that* $(0, 1)$ *is neither finite nor denumerable.*⟩ The interval $(0, 1)$ includes the subset $\left\{\dfrac{1}{2^k} : k \in \mathbb{N}\right\}$, which is infinite. Thus, by Theorem 5.1.6, $(0, 1)$ is infinite.

⟨*We now assume that* $(0, 1)$ *is denumerable and reach a contradiction.*⟩ Suppose $(0, 1)$ is denumerable. Then there is a function $f \colon \mathbb{N} \to (0, 1)$ that is one-to-one and onto $(0, 1)$. ⟨*The contradiction arises when we construct a number in* $(0, 1)$ *that is not in Rng* (f).⟩ Write the images of f, for each $n \in \mathbb{N}$, in normalized form:

$$f(1) = 0.a_{11}a_{12}a_{13}a_{14}a_{15}\ldots$$
$$f(2) = 0.a_{21}a_{22}a_{23}a_{24}a_{25}\ldots$$
$$f(3) = 0.a_{31}a_{32}a_{33}a_{34}a_{35}\ldots$$
$$f(4) = 0.a_{41}a_{42}a_{43}a_{44}a_{45}\ldots$$
$$\vdots$$
$$f(n) = 0.a_{n1}a_{n2}a_{n3}a_{n4}a_{n5}\ldots$$
$$\vdots$$

Now let b be the number $b = 0.b_1b_2b_3b_4b_5\ldots$, where

$$b_i = \begin{cases} 5 & \text{if } a_{ii} \ne 5 \\ 3 & \text{if } a_{ii} = 5 \end{cases}. \quad \langle \textit{The choices of 3 and 5 are arbitrary.} \rangle$$

Then $b \in (0, 1)$ because of the way it has been constructed. However, for each natural number n, b differs from $f(n)$ in the nth decimal place. Thus, $b \ne f(n)$ for any $n \in \mathbb{N}$, which means $b \notin$ Rng (f). Thus, f is not onto $(0, 1)$. This contradicts our assumption that f is onto $(0, 1)$. Therefore, $(0, 1)$ is not denumerable.   ■

The interval $(0, 1)$ is our first example of an uncountable set. We add it to our list of standard sets for defining cardinalities, which now consists of:

$$\varnothing, \mathbb{N}_k \text{ for every } k \in \mathbb{N}, \mathbb{N}, \text{ and } (0, 1).$$

> **DEFINITION** Let S be a set. S has **cardinal number c** iff S is equivalent to $(0, 1)$. We write $\overline{\overline{S}} = \mathbf{c}$ (which stands for **continuum**).

The cardinal number $\mathbf{c}$ is the only infinite cardinal other than $\aleph_0$ that will be identified by name. Implicit in this statement is that there are other infinite cardinals—an issue that will be addressed in Section 5.4.

There are many sets with cardinality $\mathbf{c}$. In Section 5.1 we showed that any two open intervals are equivalent. Therefore every open interval (a, b) for real numbers a and b with $a < b$, is equivalent to $(0, 1)$. Consequently:

Every open interval (a, b) has cardinality $\mathbf{c}$.

We see next that the set of all real numbers also has cardinality $\mathbf{c}$.

Theorem 5.2.5 The set $\mathbb{R}$ is uncountable and has cardinal number $\mathbf{c}$.

Proof. Define $f: (0, 1) \to \mathbb{R}$ by $f(x) = \tan\left(\pi x - \frac{\pi}{2}\right)$. See Figure 5.2.3. The function f is a contraction and translation of one branch of the tangent function and is one-to-one and onto $\mathbb{R}$. Thus $(0, 1) \approx \mathbb{R}$. ∎

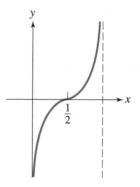

Figure 5.2.3

The proof of Theorem 5.2.5 used a trigonometric function for the one-to-one correspondence from $(0, 1)$ onto $\mathbb{R}$. Exercise 10 asks you to use a different function to show that $\mathbb{R} \approx (0, 1)$.

Example. Let C be the circle of radius $\frac{1}{2}$ with center $\left(0, \frac{1}{2}\right)$ and the point $(0, 1)$ removed, as shown in Figure 5.2.4 on the next page. For any point p in C, the line determined by $(0, 1)$ and p will intersect the x-axis in exactly one point. We define a function $f: C \to \mathbb{R}$ as follows: For each p in C let $f(p)$ be the x-coordinate of the point of intersection of the line determined by $(0, 1)$ and p. We see that different points p_1

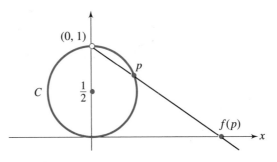

Figure 5.2.4

and p_2 will generate nonparallel lines and hence different values for $f(p_1)$ and $f(p_2)$. Thus f is one-to-one. Any point $m \in \mathbb{R}$ along the x-axis will determine a line through $(0, 1)$ that will intersect C in exactly one point q. For this point q, $f(q) = m$. Therefore, f is onto $\mathbb{R}$. Hence, f is a bijection and the set C is equivalent to $\mathbb{R}$.

Example.   The set $A = (0, 2) \cup [5, 6)$ also has cardinal number **c**. The function $f: (0, 1) \to A$, given by

$$f(x) = \begin{cases} 4x & \text{if } 0 < x < \dfrac{1}{2} \\[2mm] 2x + 4 & \text{if } \dfrac{1}{2} \le x < 1 \end{cases}$$

is a one-to-one correspondence between $(0, 1)$ and A. See Figure 5.2.5. We note that $(0, 1)$ is a proper subset of A, and A is a proper subset of $\mathbb{R}$, but all three sets have the same infinite cardinality.

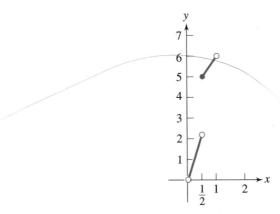

Figure 5.2.5

Exercises 5.2

1. Prove that if A is an infinite set and $A \approx B$, then B is an infinite set.

2. Prove that each of these sets is infinite.
  - **(a)** $A = \{1, \frac{1}{2}, \frac{1}{3}, \frac{1}{4}, \dots\}$
  - **(b)** $\mathbb{N} - \mathbb{N}_{15}$
  - **(c)** $(0, 0.001)$
  - **(d)** $(0, \infty)$

3. Prove that the following sets are denumerable.
  - ★ **(a)** D^+, the odd positive integers
  - **(b)** $3\mathbb{N}$, the positive integer multiples of 3
  - **(c)** $3\mathbb{Z}$, the integer multiples of 3
  - **(d)** $\{n : n \in \mathbb{N} \text{ and } n > 6\}$
  - ☆ **(e)** $\{x : x \in \mathbb{Z} \text{ and } x < -12\}$
  - **(f)** $\mathbb{N} - \{5, 6\}$
  - **(g)** $\{(x, y) \in \mathbb{N} \times \mathbb{R} : xy = 1\}$
  - **(h)** $\{x \in \mathbb{Z} : x = 1 (\text{mod } 5)\}$

4. Prove that the following sets have cardinality **c**.
  - ☆ **(a)** $(1, \infty)$
  - ☆ **(b)** (a, ∞), for any real number a
  - **(c)** $(-\infty, b)$, for any real number b
  - ☆ **(d)** $[1, 2) \cup (5, 6)$
  - **(e)** $(3, 6) \cup [10, 20)$
  - **(f)** $(0, 1] \cup (2, 3] \cup (4, 5)$
  - **(g)** $\mathbb{R} - \{0\}$

5. State whether each of the following is true or false.
  - **(a)** If a set A is countable, then A is infinite.
  - **(b)** If a set A is denumerable, then A is countable.
  - **(c)** If a set A is finite, then A is denumerable.
  - **(d)** If a set A is uncountable, then A is not denumerable.
  - **(e)** If a set A is uncountable, then A is not finite.
  - **(f)** If a set A is not denumerable, then A is uncountable.

6. ★ (a) Give an example of a bijection g from $\mathbb{N}$ to the set E^+ of positive even integers such that $g(1) = 20$.
  - **(b)** Give an example of a bijection h from $\mathbb{N}$ to E^+ such that $h(1) = 16$, $h(2) = 12$, and $h(3) = 2$.

7. Which sets have cardinal number $\aleph_0$? **c**?
  - ★ **(a)** $\mathbb{R} - [0, 1)$
  - **(b)** $(5, \infty)$
  - ★ **(c)** $\left\{\frac{1}{n} : n \in \mathbb{N}\right\}$
  - **(d)** $\{2^x : x \in \mathbb{N}\}$
  - ★ **(e)** $\{(p, q) \in \mathbb{R} \times \mathbb{R} : p + q = 1\}$
  - **(f)** $\{(p, q) \in \mathbb{R} \times \mathbb{R} : q = \sqrt{1 - p^2} \text{ and } q > 0\}$
  - **(g)** $\{(x, y) \in \mathbb{R} \times \mathbb{R} : x, y \in \mathbb{Z}\}$

· **8.** Give an example of denumerable sets A and B, neither of which is a subset of the other, such that

(a) $A \cap B$ is denumerable.

(b) $A \cap B$ is finite.

(c) $A - B$ is denumerable.

(d) $A - B$ is finite and nonempty.

9. It can be shown that the sets $[0, 1]$ and $(0, 1) \cup \mathbb{N}$ have cardinality **c**. Use these facts to show that there are sets A and B such that $\overline{\overline{A}} = \overline{\overline{B}} = \mathbf{c}$ and $A \cap B$ is

(a)   empty.   (b)   finite and nonempty.

(c)   denumerable.   (d)   uncountable.

10. Give another proof of Theorem 5.2.5 by showing that $f(x) = \dfrac{2x-1}{x(x-1)}$ is a one-to-one correspondence from $(0, 1)$ onto $\mathbb{R}$.

11. It can be shown that $\mathbb{R} \times \mathbb{R}$ has cardinality **c**. Use this fact to prove that the set $\mathbb{C}$ of complex numbers has cardinality **c**.

Proofs to Grade   · **12.** Assign a grade of A (correct), C (partially correct), or F (failure) to each. Justify assignments of grades other than A.

★   (a)   **Claim.**   Let W be the set of all natural numbers with tens digit 3 and let D^+ be the set of odd natural numbers. Then W and D^+ are equivalent.

"*Proof.*" W contains 30, 130, 230, 330, ... and many other natural numbers, so W is an infinite subset of $\mathbb{N}$. Therefore W is denumerable. D^+ is also denumerable, by Exercise 3(a). Therefore, $W \approx D^+$.   ∎

(b)   **Claim.**   If A is infinite and $x \notin A$, then $A \cup \{x\}$ is infinite.

"*Proof.*"   Let A be infinite. Then $A \approx \mathbb{N}$. Let $f: \mathbb{N} \to A$ be a one-to-one correspondence. Then $g: \mathbb{N} \to A \cup \{x\}$, defined by

$$g(t) = \begin{cases} x & \text{if } t = 1 \\ f(t-1) & \text{if } t > 1 \end{cases}$$

is one-to-one and onto $A \cup \{x\}$. Thus $\mathbb{N} \approx A \cup \{x\}$, so $A \cup \{x\}$ is infinite.   ∎

★   (c)   **Claim.**   If $A \cup B$ is infinite, then A and B are infinite.

"*Proof.*"   Assume that A and B are finite. Then by Theorem 5.1.7, $A \cup B$ is finite. Therefore if $A \cup B$ is infinite, A and B are infinite.   ∎

★   (d)   **Claim.**   If a set A is infinite, then A is equivalent to a proper subset of A.

"*Proof.*"   Let $A = \{x_1, x_2, \ldots\}$. Choose $B = \{x_2, x_3, \ldots\}$. Then B is a proper subset of A. The function $f: A \to B$ defined by $f(x_k) = x_{k+1}$ is clearly one-to-one and onto B. Thus $A \approx B$.   ∎

(e)   **Claim.**   The set $T = \{n \in \mathbb{Z}: n = 2 \,(\mathrm{mod}\,6)\}$ is denumerable.

"*Proof.*"   Define a function F on the integers by setting $F(z) = 6z + 20$. F is one-to-one because if $F(u) = F(v)$ then $6u + 20 = 6v + 20$, so $u = v$. Every element t of T has the form $6k + 2$ for some integer k, and $t = F(k - 3)$, so F maps onto T. Therefore T is equivalent to $\mathbb{Z}$, which is denumerable.   ∎

(f)   **Claim.**   The set $\mathbb{N}$ is infinite.

"*Proof.*"   The function given by $f(n) = n + 1$ is a one-to-one correspondence between $\mathbb{N}$ and $\mathbb{N} - \{1\}$, so $\mathbb{N}$ is equivalent to a proper subset of $\mathbb{N}$. Therefore, by Corollary 5.1.10, $\mathbb{N}$ is infinite.   ∎

(g)   **Claim.**   The set $5\mathbb{N} = \{5n : n \in \mathbb{N}\}$ is infinite.

"*Proof.*" $5\mathbb{N}$ is a subset of $\mathbb{N}$ and $\mathbb{N}$ is infinite. Then $5\mathbb{N}$ is infinite, because every subset of an infinite set is infinite.   ∎

(h)   **Claim.**   If A and B are denumerable, then $A \cup B$ is denumerable.

"*Proof.*"   Assume A and B are denumerable. Then there exist bijections h and g such that $h : A \rightarrow \mathbb{N}$ and $g : B \rightarrow \mathbb{N}$. Then $h \cup g : (A \cup B) \rightarrow \mathbb{N}$ is a bijection, so $A \cup B$ is denumerable.   ∎

5.3   Countable Sets

The first two sections of this chapter have presented several examples of countable sets—all of the finite sets in Section 5.1 and several denumerable sets such as $\mathbb{N}$, $\mathbb{Z}$, and $\mathbb{N} \times \mathbb{N}$ in Section 5.2. This section presents the essential facts needed by anyone who works with countable sets.

Because countable sets are those that are finite or denumerable, proofs of results about countable sets will often consider two cases.

Since there are $\aleph_0$ natural numbers and $\mathbf{c}$ real numbers, and $\mathbb{N} \subseteq \mathbb{Q} \subseteq \mathbb{R}$, we may suspect that the cardinality of $\mathbb{Q}$ is $\aleph_0$ or $\mathbf{c}$, or possibly some infinite cardinal number between them. We know that there are infinitely many rationals between any two rational numbers, so you might also suspect that $\mathbb{Q}$ is not denumerable. This is not the case. Georg Cantor* first showed that $\mathbb{Q}^+$ (the positive rationals) is indeed denumerable through a clever rearrangement of $\mathbb{Q}^+$.

Every element in $\mathbb{Q}^+$ may be expressed as $\frac{p}{q}$ for some $p, q \in \mathbb{N}$. Thus the elements of this set can be presented as in Figure 5.3.1 on the next page, where the nth row contains all the positive fractions with denominator n.

To show that $\mathbb{Q}^+$ is denumerable, Cantor listed the elements of $\mathbb{Q}^+$ in the order indicated by the arrows in Figure 5.3.1. First are all fractions in which the sum of the numerator and denominator is 2 $\left(\text{only } \frac{1}{1}\right)$, then those whose sum is 3 $\left(\frac{2}{1} \text{ and } \frac{1}{2}\right)$, then those whose sum is 4 $\left(\frac{3}{1}, \frac{2}{2}, \text{ and } \frac{1}{3}\right)$, and so on. Some rational numbers appear multiple times: For example, $\frac{2}{2}$ and $\frac{3}{3}$ are repetitions of the fraction $\frac{1}{1}$. Disregard all fractions that are not in lowest terms: $\frac{2}{2}, \frac{4}{2}, \frac{3}{3}, \frac{2}{4}, \frac{6}{2}, \ldots$. The remaining fractions have no repetitions and are circled in Figure 5.3.1.

---

* Georg Cantor (1845–1918) was a German mathematician who created set theory, primarily in papers that appeared in 1895 and 1897. This work can be seen as a revolution in mathematics, because he made it possible to think of actual infinite quantities, rather than the infinite as unattainable. He was the first to use one-to-one correspondences to describe set size, the first to show the rational numbers are countable, and the first to show that the reals are not countable. Several of his contemporaries did not accept some parts of his work.

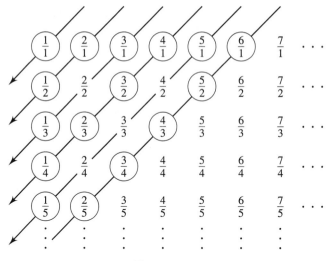

Figure 5.3.1

Starting at the top left corner of the array, traverse the array from upper right to lower left along each diagonal, assigning the natural numbers to the circled fractions, starting with $\frac{1}{1}$, then the diagonal $\frac{2}{1}, \frac{1}{2}$, and so forth. This pattern defines a one-to-one correspondence f from $\mathbb{N}$ to $\mathbb{Q}^+$ shown in Figure 5.3.2, where $f(1) = \frac{1}{1}, f(2) = \frac{2}{1}, f(3) = \frac{1}{2}, f(4) = \frac{3}{1}, f(5) = \frac{1}{3}$, etc. This correspondence can be used to establish the following theorem.* We omit the details of the proof.

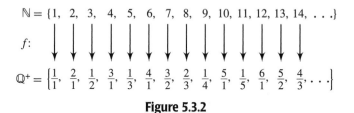

Figure 5.3.2

Theorem 5.3.1   The set $\mathbb{Q}^+$ of positive rational numbers is denumerable.

The two principal results of this section are that (1) every subset of a countable set is countable and (2) the union of countably many countable sets is countable. These two theorems are two of the most useful facts about cardinalities.

---

* Many different one-to-one correspondences are possible. For another interesting example, see N. Clakin and H. S. Wilf's article "Recounting the Rationals" in the *American Mathematical Monthly*, April 2000, pp. 360–363.

Theorem 5.3.2   Every subset of a countable set is countable.

> **Proof.**   Let A be a countable set. Assume $B \subseteq A$. If B is finite, then B is countable. Otherwise B is infinite, so A is infinite. Since A is infinite and countable, A is denumerable. Let f be a bijection from A to $\mathbb{N}$. The restriction of a one-to-one function is one-to-one, so $f|_B$ is a bijection from B to $C = \text{Rng}\,(f|_B)$. Therefore $B \approx C$.
>
> We now define a function $g \colon \mathbb{N} \to C$ by induction. ⟨*We also make use of the Well-Ordering Principle.*⟩
>
> Let $g(1)$ be the smallest integer in C. The set $C - \{g(1)\}$ is nonempty because C is infinite. For each $n \geq 1$, define $g(n + 1)$ to be the smallest element in the nonempty set $C - \{g(1), g(2), \ldots, g(n)\}$.
>
> If $r, s \in \mathbb{N}$ and $r < s$, then $g(r)$ is an element of the set $\{g(1), g(2), \ldots, g(s-1)\}$ but $g(s)$ is not. Therefore $g(r) \neq g(s)$. Thus g is an injection. Also, if $t \in C$ and there are k natural numbers less than t in C, then $g(k+1) = t$. Therefore g is onto C. Thus g is a bijection from $\mathbb{N}$ to C.
>
> Therefore, $\mathbb{N} \approx C \approx B$ and so B is denumerable.   ■

We have seen that the set $\mathbb{Q}^+$ is denumerable and therefore countable. Thus the subsets $\left\{\frac{1}{n} \colon n \in \mathbb{N}\right\}$, $\mathbb{Q} \cap (0, 1)$, and $\left\{\frac{5}{6}, \frac{6}{5}, \frac{3}{7}\right\}$ are countable sets.

Corollary 5.3.3   A set A is countable iff A is equivalent to some subset of $\mathbb{N}$.

> **Proof.**   If A is countable, then A is either finite or denumerable. Thus $A = \varnothing$, or $A \approx \mathbb{N}_k$ for some $k \in \mathbb{N}$, or $A \approx \mathbb{N}$. In each case, A is equivalent to some subset of $\mathbb{N}$.
>
> If A is equivalent to some subset of $\mathbb{N}$, then A is equivalent to a countable set, since all subsets of (countable) $\mathbb{N}$ are countable. Therefore, A is countable.   ■

We have seen (Theorem 5.1.7) that adding one or any finite number of elements to a finite set yields a finite set with larger cardinality. In the next three theorems, we consider adding elements to a denumerable set and find an important distinction between finite and denumerable sets: Adding finitely many or denumerably many elements does not change the cardinality of a denumerable set.

Theorem 5.3.4   If A is denumerable, then $A \cup \{x\}$ is denumerable.

> **Proof.**   If $x \in A$, then $A \cup \{x\} = A$, which is denumerable. Suppose that $x \notin A$. Since $\mathbb{N} \approx A$, there is a one-to-one function $f \colon \mathbb{N} \to A$ that is onto A. Define $g \colon \mathbb{N} \to A \cup \{x\}$ by
>
> $$g(n) = \begin{cases} x & \text{if } n = 1 \\ f(n-1) & \text{if } n > 1 \end{cases}.$$
>
> It is straightforward to verify that g is a one-to-one correspondence between $\mathbb{N}$ and $A \cup \{x\}$, which proves that $A \cup \{x\}$ is denumerable.   ■

Theorem 5.3.4 may be loosely restated as $\aleph_0 + 1 = \aleph_0$. Its proof is illustrated by the story of the Infinite Hotel,* attributed to the mathematician David Hilbert.[†] The Infinite Hotel has $\aleph_0$ rooms numbered $1, 2, 3, 4, \ldots$, and is full to capacity with one person in each room. You approach the desk clerk and ask for a room. When the clerk explains that each room is already occupied, you say, "There is room for me! For each n, let the person in room n move to room $n + 1$. Then I will move into room 1, and everyone will have a room as before." There are $\aleph_0 + 1$ people and they fit exactly into the $\aleph_0$ rooms. See Figure 5.3.3.

Figure 5.3.3

Rooms in the Infinite Hotel can also be found for any finite number k of additional people by asking each guest to move to room $n + k$ (Theorem 5.3.5). In the event of a fire alarm at the Grand Infinity Hotel across the street, the Infinite Hotel could even accommodate denumerably many additional guests by sending the current guest in room n to room $2n$ and assigning new guests to the odd numbered rooms (Theorem 5.3.6). Later we shall see that the clerk could find rooms if a denumerable number of additional people arrive a finite number of times (Corollary 5.3.9(c)) or even a denumerable number of times (Corollary 5.3.9(d)).

* The Infinite Hotel is one of the topics discussed in *Aha! Gotcha: Paradoxes to Puzzle and Delight* by Martin Gardner (Freeman, New York, 1981).
[†] David Hilbert (1862–1943) was a German mathematician who spent most of his career at the University of Göttingen. He is considered the most influential and creative mathematician of his time and was a staunch supporter of Cantor and his set theory. At the International Congress of Mathematicians in Paris in 1900 he proposed 23 open problems (the first one being the continuum hypothesis—see Section 5.5), which set the stage for much research in the twentieth century. Some of the 23 problems remain unsolved.

Theorem 5.3.5   If A is denumerable and B is finite, then $A \cup B$ is denumerable.

Proof.   See Exercise 3.   ∎

Theorem 5.3.6   If A and B are disjoint denumerable sets, then $A \cup B$ is denumerable.

Proof.   Let $f: \mathbb{N} \xrightarrow[\text{onto}]{1-1} A$ and $g: \mathbb{N} \xrightarrow[\text{onto}]{1-1} B$. Define $h: \mathbb{N} \to A \cup B$ via

$$
h(n) = \begin{cases} f\left(\dfrac{n+1}{2}\right) & \text{if } n \text{ is odd} \\[2mm] g\left(\dfrac{n}{2}\right) & \text{if } n \text{ is even} \end{cases}.
$$

⟨*The effect of h is to map the odd natural numbers to elements of A and the even natural numbers to B.*⟩ It is left as Exercise 4 to show that h is a one-to-one correspondence from $\mathbb{N}$ onto $A \cup B$. Therefore, $\mathbb{N} \approx A \cup B$.   ∎

We may apply the three previous theorems to produce many new examples of denumerable sets. We know from previous examples and exercises that the sets E^+ (the positive even integers) and $3\mathbb{N}$ (the integer multiples of 3) are denumerable. Therefore the sets $E^+ \cup \{5\}$, $3\mathbb{N} \cup \{1, 2, 4, 5\}$ and $E^+ \cup 3\mathbb{N}$ are denumerable.

For a more interesting example, we use the fact that $\mathbb{Q}^+$ is denumerable. By Theorem 5.3.4, $\mathbb{Q}^+ \cup \{0\}$ is denumerable. Clearly the set $\mathbb{Q}^-$ of negative rational numbers is denumerable, so by Theorem 5.3.6, $(\mathbb{Q}^+ \cup \{0\}) \cup \mathbb{Q}^-$ is denumerable. This gives us the following result.

Theorem 5.3.7   The set $\mathbb{Q}$ of all rational numbers is denumerable.

The second major theorem of this section is presented here because of its importance in dealing with countable sets. Because the proof requires the use of a new property of sets (the Axiom of Choice) that will be introduced in Section 5.5, the proof will appear in that section.

Theorem 5.3.8   Let $\mathscr{A}$ be a countable collection of countable sets. Then $\bigcup\limits_{A \in \mathscr{A}} A$ is countable.

We have already seen some theorems that are in fact special cases of Theorem 5.3.8: The union of finitely many finite sets is finite, and the union of a denumerable set with a finite set or of two disjoint denumerable sets is denumerable. Some other statements that are also immediate consequences of Theorem 5.3.8 are gathered together in the following corollary. Their placement here does not mean that each of these results requires the Axiom of Choice for a proof. In fact the first three of the four parts can be proved by methods we have already used. (See Exercises 6, 7, and 8.) If we do not refer to Theorem 5.3.8, a proof of part (d) requires the use of the Axiom of Choice.

Corollary 5.3.9   **(a)**   If $\mathscr{A}$ is a finite pairwise disjoint family of denumerable sets, then $\bigcup\limits_{A\in\mathscr{A}} A$ is countable.

(b)   If A and B are countable sets, then $A\cup B$ is countable.

(c)   If $\mathscr{A}$ is a finite collection of countable sets, then $\bigcup\limits_{A\in\mathscr{A}} A$ is countable.

(d)   If $\mathscr{A}$ is a denumerable family of countable sets, then $\bigcup\limits_{A\in\mathscr{A}} A$ is countable.

Theorem 5.3.8 provides another means to prove that the set $\mathbb{Q}^+$ of positive rationals is denumerable. $\mathbb{Q}^+$ may be written as $\bigcup\limits_{n\in\mathbb{N}} A_n$, where $A_n = \left\{\frac{a}{n}: a\in\mathbb{N}\right\}$. For each $n\in\mathbb{N}$, A_n is denumerable. By Theorem 5.3.8, $\mathbb{Q}^+$ is countable. Since $\mathbb{Q}^+$ is infinite and countable, $\mathbb{Q}^+$ is denumerable.

Example.   For each $k\in\mathbb{N}$, the set $\mathbb{N}_k$ is a finite set and therefore countable. Their union, $\bigcup\limits_{k\in\mathbb{N}}\mathbb{N}_k = \mathbb{N}$, is of course countable.

Example.   For each $n\in\mathbb{N}$, let $B_n = \{-n, n\}$. The countable union $\bigcup\limits_{n\in\mathbb{N}} B_n$ of these countable sets is $\mathbb{Z} - \{0\}$, which is countable.

We will give one more example of a denumerable set, significant for anyone with an interest in the theory of computation. A computer program is written in a given programming language and consists of a finite sequence of symbols. These symbols are selected from a finite set called an "alphabet" (typically all 26 upper and lowercase letters, the 10 digits, a blank space, certain punctuation marks, arithmetic operations, etc.). Recall, for example, that most calculators are not pre-programmed with logarithm functions other than logarithms for base 10 and base e. Here is a calculator program consisting of 80 characters to find $\log_a x$ for any base a, where $a > 0$ and $a \neq 1$:

PROGRAM:LOGBASEA	(16 symbols)
ClrHome	(7 symbols)
Input "LOG BASE?", A	(20 symbols)
Input "LOG OF?", B	(18 symbols)
ln(B)/ln(A) → C	(13 symbols)
Disp C	(6 symbols)

For each $n\in\mathbb{N}$, let P_n be the set of all programs with precisely n symbols. In most programming languages the first few P_i are empty sets. Our logarithm program above is an element of P_{80}.

Because there are only a finite number of symbols in our alphabet, there can be only a finite number of programs of length n. Therefore P_n is finite (and therefore countable) for all $n\in\mathbb{N}$. Also, since any computer program is finite in length, every program is an element of P_n for some n. Thus, the set of all possible programs is $\bigcup\limits_{n\in\mathbb{N}} P_n$.

By Theorem 5.3.8, this countable union of countable sets is countable. Hence only a countable number of programs could ever be written in a given language.

However, we saw in Section 5.1 that the set of all functions from $\mathbb{N}$ to $\{0, 1\}$ is equivalent to $\mathscr{P}(\mathbb{N})$, and we shall see in the next section that $\mathscr{P}(\mathbb{N})$ is uncountable. For a given programming language and its finite alphabet, this means that there are many functions from $\mathbb{N}$ to $\{0, 1\}$ for which there can be no computer programs to compute them. Put a different way, there are not enough solutions (programs) for all the possible problems (functions).

Exercises 5.3

1. What is the 28th term in the sequence of positive rationals produced by the counting process described in the discussion of Theorem 5.3.1?

2. Use a counting process similar to that described in the discussion of Theorem 5.3.1 to show that $\left\{\dfrac{2^x}{3^y} : x, y \in \mathbb{N}\right\}$ is denumerable.

3. Prove Theorem 5.3.5 by induction on the number of elements in the finite set B.

4. Complete the proof of Theorem 5.3.6 by showing that the function h as defined is one-to-one and onto $A \cup B$.

5. The Infinite Hotel is undergoing some remodeling, and consequently some of the rooms will be taken out of service. Show that, in a sense, this does not matter as long as only a "few" rooms are removed. That is,
    (a) prove that if A is denumerable and $x \in A$, then $A - \{x\}$ is denumerable.
    (b) prove that if A is denumerable and B is a finite subset of A, then $A - B$ is denumerable.

☆ 6. Without referring to Theorem 5.3.8, prove part (a) of Corollary 5.3.9: If $\mathscr{A} = \{A_i : i = 1, 2, 3, \ldots n\}$ is a finite pairwise disjoint family of denumerable sets, then $\bigcup_{A \in \mathscr{A}} A$ is countable.

☆ 7. Without referring to Theorem 5.3.8, prove part (b) of Corollary 5.3.9: If A and B are countable sets, then $A \cup B$ is countable.

☆ 8. Without referring to Theorem 5.3.8, prove part (c) of Corollary 5.3.9. If $\mathscr{A} = \{A_i : i = 1, 2, 3, \ldots n\}$ is a finite collection of countable sets, then $\bigcup_{A \in \mathscr{A}} A$ is countable.

9. Use the theorems of this section to prove that
    (a) an infinite subset of a denumerable set is denumerable.
    (b) if A is a countable subset of an uncountable set B, then $B - A$ is uncountable.
    (c) $\mathbb{Q} \cap (1, 2)$ is denumerable.
    (d) $\bigcup_{n=1}^{20} (\mathbb{Q} \cap (n, n + 1))$ is denumerable.
    (e) $\bigcup_{n \in \mathbb{N}} (\mathbb{Q} \cap (n, n + 1))$ is denumerable.
    (f) $\bigcup_{n \in \mathbb{N}} \left\{\dfrac{n}{2^k} : k \in \mathbb{N}\right\}$ is denumerable.

10. Prove or disprove:
  (a) If $A \subseteq B$ and B is denumerable, then A is denumerable.
  (b) If $A \subseteq B$ and A is denumerable, then B is denumerable.
  (c) $J \cup K$ is denumerable, where J is the set of all linear functions with slope 1 and rational y-intercept, and K is the set of all linear functions with slope 2 and integer y-intercept.
  (d) $\mathbb{Q} - \mathbb{Z}$ is denumerable.
  (e) If A and B are denumerable, then $A - B$ is denumerable.

☆ **11.** Prove that if $\{B_i : i \in \mathbb{N}\}$ is a denumerable family of pairwise disjoint distinct finite sets, then $\bigcup_{i \in \mathbb{N}} B_i$ is denumerable.

12. Give an example, if possible, of a family $A_1, A_2, A_3, \ldots$ of sets such that
  (a) each set A_i is finite and $\bigcup_{n=1}^{\infty} A_i$ is denumerable.
  (b) each set A_i is finite and $\bigcup_{n=1}^{\infty} A_i$ is finite.
  (c) each set A_i is finite, the family $\{A_i : i \in \mathbb{N}\}$ is pairwise disjoint, $A_i \neq A_j$ whenever $i \neq j$, and $\bigcup_{n=1}^{\infty} A_i$ is finite.

13. (a) Let S be the set of all sequences of 0's and 1's. For example, $1010101\ldots$, $1001101001\ldots$, and $011111\ldots$ are in S. Using a proof similar to that for Theorem 5.2.4, show that S is uncountable.
  ☆ **(b)** For each $n \in \mathbb{N}$, let T_n be the set of all sequences in S with exactly n 1's. Prove that T_n is denumerable for all $n \in \mathbb{N}$.
  (c) Let $T = \bigcup_{k=1}^{\infty} T_k$. Use a counting process similar to that described in the discussion of Theorem 5.3.1 to show that T is denumerable.

14. Let A be a denumerable set. Prove that
  (a) the set $\{B : B \subseteq A \text{ and } \overline{\overline{B}} = 1\}$ of all 1-element subsets of A is denumerable.
  (b) the set $\{B : B \subseteq A \text{ and } \overline{\overline{B}} = 2\}$ of all 2-element subsets of A is denumerable.
  (c) for every $k \in \mathbb{N}$, $\{B : B \subseteq A \text{ and } \overline{\overline{B}} = k\}$ is denumerable.
  (d) the set $\{B : B \subseteq A \text{ and } B \text{ is finite}\}$ of all finite subsets of A is denumerable. (Hint: Use Theorem 5.3.8.)

Proofs to Grade   **15.** Assign a grade of A (correct), C (partially correct), or F (failure) to each. Justify assignments of grades other than A.
  ★ **(a)** **Claim.** If A is denumerable, then $A - \{x\}$ is denumerable.
    "Proof." Assume A is denumerable.
    Case 1. If $x \notin A$, then $A - \{x\} = A$, which is denumerable by hypothesis.
    Case 2. Assume $x \in A$. Since A is denumerable, there exists $f : \mathbb{N} \xrightarrow[\text{onto}]{1-1} A$. Define g by setting $g(n) = f(n+1)$. Then $g : \mathbb{N} \xrightarrow[\text{onto}]{1-1} (A - \{x\})$, so $\mathbb{N} \approx A - \{x\}$. Therefore, $A - \{x\}$ is denumerable. ∎

(b) **Claim.** If A and B are denumerable, then $A \times B$ is denumerable.

   "Proof." Assume A and B are denumerable, but that $A \times B$ is not denumerable. Then $A \times B$ is finite. Since A and B are denumerable, they are not empty, so we can choose $a \in A$ and $b \in B$. Then, $A \approx A \times \{b\}$ and $B \approx \{a\} \times B$. Since $A \times B$ is finite, the subsets $A \times \{b\}$ and $\{a\} \times B$ are finite. Therefore, A and B are finite. This contradicts the statement that A and B are denumerable. We conclude that $A \times B$ is denumerable. ∎

(c) **Claim.** The set $\mathbb{Q}^+$ of positive rationals is denumerable.

   "Proof." Consider the positive rationals in the array in Figure 5.3.1. Order this set by listing all the rationals in the first row, then the second row, and so forth. Omitting fractions that are not in lowest terms, we have an ordering of $\mathbb{Q}^+$ in which every positive rational appears. Therefore, $\mathbb{Q}^+$ is denumerable. ∎

(d) **Claim.** If A and B are infinite, then $A \approx B$.

   "Proof." Suppose A and B are infinite sets. Let $A = \{a_1, a_2, a_3, \dots\}$ and $B = \{b_1, b_2, b_3, \dots\}$. Define $f: A \to B$ as shown:

$$\{a_1, a_2, a_3, a_4, \dots\}$$
$$\downarrow \ \downarrow \ \downarrow \ \downarrow$$
$$\{b_1, b_2, b_3, b_4, \dots\}.$$

   Then, since we never run out of elements in either set, f is one-to-one and onto B, so $A \approx B$. ∎

(e) **Claim.** $\mathbb{R} - \mathbb{Q}$ is uncountable.

   "Proof." $\mathbb{R}$ is uncountable and $\mathbb{R} - \mathbb{Q}$ is a subset of R. Every subset of an uncountable set is uncountable, so $\mathbb{R} - \mathbb{Q}$ is uncountable. ∎

5.4 The Ordering of Cardinal Numbers

When Georg Cantor developed set theory, he described a cardinal number of a set M as "the general concept which, with the aid of our intelligence, results from M when we abstract from the nature of its various elements and from the order of their being given." This definition was criticized as being less precise and more mystical than a definition in mathematics ought to be. Other definitions were given, and eventually the concept of cardinal number was made precise.

One way to define cardinal numbers is by choosing one fixed set from each equivalence class of sets under the relation $\approx$, and then calling this set the cardinal number of each set in the class. Under such a procedure we would think of the number 0 as being the empty set and the number 1 as being the set whose only element is the number 0. That is, $1 = \{0\}$; $2 = \{0, 1\}$; $3 = \{0, 1, 2\}$; and so on. The two infinite cardinal numbers given so far have been described by specifying a standard set, either $\mathbb{N}$ or $(0, 1)$, as the standard example for that particular cardinal.

We will not be concerned with further details of formulating a precise definition of a cardinal number. For our purposes, the essential point is that the cardinal number $\overline{\overline{S}}$ of a set S is an object associated with all sets equivalent to S and no other set. The double overbar on $\overline{\overline{S}}$ is suggestive of the double abstraction referred to by Cantor.

Intuitively, you should still think of the cardinality of a set S as the "number" of elements in S, or the "size" of the set S. So that we can compare the sizes of two sets, we make the following definitions:

DEFINITIONS   Let A and B be sets. Then

$\overline{\overline{A}} = \overline{\overline{B}}$ if and only if $A \approx B$; otherwise $\overline{\overline{A}} \neq \overline{\overline{B}}$.
$\overline{\overline{A}} \leq \overline{\overline{B}}$ if and only if there exists a one-to-one function $f: A \to B$.
$\overline{\overline{A}} < \overline{\overline{B}}$ if and only if $\overline{\overline{A}} \leq \overline{\overline{B}}$ and $\overline{\overline{A}} \neq \overline{\overline{B}}$.

We read $\overline{\overline{A}} < \overline{\overline{B}}$ as "the cardinality of A is strictly less than the cardinality of B" while $\leq$ is read "less than or equal to." In addition, we use $\overline{\overline{A}} \not< \overline{\overline{B}}$ and $\overline{\overline{A}} \not\leq \overline{\overline{B}}$ to denote the denials of $\overline{\overline{A}} < \overline{\overline{B}}$ and $\overline{\overline{A}} \leq \overline{\overline{B}}$, respectively.

A proof of $\overline{\overline{A}} \leq \overline{\overline{B}}$ will usually involve constructing a one-to-one function from A to B, while a proof of $\overline{\overline{A}} < \overline{\overline{B}}$ will have a proof of $\overline{\overline{A}} \leq \overline{\overline{B}}$ together with a proof, often by contradiction, that $\overline{\overline{A}} \neq \overline{\overline{B}}$. Once we have developed some properties of cardinal inequalities, those facts can be used to prove statements of the form $\overline{\overline{A}} \leq \overline{\overline{B}}$ without resorting to the construction of functions.

Since $1, 2, 3, \ldots$ are cardinal numbers, the natural numbers may be viewed as a subset of the collection of all cardinal numbers. In this sense the properties of $\leq$ and $<$ that we will prove for cardinal numbers in the next theorem may be viewed as extensions of those same properties of $\leq$ and $<$ that hold for $\mathbb{N}$. Proofs of parts (a), (b), (d), and (f) are left as Exercise 6.

Theorem 5.4.1   For sets A, B, and C,

(a) $\overline{\overline{A}} \leq \overline{\overline{A}}$.   (Reflexivity)
(b)   If $\overline{\overline{A}} = \overline{\overline{B}}$ and $\overline{\overline{B}} = \overline{\overline{C}}$, then $\overline{\overline{A}} = \overline{\overline{C}}$.   (Transitivity of $=$)
(c)   If $\overline{\overline{A}} \leq \overline{\overline{B}}$ and $\overline{\overline{B}} \leq \overline{\overline{C}}$, then $\overline{\overline{A}} \leq \overline{\overline{C}}$.   (Transitivity of $\leq$)
(d) $\overline{\overline{A}} \leq \overline{\overline{B}}$ iff $\overline{\overline{A}} < \overline{\overline{B}}$ or $\overline{\overline{A}} = \overline{\overline{B}}$.
(e)   If $A \subseteq B$, then $\overline{\overline{A}} \leq \overline{\overline{B}}$.
(f) $\overline{\overline{A}} \leq \overline{\overline{B}}$ iff there is a subset W of B such that $\overline{\overline{W}} = \overline{\overline{A}}$.

Proof.

(c)   Suppose $\overline{\overline{A}} \leq \overline{\overline{B}}$ and $\overline{\overline{B}} \leq \overline{\overline{C}}$. Then there exist functions $f: A \xrightarrow{1-1} B$ and $g: B \xrightarrow{1-1} C$. Since the composite $g \circ f: A \to C$ is one-to-one, we conclude $\overline{\overline{A}} \leq \overline{\overline{C}}$.

(e)   Let $A \subseteq B$. We note that the inclusion map $i: A \to B$, given by $i(a) = a$, is one-to-one, and therefore $\overline{\overline{A}} \leq \overline{\overline{B}}$.   ∎

Examples. We can show that every finite cardinal is less than $\aleph_0$ as follows. For every finite cardinal number k, $\mathbb{N}_k \subseteq \mathbb{N}$. Thus by Theorem 5.4.1(e), $k \leq \aleph_0$. Since $\mathbb{N}_k$ is not equivalent to $\mathbb{N}$, $k \neq \aleph_0$. Therefore, $k < \aleph_0$.

A similar argument shows that $\aleph_0 < \mathbf{c}$. First, $\aleph_0 \leq \mathbf{c}$ follows from $\mathbb{N} \subseteq \mathbb{R}$ and, second, $\aleph_0 \neq \mathbf{c}$ because $\mathbb{N}$ is not equivalent to $\mathbb{R}$.

We can easily show that for natural numbers m and n, if $m < n$ (in the usual sense), then $m < n$ by the definition of $<$ for cardinals. We use the inclusion map from $\mathbb{N}_m$ to $\mathbb{N}_n$ to show that $m = \overline{\overline{\mathbb{N}_m}} \leq \overline{\overline{\mathbb{N}_n}} = n$. We know that $\overline{\overline{\mathbb{N}_m}} \neq \overline{\overline{\mathbb{N}_n}}$ by the Pigeonhole Principle (Theorem 5.1.9).

Theorem 5.4.2

Cantor's Theorem
For every set A, $\overline{\overline{A}} < \overline{\overline{\mathcal{P}(A)}}$.

Proof. To show $\overline{\overline{A}} < \overline{\overline{\mathcal{P}(A)}}$, we must show that (i) $\overline{\overline{A}} \leq \overline{\overline{\mathcal{P}(A)}}$, and (ii) $\overline{\overline{A}} \neq \overline{\overline{\mathcal{P}(A)}}$. Part (i) follows from the fact that $F : A \to \mathcal{P}(A)$ defined by $F(x) = \{x\}$ is one-to-one.

To prove (ii), suppose $\overline{\overline{A}} = \overline{\overline{\mathcal{P}(A)}}$; that is, assume $A \approx \mathcal{P}(A)$. Then there exists $g : A \xrightarrow[\text{onto}]{1-1} \mathcal{P}(A)$. Let $B = \{y \in A : y \notin g(y)\}$. Since $B \subseteq A$, $B \in \mathcal{P}(A)$, and since g is onto $\mathcal{P}(A)$, $B = g(z)$ for some $z \in A$. Now either $z \in B$ or $z \notin B$. If $z \in B$, then $z \notin g(z) = B$, a contradiction. Similarly, $z \notin B$ implies $z \notin g(z)$, which implies $z \in B$, again a contradiction. We conclude that A is not equivalent to $\mathcal{P}(A)$ and hence $\overline{\overline{A}} < \overline{\overline{\mathcal{P}(A)}}$. ∎

Cantor's Theorem has some interesting consequences. First, there are infinitely many infinite cardinal numbers. We know one, $\aleph_0$, which corresponds to $\mathbb{N}$. By Cantor's Theorem, $\aleph_0 < \overline{\overline{\mathcal{P}(\mathbb{N})}}$. Since $\mathcal{P}(\mathbb{N})$ is a set, its power set $\mathcal{P}(\mathcal{P}(\mathbb{N}))$ has a strictly greater cardinality than that of $\mathcal{P}(\mathbb{N})$. In this fashion we may generate a denumerable set of cardinal numbers, each greater than its predecessor:

$$\aleph_0 < \overline{\overline{\mathcal{P}(\mathbb{N})}} < \overline{\overline{\mathcal{P}(\mathcal{P}(\mathbb{N}))}} < \overline{\overline{\mathcal{P}(\mathcal{P}(\mathcal{P}(\mathbb{N})))}} < \cdots$$

Exactly where $\mathbf{c}$, continuum, fits within this string of inequalities will be taken up later in this section. It is also an immediate consequence of Cantor's Theorem that there can be no largest cardinal number (see Exercise 7).

In Section 5.1 we showed that the set $\mathcal{F}$ of all functions from $\mathbb{N}$ to $\{0, 1\}$ is equivalent to $\mathcal{P}(\mathbb{N})$. Since $\overline{\overline{\mathbb{N}}} < \overline{\overline{\mathcal{P}(\mathbb{N})}}$, we know there are uncountably many functions from $\mathbb{N}$ to $\{0, 1\}$. Since a function from $\mathbb{N}$ to $\{0, 1\}$ is a sequence of 0's and 1's, Cantor's Theorem provides another proof for Exercise 13(a) of Section 5.3.

It appears to be obvious that if B has at least as many elements as A $(\overline{\overline{A}} \leq \overline{\overline{B}})$, and A has at least as many elements as B $(\overline{\overline{B}} \leq \overline{\overline{A}})$, then A and B are equivalent $(\overline{\overline{A}} = \overline{\overline{B}})$. The proof, however, is not obvious. The situation may be represented as in Figure 5.4.1. From $\overline{\overline{A}} \leq \overline{\overline{B}}$ and $\overline{\overline{B}} \leq \overline{\overline{A}}$, there are functions $F\colon A \xrightarrow{1-1} B$ and $G\colon B \xrightarrow{1-1} A$. The problem is to construct $H\colon A \to B$, which is both one-to-one and onto B. Cantor solved this problem in 1895, but his result was not immediately accepted because his proof used the Axiom of Choice (Section 5.5). Proofs not depending on the Axiom of Choice were given by Ernst Schröder* in 1896 and two years later by Felix Bernstein.*

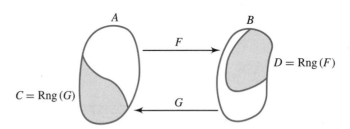

Figure 5.4.1

Theorem 5.4.3

Cantor–Schröder–Bernstein Theorem
If $\overline{\overline{A}} \leq \overline{\overline{B}}$ and $\overline{\overline{B}} \leq \overline{\overline{A}}$, then $\overline{\overline{A}} = \overline{\overline{B}}$.

Proof.   We may assume that A and B are disjoint, for otherwise we could replace A and B with the equivalent disjoint sets $A \times \{0\}$ and $B \times \{1\}$, respectively. Let $F\colon A \xrightarrow{1-1} B$ with $D = \mathrm{Rng}\,(F)$ and let $G\colon B \xrightarrow{1-1} A$ with $C = \mathrm{Rng}\,(G)$. If $B = D$ we already have $A \approx B$, so assume $B - D \neq \varnothing$.
Define a **string** to be a function $f\colon \mathbb{N} \to A \cup B$ such that

$f(1) \in B - D$,
$f(n) \in B$ implies $f(n+1) = G(f(n))$, and
$f(n) \in A$ implies $f(n+1) = F(f(n))$.

We think of a string as a sequence of elements of $A \cup B$ with first term in $B - D$, and such that thereafter the terms are alternately in C and in D. Each element of $B - D$ is the first term of some string. See Figure 5.4.2.

---

* Ernst Schröder (1841–1902) was a German mathematician known mostly for his work in logic and its applications to other areas of mathematics. He advanced the methodical use of quantifiers. The design of Schröder's proof of Theorem 5.4.3 was correct but his proof contained an error. Felix Bernstein (1878–1956), while he was still a student under Cantor, corrected the error. Bernstein made contributions in many fields, including applied mathematics, statistics and especially genetics.

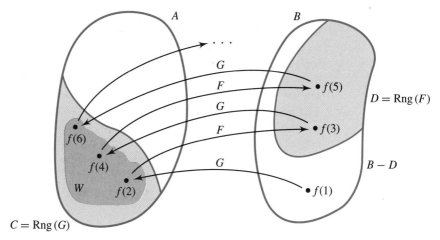

String f: $f(1), f(2), f(3), f(4), f(5), f(6), \ldots$

Figure 5.4.2

Let $W = \{x \in A : x$ is a term of some string$\}$. We note that $W \subseteq \text{Rng}\,(G)$ and that $x \in W$ iff $x = f(2n)$ for some string f and natural number n. Let $H: A \to B$ be given by

$$H(x) = \begin{cases} F(x) & \text{if } x \notin W \\ G^{-1}(x) & \text{if } x \in W \end{cases}.$$

See Figure 5.4.3. We will show that H is a one-to-one correspondence from A onto B.

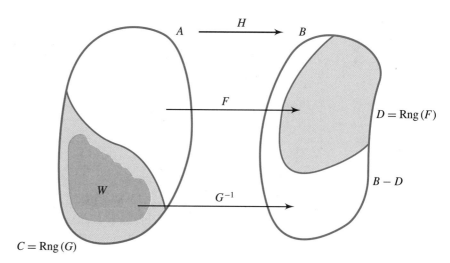

Figure 5.4.3

Suppose $x, y \in A$ and $H(x) = H(y)$. We will first show that x and y must both be in W or both in $A - W$. For a proof by contradiction, assume this is not the case.

Without loss of generality, we may assume that $x \in A - W$ and $y \in W$. ⟨*Otherwise, we could rename the elements x and y.*⟩ Then from $H(x) = H(y)$ we have $F(x) = G^{-1}(y)$, so $y = G(F(x))$. Since $y \in W$, $y = f(2n)$ for some string f and some natural number n. Therefore, $f(2n - 1) = F(x)$ ⟨*because $G(f(2n - 1)) = f(2n) = y = G(F(x))$ and G is one-to-one*⟩. If $n = 1$, then $f(1) = F(x)$, which implies $f(1) \in \text{Rng}(F)$, a contradiction to the definition of string f. Thus $n \geq 2$. But then $f(2n - 2) = x$, since $f(2n - 1) = F(x)$. This implies that x is a term in the string f, a contradiction to $x \in A - W$. Therefore, we know that x and y are either both in W or both in $A - W$.

If x and y are both in W, then $H(x) = H(y)$ implies that $G^{-1}(x) = G^{-1}(y)$. Therefore, $x = y$ since G^{-1} is one-to-one. Likewise, if x and y are both in $A - W$, then $F(x) = F(y)$ and $x = y$ because F is one-to-one. In either case, we conclude $x = y$ and H is one-to-one.

Next we show that H is onto B. Let $b \in B$. We must show that $b = H(x)$ for some $x \in A$. There are two cases:

Case 1. If $G(b) \in W$, let $x = G(b)$. Then $H(x) = H(G(b)) = G^{-1}(G(b)) = b$.

Case 2. If $G(b) \notin W$, then $b \in \text{Rng}(F)$. ⟨*If $b \notin \text{Rng}(F)$, then $b \in B - D$. Therefore, b is the first element of some string and $G(b)$ is the second element of that string, a contradiction to $G(b) \notin W$.*⟩ Since $b \in \text{Rng}(F)$, there exists $x \in A$ such that $F(x) = b$. Furthermore, $x \notin W$. ⟨*If $x \in W$, then x is a term in some string and therefore $F(x)$ and $G(F(x))$ are the next two terms of the same string. But this is a contradiction, since $G(F(x)) = G(b)$ and we have assumed that $G(b)$ is not on any string.*⟩ From $x \in A - W$ we conclude $H(x) = F(x) = b$.

In both cases, $H(x) = b$, so H is onto B. ∎

The Cantor–Schröder–Bernstein Theorem may be used to prove equivalence between sets in cases where it would be difficult to explicitly exhibit a one-to-one correspondence.

Example. We will show that $(0, 1) \approx [0, 1]$. First, note that $(0, 1) \subseteq [0, 1]$, so $\overline{\overline{(0, 1)}} \leq \overline{\overline{[0, 1]}}$. Likewise, since $[0, 1] \subseteq (-1, 2)$, we have $\overline{\overline{[0, 1]}} \subseteq \overline{\overline{(-1, 2)}}$. But we know $(0, 1) \approx (-1, 2)$ and thus $\overline{\overline{(0, 1)}} = \overline{\overline{(-1, 2)}}$. Therefore, we may write $\overline{\overline{[0, 1]}} \leq \overline{\overline{(0, 1)}}$. We conclude $\overline{\overline{(0, 1)}} = \overline{\overline{[0, 1]}}$ by the Cantor–Schröder–Bernstein Theorem and thus $(0, 1) \approx [0, 1]$.

Example. We can use the Cantor–Schröder–Bernstein Theorem to show that $\mathbb{R} \times \mathbb{R} \approx \mathbb{R}$. (See Exercise 15.) This means that there are just as many points on the real line as there are in the entire Cartesian plane.

Example. The Cantor–Schröder–Bernstein Theorem can be used to determine the relationship of $\mathbf{c}$, the cardinal number of the open interval $(0, 1)$, to the increasing sequence of cardinal numbers

$$\overline{\overline{\mathbb{N}}} < \overline{\overline{\mathscr{P}(\mathbb{N})}} < \overline{\overline{\mathscr{P}(\mathscr{P}(\mathbb{N}))}} < \overline{\overline{\mathscr{P}(\mathscr{P}(\mathscr{P}(\mathbb{N})))}} < \cdots.$$

We will show that $\overline{\overline{\mathcal{P}(\mathbb{N})}} = \mathbf{c}$.

Proof. First, recall that any real number in the interval $(0, 1)$ may be expressed in a base 2 (binary) expansion $0.b_1b_2b_3b_4\ldots$, where each b_i is either 0 or 1. If we exclude sequences that terminate with infinitely many 1's, such as $0.0101111111\ldots$ (which has the same value as $0.01100000\ldots$), then the representation is unique. Thus we may define a function $f: (0, 1) \to \mathcal{P}(\mathbb{N})$ such that for each $x \in (0, 1)$,

$$f(x) = \{n \in \mathbb{N}: b_n = 1 \text{ in the binary representation of } x\}.$$

The uniqueness of binary representations ensures that the function is defined and is one-to-one. Since f is one-to-one, $\overline{\overline{(0, 1)}} \leq \overline{\overline{\mathcal{P}(\mathbb{N})}}$.

Next, define $g: \mathcal{P}(\mathbb{N}) \to (0, 1)$ by $g(A) = 0.a_1a_2a_3a_4\ldots$, where

$$a_n = \begin{cases} 2 & \text{if } n \in A \\ 5 & \text{if } n \notin A \end{cases}.$$

For any set $A \subseteq \mathbb{N}$, $g(A)$ is a real number in $(0, 1)$ with decimal expansion consisting of 2's and 5's. ⟨*Any pair of digits not including 9 will do.*⟩ The function g is one-to-one but certainly not onto $(0, 1)$. Therefore, $\overline{\overline{\mathcal{P}(\mathbb{N})}} \leq \overline{\overline{(0, 1)}}$.

By the Cantor–Schröder–Bernstein Theorem, $\overline{\overline{\mathcal{P}(\mathbb{N})}} = \overline{\overline{(0, 1)}}$. Therefore $\overline{\overline{\mathcal{P}(\mathbb{N})}} = \mathbf{c}$. ∎

We can now identify the first two terms of the sequence $\overline{\overline{\mathbb{N}}} < \overline{\overline{\mathcal{P}(\mathbb{N})}} < \overline{\overline{\mathcal{P}(\mathcal{P}(\mathbb{N}))}} < \overline{\overline{\mathcal{P}(\mathcal{P}(\mathcal{P}(\mathbb{N})))}} < \ldots$ as being $\aleph_0$ and $\mathbf{c}$.

The Cantor–Schröder–Bernstein Theorem is another result in the extension of the familiar ordering properties of $\mathbb{N}$ to properties for all cardinal numbers. It, in turn, leads to others. In the following, parts (a) and (c) are proved; (b) and (d) are given as Exercise 13.

Corollary 5.4.4 For sets A, B, and C,

(a) if $\overline{\overline{A}} \leq \overline{\overline{B}}$, then $\overline{\overline{B}} \not< \overline{\overline{A}}$.
(b) if $\overline{\overline{A}} \leq \overline{\overline{B}}$ and $\overline{\overline{B}} < \overline{\overline{C}}$, then $\overline{\overline{A}} < \overline{\overline{C}}$.
(c) if $\overline{\overline{A}} < \overline{\overline{B}}$ and $\overline{\overline{B}} \leq \overline{\overline{C}}$, then $\overline{\overline{A}} < \overline{\overline{C}}$.
(d) if $\overline{\overline{A}} < \overline{\overline{B}}$ and $\overline{\overline{B}} < \overline{\overline{C}}$, then $\overline{\overline{A}} < \overline{\overline{C}}$.

Proof.

(a) Suppose $\overline{\overline{B}} < \overline{\overline{A}}$. Then $\overline{\overline{A}} \neq \overline{\overline{B}}$ and $\overline{\overline{B}} \leq \overline{\overline{A}}$. Combining this with the hypothesis that $\overline{\overline{A}} \leq \overline{\overline{B}}$, we conclude by the Cantor–Schröder–Bernstein Theorem that $\overline{\overline{A}} = \overline{\overline{B}}$, which is a contradiction. Therefore, $\overline{\overline{B}} \not< \overline{\overline{A}}$.

(c) Suppose $\overline{\overline{A}} < \overline{\overline{B}}$ and $\overline{\overline{B}} \leq \overline{\overline{C}}$. Then $\overline{\overline{A}} \leq \overline{\overline{B}}$; so by Theorem 5.4.1(c), $\overline{\overline{A}} \leq \overline{\overline{C}}$. Suppose $\overline{\overline{A}} \not< \overline{\overline{C}}$. Then $\overline{\overline{A}} = \overline{\overline{C}}$, which implies $\overline{\overline{C}} \leq \overline{\overline{A}}$. But $\overline{\overline{B}} \leq \overline{\overline{C}}$ and $\overline{\overline{C}} \leq \overline{\overline{A}}$ implies $\overline{\overline{B}} \leq \overline{\overline{A}}$. Combining this with $\overline{\overline{A}} \leq \overline{\overline{B}}$, we conclude by the Cantor–Schröder–Bernstein Theorem that $\overline{\overline{A}} = \overline{\overline{B}}$. Since this contradicts $\overline{\overline{A}} < \overline{\overline{B}}$, we have $\overline{\overline{A}} < \overline{\overline{C}}$. ∎

It is tempting to extend our results even further to include the converse of Corollary 5.4.4(a): "If $\overline{\overline{B}} \not< \overline{\overline{A}}$, then $\overline{\overline{A}} \leq \overline{\overline{B}}$." (As far as we know now, for two given sets

A and B, both $\overline{\overline{A}} \leq \overline{\overline{B}}$ and $\overline{\overline{B}} \leq \overline{\overline{A}}$ may be false.) The Cantor–Schröder–Bernstein Theorem turned out to be more difficult to prove than one would have guessed from its simple statement, but the situation regarding the converse of Corollary 5.4.4(a) is even more remarkable. This is discussed in Section 5.5, where "If $\overline{\overline{B}} \not< \overline{\overline{A}}$, then $\overline{\overline{A}} \leq \overline{\overline{B}}$" is rephrased as "Either $\overline{\overline{A}} < \overline{\overline{B}}$, or $\overline{\overline{A}} = \overline{\overline{B}}$, or $\overline{\overline{B}} < \overline{\overline{A}}$."

Exercises 5.4

1. Prove that for any natural number n, $n < \mathbf{c}$.

☆ 2. Prove that $\overline{\overline{\mathcal{P}(\mathbb{N})}} < \overline{\overline{\mathcal{P}(\mathbb{R})}}$.

3. Prove that if $\overline{\overline{A}} \leq \overline{\overline{B}}$ and $\overline{\overline{B}} = \overline{\overline{C}}$, then $\overline{\overline{A}} \leq \overline{\overline{C}}$.

4. Prove that if $\overline{\overline{A}} \leq \overline{\overline{B}}$ and $\overline{\overline{A}} = \overline{\overline{C}}$, then $\overline{\overline{C}} \leq \overline{\overline{B}}$.

5. State whether each of the following is true or false. For each false statement, give a counter example.
    (a) $\overline{\overline{A}} \leq \overline{\overline{B}}$ implies that $A \subseteq B$.
    ★ (b) $\overline{\overline{A \cap B}} \leq \overline{\overline{B}}$.
    (c) $\overline{\overline{A}} \leq \overline{\overline{B}}$ implies $\overline{\overline{\mathcal{P}(A)}} \leq \overline{\overline{\mathcal{P}(B)}}$.
    (d) $\overline{\overline{A}} = \overline{\overline{B}}$ implies $\overline{\overline{A}} \leq \overline{\overline{B}}$.
    (e) If $B - A$ is nonempty, then $\overline{\overline{A}} < \overline{\overline{A \cup B}}$.

6. Prove the remaining parts of Theorem 5.4.1.

7. Prove that there is no largest cardinal number.

8. Arrange the following cardinal numbers in order:
    ★ (a) $\overline{\overline{(0, 1)}}$, $\overline{\overline{[0, 1]}}$, $\overline{\overline{\{0, 1\}}}$, $\overline{\overline{\{0\}}}$, $\overline{\overline{\mathcal{P}(\mathbb{R})}}$, $\overline{\overline{\mathbb{Q}}}$, $\overline{\overline{\varnothing}}$, $\overline{\overline{\mathbb{R} - \mathbb{N}}}$, $\overline{\overline{\mathcal{P}(\mathcal{P}(\mathbb{R}))}}$, $\overline{\overline{\mathbb{R}}}$
    (b) $\overline{\overline{\{0, 5\}}}$, $\overline{\overline{[0, 5]}}$, $\overline{\overline{\{0, 3, 5\}}}$, $\overline{\overline{\mathbb{R} - \{3\}}}$, $\overline{\overline{\mathcal{P}(\{0, 5\})}}$, $\overline{\overline{\mathcal{P}((0, 5))}}$, $\overline{\overline{(0, 5) - \{3\}}}$, $\overline{\overline{\mathbb{R} - \mathbb{N}}}$
    (c) $\overline{\overline{\mathbb{Q} \cup \{\pi\}}}$, $\overline{\overline{\mathbb{R} - \{\pi\}}}$, $\overline{\overline{\mathcal{P}(\{0, 1\})}}$, $\overline{\overline{[0, 2]}}$, $\overline{\overline{(0, \infty)}}$, $\overline{\overline{\mathbb{Z}}}$, $\overline{\overline{\mathbb{R} - \mathbb{Z}}}$, $\overline{\overline{\mathcal{P}(\mathbb{R})}}$

9. Apply the proof of the Cantor–Schröder–Bernstein Theorem to this situation: $A = \{2, 3, 4, 5, \ldots\}$, $B = \left\{\frac{1}{2}, \frac{1}{3}, \frac{1}{4}, \ldots\right\}$, $F: A \to B$ where $F(x) = \dfrac{1}{x + 6}$, and $G: B \to A$ where $G(x) = \dfrac{1}{x} + 5$. Note that $\frac{1}{3}$ and $\frac{1}{4}$ are in $B - \text{Rng}(F)$. Let f be the string that begins at $\frac{1}{3}$, and let g be the string that begins at $\frac{1}{4}$.
    (a) Find $f(1), f(2), f(3), f(4)$.
    (b) Find $g(1), g(2), g(3), g(4)$.
    (c) Define H as in the proof of the Cantor–Schröder–Bernstein Theorem and find $H(2), H(8), H(13)$, and $H(20)$.

10. Suppose there exist three functions $f: A \overset{1\text{-}1}{\longrightarrow} B$, $g: B \overset{1\text{-}1}{\longrightarrow} C$, and $h: C \overset{1\text{-}1}{\longrightarrow} A$. Prove $A \approx B \approx C$. Do not assume that the functions map onto their codomains.

11. If possible, give an example of
    (a) functions f and g such that $f: \mathbb{Q} \overset{1\text{-}1}{\longrightarrow} \mathbb{N}$, $g: \mathbb{N} \overset{1\text{-}1}{\longrightarrow} \mathbb{Q}$, but neither f nor g is an onto map.
    ★ (b) a function $f: \mathbb{R} \overset{1\text{-}1}{\longrightarrow} \mathbb{N}$.

     (c)   a function $f: \mathscr{P}(\mathbb{N}) \xrightarrow{1-1} \mathbb{N}$.

     (d)   a function $f: \mathbb{R} \xrightarrow{1-1} \mathbb{Q}$.

12.   Prove that if there is a function $f: A \to \mathbb{N}$ that is one-to-one, then A is countable.

13.   Prove parts (b) and (d) of Corollary 5.4.4.

14.   Use a cardinality argument to prove that there is no universal set of all sets.

15.   Use the Cantor–Schröder–Bernstein Theorem to prove the following.

     (a)   The set of all integers whose digits are 6, 7, or 8 is denumerable.

     (b) $\mathbb{R} \times \mathbb{R} \approx \mathbb{R}$.

     (c)   If $A \subseteq \mathbb{R}$ and there exists an open interval (a, b) such that $(a, b) \subseteq A$, then $\overline{\overline{A}} = \mathbf{c}$.

16.   Consider the family $\mathscr{F} = \{f : f$ is a function from $[0, 1]$ to $[0, 1]\}$.

  ☆ **(a)**   Prove that there is no bijection from $[0, 1]$ to $\mathscr{F}$.

  ☆ **(b)**   Show that $\mathscr{F}$ is uncountable by showing that $\mathscr{F}$ has a subset equivalent to $[0, 1]$.

     (c)   What is the relationship between $\overline{\overline{[0, 1]}}$ and $\overline{\overline{\mathscr{F}}}$?

Proofs to Grade   Assign a grade of A (correct), C (partially correct), or F (failure) to each. Justify assignments of grades other than A.

17. **(a)**   **Claim.** If $\overline{\overline{A}} \le \overline{\overline{B}}$ and $\overline{\overline{A}} = \overline{\overline{C}}$, then $\overline{\overline{C}} \le \overline{\overline{B}}$.

         "Proof." Assume $\overline{\overline{A}} \le \overline{\overline{B}}$ and $\overline{\overline{A}} = \overline{\overline{C}}$. Then there exists a function f such that $f: A \xrightarrow{1-1} B$. Since $A = C$, $f: C \xrightarrow{1-1} B$. Therefore, $\overline{\overline{C}} \le \overline{\overline{B}}$. ∎

  ★ **(b)**   **Claim.** If $B \subseteq C$ and $\overline{\overline{B}} = \overline{\overline{C}}$, then $B = C$.

         "Proof." Suppose $B \ne C$. Then B is a proper subset of C. Thus $C - B \ne \varnothing$. This implies $\overline{\overline{C - B}} > 0$. But $C = B \cup (C - B)$ and, since B and $C - B$ are disjoint, $\overline{\overline{C}} = \overline{\overline{B}} + \overline{\overline{(C - B)}}$. By hypothesis, $\overline{\overline{B}} = \overline{\overline{C}}$. Thus $\overline{\overline{(C - B)}} = 0$, a contradiction. ∎

     (c)   **Claim.** If $\overline{\overline{A}} \le \overline{\overline{B}}$ and $\overline{\overline{B}} < \overline{\overline{C}}$, then $\overline{\overline{A}} < \overline{\overline{C}}$.

         "Proof." Assume $\overline{\overline{A}} \le \overline{\overline{B}}$ and $\overline{\overline{B}} < \overline{\overline{C}}$.

         Case 1. $\overline{\overline{A}} = \overline{\overline{B}}$. Then by substitution in $\overline{\overline{B}} < \overline{\overline{C}}$, $\overline{\overline{A}} < \overline{\overline{C}}$.

         Case 2. $\overline{\overline{A}} < \overline{\overline{B}}$. Then by transitivity, $\overline{\overline{A}} < \overline{\overline{C}}$. ∎

  ★ **(d)**   **Claim.** If $A \ne \varnothing$ and $\overline{\overline{A}} \le \overline{\overline{B}}$, then there exists a function $f: B \xrightarrow{\text{onto}} A$.

         "Proof." Assume $\overline{\overline{A}} \le \overline{\overline{B}}$. Then there exists a function $g: A \xrightarrow{1-1} B$. Since g is one-to-one, every b in B has exactly one pre-image in A. Thus the set $f = \{(b, y): y$ is the pre-image of b under $g\}$ is a function. This function is onto A, because for each a in A, $g(a) \in B$, and so $f(g(a)) = a$. Thus $f: B \xrightarrow{\text{onto}} A$. ∎

5.5   Comparability of Cardinal Numbers and the Axiom of Choice

One of the most useful ordering properties of $\mathbb{N}$ is the **trichotomy property:** *if m and n are any two natural numbers, then $m > n$, $m = n$, or $m < n$.* The analog for cardinal numbers is stated in the Comparability Theorem.

Theorem 5.5.1

The Comparability Theorem

If A and B are any two sets, then $\overline{\overline{A}} < \overline{\overline{B}}$, $\overline{\overline{A}} = \overline{\overline{B}}$, or $\overline{\overline{B}} < \overline{\overline{A}}$.

Surprisingly, it is impossible to prove the Comparability Theorem from the axioms and other theorems of Zermelo–Fraenkel set theory (see Section 2.1). In a formal study of set theory one can build up, starting with a few axioms specifying that certain collections are sets, to the study of the natural, rational, real, and complex numbers, polynomial, transcendental, and differentiable functions, and all the rest of mathematics. Still, comparability cannot be proved. On the other hand, it is impossible to prove in Zermelo–Fraenkel set theory that comparability is false. Theorem 5.5.1 is undecidable in our set theory; no proof of it and no proof of its negation could ever be constructed in our theory.

At this point we could choose either to assume that Theorem 5.5.1 is true (or assume true some other statement from which comparability can be proved) or else assume the truth of some statement from which we can show comparability is false. Of course, we have revealed the fact that we want comparability to be true by labeling the statement as a theorem. It has become standard practice by most mathematicians to assume the Comparability Theorem is true by assuming the truth of the following statement:

The Axiom of Choice

If $\mathscr{A}$ is any collection of nonempty sets, then there exists a function F (called a **choice function**) from $\mathscr{A}$ to $\bigcup_{A \in \mathscr{A}} A$ such that for every $A \in \mathscr{A}$, $F(A) \in A$.

The Axiom of Choice at first appears to have little significance: From a collection of nonempty sets, we can choose an element from each set. If the collection is finite, then this axiom is not needed to prove the existence of a choice function. It is only for infinite collections of sets that the result is not obvious and for which the Axiom of Choice is independent of other axioms of set theory.

Many examples and uses of the Axiom of Choice require more advanced knowledge of mathematics. The first example we present is not mathematical in content but it has become part of mathematical folklore.

A shoe store's stockroom has an infinite number of pairs of shoes and an infinite number of pairs of socks. A customer asks to see one shoe from each pair. When the clerk has an explicit rule for making a choice, he does not need to invoke the Axiom of Choice to know there is a choice function. His rule may be to choose the left shoe from each pair. If the socks in each of the infinitely many pairs are indistinguishable, and a customer asks to see one sock from each pair, then the clerk has no rule for making a choice. Without the Axiom of Choice we can't say there is a function that chooses one sock from each pair.

Example. Let $\mathscr{A} = \{A : A \subseteq \mathbb{R} \text{ and } A \neq \varnothing\}$. If we are to select one element from each set A in $\mathscr{A}$, then we will need to use the Axiom of Choice. However, if we let $\mathscr{B} = \{A : A \subseteq \mathbb{R}, A \neq \varnothing, \text{ and } A \text{ is finite}\}$, then we do not need the Axiom of Choice to select one element from each set in $\mathscr{B}$. Our choice rule might be: For each $B \in \mathscr{B}$, choose the greatest element in B. Since B is finite, such an element exists for each $B \in \mathscr{B}$.

Now that we have the Axiom of Choice available to use, we could prove the Comparability Theorem. However, we choose instead* to give a proof of Theorem 5.3.8, which was postponed in Section 5.3. We need one more preliminary result before we give that proof.

Lemma 5.5.2

Let $\{A_i : i \in \mathbb{N}\}$ be a denumerable family of sets. For each $i \in \mathbb{N}$, let $B_i = A_i - \left(\bigcup_{k=1}^{i-1} A_k \right)$. Then $\{B_i : i \in \mathbb{N}\}$ is a denumerable family of pairwise disjoint sets such that $\bigcup_{i \in \mathbb{N}} A_i = \bigcup_{i \in \mathbb{N}} B_i$.

Proof.   Let $x \in A$ and j and k be natural numbers. If $j < k$ and $x \in B_j$, then $x \notin B_k$. Thus B_j and B_k are disjoint. Thus $\{B_i : i \in \mathbb{N}\}$ is pairwise disjoint.

By definition of the B_i, $\bigcup_{n \in \mathbb{N}} B_n \subseteq \bigcup_{n \in \mathbb{N}} A_n$. If $x \in A_i$ for some $i \in \mathbb{N}$ then there is a smallest natural number k such that $x \in A_k$ and so $x \in B_k$. Thus $\bigcup_{n \in \mathbb{N}} A_n \subseteq \bigcup_{n \in \mathbb{N}} B_n$.

Therefore $\bigcup_{n \in \mathbb{N}} A_n = \bigcup_{n \in \mathbb{N}} B_n$.   ∎

Theorem 5.3.8 (restated)

Let $\mathscr{A}$ be a countable collection of countable sets. Then $\bigcup_{A \in \mathscr{A}} A$ is countable.

Proof.   Let $\mathscr{A}$ be a countable collection of countable sets. We may assume that $\mathscr{A} = \{A_1, A_2, \ldots A_m, \ldots\}$ is denumerable, because if there were only k sets in $\mathscr{A}$ we could extend $\mathscr{A}$ by defining $A_{k+1} = A_{k+2} = \cdots = \varnothing$. By Lemma 5.5.2 we may also assume that the sets in $\mathscr{A}$ are pairwise disjoint. From the fact that each set A_m is countable, we know that A_m is equivalent to a subset of $\mathbb{N}$: either to $\mathbb{N}_n$ for some $n \in \mathbb{N}$, to $\varnothing$, or to $\mathbb{N}$ itself. Thus for each A_m, there is a bijection f_m from A_m to a subset of $\mathbb{N}$.

We now define a function g from $\bigcup_{A \in \mathscr{A}} A$ to $\mathbb{N}$. Let $x \in \bigcup_{A \in \mathscr{A}} A$. Then $x \in A_m$ for exactly one natural number m. Let p_m be the mth prime number. Define $g(x) = (p_m)^{f_m(x)}$. ⟨*See the example below.*⟩ We claim that g is one-to-one: Suppose $g(a) = g(b)$ for some $a, b \in \bigcup_{A \in \mathscr{A}} A$, where $a \in A_i$ and $b \in A_j$. Then $(p_i)^{f_i(a)} = (p_j)^{f_j(b)}$ so by the Fundamental Theorem of Arithmetic, $i = j$ and $f_i(a) = f_i(b)$. Since f_i is one-to-one, $a = b$.

Since $g \colon \bigcup_{A \in \mathscr{A}} A \xrightarrow{1-1} \mathbb{N}$, $\bigcup_{A \in \mathscr{A}} A$ is equivalent to a subset of $\mathbb{N}$. Since every subset of $\mathbb{N}$ is countable, $\bigcup_{A \in \mathscr{A}} A$ is countable.   ∎

As an example of how the function g in the proof above works, suppose $A_5 = \{r, s, t\}$ and $f_5 = \{(r, 1), (s, 3), (t, 2)\}$. Since 11 is the fifth prime, $p_5 = 11$. We compute $g(r) = 11^1$, $g(s) = 11^3$, and $g(t) = 11^2$. Elements of the set A_6 would be mapped to distinct powers of 13, and so forth. If A_6 happens to be denumerable, then every power of 13 will be in Rng(g).

---

* For a proof of the Comparability Theorem, see Paul R. Halmos, *Naive Set Theory* (Undergraduate Texts in Mathematics), Springer-Verlag, 1998.

Where was the Axiom of Choice used in the proof of Theorem 5.3.8? For each $m \in \mathbb{N}$, A_m is countable, and there are generally *many* bijections from A_m to a subset of $\mathbb{N}$. In our proof, we select one such bijection and call it f_m. We do this infinitely many times, once for each $m \in \mathbb{N}$. Our collection consisted of sets of bijections, and we needed one bijection from each set of bijections. There is no way to select the f_m without the Axiom of Choice.

Many other important theorems, in many areas of mathematics, cannot be proved without the use of the Axiom of Choice. In fact, several crucial results are equivalent to it.* Some of the consequences of the axiom are not as natural as the Comparability Theorem, however, and some of them are extremely difficult to believe. One of these is that the real numbers can be rearranged in such a way that every nonempty subset of R has a smallest element—in other words, that the reals can be well ordered. Another, called the Banach–Tarski paradox, states that a ball can be cut into a finite number of pieces that can be rearranged to form two balls the same size as the original ball. Actually, this "paradox" is hardly more surprising than the result in Section 5.4 that $\mathbb{R} \times \mathbb{R} \approx \mathbb{R}$, although that theorem can be proved without the Axiom of Choice.

The Axiom of Choice has been objected to because of such consequences, and also because of a lack of precision in the statement of the axiom, which does not provide any hint of a rule for constructing the choice function F. Because of these objections, it is common practice to call attention to the fact that the Axiom of Choice has been used in a proof, so that anyone who is interested can attempt to find an alternate proof that does not use the axiom.

We conclude this chapter with three more results whose proofs rely on the Axiom of Choice. The first theorem says that if there is a function from a set A onto a set B, then A must have at least as many elements as B. The proof uses the Axiom of Choice to choose, for every $b \in B$, an $a \in A$ such that $f(a) = b$.

Theorem 5.5.3

If there exists a function from a set A onto a set B, then $\overline{\overline{B}} \leq \overline{\overline{A}}$.

Proof. If $B = \varnothing$, then $B \subseteq A$. Therefore, in this case $\overline{\overline{B}} \leq \overline{\overline{A}}$. Suppose $B \neq \varnothing$, and suppose $f: A \to B$ is onto B. ⟨*To show that* $\overline{\overline{B}} \leq \overline{\overline{A}}$, *we must construct a function* $h: B \to A$ *that is one-to-one.*⟩ Let $b \in B$. Since f is onto B, b is in Rng (f). Therefore the set $C_b = \{a \in A: f(a) = b\}$ is nonempty. See Figure 5.5.1. Thus $\mathcal{A} = \{C_b: b \in B\}$ is a nonempty collection of nonempty sets.

By the Axiom of Choice, there is a function $g: \mathcal{A} \to \bigcup_{b \in B} C_b$ such that $g(C_b) \in C_b$ for every $b \in B$. ⟨*Since f is a function with domain A,* $\bigcup_{b \in B} C_b = A$. *Therefore g is a function from $\mathcal{A}$ to A.*⟩

Define $h: B \to A$ by $h(b) = g(C_b)$ for every $b \in B$. We will show that h is one-to-one. Let r and s be elements of B and suppose that $h(r) = h(s)$. Then

---

* Paul Howard and Jean E. Rubin, *Consequences of the Axiom of Choice*, American Mathematical Society, Mathematical Surveys and Monographs, v. 59, 1998.

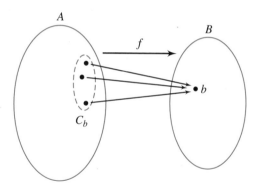

Figure 5.5.1

$g(C_r) = g(C_s)$. Call this object x. By definition of g, $x \in C_r$ and $x \in C_s$. Then $f(x) = r$ and $f(x) = s$. But f is a function, so $r = s$. Therefore h is one-to-one. We conclude that $\overline{\overline{B}} \leq \overline{\overline{A}}$.   ∎

Theorem 5.5.4   Every infinite set A has a denumerable subset.

Proof.   Suppose A is infinite. We inductively define a denumerable subset of A. First, since A is infinite, $A \neq \varnothing$. Choose $a_1 \in A$. Then $A - \{a_1\}$ is infinite, hence non-empty. Choose $a_2 \in A - \{a_1\}$. Note that $a_2 \neq a_1$ and $a_2 \in A$. Continuing in this fashion, suppose $a_1, \ldots, a_k$ have been defined. Then $A - \{a_1, \ldots, a_k\} \neq \varnothing$, so select any a_{k+1} from this set. By the Axiom of Choice, a_n is defined for all $n \in \mathbb{N}$. The a_n have been constructed so that each $a_n \in A$ and $a_i \neq a_j$ for $i \neq j$. Thus $B = \{a_n : n \in \mathbb{N}\}$ is a subset of A, and the function f given by $f(n) = a_n$ is a one-to-one correspondence from $\mathbb{N}$ to B. Thus B is denumerable.   ∎

Theorem 5.5.4 can be used to prove that every infinite set is equivalent to one of its proper subsets. (See Exercise 8.) This result characterizes infinite sets because, as we saw in Section 5.1, no finite set is equivalent to any of its proper subsets.

Theorem 5.5.4 also confirms that $\aleph_0$ is the smallest infinite cardinal number. For any set A with infinite cardinality, there is a denumerable subset B of A. Therefore, $\aleph_0 = \overline{\overline{B}} \leq \overline{\overline{A}}$.

Corollary 5.5.5   A nonempty set A is countable iff there exists a function $f : \mathbb{N} \xrightarrow{\text{onto}} A$.

Proof.   Exercise 9.   ∎

We have seen that $\aleph_0 = \overline{\overline{\mathbb{N}}} < \mathbf{c} = \overline{\overline{\mathcal{P}(\mathbb{N})}} < \overline{\overline{\mathcal{P}(\mathcal{P}(\mathbb{N}))}} < \overline{\overline{\mathcal{P}(\mathcal{P}(\mathcal{P}(\mathbb{N})))}} < \ldots$. The fact that $\aleph_0$ and $\mathbf{c}$ are the first two cardinal numbers in this sequence does not necessarily mean that $\mathbf{c}$ is the next largest cardinal number after $\aleph_0$. Cantor conjectured that this is so: That is, no set X exists such that $\aleph_0 < \overline{\overline{X}} < \mathbf{c}$. This conjecture, called the **continuum hypothesis,** is one of the most famous problems in modern mathematics. The combined work of Kurt Gödel* in the 1930s and Paul Cohen[†] in 1963 shows that the continuum hypothesis can neither be proved nor disproved in Zermelo–Fraenkel set theory. Like the Axiom of Choice, the continuum hypothesis is undecidable.

Exercises 5.5

1. Indicate whether the Axiom of Choice must be employed to select one element from each set in the following collections.
    (a) an infinite collection of sets, each set containing one odd and one even integer
    ★ (b) a finite collection of sets such that each set is uncountable
    (c) an infinite collection of sets, each set containing exactly one integer
    (d) a denumerable collection of uncountable sets
    (e) $\{(a, \infty): a \in \mathbb{R}\}$
    (f) $\{A: A \subseteq \mathbb{N}$ and both A and $\mathbb{N} - A$ are infinite$\}$
    (g) $\{A: A \subseteq \mathbb{R}$ and both A and $\mathbb{R} - A$ are infinite$\}$
    ★ (h) $\{A: A \subseteq \mathbb{R}$ and A is denumerable$\}$

2. (a) Prove this partial converse of Theorem 5.5.3 without using the Axiom of Choice. Let A and B be sets with $B \neq \varnothing$. If $\overline{\overline{B}} \leq \overline{\overline{A}}$ then there exists $g: A \to B$ that is onto B.

    (b) Use the Axiom of Choice to prove that if there exists $f: A \xrightarrow{\text{onto}} B$ then there exists a function $g: B \xrightarrow{1-1} A$.

3. Let A and B be any two nonempty sets. Use the results of this section to prove that there exists $f: A \to B$ that has at least one of these properties:
    (i) f is one-to-one   or
    (ii) f is onto B.

4. Prove that if $f: A \to B$, then $\overline{\overline{\text{Rng}(f)}} \leq \overline{\overline{A}}$.

★ 5. Suppose A is a denumerable set and B is an infinite subset of A. Prove $A \approx B$.

6. Suppose $\overline{\overline{B}} < \overline{\overline{C}}$ and $\overline{\overline{B}} \nleq \overline{\overline{A}}$. Prove that $\overline{\overline{A}} < \overline{\overline{C}}$.

7. Let $\{A_i: i \in \mathbb{N}\}$ be a collection of distinct pairwise disjoint nonempty sets. That is, if i and j are in $\mathbb{N}$ and $i \neq j$, then $A_i \neq A_j$ and $A_i \cap A_j = \varnothing$. Prove that $\bigcup_{i \in \mathbb{N}} A_i$ includes a denumerable subset.

☆ 8. Let A be an infinite set. Prove that A is equivalent to a proper subset of A.

9. Prove Corollary 5.5.5: A nonempty set A is countable iff there is a function $f: \mathbb{N} \to A$ that is onto A.

---

* Kurt Gödel (1906–1978) was an Austrian–American logician best known for his Incompleteness Theorem, which says (roughly) that in any logical system rich enough to include the theory of the natural numbers, there will always be true statements that are unprovable.
[†] Paul Cohen (1934–2007), an American logician, created a method of proof that he used to show that neither the Axiom of Choice nor the continuum hypothesis can be proved in the set theory based on the Zermelo–Fraenkel axioms.

Proofs to Grade   **10.** Assign a grade of A (correct), C (partially correct), or F (failure) to each. Justify assignments of grades other than A.

★ **(a)** **Claim.** There is a denumerable set B of irrational numbers such that any two elements of the set differ by an integer.
"*Proof.*" Define the sequence s by setting the nth term to be $s_n = \pi + n$. Then s is a function with domain $\mathbb{N}$. Let B be the range of this function. By Theorem 5.3.3, $\overline{\overline{B}} \le \overline{\overline{\mathbb{N}}}$. Since B is infinite, B is denumerable. The difference between any two elements $\pi + m$ and $\pi + n$ of B is $m - n$, an integer. ∎

(b) **Claim.** Every infinite set A has a denumerable subset.
"*Proof.*" Suppose no subset of A is denumerable. Then all subsets of A must be finite. In particular $A \subseteq A$. Thus A is finite, contradicting the assumption. ∎

★ **(c)** **Claim.** Every infinite set A has a denumerable subset B.
"*Proof.*" If A is denumerable, let $B = A$, and we are done. Otherwise, A is uncountable. Choose $x_1 \in A$. If $A - \{x_1\}$ is denumerable, let $B = A - \{x_1\}$. Otherwise, choose $x_2 \in A - \{x_1\}$. If $A - \{x_1, x_2\}$ is denumerable, let $B = A - \{x_1, x_2\}$. Continuing in this manner, using the Axiom of Choice, we obtain a subset $C = \{x_1, x_2, \ldots\}$ such that $B = A - C$ is denumerable. ∎

(d) **Claim.** Every infinite set has two disjoint denumerable subsets.
"*Proof.*" Let A be an infinite set. By Theorem 5.5.4, A has a denumerable subset B. Then $A - B$ is infinite, because A is infinite, and is disjoint from B. By Theorem 5.5.4, $A - B$ has a denumerable subset C. Then B and C are disjoint denumerable subsets of A. ∎

(e) **Claim.** If $A \ne \varnothing$ and $\overline{\overline{A}} \le \overline{\overline{B}}$, then there exists a function $f : B \xrightarrow{\text{onto}} A$.
"*Proof.*" Assume $\overline{\overline{A}} \le \overline{\overline{B}}$. Then there exists a function $g : A \xrightarrow{1-1} B$. Then g^{-1} is a function that maps $\text{Rng}(g)$ onto A. Let a^* be some fixed element of A and define $f = g^{-1} \cup \{(b, a^*) : b \in B - \text{Rng}(g)\}$. Then $f : B \xrightarrow{\text{onto}} A$. ∎

(f) **Claim.** Every infinite set has two disjoint denumerable subsets.
"*Proof.*" Let A be an infinite set. By Theorem 5.5.4, A has a denumerable subset B. Since B is denumerable, there is a function $f : \mathbb{N} \xrightarrow[\text{onto}]{1-1} B$. Let $C = \{f(2n) : n \in \mathbb{N}\}$ and $D = \{f(2n - 1) : n \in \mathbb{N}\}$. Then $C = \{f(2), f(4), f(6), \ldots\}$ and $D = \{f(1), f(3), f(5), \ldots\}$ are disjoint denumerable subsets of A. ∎

(g) **Claim.** Every subset of a countable set is countable.
"*Proof.*" Let A be a countable set and let $B \subseteq A$. If B is finite, then B is countable by definition. If B is infinite, since $B \subseteq A$, A is infinite. Thus A is denumerable. By Theorem 5.5.4, B has a denumerable subset C. Thus $C \subseteq B \subseteq A$, which implies $\aleph_0 = \overline{\overline{C}}$ and $\overline{\overline{C}} \le \overline{\overline{B}} \le \overline{\overline{A}} = \aleph_0$. Therefore $\overline{\overline{A}} = \overline{\overline{B}} = \aleph_0$. Thus B is denumerable and hence countable. ∎

Concepts of Algebra

The broad meaning of algebra refers to systems of computation and the study of properties of such systems. In this chapter we make precise the idea of an algebraic system and introduce several different types of systems. The goal is to make available additional opportunities to sharpen your proof writing skills while providing a first experience with some of the topics in this important field.

6.1   Algebraic Structures

We start with the notion of a computation. For example, the natural number system includes the operation of addition, which provides a structure to compute sums. We write the familiar $5 + 3 = 8$, which means: Given 5 and 3, the result is 8. Thus, addition of natural numbers is a function that acts on a pair of elements of $\mathbb{N}$ to produce another element of $\mathbb{N}$.

> **DEFINITION**   Let A be a nonempty set. A **binary operation on A** is a function from $A \times A$ to A.

We will usually denote an operation by one of the symbols $+$, $\cdot$, $\circ$, or $*$.

If $\circ$ is an operation on A and (x, y) is in the domain of $\circ$, we usually write $x \circ y$ in place of the standard function notation $\circ (x, y)$ for the image of (x, y). This notation is familiar from the operations of addition and multiplication on the real numbers, where we write $5 + 3 = 8$ and $8 \cdot 3 = 24$ instead of $+(5, 3) = 8$ and $\cdot (8, 3) = 24$, respectively. For some operations we omit the operation symbol completely and write $xy = z$, as is done with multiplication.

The images $x \circ y$, $x * y$, and xy are called *products,* regardless of whether the operations have anything to do with multiplication. Similarly, $x + y$ is referred to as the sum of x and y, even when the operation $+$ does not involve adding numbers.

In addition to the usual arithmetic operations on sets of numbers, you are already familiar with many other binary operations. Matrix multiplication, for example, is a binary operation on the set of all 2×2 matrices. For a set A, the operations of union $\cup$ and intersection $\cap$ are binary operations on $\mathcal{P}(A)$, the power set of A.

There are operations other than binary operations. Ternary operations map $A \times A \times A$ to A, and unary operations map A to A. "Inverse," for example, is a unary operation on $\mathbb{Z}$ that assigns to each integer x its additive inverse $-x$. In this chapter, when we say "operation" we mean a binary operation.

DEFINITION An **algebraic system** or **algebraic structure** is a non-empty set A with a collection of one or more operations on A and a (possibly empty) collection of relations on A.

The system $\mathbb{R}$ of real numbers together with operations of addition and multiplication and the relation "less than" make up a familiar algebraic system. The rational numbers with the operation of multiplication make up a different algebraic system, as does the set of natural numbers with addition and "less than."

Except for the last section of this chapter, *algebraic system* will refer to a structure with one binary operation and no relations. The notation for the algebraic system A with operation $*$ is $(A, *)$.

DEFINITION Let $(A, *)$ be an algebraic system. Let B be a subset of A. We say B is **closed under the operation** $*$ iff for all $x, y \in B$, $x * y \in B$.

For an algebraic system $(A, *)$ the set A is of course closed under $*$ (because the operation $*$ is a function that maps to A). For any proper subset B of A that is closed under $*$, we use the same operation symbol, $*$, to denote the restriction of the operation to $B \times B$. The three statements "B is closed under $*$," "$*$ is an operation on B," and "$(B, *)$ is an algebraic system" are all equivalent.

The algebraic system $(\mathbb{R}, \cdot)$ of real numbers with the usual multiplication has many subsets that are closed under multiplication. The set $\mathbb{Q}$ of rational numbers is closed under multiplication because the product of any two rational numbers is rational. The set of even integers, the open interval $(0, 1)$, and the set $\{-1, 1\}$ are all subsets of $\mathbb{R}$ that are closed under multiplication. When we consider the algebraic system consisting of the real numbers with the addition operation instead of multiplication, we find that $\mathbb{Q}$ and the set of even integers are closed under $+$, but the set $\{-1, 1\}$ is not. The interval $(0, 1)$ is not closed under addition because $0.59 + 0.43$ is not in $(0, 1)$.

If A is a finite set, the **order** of the algebraic system $(A, *)$ is the number of elements in A. When A is infinite, we simply say $(A, *)$ has infinite order.

A convenient way to display information about a binary operation, at least for a system of small finite order, is by means of its **operation table,** or **Cayley**[†] **table.** An operation table for a system $(A, *)$ of order n is an $n \times n$ array of products such that $x * y$ appears in row x and column y. Table 1 represents a system $(A, *)$ with $A = \{1, 2, 3\}$ in which, for example, $2 * 1 = 3$. As an example of computation in this system of order 3, notice that $(3 * 2) * (1 * 3) = 3 * 1 = 2$.

Table 1

$*$	1	2	3
1	3	2	1
2	3	1	3
3	2	3	3

Cayley tables are impractical for algebraic systems of large order and impossible to construct for infinite algebraic systems. In these cases, the operation must be described by a rule or algorithm.

We add more structure to an algebraic system by imposing additional properties on the operation.

DEFINITIONS Let $(A, *)$ be an algebraic system. Then

(i) $*$ is **commutative** on A iff for all $x, y \in A$, $x * y = y * x$.
(ii) $*$ is **associative** on A iff for all $x, y, z \in A$, $(x * y) * z = x * (y * z)$.
(iii) an element e of A is an **identity element** for $*$ iff for all $x \in A$, $x * e = e * x = x$.
(iv) if A has an identity element e, and a and b are in A, then b is an **inverse** of a iff $a * b = b * a = e$. In this case a would also be an inverse of b.

You are familiar with the fact that the system $(\mathbb{Z}, \cdot)$, with the usual multiplication of integers, is commutative and associative. In this system, 1 is the identity element, and only the elements 1 and -1 have inverses. For the system consisting of the real numbers with addition, the operation is commutative and associative, 0 is the identity, and every element has an inverse (its negative).

When the group operation is defined by a Cayley table, it is easy to see whether the operation is commutative—the table is symmetric about its *main diagonal,* from the upper left to the lower right. A Cayley table will have an identity element e if the row labeled e is identical to the row header and the column labeled e is identical to the column header. Elements x and y will be inverses iff e is the entry in both row x, column y and row y, column x.

The operation $*$ of Table 1 is not commutative because the table is not symmetric about the main diagonal. We see, for example, $1 * 3 = 1$ but $3 * 1 = 2$. Because

---

[†] Arthur Cayley (1821–1895) was the leader of the British school of pure mathematics in the 19th century. To earn a living, Cayley was a lawyer the first 14 years of his adult life, specializing in property law. During that time he wrote nearly 300 mathematical papers. His work included many contributions to the algebra of matrices, non-Euclidean geometry, and n-dimensional geometry.

Table 2			
$\circ$	1	2	3
1	1	2	3
2	1	2	3
3	1	2	3

Table 3			
$\cdot$	1	2	3
1	3	1	2
2	1	2	3
3	2	3	1

Table 4			
$+$	1	2	3
1	3	3	1
2	1	1	2
3	1	2	3

no row duplicates the row header there is no identity element for Table 1. As a result, the question of inverses does not arise.

Three different operations on $A = \{1, 2, 3\}$ are shown in Tables 2, 3, and 4. The operation of Table 3 is commutative (note the symmetry about the main diagonal), but the operations of Tables 2 and 4 are not.

The element 2 is an identity for $\cdot$ of Table 3 and in this system every element has an inverse: 3 is an inverse of 1, 2 is an inverse of 2, and 1 is an inverse of 3. Table 2 has no identity element. In Table 4, where 3 is the identity element, only 1 and 3 have inverses.

It is not easy to tell by looking at a table whether an operation is associative. For a system of order n, verification of associativity may require checking n^3 products of three elements, each grouped two ways. The operation in Table 1 is not associative because $(1 * 1) * 2 \neq 1 * (1 * 2)$. The operations in Tables 2 and 3 are associative, but $+$ (Table 4) is not associative on A. You should find elements a, b, and c, not necessarily distinct, for which $(a + b) + c \neq a + (b + c)$.

The associative property is a great convenience in computing products. First, it means that so long as factors appear in the same order, we need no parentheses. For both $x(yz)$ and $(xy)z$ we can write xyz. This can be extended inductively to products of four or more factors: $(xy)(zw) = (x(yz))w = (xy(zw))$, and so forth. Second, for an associative operation, we can define powers. Without associativity, $(xx)x$ might be different from $x(xx)$, but with associativity they are equal, and both can be denoted by x^3.

Theorem 6.1.1   Let $(A, \circ)$ be an algebraic structure.

(a) $(A, \circ)$ has at most one identity element.

(b)   Suppose $\circ$ is associative with identity e. If $a \in A$ has an inverse, then a has only one inverse.

Proof.

(a)   ⟨*We need to show that if e and f are both identities for* $\circ$, *then* $e = f$.⟩ Suppose that e and f are both identities for $\circ$. Then since e is an identity, the product $ef = f$. Likewise, since f is an identity, $ef = e$. Therefore, $e = f$.

(b)   See Exercise 9.   ■

Other important algebraic structures are based on the equivalence relation $\equiv_m$ of congruence modulo m on the set of integers. We saw in Section 3.2 that for each natural number m there are exactly m equivalence classes, which are denoted by $\overline{0}, \overline{1}, \overline{2}, \overline{3}, \ldots, \overline{m-1}$. It seems natural to define operations of addition and multiplication on the set $\mathbb{Z}_m$ of equivalence classes as follows:

> **DEFINITION**   For $\bar{a}$ and $\bar{b}$ in $\mathbb{Z}_m$,
>
> $$\bar{a} +_m \bar{b} = \overline{a + b} \quad \text{and} \quad \bar{a} \cdot_m \bar{b} = \overline{a \cdot b}.$$

That is, the sum of two equivalence classes is the class of the sum, and the product of two equivalence classes is the class of the product.

Before we can say that $(\mathbb{Z}_m, +_m)$ and $(\mathbb{Z}_m, \cdot_m)$ are algebraic systems, we need to make certain that $+_m$ and $\cdot_m$ defined in this way are truly operations on $\mathbb{Z}_m$, that is, to verify that both are functions from $\mathbb{Z}_m \times \mathbb{Z}_m$ to $\mathbb{Z}_m$. Take for example $\mathbb{Z}_5$, with $\bar{a} = \bar{1}$ and $\bar{b} = \bar{3}$. Then $\bar{a} + \bar{b}$ is $\overline{1 + 3} = \bar{4}$. But there are other ways to represent each of these classes. Because $\overline{31}$ is the same as $\bar{1}$ and $\overline{-12}$ is the same as $\bar{3}$, $\bar{a} + \bar{b}$ must be $\overline{31 + (-12)}$. Fortunately, this answer is $\overline{19} = \bar{4}$, the same result as our first computation. In order for $+_m$ and $\cdot_m$ to be binary operations, we need to know that $\overline{a + b} = \overline{c + d}$ and $\overline{a \cdot b} = \overline{c \cdot d}$ whenever $\bar{a} = \bar{c}$ and $\bar{b} = \bar{d}$.

Theorem 6.1.2   Let a, b, c, and d be integers. If $a = c \pmod{m}$ and $b = d \pmod{m}$, then

(i) $a + b = c + d \pmod{m}$.
(ii) $a \cdot b = c \cdot d \pmod{m}$.

Proof.   See Exercise 12.   ■

Examples. $(\mathbb{Z}_{12}, +_{12})$ is an algebraic system with 12 elements, $\bar{0}, \bar{1}, \bar{2}, \bar{3}, \ldots, \overline{11}$. We have

$$\bar{3} +_{12} \bar{4} = \overline{3 + 4} = \bar{7}, \text{ and}$$
$$\bar{9} +_{12} \bar{8} = \bar{5}, \text{ because } 9 + 8 = 17 \text{ and } \overline{17} = \bar{5}.$$

$(\mathbb{Z}_{12}, +_{12})$ is associative and commutative. $\bar{0}$ is the identity in $(\mathbb{Z}_{12}, +_{12})$ and every element has an inverse: The inverse of $\bar{a}$ is $\overline{-a}$. For example, the inverse of $\bar{2}$ is $\overline{10}$ because $\overline{-2} = \overline{10}$.

For $(\mathbb{Z}_{12}, \cdot_{12})$ we have

$$\bar{3} \cdot_{12} \bar{2} = \overline{3 \cdot 2} = \bar{6},$$
$$\bar{4} \cdot_{12} \bar{5} = \bar{8}, \text{ because } 4 \cdot 5 = 20 \text{ and } \overline{20} = \bar{8}, \text{ and}$$
$$\bar{3} \cdot_{12} \bar{8} = \bar{0}.$$

$(\mathbb{Z}_{12}, \cdot_{12})$ is associative and commutative. $\bar{1}$ is the identity in $(\mathbb{Z}_{12}, \cdot_{12})$ and *some* elements have an inverse. We note that $\bar{7}$ is its own inverse because $\bar{7} \cdot_{12} \bar{7} = \overline{49} = \bar{1}$. The element $\bar{3}$, however, has no inverse in $(\mathbb{Z}_{12}, \cdot_{12})$.

Theorem 6.1.3   For every natural number m,

(a) $(\mathbb{Z}_m, +_m)$ is an algebraic system that is associative and commutative with identity element $\overline{0}$. Every element has an inverse.

(b) $(\mathbb{Z}_m, \cdot_m)$ is an algebraic system that is associative and commutative. If $m > 1$, the system has identity element $\overline{1}$.

Proof.   See Exercise 13.   ∎

The notation for computations in $\mathbb{Z}_m$ is usually simplified by writing a instead of $\overline{a}$ for the elements of $\mathbb{Z}_m$ and omitting the subscript from the operation symbols $+_m$ and $\cdot_m$. Operation tables are given below for $(\mathbb{Z}_2, +)$, $(\mathbb{Z}_6, +)$, and $(\mathbb{Z}_6, \cdot)$.

$(\mathbb{Z}_2, +)$

+	0	1
0	0	1
1	1	0

$(\mathbb{Z}_6, +)$

+	0	1	2	3	4	5
0	0	1	2	3	4	5
1	1	2	3	4	5	0
2	2	3	4	5	0	1
3	3	4	5	0	1	2
4	4	5	0	1	2	3
5	5	0	1	2	3	4

$(\mathbb{Z}_6, \cdot)$

·	0	1	2	3	4	5
0	0	0	0	0	0	0
1	0	1	2	3	4	5
2	0	2	4	0	2	4
3	0	3	0	3	0	3
4	0	4	2	0	4	2
5	0	5	4	3	2	1

Multiplication in $\mathbb{Z}_m$ can produce a result that we never find with multiplication of integers, or any real numbers. For real numbers a and b, if $ab = 0$, then $a = 0$ or $b = 0$. In the table above for $\mathbb{Z}_6$, we see that $3 \cdot 4 = 0$. Another way to say this is that 3 and 4 divide 0.

DEFINITION   In $\mathbb{Z}_m$, if $a \neq 0$ is an element such that $a \cdot b = 0$ for some $b \neq 0$, we say that a is a **divisor of 0**.

In $\mathbb{Z}_{12}$, the divisors of 0 are 2, 3, 4, 6, 8, 9, and 10. In $\mathbb{Z}_2$ and $\mathbb{Z}_3$, there are no divisors of 0. The only divisor of 0 in $\mathbb{Z}_4$ is 2.

Solving equations requires special care when $\mathbb{Z}_m$ has divisors of zero. For example, in $\mathbb{Z}_{12}$ a simple linear equation such as $2x = 10$ has two solutions, $x = 5$ and $x = 11$, because $2 \cdot 11 = 22 = 10(\bmod 12)$. Other equations such as $6x = 2$ have no solutions in $\mathbb{Z}_{12}$.

Exercises 6.1

1.   Which of the following are algebraic structures? (The operation symbols have their usual meanings.)

⋆ (a) $(\mathbb{Z}, -)$ (b) $(\mathbb{Z}, \div)$ (c) $(\mathbb{R}, -)$

   (d) $(\mathbb{R}, \div)$ ⋆ (e) $(\mathbb{N}, -)$ (f) $(\mathbb{Q}, \div)$

(g) $(\mathbb{Q} - \{0\}, \div)$ **(h)** $(\mathscr{P}(A), \cap)$ **(i)** $(\mathscr{P}(A), \cup)$

(j) $(\mathscr{P}(A) - \{\varnothing\}, -)$ **(k)** $(\{0, 1\}, \cdot)$ **(l)** $(\{0, 1\}, +)$

☆ **2.** Which of the operations in Exercise 1 are commutative?

☆ **3.** Which of the operations in Exercise 1 are associative?

★ **4.** Consider the set $A = \{a, b, c, d\}$ with operation $\circ$ given by the Cayley table at the right.

(a) Name the identity element of this system.

(b) Is the operation $\circ$ associative on A?

(c) Is the operation $\circ$ commutative on A?

(d) For each element of A that has an inverse, name the inverse.

(e) Is $B_1 = \{a, b, c\}$ closed under $\circ$?

(f) Is $B_2 = \{a, c\}$ closed under $\circ$?

(g) Name all subsets of A that are closed under $\circ$.

(h) True or False? For all $x, y \in A$, $x \circ x = y \circ y$.

$\circ$	a	b	c	d
a	a	b	c	d
b	b	a	d	c
c	c	d	a	b
d	d	c	b	a

5. Repeat Exercise 4 with the operation $*$ given by the table on the right.

$*$	a	b	c	d
a	c	d	a	b
b	d	a	b	c
c	a	b	c	d
d	b	c	d	a

6. The Cayley tables for operations $\circ$, $*$, $+$, and $\times$ are listed below.

$\circ$	a	b
a	a	b
b	b	a

$*$	a	b	c
a	c	a	c
b	a	b	c
c	b	c	b

$+$	a	b
a	a	a
b	a	a

$\times$	a	b	c
a	a	c	b
b	c	b	a
c	b	a	c

(a) Which of the operations are commutative?

(b) Which of the operations are associative?

(c) Which systems have an identity? What is the identity element?

(d) For those systems that have an identity, which elements have inverses?

7. Let $m, n \in \mathbb{N}$ and $\mathcal{M} = \{A: A$ is an $m \times n$ matrix with real number entries$\}$.

(a) Let $\cdot$ be matrix multiplication. Under what conditions on m and n is $(\mathcal{M}, \cdot)$ an algebraic system?

(b) Let $+$ be matrix addition. Under what conditions on m and n is $(\mathcal{M}, +)$ an algebraic system?

☆ **8.** Let $\cdot$ be an associative operation on nonempty set A with identity e. Suppose $a, b, c,$ and d are elements of A, b is the inverse of a, and d is the inverse of c. Prove that db is the inverse of ac.

9. Let $(A, \circ)$ be an algebraic structure, $a \in A$, and e the identity for $\circ$.

☆ **(a)** Prove that if $\circ$ is associative, and x and y are inverses of a, then $x = y$.

(b) Give an example of a nonassociative structure in which inverses are not unique.

10. Suppose $(A, *)$ is an algebraic system and $*$ is associative on A.
    (a) Prove that if a_1, a_2, a_3, a_4 are in A, then

$$(a_1 * a_2) * (a_3 * a_4) = a_1 * ((a_2 * a_3) * a_4).$$

   ☆ (b) Use complete induction to prove that any product of n factors a_1, $a_2, a_3, \ldots, a_n$ in that order is equal to the left-associated product $(\ldots ((a_1 * a_2) * a_3) \ldots) * a_n$. Thus the product of n factors is always the same, no matter how they are grouped by parentheses, as long as the order of the factors is not changed.

11. Let $(A, \circ)$ be an algebra structure. An element $l \in A$ is a **left identity** for $\circ$ iff $l \circ a = a$ for every $a \in A$.
    (a) Give an example of a structure of order 3 with exactly two left identities.
    (b) Define a **right identity** for $(A, \circ)$.
    (c) Prove that if $(A, \circ)$ has a right identity r and left identity l, then $r = l$, and that $r = l$ is an identity for $\circ$.

   ☆ 12. Prove Theorem 6.1.2.

13. Prove Theorem 6.1.3.

14. Construct the operation table for each of the following:
    (a) $(\mathbb{Z}_8, +)$ (b) $(\mathbb{Z}_5, +)$
    (c) $(\mathbb{Z}_8, \cdot)$ (d) $(\mathbb{Z}_5, \cdot)$

15. Find all the divisors of zero
   ★ (a) in $\mathbb{Z}_8$.
    (b) in $\mathbb{Z}_{12}$.
   ★ (c) in $\mathbb{Z}_7$.
    (d) in $\mathbb{Z}_{11}$.

Proofs to Grade

16. Assign a grade of A (correct), C (partially correct), or F (failure) to each. Justify assignments of grades other than A.
    (a) **Claim.**   Let $(A, \circ)$ be an algebraic structure. If e is an identity for $\circ$, and if x and y are both inverses of a, then $x = y$.
    "Proof."   Since x and y are inverses of a, $x \circ a = e$ and $y \circ a = e$. Thus $x \circ a = y \circ a$. By cancellation, $x = y$.   ∎

   ★ (b) **Claim.** If every element of a structure $(A, \circ)$ has an inverse, then $\circ$ is commutative.
    "Proof."   Let x and y be in A. The element y has an inverse, which we will call y'. Then $y \circ y' = e$, so y is the inverse of y'. Now $x = x$ and multiplying both sides of the equation by the inverse of y', we have $y \circ x = x \circ y$. Therefore, $\circ$ is commutative.   ∎

    (c) **Claim.**   If a and b are zero divisors in $(\mathbb{Z}_m, \cdot)$, then ab is a zero divisor.
    "Proof."   If a and b are zero divisors, then $ab = 0$. Thus $(ab)(ab) = 0 \cdot 0 = 0$ and ab is a zero divisor.   ∎

★ **(d)** **Claim.** If a and b are zero divisors in $(\mathbb{Z}_m, \cdot)$ and $ab \neq 0$, then ab is a zero divisor.

"Proof." Since a is a zero divisor, $ax = 0$ for some $x \neq 0$ in $\mathbb{Z}_m$. Likewise, $by = 0$ for some $y \neq 0$ in $\mathbb{Z}_m$. Therefore, $(ab)(xy) = (ax)(by) = 0 \cdot 0 = 0$. Thus ab is a zero divisor. ∎

6.2   Groups

In this section, we focus on one particularly important algebraic structure, the *group*. It was the work of Evariste Galois* on polynomial equations that led to the study of groups as an aid to solving equations. The concept of a group has influenced and enriched many other areas of mathematics. In geometry, for example, the ideas of Euclidean and non-Euclidean geometries are unified by the notion of a group. Group theory has applications outside of mathematics, too, in fields such as nuclear physics and crystallography.

The properties of associativity, the existence of an identity, and the existence of an inverse for each element are just the properties needed to define a group. Our approach to defining a group is **axiomatic,** in the sense that we shall list the desired properties (axioms) of a structure, and any system satisfying these properties is called a group.

> **DEFINITION** $(G, \circ)$ is a **group** iff $(G, \circ)$ is an algebraic system such that
>
> **(i)**   the operation $\circ$ is associative on G.
> **(ii)**   there is an identity element e in G for $\circ$.
> **(iii)**   every $x \in G$ has an inverse x^{-1} in G.

The systems $(\mathbb{R}, +)$, $(\mathbb{Q}, +)$, and $(\mathbb{Z}, +)$ are all groups with identity 0. The algebraic system $(\mathbb{R}, \cdot)$ is not a group because 0 has no multiplicative inverse. The system $(\mathbb{R}^+, \cdot)$, where $\mathbb{R}^+$ denotes the positive real numbers, is a group with identity 1. The system $(\{0\}, +)$ is the smallest group.

The system $(\mathbb{N}, +)$ is not a group because it fails to satisfy group axioms (ii) and (iii). There is no identity and, therefore, it makes no sense to discuss inverses. The algebraic structure $(\mathbb{Z} - \{0\}, \cdot)$ is not a group because although multiplication is associative and the number 1 is an identity, only the elements 1 and -1 have multiplicative inverses in $\mathbb{Z}$.

---

*Evariste Galois (1811–1832) was a French mathematician who discovered elegant necessary and sufficient conditions for a polynomial equation to be solvable by radicals. He introduced the concept of a finite field, and was the first to use the word "group," in reference to a group of permutations. He died at age 20 as the result of a duel, but his work led to the development of an area of algebra that is known as Galois Theory.

When we refer to "the group G" without specifying the operation, we call the result of the group operation on x and y the product of x and y, and write the product as xy.

Example.   Let $G = \{e, a, b, c, d\}$ with operation given by the Cayley table shown below. Then G is a group of order 5.

	e	a	b	c	d
e	e	a	b	c	d
a	a	b	c	d	e
b	b	c	d	e	a
c	c	d	e	a	b
d	d	e	a	b	c

Proof.   To show that G is a group of order 5, we first note that G has five elements and is therefore nonempty. The table defines an operation on G because the product of every pair of elements of G is specified by the table.

The verification that G satisfies $(xy)z = x(yz)$ may be done by considering all $5^3 = 125$ possible assignments of values to x, y, and z. This work can be shortened considerably by noting that the equation is clearly true when any of x, y, or z is the identity. As for the remaining 64 cases, we see that, for example, $(bd)a = b(da)$ because both expressions have value b, and $(ca)b = c(ab)$ because both expressions have value a.

We see that e is the identity for G because $ex = e$ (examine the first row of the table) and $xe = x$ (examine the first column) for all $x \in G$.

Finally, every element of G has an inverse. The inverse of a is d because $ad = e$ and $da = e$. These equations also prove that a is the inverse of d. The elements b and c are inverses because $bc = cb = e$, and e is its own inverse because $ee = e$.

Therefore G is a group.   ∎

In Section 6.1 we defined the operation $+$ on the set $\mathbb{Z}_m$ and proved that $(\mathbb{Z}_m, +)$ is an algebraic system that is associative and has identity element 0. Every element of $\mathbb{Z}_m$ has an additive inverse. Thus we may restate Theorem 6.1.3(a) as follows:

Theorem 6.2.1   For every natural number m, $(\mathbb{Z}_m, +)$ is a group of order m.

The algebra for multiplication in $\mathbb{Z}_m$ is more complicated. For $m = 1$, $(\mathbb{Z}_1, \cdot)$ is the group $(\{0\}, \cdot)$. By Theorem 6.1.3(b), for every $m > 1$, the system $(\mathbb{Z}_m, \cdot)$ is associative and has identity 1. However, 0 has no multiplicative inverse because there is no x in $\mathbb{Z}_m$ such that $x \cdot 0 = 1$. Therefore $(\mathbb{Z}_m, \cdot)$ is *not* a group when $m > 1$.

If we remove the element 0, then $(\mathbb{Z}_m - \{0\}, \cdot)$ may not be an algebraic system, because the set may not be closed under the operation $\cdot$. For example, the product $3 \cdot 4 = 0$ is not in the set $\mathbb{Z}_{12} - \{0\}$. When m is prime, $(\mathbb{Z}_m - \{0\}, \cdot)$ has not only the closure and associative properties, but 1 is the identity element, and every element has an inverse.

Theorem 6.2.2   The system $(\mathbb{Z}_m - \{0\}, \cdot)$ is a group iff m is prime.

Proof.   See Exercise 18.   ∎

You may have noticed that the commutative property is not included among the group axioms. All of the groups considered so far in this section, including $(\mathbb{R}, +)$, $(\mathbb{Q}, +)$, $(\mathbb{R} - \{0\}, \cdot)$, and $(\mathbb{Z}_m, +)$ have operations that are commutative, but not all groups have commutative operations. Commutative groups are called abelian groups, and are so named in honor of Niels Abel.*

> **DEFINITION**   A group G is **abelian** iff the group operation is commutative.

The abelian property is independent of the group axioms; that is, it cannot be proved from those axioms. It could have been considered as another axiom for defining a group. Because there are many important algebraic structures that are groups but do not satisfy the commutative property, mathematicians choose to not include commutativity in the definition of a group.

Our work with one-to-one correspondences revealed algebraic properties that we can now use to form *groups whose elements are functions*. In this way we encounter our first examples of nonabelian groups.

A **permutation** on a nonempty set A was defined in Section 4.4 as a function $f: A \rightarrow A$ that is both one-to-one and onto A. If the elements of A are listed in order, the effect of the permutation is rearranging (or permuting) the elements of A.

We use a simplified notation (see Section 4.4) to describe a permutation f on the set $\mathbb{N}_k = \{1, 2, 3, \ldots, k\}$ by listing the images of $1, 2, 3, \ldots, k$ in order within brackets, as follows:

$$[f(1)f(2)f(3)\ldots f(k)].$$

For example, the permutation $g = \{(1, 3), (2, 1), (3, 4), (4, 5), (5, 2)\}$ on the set $A = \{1, 2, 3, 4, 5\}$ is written as $g = [3\ 1\ 4\ 5\ 2]$. The identity permutation on A is $I_A = [1\ 2\ 3\ 4\ 5]$ because it maps 1 to 1, 2 to 2, and so on. The permutation $h = [5\ 4\ 3\ 1\ 2]$ is the function given by $h(1) = 5, h(2) = 4, h(3) = 3, h(4) = 1$, and $h(5) = 2$.

From Theorem 4.4.5 (b) we know that the composite of two permutations on a set A is again a permutation of A. Therefore, the set of all permutations on A, with composition as the operation, is an algebraic structure. We know by Theorem 4.2.1 that composition is associative. Theorem 4.4.5 goes on to say that I_A is an identity (parts (a) and

---

* Niels Abel (1802–1829) was a Norwegian mathematician who made fundamental contributions to the theory of functions and proved that no general solution involving radicals exists for 5th degree polynomial equations. Tuberculosis ended his brilliant career at age 26.

(d) of the theorem) and every permutation has an inverse (parts (c) and (e)). Combining these results, we have established:

Theorem 6.2.3
Let A be a nonempty set. The set of all permutations on A with the operation of function composition is a group, called the **group of permutations of A**.

The group of permutations on the set $\{1, 2, 3, \ldots, n\}$ is given a special name.

> **DEFINITION** Let n be a natural number. The group of all permutations of $\mathbb{N}_n = \{1, 2, 3, \ldots, n\}$ is called the **symmetric group on n symbols** and is designated by S_n.

There are exactly $n!$ arrangements of the elements of a set A with n elements, so the order of S_n is $n!$

Example. For the set $\mathbb{N}_3 = \{1, 2, 3\}$, there are six permutations in S_3: $I_A = [1\ 2\ 3]$, $f_1 = [1\ 3\ 2]$, $f_2 = [3\ 2\ 1]$, $f_3 = [2\ 1\ 3]$, $g = [2\ 3\ 1]$, and $h = [3\ 1\ 2]$. Remember that we compute products as function composites. For example, $[3\ 2\ 1][1\ 3\ 2] = f_2 \circ f_1 = h = [3\ 1\ 2]$ because

$$(f_2 \circ f_1)(1) = f_2(f_1(1)) = f_2(1) = 3,$$
$$(f_2 \circ f_1)(2) = f_2(f_1(2)) = f_2(3) = 1, \text{ and}$$
$$(f_2 \circ f_1)(3) = f_2(f_1(3)) = f_2(2) = 2.$$

Likewise, $[3\ 1\ 2][2\ 3\ 1] = h \circ g = [1\ 2\ 3] = I_A$. The complete Cayley table for S_3 is

	I_A [1 2 3]	f_1 [1 3 2]	f_2 [3 2 1]	f_3 [2 1 3]	g [2 3 1]	h [3 1 2]
I_A [1 2 3]	[1 2 3]	[1 3 2]	[3 2 1]	[2 1 3]	[2 3 1]	[3 1 2]
f_1 [1 3 2]	[1 3 2]	[1 2 3]	[2 3 1]	[3 1 2]	[3 2 1]	[2 1 3]
f_2 [3 2 1]	[3 2 1]	[3 1 2]	[1 2 3]	[2 3 1]	[2 1 3]	[1 3 2]
f_3 [2 1 3]	[2 1 3]	[2 3 1]	[3 1 2]	[1 2 3]	[1 3 2]	[3 2 1]
g [2 3 1]	[2 3 1]	[2 1 3]	[1 3 2]	[3 2 1]	[3 1 2]	[1 2 3]
h [3 1 2]	[3 1 2]	[3 2 1]	[2 1 3]	[1 3 2]	[1 2 3]	[2 3 1]

It is important to note that the two products $[2\ 1\ 3][3\ 1\ 2]$ and $[3\ 1\ 2][2\ 1\ 3]$ are different. Since $[2\ 1\ 3][3\ 1\ 2] = [3\ 2\ 1] \neq [1\ 3\ 2] = [3\ 1\ 2][2\ 1\ 3]$, the group S_3 is *not* abelian.

Groups whose elements are some (but not necessarily all) permutations of a set are called **permutation groups.** The reason for the importance of permutation groups is that, for every group of elements of any kind (numbers, sets, functions, …), there is a corresponding permutation group with the same number of elements and the same structure. This fact is known as Cayley's Theorem, and appears in Section 6.4.

The next results are consequences of the group axioms and facilitate calculations involving elements of a group. Notice that proving a statement like $x = y^{-1}$ is not like proving, say, a trigonometric identity. The statement $x = y^{-1}$ is read "x is the inverse of y," and is proved by showing that x plays the role of an inverse for y, i.e., that the product of x and y is the identity.

Theorem 6.2.4   Let G be a group with identity e. For all a, b, and c in G,

(a) $(a^{-1})^{-1} = a$.
(b) $(ab)^{-1} = b^{-1}a^{-1}$.
(c) If $ac = bc$, then $a = b$ (Right Cancellation Law).
(d) If $ca = cb$, then $a = b$ (Left Cancellation Law).

Proof.

(a)   Because a^{-1} is the inverse of a, $a^{-1}a = aa^{-1} = e$. Therefore, a acts as the $\langle unique \rangle$ inverse of a^{-1}, so $(a^{-1})^{-1} = a$.

(b)   We know $(ab)^{-1}$ is the unique element x of G such that $(ab)x = x(ab) = e$. We see that $b^{-1}a^{-1}$ meets this criterion by computing

$$(ab)(b^{-1}a^{-1}) = a(bb^{-1})a^{-1} = a(e)a^{-1} = aa^{-1} = e.$$

Similarly, $(b^{-1}a^{-1})(ab) = e$, so $b^{-1}a^{-1}$ is the inverse of ab.

(c)   Suppose $ac = bc$ in the group G. Then c^{-1} is in G and $(ac)c^{-1} = (bc)c^{-1}$. Using the associative, inverse, and identity properties, we see that

$$(ac)c^{-1} = a(c\,c^{-1}) = ae = a \quad \text{and}$$
$$(bc)c^{-1} = b(c\,c^{-1}) = be = b.$$

Therefore $a = b$.

(d)   Exercise 10. ∎

Theorem 6.2.5   Let G be a finite group with identity e. For every $a \in G$,

(a)   The function $\lambda_a: G \rightarrow G$, where $\lambda_a(x) = ax$ for each $x \in G$, is a permutation of G.

(b)   The function $\rho_a: G \rightarrow G$, where $\rho_a(x) = xa$ for each $x \in G$, is a permutation of G.

Proof.

(a)   To show that λ_a is one-to-one, let $x, y \in G$ and suppose $\lambda_a(x) = \lambda_a(y)$. Then $ax = ay$, so by the Left Cancellation Law, $x = y$.
To show that λ_a is onto G, let $b \in G$. $\langle We\ need\ to\ find\ x\ in\ G\ so\ that\ \lambda_a(x) = b.\rangle$ Choose $x = a^{-1}b$. Then $\lambda_a(x) = \lambda_a(a^{-1}b) = a(a^{-1}b) = (a\,a^{-1})b = eb = b$.

(b)   The proof of part (b) is similar and is Exercise 11. ∎

Lambda (λ_a) is the function that performs left multiplication by a, while rho (ρ_a) is the function that performs right multiplication by a. For a finite group G, Theorem 6.2.4 says that for any row (specified by $a \in G$) in the Cayley table, the row is a permutation of the list of elements in G in the order presented by the table's row heading. The same is true for every element column in the table. Therefore,

> *If G is a finite group, then every element of G occurs exactly once in every row and exactly once every column of the Cayley table.*

The converse of this statement is false. It is possible to have each element occur exactly once in every row and once in every column of an operation table for a structure that is not a group. See Exercise 5.

If G is a group, it is convenient to have notation for powers of elements of G. Let $a \in G$ and $n \in \mathbb{N}$. We define

$$a^0 = e \text{ and } a^{n+1} = a^n a.$$

Thus a^n is defined inductively for all $n \geq 0$. Define a^n for $n < 0$ by

$$a^n = (a^{-1})^{-n}.$$

For all integers m and n, these familiar laws of exponents hold in a group:

$$a^m a^n = a^{m+n}$$
$$(a^m)^n = a^{mn}, \text{ and}$$
$$(a^n)^{-1} = a^{-n}, \text{ for } n > 0.$$

When the group operation is $+$, we use different words for the concepts we have termed "product," ab, a^{-1}, a^2, a^n and a^{-n}:

The operation is called **addition**, and $a + b$ is the **sum** of a and b;
the additive inverse of a is $-a$, called the **negative** of a;
$a - b$ is an abbreviation for $a + (-b)$, the **difference** of a and b;
$a + a$ is $2a$, $a + 2a$ is $3a$, and for $n \in \mathbb{N}$, $a + a + \cdots + a$ (n times) is na, the **nth multiple** of a; and
$(-n)a$ is $-(na)$.

Example.   Prove that for every natural number t, the set $t\mathbb{Z}$ of all integer multiples of t is an additive group.

Proof.

(i)   If $x, y \in t\mathbb{Z}$, then $x = k_1 t$ and $y = k_2 t$ for some integers k_1 and k_2. The set $t\mathbb{Z}$ is closed under addition because the sum $x + y = (k_1 + k_2)t \in t\mathbb{Z}$.

(ii)   Addition is associative in $\mathbb{Z}$ and therefore in $t\mathbb{Z}$.

(iii)   The additive identity element in $t\mathbb{Z}$ is 0.

(iv)   The additive inverse (negative) of an element $x = kt$ in $t\mathbb{Z}$ is $(-k)t$.   ∎

Examples. In the group $(5\mathbb{Z}, +)$, the third multiples of the elements 20 and -35 are $3(20) = 60$ and $3(-35) = -105$. The negative of the element $(6) \cdot 5$ is $(-(6)) \cdot 5 = -30$.

We close this section with some important observations about the axiomatic approach used to define a group. First, a small set of axioms is advantageous, although challenging to produce, because a small set means that fewer properties need to be checked to be sure a given structure satisfies the axioms. The definition of a group uses just three axioms.

Second, it may be best to leave a desired property out of the axioms if it can be deduced from the remaining axioms. Theorem 6.1.1 tells us that the identity and inverses of elements in a group are unique. In the definition of a group, we could have said "there is a unique identity element e in G for $\circ$." Verifying this property requires showing the existence and uniqueness of the identity rather than only the existence. We stated the definition of a group as we did to make it easier to verify that a structure is a group.

Finally, the fact that axioms may be altered by adding or deleting specific axioms does not mean that the axioms are chosen at random, or that all of the axioms are equally worthy of study. The group axioms are chosen because the structures they describe are so important to mathematics and its applications. Treating an additional property, such as commutativity, as a property that holds for many but not all groups, allows us to keep the basic axioms for a group minimal.

Exercises 6.2

1. Show that each of the following algebraic structures is a group.
   (a) $(\{1, -1\}, \cdot)$ where $\cdot$ is multiplication of integers.
   (b) $(\{1, \alpha, \beta\}, \cdot)$ where $\alpha = \dfrac{-1 + i\sqrt{3}}{2}, \beta = \dfrac{-1 - i\sqrt{3}}{2}$, and $\cdot$ is complex number multiplication.
   ★ (c) $(\{1, -1, i, -i\}, \cdot)$ where $\cdot$ is complex number multiplication.
   ☆ (d) $(\mathscr{P}(X), \triangle)$ where X is a nonempty set and $\triangle$ is the symmetric difference operation $A \triangle B = (A - B) \cup (B - A)$.

★ 2. Given that $G = \{e, u, v, w\}$ is a group of order 4 with identity e and $u^2 = v, v^2 = e$, construct the operation table for G.

3. Given that $G = \{e, u, v, w\}$ is a group of order 4 with identity e and $u^2 = v^2 = w^2 = e$, construct the operation table for G.

☆ 4. Which of the groups of Exercise 1 are abelian?

5. Give an example of an algebraic system $(G, \circ)$ that is not a group such that in the operation table for $\circ$, every element of G appears exactly once in every row and once in every column. This can be done with as few as three elements in G.

6. Construct the operation table for each of the following groups.
   (a) $(\mathbb{Z}_6, +)$ (b) $(\mathbb{Z}_7, +)$
   (c) $(\mathbb{Z}_5 - \{0\}, \cdot)$ (d) $(\mathbb{Z}_7 - \{0\}, \cdot)$

7. Construct the operation table for S_2, the symmetric group on 2 elements. Is S_2 abelian?

8. **(a)** What is the order of S_4, the symmetric group on 4 elements?
   ⋆ **(b)** Compute these products in S_4: [1 2 4 3][4 2 1 3], [4 3 2 1][4 3 2 1], and [2 1 4 3][1 3 2 4].
   (c) Compute these products in S_4: [3 1 2 4][3 2 1 4], [4 3 2 1][3 1 2 4], and [1 4 3 2][1 4 3 2].
   (d) Show that S_4 is not abelian.

9. Let G be a group, and $a_i \in G$ for all $n \in \mathbb{N}$.
   (a) Prove that $(a_1 a_2 a_3)^{-1} = a_3^{-1} a_2^{-1} a_1^{-1}$.
   (b) State and prove a result similar to part (a) for n elements of G, for all $n \in \mathbb{N}$.

10. Prove part (d) of Theorem 6.2.4. That is, prove that if G is a group, a, b, and c are elements of G, and $ca = cb$, then $a = b$.

11. Prove part (b) of Theorem 6.2.5.

☆ 12. Let G be a group. Prove that if $g^2 = e$ for all $g \in G$, then G is abelian.

☆ 13. Give an example of an algebraic structure of order 4 that has both right and left cancellation but that is not a group.

14. Let G be a group. Prove that
   (a) G is abelian iff $a^2 b^2 = (ab)^2$ for all $a, b \in G$.
   (b) G is abelian iff $a^n b^n = (ab)^n$ for all $n \in \mathbb{N}$ and $a, b \in G$.

15. Show that the structure $(\mathbb{R} - \{1\}, \circ)$, with operation $\circ$ defined by $a \circ b = a + b - ab$, is an abelian group. You should first show that $(\mathbb{R} - \{1\}, \circ)$ is a structure.

16. ⋆ **(a)** In the group G of Exercise 2, find x such that $v \circ x = e$; x such that $v \circ x = u$; x such that $v \circ x = v$; and x such that $v \circ x = w$.
   ☆ **(b)** Let $(G, *)$ be a group and $a, b \in G$. Show that there exist unique elements x and y in G such that $a * x = b$ and $y * a = b$.

17. Show that $(\mathbb{Z}, \#)$, with operation $\#$ defined by $a \# b = a + b + 1$, is a group. Find x such that $50 \# x = 100$.

18. **(a)** Prove that if m is composite, then the set $\mathbb{Z}_m - \{0\}$ is not closed under multiplication.
   (b) Let p be a prime natural number. Prove that $(\mathbb{Z}_p - \{0\}, \cdot)$ is an associative algebraic system with identity 1. *Hint:* Use Euclid's Lemma.
   (c) Prove that if p is prime, then every element of $(\mathbb{Z}_p - \{0\}, \cdot)$ has an inverse. *Hint:* Suppose $x \in \mathbb{Z}_p - \{0\}$. Then x and p are relatively prime, so there exist integers r and s such that $rx + sp = 1$.
   (d) Conclude that $\mathbb{Z}_m - \{0\}$ with multiplication is a group iff m is prime.

⋆ 19. Let p be a prime natural number. Show that $(p - 1)^{-1} = p - 1$ in $(\mathbb{Z}_p - \{0\}, \cdot)$.

20. Find all solutions in $(\mathbb{Z}_{20}, \cdot)$ for the following equations.
   ⋆ **(a)** $5x = 0$ **(b)** $3x = 0$
   (c) $x^2 = 0$ **(d)** $x^2 = 9$

21. In $(\mathbb{Z}_8, +)$, find all solutions for the following equations.
    - (a) $4 + x = 6$ (b) $x + 7 = 3$ ★ (c) $3 + x = 1$
    - (d) $2x = 4$ ★ (e) $2x = 3$ (f) $2x + 3 = 1$

22. Galois discovered a connection between certain groups and the solutions to polynomial equations. Refer to the finite sets in Exercises 1(a), (b) and (c). Each of these sets forms a group (although these groups do not represent the general case of Galois' work). Find a polynomial equation, and verify that your equation has integer coefficients, such that
    - ★ (a) the equation has degree 2 and $\{1, -1\}$ is the solution set.
    - (b) the equation has degree 3 and $\{1, \alpha, \beta\}$ is the solution set.
    - (c) the equation has degree 4 and $\{1, -1, i, -i\}$ is the solution set.

Proofs to Grade

23. Assign a grade of A (correct), C (partially correct), or F (failure) to each. Justify assignments of grades other than A.
    - (a) **Claim.** If G is a group with identity e, then G is abelian.
      "Proof." Let a and b be elements of G. Then

$$
\begin{aligned}
ab &= aeb \\
&= a(ab)(ab)^{-1}b \\
&= a(ab)(b^{-1}a^{-1})b \\
&= (aa)(bb^{-1})a^{-1}b \\
&= (aa)a^{-1}b \\
&= (aa)(b^{-1}a)^{-1} \\
&= a(a(b^{-1}a)^{-1}) \\
&= a((b^{-1}a)^{-1}a) \\
&= a((a^{-1}b)a) \\
&= (aa^{-1})(ba) \\
&= e(ba) \\
&= ba.
\end{aligned}
$$

      Therefore $ab = ba$ and G is abelian. ∎

    - ★ (b) **Claim.** If G is a group with elements x, y, and z, and if $xz = yz$, then $x = y$.
      "Proof." If $z = e$, then $xz = yz$ implies that $xe = ye$, so $x = y$. If $z \neq e$, then the inverse of z exists, and $xz = yz$ implies $\frac{xz}{z} = \frac{yz}{z}$ and $x = y$. Hence in all cases, if $xz = yz$, then $x = y$. ∎

    - ★ (c) **Claim.** The set $\mathbb{Q}^+$ of positive rationals with the operation of multiplication is a group.
      "Proof." The product of two positive rationals is a positive rational, so $\mathbb{Q}^+$ is closed under multiplication. Since $1 \cdot r = r = r \cdot 1$ for every $r \in \mathbb{Q}$, 1 is the identity. The inverse of the positive rational $\frac{a}{b}$ is the positive rational $\frac{b}{a}$. The rationals are associative under multiplication because the reals are associative under multiplication. ∎

(d)   Claim.   If m is prime, then $(\mathbb{Z}_m - \{0\}, \cdot)$ has no divisors of zero.
"Proof."   Suppose a is a divisor of zero in $(\mathbb{Z}_m - \{0\}, \cdot)$. Then $a \neq 0$, and there exists $b \neq 0$ in $\mathbb{Z}_m$ such that $ab = 0$. Then $ab = m \,(\text{mod } m)$, so $ab = m$. This contradicts the assumption that m is prime.   ∎

(e)   Claim.   If m is prime, then $(\mathbb{Z}_m - \{0\}, \cdot)$ has no divisors of zero.
"Proof."   Suppose a is a divisor of zero in $(\mathbb{Z}_m - \{0\}, \cdot)$. Then $a \neq 0$, and there exists $b \neq 0$ in $\mathbb{Z}_m$ such that $ab = 0$. Then $ab = 0 \,(\text{mod } m)$, so m divides ab. Since m is prime, m divides a or m divides b. But since a and b are elements of $\mathbb{Z}_m$, both are less than m. This is impossible.   ∎

(f)   Claim.   For every natural number m, $(\mathbb{Z}_m - \{0\}, \cdot)$ is a group.
"Proof."   We know that $(\mathbb{Z}, \cdot)$ is associative with identity element 1. Therefore, $(\mathbb{Z}_m - \{0\}, \cdot)$ is associative with identity element 1. It remains to show every element has an inverse. For $x \in \mathbb{Z}_m - \{0\}, x \neq 0$. Therefore, $1/x \in \mathbb{Z}_m - \{0\}$ and $x \cdot 1/x = x(1/x) = 1$. Therefore, every element of $\mathbb{Z}_m - \{0\}$ has an inverse.   ∎

(g)   Claim.   If $(\mathbb{Z}_m - \{0\}, \cdot)$ is a group, then m is prime.
"Proof."   Assume that $(\mathbb{Z}_m - \{0\}, \cdot)$ is a group. Suppose m is not prime. Let $m = rs$, where r and s are integers greater than 1 and less than m. Then $r \cdot s = m = 0 \,(\text{mod } m)$. Since r has an inverse t in $\mathbb{Z}_m - \{0\}, t \cdot r = 1$. Then $s = 1 \cdot s = (t \cdot r) \cdot s = t \cdot (r \cdot s) = t \cdot 0 = 0$. That is, $s = 0 \,(\text{mod } m)$. This is impossible, because $1 < s < m$.   ∎

6.3   Subgroups

A substructure of an algebraic system $(A, *)$ consists of a subset of A together with all the operations and relations in the original structure, *provided* that this is an algebraic structure. This proviso is necessary, for it may happen that a subset of A is not closed under an operation. For example, the subset of $\mathbb{R}$ consisting of the irrationals is not closed under multiplication. The idea of substructures is a natural one, and we can't fully understand a group until we understand its subgroups.

> **DEFINITION**   Let $(G, \circ)$ be a group and H a subset of G. Then $(H, \circ)$ is a **subgroup** of G iff $(H, \circ)$ is a group.

It is understood that the operation $\circ$ on H agrees with the operation $\circ$ on G. That is, the operation on H is the function $\circ$ restricted to $H \times H$.

Suppose H is a nonempty subset of G and $(G, \circ)$ is a group. What must we do to prove that H is a subgroup of G? The first answer that comes to mind is to prove that H is closed under $\circ$ and to verify all three of the group properties. As a first step in shortening this process, we observe that it is not necessary to check associativity: this property is "inherited" from the group.

Theorem 6.3.1   Let G be a group and H be a nonempty subset of G. Then H is a subgroup of G if

    (i) H is closed under $\circ$,

    (ii)   the identity e of G is in H,

    (iii) every $x \in H$ has an inverse x^{-1} in H.

Proof.   Suppose H is a nonempty subset of the group G that satisfies conditions (i), (ii), and (iii). Let x, y, and z be in H. Then x, y, and z are in G, so by associativity for G, $(xy)z = x(yz)$. Thus H is a subgroup of G. ∎

    The set E of even numbers is a subgroup of $(\mathbb{Z}, +)$ because E is nonempty, E is closed under addition, the identity $0 \in E$ and the negative of an even integer is even. We note that E is the set of all multiples of 2. In general, for every integer t the set $t\mathbb{Z}$ of all multiples of t is a subgroup of $(\mathbb{Z}, +)$.

    Two subsets of $\mathbb{Z}_6$ that are closed under $+$ can be seen in the following tables. It is easy to check that both $H = \{0, 3\}$ and $K = \{0, 2, 4\}$ are subgroups of $(\mathbb{Z}_6, +)$.

$(\mathbb{Z}_6, +)$	+	0	1	2	3	4	5
	0	0	1	2	3	4	5
	1	1	2	3	4	5	0
	2	2	3	4	5	0	1
	3	3	4	5	0	1	2
	4	4	5	0	1	2	3
	5	5	0	1	2	3	4

$(\mathbb{Z}_6, +)$	+	0	1	2	3	4	5
	0	0	1	2	3	4	5
	1	1	2	3	4	5	0
	2	2	3	4	5	0	1
	3	3	4	5	0	1	2
	4	4	5	0	1	2	3
	5	5	0	1	2	3	4

$(H, +)$	+	0	3
	0	0	3
	3	3	0

$(K, +)$	+	0	2	4
	0	0	2	4
	2	2	4	0
	4	4	0	2

    For every group $(G, \circ)$ with identity e, $(\{e\}, \circ)$ is a group called the **identity subgroup,** or **trivial subgroup** of G. Also, every group is a subgroup of itself. All subgroups of G other than G are called **proper subgroups.**

    The symmetric group S_3 with 6 elements has six subgroups, two of which are the trivial subgroup $\{[1\ 2\ 3]\}$ and S_3 itself. Let $J = \{[1\ 2\ 3], [2\ 1\ 3]\}$. Then J contains the identity element $[1\ 2\ 3]$ of S_3. By computing

$$[2\ 1\ 3][2\ 1\ 3] = [1\ 2\ 3]$$
$$[2\ 1\ 3][1\ 2\ 3] = [2\ 1\ 3]$$
$$[1\ 2\ 3][2\ 1\ 3] = [2\ 1\ 3]$$
$$[1\ 2\ 3][1\ 2\ 3] = [1\ 2\ 3]$$

we see that the inverse of $[2\ 1\ 3]$ is $[2\ 1\ 3]$ and J is closed under composition, so J satisfies the conditions of Theorem 6.3.1. Therefore, J is a subgroup of S_3. Similar computations show that $K = \{[1\ 2\ 3], [3\ 2\ 1]\}$ and $L = \{[1\ 2\ 3], [1\ 3\ 2]\}$ are also subgroups of S_3. The only subgroup of S_3 of order three is $M = \{[1\ 2\ 3], [2\ 3\ 1], [3\ 1\ 2]\}$.

Important questions to be answered are whether the identity element in a subgroup can be different from the identity element of the original group, and whether the inverse of an element in H could be different from its inverse in G. The answers are "no" and "no."

Lemma 6.3.2

Let H be a subgroup of G. Then

(a)   The identity of H is the identity e of G.
(b)   If $x \in H$, the inverse of x in H is its inverse in G.

Proof.

(a)   If i is the identity element of H, then $ii = i$. But in G, $ie = i$, so $ii = ie$ and, by cancellation, $i = e$.
(b)   See Exercise 3.   ∎

The next theorem makes it easier to prove that a subset of a group is a subgroup. It is given in "iff" form for completeness, but the important result is that only two properties must be checked to show that H is a subgroup of G. The first is that H is nonempty. This is usually done by showing that the identity e of G is in H. The other is to show that $ab^{-1} \in H$ whenever a and b are in H. This is usually less work than showing both that H is closed under the group operation and that $b \in H$ implies $b^{-1} \in H$.

Theorem 6.3.3

Let G be a group. A subset H of G is a group iff H is nonempty and for all $a, b \in H$, $ab^{-1} \in H$.

Proof.   First, suppose H is a subgroup of G. Then H is a group, so by Lemma 6.3.2 (a) H contains the identity e. Therefore $H \neq \varnothing$. Also, if a and b are in H, then $b^{-1} \in H$ (by the inverse property) and $ab^{-1} \in H$ (by the closure property).

Now suppose $H \neq \varnothing$ and for all $a, b \in H$, $ab^{-1} \in H$. *(We show that H is a subgroup by showing that H satisfies the conditions of Theorem 6.3.1. It is best to proceed in the order that follows.)*

(i) $H \neq \varnothing$, so there is some $a \in H$. Then $aa^{-1} = e \in H$.
(ii)   Suppose $x \in H$. Then e and x are in H, so by hypothesis, $ex^{-1} = x^{-1} \in H$.
(iii)   Let x and y be in H. Then by (ii), $y^{-1} \in H$. Then x and y^{-1} are in H, so by hypothesis, $x(y^{-1})^{-1} = xy \in H$.   ∎

If a is a member of G, then by the closure property, all powers of a are in G. The next theorem shows that the set $\{\ldots, a^{-2}, a^{-1}, a^0 = e, a, a^2, \ldots\}$ is a subgroup of G.

Theorem 6.3.4

If G is a group and $a \in G$, then the set of all powers of a is an abelian subgroup of G.

Proof.   Since $a^1 = a$ is a power of a, the set of all powers of a is nonempty. Suppose x and y are powers of a. Then $x = a^m$ and $y = a^n$ for some integers m and n. Thus $xy^{-1} = a^m(a^n)^{-1} = a^m a^{-n} = a^{m-n}$ is a power of a. Therefore, by Theorem 6.3.3, $\{a^n: n \in \mathbb{Z}\}$ is a subgroup of G. The subgroup is abelian because $a^m a^n = a^{mn} = a^{nm} = a^n a^m$.   ∎

DEFINITIONS   Let G be a group and $a \in G$. Then $(a) = \{a^n : n \in \mathbb{Z}\}$ is
called the **cyclic subgroup generated by a**.

The **order of the element a** is the order of (number of elements in)
the group (a). If (a) is an infinite set, we say a has **infinite order**.

Examples.   For the group $(\mathbb{R} - \{0\}, \cdot)$, the element 2 generates the infinite cyclic
subgroup $(2) = \{2^n : n \in \mathbb{Z}\} = \left\{ \ldots, -\frac{1}{8}, -\frac{1}{4}, -\frac{1}{2}, 1, 2, 3, 4, 8, \ldots \right\}$. Thus 2 has
infinite order. We note that $\left(\frac{1}{2}\right)$ is the same subgroup as (2).

The element -1 generates the cyclic subgroup $\{-1, 1\}$. Since this subgroup
has 2 elements, we say -1 has order 2.

Examples.   The group $(\mathbb{Z}_7 - \{0\}, \cdot)$ has 6 elements: 1, 2, 3, 4, 5, 6. The cyclic
subgroup generated by 1 is $\{1\}$, so the order of 1 is 1.

The cyclic subgroup $(2) = \{1, 2, 4\}$ since $2^0 = 1, 2^1 = 2, 2^2 = 4$, and
$2^3 = 8 = 1$ again. The order of 2 is 3.

Computing modulo 7, we have $3^2 = 9 = 2, 3^3 = 27 = 6, 3^4 = 81 = 4, 3^5 =$
$243 = 5$, and $3^6 = 729 = 1$, so the nonnegative powers of 3 are, in order, 1, 3, 2, 6,
4, 5, and then 1 again. The element 3 has order 6 because (3) has 6 elements. We can
also show that 4 has order 3; 5 has order 6; and 6 has order 2.

DEFINITIONS   Let G be a group. If there is an element $a \in G$ such that
$(a) = G$, then we say G is a **cyclic group**. Any element a of G such that
$(a) = G$ is called a **generator for G**.

In the $\mathbb{Z}_7 - \{0\}$ example above, $\mathbb{Z}_7 - \{0\} = (3)$, so the element 3 is a generator. Since 5 has order 6, $(5) = \mathbb{Z}_7 - \{0\}$ and 5 is another generator. The elements 3
and 5 are the only generators of $\mathbb{Z}_7 - \{0\}$.

When the group operation is addition, the cyclic subgroup generated by the
element a is the set of all multiples of a. For example, $(\mathbb{Z}, +)$ is cyclic with generators 1 and -1. For $m > 1$, every group $(\mathbb{Z}_m, +)$ is cyclic with generators 1 and
$m - 1$. See Exercise 16.

The cyclic group $(\mathbb{Z}_4, +)$ has only two generators, 1 and 3. This is because the
multiples of 1 and 3 are:

$$
\begin{array}{lll}
1 \cdot 1 = 1 & \text{and} & 1 \cdot 3 = 3 \\
2 \cdot 1 = 1 + 1 = 2 & & 2 \cdot 3 = 3 + 3 = 2 \\
3 \cdot 1 = 2 + 1 = 3 & & 3 \cdot 3 = 2 + 3 = 1 \\
4 \cdot 1 = 3 + 1 = 0 & & 4 \cdot 3 = 1 + 3 = 0.
\end{array}
$$

The element 2 does not generate $\mathbb{Z}_4$; it has order 2 and generates the subgroup
$\{0, 2\}$.

Example. The group S_3 is not cyclic because none of its elements generates the entire group. For example, for the element $[312]$

$$[3\ 1\ 2]^1 = [3\ 1\ 2]$$
$$[3\ 1\ 2]^2 = [2\ 3\ 1]$$
$$[3\ 1\ 2]^3 = [1\ 2\ 3], \quad \text{the identity}$$
$$[3\ 1\ 2]^4 = [3\ 1\ 2]^1 = [3\ 1\ 2]$$
$$[3\ 1\ 2]^5 = [3\ 1\ 2]^2 = [2\ 3\ 1]$$
$$[3\ 1\ 2]^6 = [3\ 1\ 2]^3 = [1\ 2\ 3], \quad \text{and so forth.}$$

All other powers of $[3\ 1\ 2]$ are equal to one of these three elements, so the cyclic subgroup generated by $[3\ 1\ 2]$ is $\{[1\ 2\ 3], [3\ 1\ 2], [2\ 3\ 1]\}$. Similar calculations show that none of the other elements of S_3 generate S_3 (Exercises 9(a) and 10(a)). Of course, the fact that S_3 is not abelian is sufficient to conclude that S_3 is not cyclic, because every cyclic group is abelian.

Theorem 6.3.5   Let G be a group and a be an element of G with order r. Then r is the smallest positive integer such that $a^r = e$, the identity, and $(a) = \{e, a, a^2, \ldots, a^{r-1}\}$.

Proof.   Since the order of a is finite, the powers of a are not all distinct. Let $a^m = a^n$ with $0 \le m < n$. Then $a^{n-m} = e$ with $n - m > 0$. Therefore, the set of positive integers p such that $a^p = e$ is nonempty. Let k be the smallest such integer. $\langle$*This k exists by the Well-Ordering Principle.*$\rangle$ We prove that $k = r$ by showing that the elements of (a) are exactly $a^0 = e, a^1, a^2, \ldots, a^{k-1}$.

First, we show that the elements $e, a^1, a^2, \ldots, a^{k-1}$ are distinct. If $a^s = a^t$ with $0 \le s < t < k$, then $a^{t-s} = e$ and $0 < t - s < k$, contradicting the definition of k.

Second, we show that every element of (a) is one of $e, a^1, a^2, \ldots, a^{k-1}$. Consider a^t for $t \in \mathbb{Z}$. By the Division Algorithm, $t = mk + s$ with $0 \le s < k$. Thus $a^t = a^{mk+s} = a^{mk}a^s = (a^k)^m a^s = e^m a^s = ea^s = a^s$, so that $a^t = a^s$ with $0 \le s < k$.

We have shown that the elements a^s for $0 \le s < k$ are all distinct and that every power of a is equal to one of these. Since (a) has exactly r elements, $r = k$ and $a^r = e$. ∎

If $a \in G$ has infinite order, then all the powers of a are distinct and

$$(a) = \{\ldots, a^{-2}, a^{-1}, a^0 = e, a^1, a^2, \ldots\}.$$

Exercises 6.3

---

1. By looking for subsets closed under the group operation, then checking the group axioms, find all subgroups of
   ★ (a) $(\mathbb{Z}_8, +)$.   (b) $(\mathbb{Z}_7 - \{0\}, \cdot)$.
   (c) $(\mathbb{Z}_5, +)$.

☆ **(d)** $(J, *)$, with $J = \{a, b, c, d, e, f\}$ and the table for $*$ shown at the right.

$*$	a	b	c	d	e	f
a	a	b	c	d	e	f
b	b	a	f	e	d	c
c	c	e	a	f	b	d
d	d	f	e	a	c	b
e	e	c	d	b	f	a
f	f	d	b	c	a	e

2. In the group S_4,
   (a) find two different subgroups that have 3 elements.
   (b) find two different subgroups that have 4 elements.
   (c) [2 3 1 4][3 1 2 4] = [1 2 3 4]. Is there a subgroup of S_4 that contains [2 3 1 4] but not [3 1 2 4]? Explain.
   (d) find the smallest subgroup that contains [4 2 1 3] and [3 2 4 1]. (*Hint:* Use Theorem 6.3.1.)

3. Prove that if G is a group and H is a subgroup of G, then the inverse of an element $x \in H$ is the same as its inverse in G (Theorem 6.3.2).

4. Prove that if H and K are subgroups of a group G, then $H \cap K$ is a subgroup of G.

5. Prove that if $\{H_\alpha: \alpha \in \Delta\}$ is a family of subgroups of a group G, then $\bigcap_{\alpha \in \Delta} H_\alpha$ is a subgroup of G.

6. Give an example of a group G and subgroups H and K of G such that $H \cup K$ is not a subgroup of G.

7. Let G be a group and H be a subgroup of G.
★ **(a)** If G is abelian, must H be abelian? Explain.
   (b) If H is abelian, must G be abelian? Explain.

8. Let G be a group. If H is a subgroup of G and K is a subgroup of H, prove that K is a subgroup of G.

9. Find the order of each element of the group
   (a) S_3.   **(b)** $(\mathbb{Z}_7, +)$.
★ **(c)** $(\mathbb{Z}_8, +)$.   **(d)** $(\mathbb{Z}_{11} - \{0\}, \cdot)$.

10. List all generators of each cyclic group in Exercise 9.

★ 11. Let G be a group with identity e and let $a \in G$. Prove that the set $C_a = \{x \in G: xa = ax\}$, called the **centralizer** of a in G, is a subgroup of G.

12. Let G be a group and let $C = \{x \in G: \text{for all } y \in G, xy = yx\}$. Prove that C, the **center** of G, is a subgroup of G.

13. Prove that if G is a group and $a \in G$, then the center of G is a subgroup of the centralizer of a in G.

★ 14. Let G be a group and let H be a subgroup of G. Let a be a fixed element of G. Prove that $K = \{a^{-1}ha: h \in H\}$ is a subgroup of G.

15. Let $(\mathbb{C} - \{0\}, \cdot)$ be the group of nonzero complex numbers with complex number multiplication. Let $\alpha = \dfrac{1 + i\sqrt{3}}{2}$.
   (a) Find (α).
   (b) Find a generator of (α) other than α.

16. Prove that for every natural number m greater than 1, the group $(\mathbb{Z}_m, +)$ is cyclic with generators 1 and $m - 1$.

☆ 17. Prove that every subgroup of a cyclic group is cyclic.

18. Let $G = (a)$ be a cyclic group of order 30.
   (a) What is the order of a^6?   (b) List all elements of order 2.
   ★ (c) List all elements of order 3.   (d) List all elements of order 10.

Proofs to Grade 19. Assign a grade of A (correct), C (partially correct), or F (failure) to each. Justify assignments of grades other than A.
   ★ (a) **Claim.** If H and K are subgroups of a group G, then $H \cap K$ is a subgroup of G.
      "Proof." Let $a, b \in H \cap K$. Then $a, b \in H$ and $a, b \in K$. Since H and K are subgroups, $ab^{-1} \in H$ and $ab^{-1} \in K$. Therefore, $ab^{-1} \in H \cap K$. ∎
   (b) **Claim.** If H is a subgroup of a group G and $x \in H$, then $xH = \{xh : h \in H\}$ is a subgroup of G.
      "Proof." First, the identity $e \in H$. Thus, $x = xe \in xH$. Therefore, $xH \neq \varnothing$. Second, let $a, b \in xH$. Then $a = xh$ and $b = xk$, for some $h, k \in H$. Then we have $ab^{-1} = (xh)(xk)^{-1} = (xh)(k^{-1})(x^{-1}) = x(hk^{-1}x^{-1}) \in xH$. Therefore, xH is a subgroup of G. ∎

6.4   Operation Preserving Maps

One of the most important concepts in algebra involves mappings between systems. In particular, we are interested in those functions from one algebraic system $(A, \circ)$ to another $(B, *)$ that preserves structure; that is, functions that align the structure of A with that of B. For example, we will see that under an operation preserving map the identity element for A corresponds to the identity for B, and if x has an inverse in A, then the image of x has an inverse in B.

> **DEFINITION**   Let $(A, \circ)$ and $(B, *)$ be algebraic systems and f be a function from A to B. Then f is **operation preserving (OP)** iff for all $x, y \in A$,
>
> $$f(x \circ y) = f(x) * f(y).$$

Because $f(x \circ y)$ and $f(x) * f(y)$ are elements in B, the definition of operation preserving is a statement about equality of elements in B. Calculating each of $f(x \circ y)$ and $f(x) * f(y)$ requires two steps: performing an operation and applying the function. If f is an OP map, the equation $f(x \circ y) = f(x) * f(y)$ means that the result is the same whether the operation $x \circ y$ is performed first or the images $f(x)$ and $f(y)$ are determined first. Another way to say a mapping is operation preserving is "the image of the product is the product of the images."

You have already encountered OP mappings in previous courses. For example, the familiar equation $\log(x \cdot y) = \log x + \log y$ tells us that the logarithm function from $((0, \infty), \cdot)$ to $(\mathbb{R}, +)$ is operation preserving.

Examples.   Let $(\mathbb{R}[x], +)$ be the set of all polynomials with real coefficients with the operation of polynomial addition. Let D be the differentiation mapping $D: (\mathbb{R}[x], +) \to (\mathbb{R}[x], +)$, where for each $f \in \mathbb{R}[x]$, $D(f) = \frac{df}{dx}$, the first derivative of the polynomial f. The function D is an OP map because we know that

$$D(f + g) = \frac{d}{dx}(f + g) = \frac{df}{dx} + \frac{dg}{dx} = D(f) + D(g).$$

If we change the operation on $\mathbb{R}[x]$ to polynomial multiplication, then $D: (\mathbb{R}[x], \cdot) \to (\mathbb{R}[x], \cdot)$ is not operation preserving because the derivative of a product is not always equal to the product of the derivatives.

Example.   Let the operation $\circ$ on $\mathbb{R} \times \mathbb{R}$ be defined by setting $(a, b) \circ (c, d) = (a + c, b + d)$ and let $\cdot$ be the usual multiplication on $\mathbb{R}$. Then the function $f: (\mathbb{R} \times \mathbb{R}, \circ) \to (\mathbb{R}, \cdot)$ given by $f(a, b) = 2^a 3^b$ is operation preserving.

Proof.   Let (x, y) and (u, v) be elements of $\mathbb{R} \times \mathbb{R}$. Then

$$f((x, y) \circ (u, v)) = f(x + u, y + v) = 2^{x+u} \cdot 3^{y+v}.$$

Also

$$f(x, y) \cdot f(u, v) = (2^x \cdot 3^y) \cdot (2^u \cdot 3^v) = 2^{x+u} \cdot 3^{y+v}.$$

Therefore, f is operation preserving.   ∎

The next three theorems explain in more detail what we mean by saying that an OP map $f: (A, \circ) \to (B, *)$ preserves the structure of $(A, \circ)$.

Theorem 6.4.1   Let f be an OP map from $(A, \circ)$ to $(B, *)$. Then $(\text{Rng}(f), *)$ is an algebraic structure.

Proof.   ⟨*What we must show is that* $\text{Rng}(f)$ *is closed under the operation* $*$*; that is, if* $u, v \in \text{Rng}(f)$, *then* $u * v \in \text{Rng}(f)$.⟩

First, note that because A is nonempty, $\text{Rng}(f)$ is nonempty. Assume that $u, v \in \text{Rng}(f)$. Then there exist elements x and y of A such that $f(x) = u$ and $f(y) = v$. Then $u * v = f(x) * f(y) = f(x \circ y)$, so $u * v$ is the image of $x \circ y$, which is in A. Therefore, $u * v \in \text{Rng}(f)$.   ∎

Theorem 6.4.2   Let f be an OP map from $(A, \circ)$ onto $(B, *)$. If $\circ$ is commutative on A, then $*$ is commutative on B.

Proof.   Assume that f is OP, f is onto B, and $\circ$ is commutative on A. Let u and v be elements of B. Then there are x and y in A such that $u = f(x)$ and $v = f(y)$. Then

$$
\begin{aligned}
u * v &= f(x) * f(y) \\
&= f(x \circ y) \\
&= f(y \circ x) \\
&= f(y) * f(x) \\
&= v * u.
\end{aligned}
$$

Therefore $u * v = v * u$.   ∎

The properties of associativity, existence of identities, and existence of inverses are all preserved by an OP mapping.

Theorem 6.4.3   Let f be an OP map from $(A, \circ)$ onto $(B, *)$.

(a)   If $\circ$ is associative on A, then $*$ is associative on B.
(b)   If e is the identity for A, then $f(e)$ is the identity for B.
(c)   If x^{-1} is the inverse for x in A, then $f(x^{-1})$ is the inverse for $f(x)$ in B.

Proof.   See Exercise 9.   ∎

Example.   For each natural number m, the canonical map is an important operation preserving map from $(\mathbb{Z}, +)$ onto $(\mathbb{Z}_m, +)$. We define $H: \mathbb{Z} \to \mathbb{Z}_m$ by setting $H(x) = \bar{x}$, the equivalence class determined by x. The canonical map is operation preserving because, by Theorem 6.1.2(a),

$$
\text{for all } a, b \in \mathbb{Z}, H(a + b) = \overline{a + b} = \bar{a} + \bar{b} = H(a) + H(b).
$$

The canonical map H is a surjection because for every $\bar{k} \in \mathbb{Z}_m$, $k \in \mathbb{Z}$ and $H(k) = \bar{k}$.

We note that operation preserving mappings need not be limited to an algebraic structure with a single operation. Since

$$
H(ab) = \overline{ab} = \bar{a} \cdot \bar{b} = H(a) \cdot H(b),
$$

the canonical map H also preserves multiplication. Thus, $H: (\mathbb{Z}, +, \cdot) \to (\mathbb{Z}_m, +, \cdot)$ is OP for both multiplication and addition.

Special terminology is used for operation preserving mappings where the algebraic structures involved are groups.

DEFINITIONS   Let $(G, \circ)$ and $(H, *)$ be groups. An OP mapping $h: (G, \circ) \to (H, *)$ is called a **homomorphism** from $(G, \circ)$ to $(H, *)$. The range of h is called the **homomorphic image** of $(G, \circ)$ under h.

The function $f: (\mathbb{Z}, +) \to 5\mathbb{Z}, +)$ where $5\mathbb{Z}$ is the set of all integer multiples of 5 and $f(x) = 5x$, is an example of a homomorphism. To verify this, we note that for all $x, y \in \mathbb{Z}$, $f(x + y) = 5(x + y) = 5x + 5y = f(x) + f(y)$. The homomorphism f maps onto $5\mathbb{Z}$, because for each $w \in 5\mathbb{Z}$, $w = 5k$ for some integer k and therefore, $f(k) = 5k = w$. Thus $(5\mathbb{Z}, +)$ is the homomorphic image of $(\mathbb{Z}, +)$ under f.

Example. Let $\mathbb{Z}_6 = \{\overline{0}, \overline{1}, \overline{2}, \overline{3}, \overline{4}, \overline{5}\}$ and $\mathbb{Z}_3 = \{[0], [1], [2]\}$. Define $T: \mathbb{Z}_6 \to \mathbb{Z}_3$ by $T(\overline{0}) = T(\overline{3}) = [0]$, $T(\overline{1}) = T(\overline{4}) = [1]$, and $T(\overline{2}) = T(\overline{5}) = [2]$. See Figure 6.4.1. It can be verified by checking all cases that $T: (\mathbb{Z}_6, +) \to (\mathbb{Z}_3, +)$ is a homomorphism onto $(\mathbb{Z}_3, +)$.

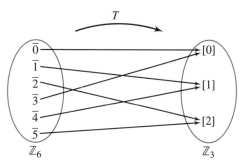

Figure 6.4.1

Theorem 6.4.4   Let $(G, \cdot)$ and $(K, *)$ be groups. If $f: (G, \cdot) \to (K, *)$ is a homomorphism, then $(\text{Rng}(f), *)$ is a subgroup of $(K, *)$. Furthermore, if the operation $\cdot$ is commutative, then $*$ is a commutative operation on $(\text{Rng}(f)$. In other words,

(a)   the homomorphic image of a group is a group, and
(b)   the homomorphic image of an abelian group is an abelian group.

Proof.   This theorem is a restatement of previous results using our new terminology. The image of a group under a homomorphism is an algebraic system (Theorem 6.4.1); the image is associative, has an identity element, and every element has an inverse (Theorem 6.4.3); and, if a group is abelian, then its image is abelian (Theorem 6.4.2). ∎

Example.   Let $(G, \circ)$ be any group with identity e. The mapping $F: G \to G$ given by $F(x) = e$ for all $x \in G$ is a homomorphism, called the *trivial homomorphism*. We can verify this by observing that for $x, y \in G$, both $F(x \circ y) = e$ and $F(x) \circ F(y) = e \circ e = e$. Therefore, $F(x \circ y) = F(x) \circ F(y)$. In this case the homomorphic image of $(G, \circ)$ is $(\{e\}, \circ)$.

Example.   The function $T: \mathbb{Z}_3 \to \mathbb{Z}_6$ given by $T([x]) = \overline{2x}$ is a homomorphism since for all $[x]$ and $[y]$ in $\mathbb{Z}_3$,

$$
\begin{aligned}
T([x] + [y]) &= T([x + y]) \\
&= \overline{2(x + y)} \\
&= \overline{2x + 2y} \\
&= \overline{2x} + \overline{2y} \\
&= T([x]) + T([y]).
\end{aligned}
$$

However, $\text{Rng}(T) = \{\bar{0}, \bar{2}, \bar{4}\}$, so T is not onto $\mathbb{Z}_6$. We note that $(\{\bar{0}, \bar{2}, \bar{4}\}, +)$ is a subgroup of $\mathbb{Z}_6$, in agreement with Theorem 6.4.4.

DEFINITIONS   Let $(G, \circ)$ and $(H, *)$ be groups. A homomorphism $h: (G, \circ) \to (H, *)$ that is one-to-one and onto H is called an **isomorphism**. If h is an isomorphism, we say $(G, \circ)$ and $(H, *)$ are **isomorphic**.

The word *isomorphic* comes from the Greek words *isos* (equal) and *morphe* (form), literally meaning "equal form" because two isomorphic groups will differ only in the names or nature of their elements. *All their algebraic properties are identical.* Inverses and composites of isomorphisms are also isomorphisms. Thus, the relation of being isomorphic is an equivalence relation on the class of all groups. (See Exercise 19.)

The three groups $(\mathbb{Z}_2, +), (\{1, -1\}, \cdot)$, and $(\{\varnothing, A\}, \triangle)$ of order 2 are isomorphic, where A is a nonempty set and $\triangle$ is the symmetric difference operation defined by $X \triangle Y = (X - Y) \cup (Y - X)$. The Cayley tables for the three groups are

+	0	1
0	0	1
1	1	0

$\cdot$	1	-1
1	1	-1
-1	-1	1

$\triangle$	$\varnothing$	A
$\varnothing$	$\varnothing$	A
A	A	$\varnothing$

In fact, any two groups of order two are isomorphic (see Exercise 10(a)).

Example.   Let $K = \{2^x : x \in \mathbb{Z}\}$ and $\cdot$ be multiplication. Then $(K, \cdot)$ is a group isomorphic to $(\mathbb{Z}, +)$. The one-to-one correspondence $f: \mathbb{Z} \to K$, where $f(x) = 2^x$, is OP because $f(x + y) = 2^{x+y} = 2^x 2^y = f(x) f(y)$.

Example.   The groups $(\mathbb{Z}_4, +)$ and $(\mathbb{Z}_5 - \{0\}, \cdot)$, are isomorphic. We define an isomorphism $h: (\mathbb{Z}_4, +) \to (\mathbb{Z}_5 - \{0\}, \cdot)$ as follows, keeping in mind that elements of $\mathbb{Z}_4$ are equivalence classes mod 4, while elements of $\mathbb{Z}_5$ are equivalence classes mod 5:

$$h(0) = 1,$$
$$h(1) = 2,$$
$$h(2) = 4, \quad \text{and}$$
$$h(3) = 3.$$

The Cayley tables for the groups are shown next. The elements of $\mathbb{Z}_5 - \{0\}$ are listed in the order determined by their preimages. This makes it easier to see that the two groups have the same structure.

+	0	1	2	3
0	0	1	2	3
1	1	2	3	0
2	2	3	0	1
3	3	0	1	2

$(\mathbb{Z}_4, +)$

·	1	2	4	3
1	1	2	4	3
2	2	4	3	1
4	4	3	1	2
3	3	1	2	4

$(\mathbb{Z}_5 - \{0\}, \cdot)$

We claimed in Section 6.2 that for every group there is a permutation group with the same structure. This result, due to Arthur Cayley, is proved in the next theorem.

Theorem 6.4.5

Cayley's Theorem

Every group G is isomorphic to a permutation group.

Proof.   We choose the set G itself to be the set of objects to be permuted. By Theorem 6.2.5(a) for every a in G the function $\lambda_a\colon G \to G$, where $\lambda_a(x) = ax$ for each $x \in G$, is a permutation of G. Let $\mathcal{H} = \{\lambda_a\colon a \in G\}$. By Theorem 6.2.3, the set $\mathcal{G}$ of all permutations of the elements of G is a group with the operation of function composition. We claim that $\mathcal{H}$ is the image of a one-to-one homomorphism from G to $\mathcal{G}$, and conclude that G is isomorphic to the permutation group $\mathcal{H}$.

Let $f\colon G \to \mathcal{G}$ be given by $f(a) = \lambda_a$. Let $a, b \in G$. ⟨*We will prove that f is a homomorphism by showing that $f(ab) = f(a) \circ f(b)$.*⟩ Let $x \in G$. Then,

$$\begin{aligned}
f(ab)(x) &= \lambda_{ab}(x)\\
&= (ab)x\\
&= a(bx)\\
&= \lambda_a(\lambda_b(x))\\
&= (\lambda_a \circ \lambda_b)(x)\\
&= (f(a) \circ f(b))(x).
\end{aligned}$$

Thus $f(ab) = f(a) \circ f(b)$ for all $a, b \in G$. Therefore, f is a homomorphism.

To show that f is one-to-one, suppose that $f(a) = f(b)$. Then $\lambda_a = \lambda_b$. Therefore, $ax = bx$ for every $x \in G$. In particular, if e is the identity for G, $ae = be$, so $a = b$.

By definition, every permutation in $\mathcal{H}$ is λ_a for some $a \in G$. Therefore, f maps onto $\mathcal{H}$.

We have shown that $\mathcal{H}$ is the homomorphic image of G, so by Theorem 6.4.4, $\mathcal{H}$ is a group. Since f is one-to-one, f is an isomorphism. Therefore G and $\mathcal{H}$ are isomorphic groups.   ■

Example.   Let $(G, \cdot)$ be the group $\{1, -1, i, -i\}$ of four complex numbers, where $\cdot$ is the usual multiplication of complex numbers. The corresponding group $\mathcal{H}$ of permutations, as described above, consists of the four left translations by the elements of G. For example, λ_i is the mapping that multiplies each element of G on the left by i:

$$\begin{aligned}
\lambda_i(1) &= i \cdot 1 = i,\\
\lambda_i(-1) &= i \cdot (-1) = -i,\\
\lambda_i(i) &= i \cdot i = -1, \text{ and}\\
\lambda_i(-i) &= i \cdot (-i) = 1.
\end{aligned}$$

Thus, $\lambda_i = [i \ -i \ -1 \ 1]$. The other three permutations are $\lambda_1 = [1 \ -1 \ i \ -i]$, $\lambda_{-1} = [-1 \ 1 \ -i \ i]$, and $\lambda_{-i} = [-i \ i \ 1 \ -1]$. The tables for G and $\mathcal{H}$ show that they have the same structure.

$(G, \cdot)$

$\cdot$	1	-1	i	$-i$
1	1	-1	i	$-i$
-1	-1	1	$-i$	i
i	i	$-i$	-1	1
$-i$	$-i$	i	1	-1

$(\mathcal{H}, \circ)$

$\circ$	λ_1	λ_{-1}	λ_i	λ_{-i}
λ_1	λ_1	λ_{-1}	λ_i	λ_{-i}
λ_{-1}	λ_{-1}	λ_1	λ_{-i}	λ_i
λ_i	λ_i	λ_{-i}	λ_{-1}	λ_1
λ_{-i}	λ_{-i}	λ_i	λ_1	λ_{-1}

Finally, we note that G (and therefore $\mathcal{H}$) is isomorphic to $(\mathbb{Z}_4, +)$. (See Exercise 17.) This fact suggests the possibility that all groups of order 4 are isomorphic to $(\mathbb{Z}_4, +)$. See Exercise 10(c) for a counterexample to this conjecture.

Exercises 6.4

1. Define $SQRT: \mathbb{R}^+ \rightarrow \mathbb{R}^+$ by $SQRT(x) = \sqrt{x}$.
   (a) Is $SQRT: (\mathbb{R}^+, +) \rightarrow (\mathbb{R}^+, +)$ operation preserving?
   (b) Is $SQRT: (\mathbb{R}^+, \cdot) \rightarrow (\mathbb{R}^+, \cdot)$ operation preserving?

2. Define $SQR: \mathbb{R} \rightarrow \mathbb{R}$ by $SQR(x) = x^2$.
   (a) Is $SQR: (\mathbb{R}, +) \rightarrow (\mathbb{R}, +)$ operation preserving?
   (b) Is $SQR: (\mathbb{R}, \cdot) \rightarrow (\mathbb{R}, \cdot)$ operation preserving?

3. Define $\otimes$ on $\mathbb{R} \times \mathbb{R}$ by setting $(a, b) \otimes (c, d) = (ac - bd, ad + bc)$.
   (a) Show that $(\mathbb{R} \times \mathbb{R}, \otimes)$ is an algebraic system.
   (b) Show that the function h from the system $(\mathbb{C}, \cdot)$ to $(\mathbb{R} \times \mathbb{R}, \otimes)$ given by $h(a + bi) = (a, b)$ is a one-to-one function from the set of complex numbers that is onto $\mathbb{R} \times \mathbb{R}$ and is operation preserving.

★ 4. Let $\mathcal{F}$ be the set of all real-valued integrable functions defined on the interval $[a, b]$. Then $(\mathcal{F}, +)$ is an algebraic structure, where $+$ is the addition of functions. Define $I: (\mathcal{F}, +) \rightarrow (\mathbb{R}, +)$ by $I(f) = \int_a^b f(x)\,dx$. Use your knowledge of calculus to verify that I is an OP map.

5. Let $f: (A, \cdot) \rightarrow (B, *)$ and $g: (B, *) \rightarrow (C, \times)$ be OP maps.
   (a) Prove that $g \circ f$ is an OP map.
   (b) Prove that if f^{-1} is a function, then f^{-1} is an OP map.

6. Let $\mathcal{M}$ be the set of all 2×2 matrices with real entries. Define Det: $\mathcal{M} \rightarrow \mathbb{R}$ by
$$\text{Det} \begin{bmatrix} a & b \\ c & d \end{bmatrix} = ad - bc.$$

    (a) Prove that Det: $(\mathcal{M}, \cdot) \rightarrow (\mathbb{R}, \cdot)$ is operation preserving, where $(\mathcal{M}, \cdot)$ denotes $\mathcal{M}$ with matrix multiplication.

    (b) Prove that Det: $(\mathcal{M}, +) \rightarrow (\mathbb{R}, +)$ is not operation preserving, where $(\mathcal{M}, +)$ denotes $\mathcal{M}$ with matrix addition.

7. Let Conj: $\mathbb{C} \rightarrow \mathbb{C}$ be the conjugate mapping for complex numbers given by Conj $(a + bi) = a - bi$.

    (a) Prove that Conj: $(\mathbb{C}, +) \rightarrow (\mathbb{C}, +)$ is operation preserving, where $(\mathbb{C}, +)$ denotes the complex numbers with addition.

    (b) Prove that Conj: $(\mathbb{C}, \cdot) \rightarrow (\mathbb{C}, \cdot)$ is operation preserving, where $(\mathbb{C}, \cdot)$ denotes the complex numbers with multiplication.

8. Let f be a function from set A to set B. Let f and f^{-1} be the induced functions on $\mathcal{P}(A)$ as defined in Section 4.5.

★   **(a)** Prove that the induced function $f: (\mathcal{P}(A), \cup) \rightarrow (\mathcal{P}(B), \cup)$ is an OP map.

    (b) Prove that the induced function $f^{-1}: (\mathcal{P}(B), \cap) \rightarrow (\mathcal{P}(A), \cap)$ is an OP map.

    (c) Prove that the induced function $f^{-1}: (\mathcal{P}(B), \cup) \rightarrow (\mathcal{P}(A), \cup)$ is an OP map.

9. Prove Theorem 6.4.3.

10. ☆ **(a)** Show that any two groups of order 2 are isomorphic.

    (b) Show that any two groups of order 3 are isomorphic.

☆   **(c)** Prove that there exist two groups of order 4 that are not isomorphic.

11. Let $3\mathbb{Z}$ and $6\mathbb{Z}$ be the sets of integer multiples of 3 and 6, respectively. Let f be the function from $(3\mathbb{Z}, +)$ to $(6\mathbb{Z}, +)$ given by $f(x) = 4x$.

    (a) Prove that f is a homomorphism.

    (b) What group is the homomorphic image of $(3\mathbb{Z}, +)$ under f?

12. Let $(3\mathbb{Z}, +)$ and $(6\mathbb{Z}, +)$ be the groups in Exercise 12 and let g be the function from $3\mathbb{Z}$ to $6\mathbb{Z}$ given by $g(x) = x + 3$. Is g a homomorphism? Explain.

13. Let $(\{a, b, c\}, \circ)$ be the group with the operation table shown here.

$\circ$	a	b	c
a	a	b	c
b	b	c	a
c	c	a	b

    Verify that the mapping $g: (\mathbb{Z}_6, +) \rightarrow (\{a, b, c\}, \circ)$ defined by $g(0) = g(3) = a$, $g(1) = g(4) = b$, and $g(2) = g(5) = c$ is a homomorphism.

14. Let $\{\bar{0}, \bar{1}, \ldots, \overline{17}\} = \mathbb{Z}_{18}$ and $\{[0], [1], \ldots, [23]\} = \mathbb{Z}_{24}$.

★   **(a)** Prove that the function $f: \mathbb{Z}_{18} \rightarrow \mathbb{Z}_{24}$ given by $f(\bar{x}) = [4x]$ is well defined and is a homomorphism from $(\mathbb{Z}_{18}, +)$ to $(\mathbb{Z}_{24}, +)$.

★   **(b)** Find Rng (f) and give the operation table for the subgroup Rng (f) of $\mathbb{Z}_{24}$.

15. Let $\{\bar{0}, \bar{1}, \ldots, \overline{14}\} = \mathbb{Z}_{15}$ and $\{[0], [1], \ldots, [11]\} = \mathbb{Z}_{12}$. Define $f: \mathbb{Z}_{15} \rightarrow \mathbb{Z}_{12}$ by $f(\bar{x}) = [4x]$.

(a)   Prove that f is a well-defined function and a homomorphism from $(\mathbb{Z}_{15}, +)$ to $(\mathbb{Z}_{12}, +)$.

(b)   Find Rng (f) and give the operation table for this subgroup of $\mathbb{Z}_{12}$.

16.   Let $(G, \circ)$ and $(H, *)$ be groups, i be the identity element for H, and $h\colon (G, \circ) \to (H, *)$ be a homomorphism. The **kernel** of f is ker $(f) = \{x \in G\colon f(x) = i\}$. ker (f) is all the elements of G that map to the identity in H. Show that ker (f) is a subgroup of G.

17.   Show that $(\mathbb{Z}_4, +)$ and $(\{1, -1, i, -i\}, \cdot)$ are isomorphic.

18.   Is S_3 isomorphic to $(\mathbb{Z}_6, +)$? Explain.

19.   Prove that the relation of isomorphism is an equivalence relation. That is, prove that
(a)   if $(G, \cdot)$ is a group, then $(G, \cdot)$ is isomorphic to $(G, \cdot)$.
(b)   if $(G, \cdot)$ is isomorphic to $(H, *)$, then $(H, *)$ is isomorphic to $(G, \cdot)$.
(c)   if $(G, \cdot)$ is isomorphic to $(H, *)$ and $(H, *)$ is isomorphic to $(K, \otimes)$, then $(G, \cdot)$ is isomorphic to $(K, \otimes)$.

20.   Use the method of proof of Cayley's Theorem to find a group of permutations isomorphic to
(a) $(\mathbb{Z}_3, +)$.
(b) $(\mathbb{Z}_5, +)$.
(c) $(\mathbb{R}, +)$.

Proofs to Grade   21.   Assign a grade of A (correct), C (partially correct), or F (failure). Justify assignments of grades other than A.
(a)   **Claim.**   Let $\circ$ be the operation on $\mathbb{R} \times \mathbb{R}$ defined by setting $(a, b) \circ (c, d) = (a + c, b + d)$ and let $-$ be the usual subtraction on $\mathbb{R}$. Then the function f given by $f(a, b) = a - 3b$ is an OP map from $(\mathbb{R} \times \mathbb{R}, \circ)$ to $(\mathbb{R}, -)$.

"*Proof.*" $(4, 2)$ and $(3, 1)$ are in $\mathbb{R} \times \mathbb{R}$. Then $f((4, 2) \circ (3, 1)) = f(7, 3) = 7 - 3 \cdot 3 = -2$, whereas $f(4, 2) - f(3, 1) = -2 - 0 = -2$, so f is operation preserving.   ∎

(b)   **Claim.**   Let $f\colon (G, *) \to (H, \cdot)$ and $g\colon (H, \cdot) \to (K, \otimes)$ be OP maps. Then the composite $g \circ f\colon (G, *) \to (K, \otimes)$ is an OP map.

"*Proof.*" $g \circ f(ab) = g(f(ab)) = g(f(a)f(b)) = g(f(a))g(f(b)) = (g \circ f(a))(g \circ f(b))$.   ∎

6.5   Rings and Fields

Thus far we have considered algebraic structures with exactly one binary operation, and in this setting we have explored the derivation of structural properties (such as uniqueness of the identity element and cancellation) and the concepts of substructure and isomorphism. We have considered systems such as $(\mathbb{N}, +)$ and $(\mathbb{N}, \cdot)$ as distinct algebraic systems, ignoring any interaction between the two operations. In this section we extend our study of algebraic structures by investigating systems with two binary operations (addition and multiplication). We will identify those algebraic properties of $\mathbb{Z}$, $\mathbb{Q}$, and $\mathbb{R}$ that distinguish them from the natural numbers and from one another.

The two common properties of all these systems that describe the interaction between addition and multiplication are the distributive laws. Thus we begin by making the distributive laws part of the definition for the algebraic structure called a ring.

DEFINITION   A **ring** $(R, +, \cdot)$ is a set R together with two binary operations $+$ and $\cdot$ that satisfy the following axioms:

(1) $(R, +)$ is an abelian group. Thus for all $a, b, c \in R$,
  (a)   there is an identity element $0 \in R$ such that $a + 0 = 0 + a = a$.
  (b)   for every $a \in R$, there is an additive inverse $-a \in R$ such that $a + (-a) = (-a) + a = 0$.
  (c) $a + (b + c) = (a + b) + c$.
  (d) $a + b = b + a$.

(2) The operation $\cdot$ is associative. Thus for all $a, b, c \in R$,
$$a \cdot (b \cdot c) = (a \cdot b) \cdot c.$$

(3) The multiplication operation is distributive over addition. Thus for all $a, b, c \in R$,
$$a \cdot (b + c) = (a \cdot b) + (a \cdot c) \quad \text{and}$$
$$(a + b) \cdot c = (a \cdot c) + (b \cdot c).$$

The definition of a ring says a great deal about addition but does not require that multiplication satisfy any of the properties for a group operation except associativity, nor that the multiplication operation be commutative. As we did for groups, we often write ab instead of $a \cdot b$, and we write $a - b$ for $a + (-b)$.

Examples.   The real number system $(\mathbb{R}, +, \cdot)$ with addition and multiplication is a ring. The systems $(\mathbb{Z}, +, \cdot)$ and $(\mathbb{Q}, +, \cdot)$ of integers and rational numbers also form rings. If E is the set of even integers, then $(E, +, \cdot)$ is a ring.

The number system $(\mathbb{N} \cup \{0\}, +, \cdot)$ is *not* a ring because $(\mathbb{N} \cup \{0\}, +)$ is not a group—only the element 0 has an additive inverse.

Example.   Let $\mathcal{M}_2$ be the set of all 2 by 2 matrices with real number entries. Then $(\mathcal{M}_2, +, \cdot)$ is a ring because matrix addition is associative and commutative, matrix multiplication is associative, and multiplication distributes over addition. The additive identity is the zero matrix $\begin{bmatrix} 0 & 0 \\ 0 & 0 \end{bmatrix}$ and the additive inverse

of $\begin{bmatrix} a & b \\ c & d \end{bmatrix}$ is the matrix $-\begin{bmatrix} a & b \\ c & d \end{bmatrix} = \begin{bmatrix} -a & -b \\ -c & -d \end{bmatrix}$. Multiplication in this ring is not commutative.

Example.   Let $\mathcal{F}(\mathbb{R})$ be the set of all functions from $\mathbb{R}$ to $\mathbb{R}$. Then $(\mathcal{F}(\mathbb{R}), +, \cdot)$ is a ring. (See Exercise 5.)

Theorem 6.5.1   For every $m \in \mathbb{N}$, $(\mathbb{Z}_m, +, \cdot)$ is a ring.

Proof.   We know by Theorem 6.1.3(a) that for every natural number m, $(\mathbb{Z}_m, +)$ is an abelian group, and the operation $\cdot$ is associative on $\mathbb{Z}_m$ by Theorem 6.1.3(b). We need only verify the distributive axioms to show that $(\mathbb{Z}_m, +, \cdot)$ is a ring. Let a, b, c be integers. Then

$$
\begin{aligned}
\overline{a} \cdot (\overline{b} + \overline{c}) = \overline{a} \cdot \overline{(b + c)} \qquad &\langle \textit{by definition of addition in } \mathbb{Z}_m \rangle \\
= \overline{a(b + c)} \qquad &\langle \textit{by definition of multiplication in } \mathbb{Z}_m \rangle \\
= \overline{(ab + ac)} \qquad &\langle \textit{by distributivity of } + \textit{ and } \cdot \textit{ in } \mathbb{Z} \rangle \\
= \overline{ab} + \overline{ac} \qquad &\langle \textit{by definition of addition in } \mathbb{Z}_m \rangle \\
= \overline{a} \cdot \overline{b} + \overline{a} \cdot \overline{c} \qquad &\langle \textit{by definition of multiplication in } \mathbb{Z}_m \rangle.
\end{aligned}
$$

The proof of the other distributive axiom is an exercise. Therefore, $(\mathbb{Z}_m, +, \cdot)$ is a ring. ■

We next consider properties that are shared by every ring $(R, +, \cdot)$. As a first step, we note that since $(R, +)$ is an abelian group, properties that hold for every abelian group certainly hold for $(R, +)$.

Theorem 6.5.2   Let $(R, +, \cdot)$ be a ring, and $a, b, c \in R$. Then

(a)   the additive identity (zero) of R is unique.
(b)   additive inverses (negatives) of elements of R are unique.
(c)   left and right cancellation hold in R. That is,

$$\text{if } a + b = a + c, \text{ then } b = c, \text{ and}$$
$$\text{if } b + a = c + a, \text{ then } b = c.$$

(d) $-(-a) = a$ and $-(a + b) = (-a) + (-b)$.
(e)   for all integers m and n, $m(a + b) = ma + mb$, $(m + n)a = ma + na$, and $m(na) = (mn)a$.

Proof.   All the above are restatements of properties of abelian groups developed in Sections 6.2 and 6.3.   ∎

The operations in property (e) of Theorem 6.5.2 must be interpreted carefully. Addition in the expression $m(a + b) = ma + mb$ involves the ring addition operation $+$ for both sides of the equation. The terms ma, mb and $m(a + b)$ do not represent multiplication of ring elements, but instead are expressions for multiples of a, b, and $a + b$. The expression $(m + n)a = ma + na$, however, involves two different addition operations. The term $m + n$ adds two integers, but the $+$ sign in the term $ma + mb$ means the (ring) addition of two ring elements. Likewise, in the equation $m(na) = (mn)a$, mn refers to multiplication of integers, whereas $m(na)$ and $(mn)a$ represent the ring sums $na + na + \cdots + na$ (m times) and $a + a + \cdots + a$ (mn times), respectively.

The distributive axioms allow us to derive properties that relate multiplication to the zero and negatives in a ring.

Theorem 6.5.3

Let $(R, +, \cdot)$ be a ring with zero element 0. Then, for all $a, b, c \in R$,

(a) $0 \cdot a = a \cdot 0 = 0$.

(b) $a \cdot (-b) = (-a) \cdot b = -(a \cdot b)$.

(c) $(-a) \cdot (-b) = a \cdot b$.

(d) $a \cdot (b - c) = a \cdot b - a \cdot c$ and $(a - b) \cdot c = a \cdot c - b \cdot c$.

Proof.

(a)   Let $a \in R$. ⟨*We use the fact that 0 is the additive identity in two different ways.*⟩ Then $(0 \cdot a) + 0 = 0 \cdot a = (0 + 0) \cdot a = (0 \cdot a) + (0 \cdot a)$. By the left cancellation property, $0 = 0 \cdot a$. The proof that $a \cdot 0 = 0$ is an exercise.

(b)   Let $a, b \in R$. ⟨*To show that $a \cdot (-b) = -(a \cdot b)$, we must show that $a \cdot (-b)$ plays the role of the inverse of $a \cdot b$.*⟩ Then $a \cdot (-b) + (a \cdot b) = a \cdot (-b + b) = a \cdot 0 = 0$. Since $a \cdot (-b)$ is a negative of $(a \cdot b)$ and negatives are unique, $a \cdot (-b) = -(a \cdot b)$. The proof that $(-a) \cdot b = -(a \cdot b)$ is an exercise.

(c)   Let $a, b \in R$. Then $-a$ and b are in R, so by part (b), $(-a) \cdot (-b) = -[(-a) \cdot b] = -[-(a \cdot b)]$. By Theorem 6.5.2(d), $-[-(a \cdot b)] = a \cdot b$, so $(-a) \cdot (-b) = a \cdot b$.

(d)   Let $a, b, c \in R$. Then $a \cdot (b - c) = a \cdot [b + (-c)] = a \cdot b + a \cdot (-c) = a \cdot b + [-(a \cdot c)] = a \cdot b - a \cdot c$. The proof that $(a - b) \cdot c = a \cdot c - b \cdot c$ is an exercise.   ∎

In Section 6.4 we discussed operation preserving maps and defined a group homomorphism as a function that preserves the operation. A ring homomorphism must be operation preserving for both addition and multiplication.

> **DEFINITIONS** Let $(R, +, \cdot)$ and $(S, \oplus, \otimes)$ be rings. A function $h \colon R \to S$ is **a ring homomorphism** iff for all $a, b, \in R$,
>
> $$h(a + b) = h(a) \oplus h(b) \quad \text{and}$$
> $$h(a \cdot b) = h(a) \otimes h(b).$$
>
> If h is one-to-one and onto S, then h is a **ring isomorphism**.

Example. Let $h \colon (\mathbb{Z}, +, \cdot) \to (\mathbb{Z}_m, +, \cdot)$ be the canonical function given by $h(x) = \bar{x}$. We have seen in Section 6.4 that $h \colon (\mathbb{Z}, +) \to (\mathbb{Z}_m, +)$ and $h \colon (\mathbb{Z}, \cdot) \to (\mathbb{Z}_m, \cdot)$ are both operation preserving. Therefore h is a ring homomorphism.

Example. The function $g \colon \mathbb{Z}_6 \to \mathbb{Z}_6$ defined by $g(x) = 3x$ is a ring homomorphism.

Proof. Let $x, y \in \mathbb{Z}_6$. Then $g(x + y) = 3(x + y) = 3x + 3y = g(x) + g(y)$. Also, $g(xy) = 3xy = 9xy = 3x \cdot 3y = g(x) \cdot g(y)$. $\langle$*We used the fact that* $3 = 9 \pmod{6}$.$\rangle$ ∎

The ring $(\mathbb{Z}, +, \cdot)$ of integers has several properties beyond those required to be a ring. For example, the element 1 is a multiplicative identity, also called the unity element. Multiplication in the integers is commutative. The integers also have the property that for any two integers a and b, whenever $ab = 0$, either $a = 0$ or $b = 0$. That is, in the ring of integers there are no divisors of zero. These properties are collected in the next definitions.

> **DEFINITIONS** Let $(R, +, \cdot)$ be a ring.
>
> $(R, +, \cdot)$ is a **ring with unity** iff there is an element $1 \in R$ such that for all $a \in R$, $a \cdot 1 = 1 \cdot a = a$.
> $(R, +, \cdot)$ is a **commutative ring** iff for all $a, b, \in R$, $a \cdot b = b \cdot a$.
> $(R, +, \cdot)$ is an **integral domain** iff R is a commutative ring with unity element 1, $1 \neq 0$, and R has no divisors of zero.

Examples. All three of $(\mathbb{Z}, +, \cdot)$, $(\mathbb{R}, +, \cdot)$, and $(\mathbb{Q}, +, \cdot)$ are integral domains. For every natural number m, the system $(\mathbb{Z}_m, +, \cdot)$ is a commutative ring with unity. The next theorem reveals which of the modular arithmetic rings are integral domains.

Theorem 6.5.4   For $m \in \mathbb{N}$, the ring $(\mathbb{Z}_m, +, \cdot)$ has no zero divisiors iff m is a prime.

Proof. See Exercise 13. ∎

Examples. The ring of even integers is a commutative ring with no divisors of zero, but it is not an integral domain because it has no unity element. The ring of 2 by 2 matrices is a ring with unity because the identity matrix $I = \begin{bmatrix} 1 & 0 \\ 0 & 1 \end{bmatrix}$ is the multiplicative identity, but it is not an integral domain because matrix multiplication is not commutative and the ring has zero divisors. The matrices $A = \begin{bmatrix} 2 & 1 \\ 6 & 3 \end{bmatrix}$ and $B = \begin{bmatrix} 1 & 1 \\ -2 & -2 \end{bmatrix}$ are zero divisors because $AB = \begin{bmatrix} 0 & 0 \\ 0 & 0 \end{bmatrix}$.

The ring $(\mathscr{F}(\mathbb{R}), +, \cdot)$ of functions from $\mathbb{R}$ to $\mathbb{R}$ is a commutative ring with unity but is also not an integral domain. The constant function C_1, with range $\{1\}$, is the multiplicative identity. To construct a pair of zero divisors in the ring $\mathscr{F}(\mathbb{R})$, we let $A = [0, \infty)$, $B = (-\infty, 0)$, and X_A and X_B be the characteristic functions of the sets A and B, respectively. Then $X_A \cdot X_B = 0$ because $X_A \cdot X_B(x) = X_A(x) \cdot X_B(x) = 0$ for all $x \in \mathbb{R}$.

In an integral domain we can apply cancellation laws to simplify products with a nonzero common factor.

Theorem 6.5.5   Let $(R, +, \cdot)$ be an integral domain, and $a, b, c \in R$ with $a \neq 0$. If $a \cdot b = a \cdot c$, then $b = c$, and if $b \cdot a = c \cdot a$, then $b = c$.

Proof.   Assume $a \neq 0$ and $a \cdot b = a \cdot c$. Then $a \cdot b - a \cdot c = 0$, so by Theorem 6.5.4(d), $a \cdot (b - c) = 0$. Since there are no divisors of zero in R and $a \neq 0$, $b - c = 0$. Therefore, $b = c$. The proof that $b \cdot a = c \cdot a$ implies $b = c$ is an exercise.   ■

Let $(R, +, \cdot)$ be an integral domain. One might hope that $(R, \cdot)$ would be an abelian group, since $\cdot$ is associative and commutative and 1 is the multiplicative identity. On second thought this is impossible, because 0 cannot have a multiplicative inverse. (See Exercise 12.) It is possible, however, that the nonzero elements of R all have inverses. An integral domain with this property is called a **field**.

DEFINITION   The ring $(R, +, \cdot)$ is a **field** iff $(R, +, \cdot)$ is an integral domain and $(R - \{0\}, \cdot)$ is an abelian group.

Examples.   The rings of rational numbers and real numbers are fields. The ring of integers is not a field, because no element of $\mathbb{Z}$ (other than 1 and -1) has a multiplicative inverse in $\mathbb{Z}$. Sets with only finitely many elements can also be fields: the ring $(\mathbb{Z}_m, +, \cdot)$ is a field if and only if m is prime. (See Exercise 15.)

A field can also be described as an algebraic structure $(R, +, \cdot)$ such that:

(i) $(R, +)$ is an abelian group with identity 0.
(ii) $(R - \{0\}, \cdot)$ is an abelian group with identity 1.

(iii)   For all $a, b, c \in R$, $a \cdot (b + c) = (a \cdot b) + (a \cdot c)$.

(iv) $0 \neq 1$.

The proof of this fact is an exercise. All that is required is to verify that the second distributive axiom holds and that R has no divisors of zero.

Rings, rings with unity, commutative rings, integral domains, and fields are all objects of important study in mathematics. In Chapter 7 we will consider properties that distinguish the field $\mathbb{R}$ from all other fields.

Exercises 6.5

1.   Which of the following is a ring with the usual operations of addition and multiplication? For each structure that is not a ring, list the ring axioms that are not satisfied.

  (a) $\mathbb{N}$

★ (b)   the closed interval $[-1, 1]$

  (c) $\{a + bi \colon a, b \in \mathbb{Z}\}$, where $i^2 = -1$

  (d) $\{bi \colon b \in \mathbb{Z}\}$, where $i^2 = -1$

2.   Let $\mathbb{Z}[\sqrt{2}]$ be the set $\{a + b\sqrt{2} \colon a, b \in \mathbb{Z}\}$. $\mathbb{Z}[\sqrt{2}]$ is called "$\mathbb{Z}$ adjoin $\sqrt{2}$." Define addition and multiplication on $\mathbb{Z}[\sqrt{2}]$ in the usual way. That is,

$$(a + b\sqrt{2}) + (c + d\sqrt{2}) = (a + c) + (b + d)\sqrt{2} \quad \text{and}$$
$$(a + b\sqrt{2}) \cdot (c + d\sqrt{2}) = ac + ad\sqrt{2} + bc\sqrt{2} + bd(\sqrt{2})^2$$
$$= ac + 2bd + (ad + bc)\sqrt{2}.$$

  Prove that $(\mathbb{Z}[\sqrt{2}], +, \cdot)$ is a ring.

3.   Complete the proof that for every $m \in \mathbb{N}$, $(\mathbb{Z}_m, +, \cdot)$ is a ring (Theorem 6.5.1) by showing that $(b + c)a = ba + ca \pmod{m}$ for all integers a, b, and c.

☆ 4.   Define addition $\oplus$ and multiplication $\otimes$ on the set $\mathbb{Z} \times \mathbb{Z}$ as follows. For $a, b, c, d \in \mathbb{Z}$, $(a, b) \oplus (c, d) = (a + c, b + d)$ and $(a, b) \otimes (c, d) = (ac, bd)$. Prove that $(\mathbb{Z} \times \mathbb{Z}, \oplus, \otimes)$ is a ring.

5.   Let $\mathscr{F}(\mathbb{R})$ be the set of all functions from $\mathbb{R}$ to $\mathbb{R}$ and define addition and multiplication operations on $\mathscr{F}(\mathbb{R})$ as follows. For $f, g \in \mathscr{F}(\mathbb{R})$ and $x \in \mathbb{R}$, $(f + g)(x) = f(x) + g(x)$ and $(f \cdot g)(x) = f(x) \cdot g(x)$. Prove that $(\mathscr{F}(\mathbb{R}), +, \cdot)$ is a ring.

6.   Let $(R, +, \cdot)$ be a ring and $a, b \in R$. Prove that $b + (-a)$ is the unique solution to the equation $x + a = b$.

7.   Prove the remaining parts of Theorem 6.5.3: for all $a, b, c \in \mathbb{R}$,

  (a) $a \cdot 0 = 0$.

  (b) $(-a) \cdot b = -(a \cdot b)$.

  (c) $(a - b) \cdot c = (a \cdot c) - (b \cdot c)$.

8. We define a *subring* of a ring in the same way we defined a subgroup of a group: $(S, +, \cdot)$ is a subring of $(R, +, \cdot)$ if and only if $(R, +, \cdot)$ is a ring, $S \subseteq R$, and $(S, +, \cdot)$ is a ring with the same operations. For example, the ring of even integers is a subring of the ring of integers, and both are subrings of the ring of rational numbers.

   (a) Prove that the ring $(\{0\}, +, \cdot)$ is a subring of any ring $(R, +, \cdot)$ (called the *trivial* subring).

   (b) (Subring Test) Prove that if $(R, +, \cdot)$ is a ring, T is a nonempty subset of R, and T is closed under subtraction and multiplication, then $(T, +, \cdot)$ is a subring.

9. Let $3\mathbb{Z} = \{3k : k \in \mathbb{Z}\}$. Apply the Subring Test (Exercise 8(b)) to show that $(3\mathbb{Z}, +, \cdot)$ is a subring of $(\mathbb{Z}, +, \cdot)$.

10. **(a)** Show that the function $h: \mathbb{Z} \to \mathbb{Z}$ defined by $h(x) = 3x$ is not a ring homomorphism.

   (b) Show that the function $h: \mathbb{Z}_6 \to \mathbb{Z}_6$ defined by $h(x) = 2x$ is not a ring homomorphism.

   (c) Let $\mathbb{Z}[x]$ be the set of all polynomials $p(x)$ in the variable x with integer coefficients. Show that the function $g: \mathbb{Z}[x] \to \mathbb{Z}$ defined by $g(p(x)) = p(0)$ is a ring homomorphism.

11. Suppose P is a set of ordered pairs of integers, and that $(P, \oplus, \otimes)$ is a ring, where

$$(a, b) \oplus (c, d) = (ad + bc, bd) \quad \text{and} \quad (a, b) \otimes (c, d) = (ac, bd).$$

Suppose $f: (\mathbb{Q}, +, \cdot) \to (P, \oplus, \otimes)$ is given by $f(p/q) = (p, q)$. Prove that f is a ring homomorphism.

12. Let $(R, +, \cdot)$ be an integral domain. Prove that 0 has no multiplicative inverse.

13. Let $m \in \mathbb{N}$.

  ☆ **(a)** Prove that m is prime iff $(\mathbb{Z}_m, +, \cdot)$ has no zero divisors. (Theorem 6.5.4).

   (b) Deduce that $(\mathbb{Z}_m, +, \cdot)$ is a field iff m is prime.

14. Complete the proof of Theorem 6.5.5. That is, prove that if $(R, +, \cdot)$ is an integral domain, $a, b, c \in R$ and $a \neq 0$, then $b \cdot a = c \cdot a$ implies $b = c$.

15. Let $(R, +, \cdot)$ be an algebraic structure such that

   (i) $(R, +)$ is an abelian group with identity 0.

   (ii) $(R - \{0\}, \cdot)$ is an abelian group with identity 1.

   (iii) For all $a, b, c \in R$, $a \cdot (b + c) = (a \cdot b) + (a \cdot c)$.

   (iv) $0 \neq 1$.

Prove that $(R, +, \cdot)$ is a field by showing that

   (a) for all $a, b, c \in R$, $(a + b) \cdot c = (a \cdot c) + (b \cdot c)$.

   (b) R has no divisors of zero.

Proofs to Grade   **16.** Assign a grade of A (correct), C (partially correct), or F (failure). Justify assignments of grades other than A.

  (a) Claim. If $(R, +, \cdot)$ is a ring, $a, b \in R$ and $a \neq 0$, then the equation $ax = b$ has a unique solution.

  "Proof." Suppose p and q are two solutions to $ax = b$. Then $ap = b$ and $aq = b$. Therefore, $ap = aq$. Therefore, $p = q$. ∎

  (b) Claim. If $(R, +, \cdot)$ is a finite integral domain, then $(R, +, \cdot)$ is a field.

  "Proof." Suppose R has n elements. Let $x \in R$. Then the $n + 1$ powers of x: $e = x^0, x, x^2, x^3, \ldots, x^n$ are not all distinct. Therefore, $x^t = x^r$ for integers t, r, where we may assume that $t < r$. Then $x^{-t}x^t = x^{-t}x^r$ and therefore, $e = x^{r-t}$. Thus, $e = x \cdot x^{r-t-1}$. Therefore, x has an inverse. Hence, R is a field. ∎

CHAPTER 7

Concepts of Analysis

In this chapter we give an introduction to the analyst's point of view of the real numbers. We consider the reals as a *field* (a set of numbers with operations of addition and multiplication) that is *ordered* (so that all the real numbers may be thought of as forming a line) and *complete* (so that there are no missing numbers anywhere along the line). We examine in depth just what we mean when we say that there are no missing numbers along the real line; each section of this chapter addresses the concept of completeness from a different perspective.

In Section 7.1 we begin with the idea that there are enough real numbers so that there is always a "best" bound for every bounded set. Section 7.2 considers "open," "closed," and "compact" subsets of the reals and establishes an important relationship among these concepts: the Heine–Borel Theorem, whose proof is based on the completeness of the reals. The Bolzano–Weierstrass Theorem of Section 7.3 says that in sets meeting certain conditions, there will always be some element or elements of the set for which there are infinitely many other nearby elements. This fact is proved using the Heine–Borel Theorem. The Bounded Monotone Sequence Theorem of Section 7.4 is derived from the Bolzano–Weierstrass Theorem. It says that if a sequence is increasing (or decreasing) and bounded, then there are enough real numbers so that the limit of the sequence exists. Finally, Section 7.5 shows how the Bounded Monotone Sequence Theorem implies that the real number system is complete.

The sequence of deductions outlined here is circular—we start by assuming completeness of the reals in Section 7.1 and eventually (in Section 7.5) return to the fact that the real numbers are complete. While we will not have proved completeness, we will have seen different ways of understanding completeness and we will have proved that completeness is equivalent to each of the three theorems named. For the purposes of this text, that is sufficient. A separate proof that the reals are complete involves a more careful definition of the real numbers. More will be said about this at the conclusion of Section 7.5.

7.1   Completeness of the Real Numbers

This section assumes a general knowledge of the properties of the rational and real number systems commonly obtained in a calculus class. You should think of the set $\mathbb{R}$ as the set of all decimal numbers along the number line, and the set $\mathbb{Q}$ of rationals as the subset of $\mathbb{R}$ consisting of the repeating or terminating decimals.

Both $\mathbb{R}$ and $\mathbb{Q}$ have algebraic properties of addition and multiplication (listed at the end of this section for reference) that make them *fields*.

Algebraic structures are studied in Chapter 6 but it is not necessary for you to have studied the material on fields before reading this chapter.

Both $\mathbb{R}$ and $\mathbb{Q}$ also have ordering properties for the relation "less than" that make them *ordered fields*. For convenience, the ordering concepts of bounded sets, supremum and infimum from Section 3.4 are repeated at the end of this section, but again it is not necessary to have studied that section. You will need to have studied Section 4.6 on sequences before starting Section 7.4.

It follows from the properties of an ordered field that

Between any two distinct elements there is a third element.

That is, if $a < b$, then there is a third element c such that $a < c < b$. We reason as follows: if $a < b$, then

$$a + a < a + b < b + b$$
$$(1 + 1)a < a + b < (1 + 1)b$$
$$2a < a + b < 2b$$
$$a < \frac{1}{2}(a + b) < b.$$

(Here we have used 2 for the element $1 + 1$ and $\frac{1}{2}$ as the symbol for the multiplicative inverse of 2—see field property 2(b).) Therefore, $c = \frac{1}{2}(a + b)$ is an element of the field that is between the elements a and b. We can repeat this strategy to produce a different element between a and c, and another element between c and b. In fact, by similar reasoning we can produce infinitely many elements between any two elements of an ordered field. The important observation about any ordered field is that

There are never any empty spaces between elements.

Another way to think of this property is that—unlike the integers, where every integer is followed by the next (successor) integer—*nowhere on the number line is there a number that is followed by "the next rational number" or "the next real number."* In particular, there is no positive real number that is the "next" number after 0 because in between 0 and *any* given positive real number, there are always infinitely many real numbers.

The property of having no empty spaces between numbers is a property that holds for both the rational number system and the real number system, because both are ordered fields. Our focus throughout this chapter will be on a different property—a property of the reals that is not shared by the rational numbers.

Although the rational number system has no empty spaces between numbers, it is missing some numbers that in some sense ought to be there. For example, every rational number in the set $B = \{x \in \mathbb{Q}: x^2 < 2\}$ is less than $\sqrt{2}$. There are many rational numbers that are larger than any number in B: the rational number 1.5 is larger than every element of B; the rational number 1.42 is larger than every element of B, and so is 1.415, and 1.4143, and 1.41422, and 1.414214. We could continue this list by finding smaller and smaller rational numbers each of which is larger than every element of B. Ideally, we'd like to find the *smallest rational* number that is larger than every element of B. We will soon prove that there is no such *rational* number. Because the field of rational numbers is missing numbers like these, we say that $\mathbb{Q}$ is not complete.

The goals of this section are to gain a clear understanding of what it means for an ordered field to be complete and to prove that the field of rationals is not complete. We begin with the idea of bounds for a set.

DEFINITIONS   Let A be a subset of an ordered field F. Then
  $u \in F$ is an **upper bound** for A iff $a \leq u$ for every $a \in A$. If A has an upper bound, we say A is **bounded above**.
  $l \in F$ is a **lower bound** for A iff $l \leq a$ for every $a \in A$. If A has a lower bound, we say A is **bounded below**.
  If A has an upper bound and a lower bound, we say A is **bounded**.

In $\mathbb{R}$, the half-open interval $[0, 3)$ has 3 as an upper bound. In fact, π, 18, and 206 are also upper bounds for $[0, 3)$. Both -0.5 and 0 are lower bounds for $[0, 3)$. We note that some bounds for sets are elements of the set while other bounds are not.

Any finite nonempty subset $A = \{x_1, x_2, x_3, \ldots, x_n\}$ of $\mathbb{R}$ is both bounded above and below:

   $u = \max \{x_i : x_i \in A\}$ is an upper bound for A and
   $l = \min \{x_i : x_i \in A\}$ is a lower bound for A.

In $\mathbb{R}$, the subset $\mathbb{N}$ is bounded below but not above, while the sets $\mathbb{Q}$ and $\mathbb{Z}$ are neither bounded above nor bounded below.

In $\mathbb{Q}$, the set $\left\{\frac{1}{2}, \frac{1}{4}, \frac{1}{8}, \frac{1}{16}, \frac{1}{32}, \ldots\right\}$ of negative integer powers of 2 is bounded above by $\frac{1}{2}$ and below by 0. The set $A = \{x \in \mathbb{Q}: x^3 < 4\}$ has many upper bounds: 8, 1.6, 1.59, and so on. However, A has no lower bounds. The set $B = \{x \in \mathbb{Q}: x^2 < 2\}$ is bounded above by 3 and below by -3.

The best (smallest) possible upper bound for a set A is called the supremum of A.

DEFINITIONS   Let A be a subset of an ordered field F. Then $s \in F$ is a **least upper bound** for A (or **supremum** for A) iff

(i) s is an upper bound for A and
(ii) $s \leq x$ for every upper bound x for A.

$i \in F$ is a **greatest lower bound** for A (or **infimum** for A) iff

(i) i is a lower bound for A and
(ii) $x \leq i$ for every lower bound x for A.

We write **sup**(A) to denote a supremum of A. An infimum of A is denoted **inf**(A).

While a set A may have many upper bounds, when $\sup(A)$ exists it is unique. Likewise, $\inf(A)$ is unique if it exists. See Exercise 8.

Examples.   In the ordered field $\mathbb{R}$,

$\inf([0, 3)) = 0$ and $\sup([0, 3)) = 3$.
$\inf(\mathbb{N}) = 1$, but $\sup(\mathbb{N})$ does not exist because $\mathbb{N}$ is not bounded above.
For $A = \{2^{-k} : k \in \mathbb{N}\}$, $\inf(A) = 0$ and $\sup(A) = \frac{1}{2}$.
For $B = \{x \in \mathbb{Q} : x^2 < 2\}$, $\inf(B) = -\sqrt{2}$ and $\sup(B) = \sqrt{2}$.

In the ordered field $\mathbb{Q}$,
For $A = \{2^{-k} : k \in \mathbb{N}\}$, $\inf(A) = 0$ and $\sup(A) = \frac{1}{2}$.

For $B = \{x \in \mathbb{Q} : x^2 < 2\}$, B is bounded, but as we shall see, $\inf(B)$ and $\sup(B)$ do not exist *in the field* $\mathbb{Q}$.

The following theorem provides a characterization of the supremum of a set. Its interpretation, which depends on taking the view that ε may be a very small positive number, is that every element of A is strictly less than every number that is larger than $\sup(A)$, and every number that is smaller than $\sup(A)$ is exceeded by some element of A. Thus *every number larger than $\sup(A)$ is an upper bound for A*, and *every number smaller than $\sup(A)$ is not an upper bound for A*.

Theorem 7.1.1   Let A be a subset of an ordered field F. Then $s = \sup(A)$ iff

(i)   for all $\varepsilon > 0$, if $x \in A$, then $x < s + \varepsilon$.
(ii)   for all $\varepsilon > 0$, there exists $y \in A$ such that $y > s - \varepsilon$.

Proof.   First, suppose $s = \sup(A)$. Let $\varepsilon > 0$ be given. Then $x \leq s < s + \varepsilon$ for all $x \in A$, which establishes property (i).

To verify property (ii), suppose $\varepsilon > 0$ and there is no $y \in A$ such that $y > s - \varepsilon$. Then $s - \varepsilon$ is an upper bound for A less than the least upper bound of A, a contradiction.

Suppose now that s is a number that satisfies conditions (i) and (ii). To show that $s = \sup(A)$, we must first show that s is an upper bound for A. Suppose there is

$y \in A$ such that $y > s$. See Figure 7.1.1(a). If we let $\varepsilon = \dfrac{y - s}{2}$, then $y > s + \varepsilon$, which violates condition (i). ⟨*The idea here is that ε is only half the distance from s to the larger number y, so $s + \varepsilon$ must still be less than y. To verify this algebraically, we have $2y > s + y$, so $y > \dfrac{s + y}{2} = \dfrac{2s + y - s}{2} = s + \dfrac{y - s}{2} = s + \varepsilon$.*⟩ We conclude that $y \le s$ for all $y \in A$, so s is an upper bound for A.

To show s is the least of all upper bounds, suppose that there is another upper bound t such that $t < s$. ⟨*We will show that t cannot be an upper bound.*⟩ If we let $\varepsilon = s - t$, then by condition (ii), there is a number $z \in A$ such that $z > s - \varepsilon$. See Figure 7.1.1(b). Thus $z > s - \varepsilon = s - (s - t) = t$. This contradicts the assumption that t is an upper bound for A. Therefore s is indeed the least upper bound for A. ∎

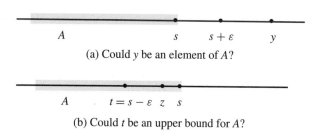

(a) Could y be an element of A?

(b) Could t be an upper bound for A?

Figure 7.1.1

For example, in the field $\mathbb{R}$, the supremum of the set $(2, 4)$ is 4. Even if we take ε to a very small positive number, say $\varepsilon = 0.0001$, every element of $(2, 4)$ is less than $4 + \varepsilon = 4.0001$. Furthermore, $4 - \varepsilon = 3.9999$ is not quite big enough to be the supremum, because 3.99995 is an element of $(2, 4)$ and is greater than $4 - \varepsilon$.

We said earlier that the ordered field $\mathbb{Q}$ is missing some numbers. To be precise, there are subsets of $\mathbb{Q}$ that have upper bounds in $\mathbb{Q}$, but for which there is no least upper bound in $\mathbb{Q}$—the suprema for these sets are missing from $\mathbb{Q}$. As an example, we prove that the subset $B = \{x \in \mathbb{Q} : x^2 < 2\}$ is one such subset. The proof uses a version of the Archimedean Principle (see Theorem 2.4.2) for the real numbers, which states that for every positive real number r there is a natural number K that is so large that $\dfrac{1}{K} < r$.

Example. The set $B = \{x \in \mathbb{Q} : x^2 < 2\}$ is bounded above in the field $\mathbb{Q}$ but has no supremum in $\mathbb{Q}$.

Proof. There are many rational numbers, such as $\dfrac{3}{2}$ and 1.44, that are upper bounds for B, so B is bounded above in $\mathbb{Q}$.

Now suppose there is a rational number s such that $s = \sup (B)$. We will first show that both $s < \sqrt{2}$ and $s > \sqrt{2}$ are false.

If $s < \sqrt{2}$, then $\sqrt{2} - s$ is positive. Choose a natural number K such that $\dfrac{1}{K} < \sqrt{2} - s$. Then $s + \dfrac{1}{K} < \sqrt{2}$ and $s + \dfrac{1}{K}$ is rational ⟨*because both s and $\dfrac{1}{K}$ are rational*⟩. Thus $s + \dfrac{1}{K}$ is an element of B, which contradicts s being an upper bound for B.

If $s > \sqrt{2}$, then $s - \sqrt{2}$ is positive. Choose a natural number M such that $\frac{1}{M} < s - \sqrt{2}$. Then $s - \frac{1}{M} > \sqrt{2}$ and $s - \frac{1}{M}$ is rational. Then for all x in B, $x < \sqrt{2} < s - \frac{1}{M} < s$, so $s - \frac{1}{M}$ is an upper bound for B that is less than s. This contradicts the assumption that s is the least upper bound for B in $\mathbb{Q}$.

Because both $s < \sqrt{2}$ and $s > \sqrt{2}$ are false, we conclude that $s = \sqrt{2}$. But this is impossible, because s is a rational number. Therefore $\sup(B)$ does not exist in the ordered field $\mathbb{Q}$. ∎

Remember that the field $(\mathbb{Q}, +)$ has no gaps—between any two rational numbers there is always another rational number. Nevertheless, this example shows that $\mathbb{Q}$ has "pinholes" (points missing from $\mathbb{Q}$) where the suprema (and infima) of some bounded subsets of $\mathbb{Q}$ "ought to be" but are not. (Exercise 18 shows that $\mathbb{Q}$ has many missing points.) We have identified a property that we can use to distinguish the real number system from $\mathbb{Q}$.

> **DEFINITION** An ordered field F is **complete** iff every nonempty subset of F that has an upper bound in F has a supremum that is an element of F.

The fact that the bounded set $B = \{x \in \mathbb{Q}: x^2 < 2\}$ has no supremum in $\mathbb{Q}$ means that the field $\mathbb{Q}$ is not complete. Note, however, that when B is considered a subset of $\mathbb{R}$, it *does* have a least upper bound in $\mathbb{R}$: $\sup(B) = \sqrt{2}$. In fact, every set of real numbers that is bounded above has a supremum in $\mathbb{R}$.

Theorem 7.1.2   The field $(\mathbb{R}, +, \cdot)$ is a complete ordered field.

We state this fact without proof. A proof requires considerable preliminary study of the nature of the real number system and is beyond the goals of this text. Section 7.5 includes a brief description of how the real numbers may be built up from the rationals to achieve this result.

For everything that has been said about upper bounds and suprema there is a corresponding statement about lower bounds and infima. In particular, the definition of completeness could have been stated in terms of lower bounds and infima. That is, an ordered field F is complete iff every nonempty subset of F that has a lower bound in F has an infimum in F. See Exercise 19.

As promised early in this section, we present here for your reference the formal definitions of "field" and "ordered". First are the algebraic properties.

> **DEFINITION** A **field** $(F, +, \cdot)$ is a set F with two operations $+$ and $\cdot$ such that
>
> **(1)** $+$ is an operation on F such that for all $x, y, z \in F$,
> - **(a)** $(x + y) + z = x + (y + z)$.
> - **(b)** there is an additive identity 0 such that $x + 0 = 0 + x = x$.
> - **(c)** for every $x \in F$, there is an additive inverse $-x \in F$ such that $x + (-x) = (-x) + x = 0$.
> - **(d)** $x + y = y + x$.

(2) $\cdot$ is an operation on F so that for all $x, y, z \in F$,

    (a) $(x \cdot y) \cdot z = x \cdot (y \cdot z)$.

    (b) there is a multiplicative identity 1 such that $x \cdot 1 = 1 \cdot x = x$.

    (c) for every $x \in F - \{0\}$, there exists a multiplicative inverse
$x^{-1} \in F - \{0\}$ such that $x \cdot x^{-1} = x^{-1} \cdot x = 1$.

    (d) $x \cdot y = y \cdot x$.

(3) For all x, y, z in F, $x \cdot (y + z) = x \cdot y + x \cdot z$.

(4) $0 \neq 1$.

If you are familiar with the terminology of Chapter 6, part (1) of the definition says that $(F, +)$ is an abelian group, part (2) says that $(F - \{0\}, \cdot)$ is an abelian group, and part (3) is one of the distributive laws. Each of the rationals, the reals, and the complex numbers with the familiar operations of addition and multiplication is a field. Another result from Chapter 6 is that if p is a prime, then the modular arithmetic structure $(\mathbb{Z}_p, +, \cdot)$ is a field with p elements.

The definition of an ordered field is:

DEFINITION A field $(F, +, \cdot)$ is **ordered** iff there is a relation $<$ on F such that for all $x, y, z \in F$,

(1) $x \not< x$ (irreflexivity).

(2) if $x < y$ and $y < z$, then $x < z$ (transitivity).

(3) either $x < y$, $x = y$, or $y < x$ (trichotomy).

(4) if $x < y$, then $x + z < y + z$.

(5) if $x < y$ and $0 < z$, then $x \cdot z < y \cdot z$.

Taken together, these properties ensure that the field elements are linearly arranged, and that the ordering is compatible with the operations of addition and multiplication.

All the order properties of $\mathbb{R}$ and $\mathbb{Q}$ can be derived from these definitions. For example, we can show that $0 < 1$, $-1 < 0$, and $-y < -x$ whenever $x < y$. We can also prove that if $x < y$ and $z < 0$, then $x \cdot z > y \cdot z$. See Exercise 20.

Not all fields are ordered. The fields $(\mathbb{C}, +, \cdot)$ and $(\mathbb{Z}_p, +, \cdot)$ where p is a prime are not ordered.

Exercises 7.1

1. Find four upper bounds (if any exist) for each of the following sets.

    (a) $\{x \in \mathbb{R}: x^2 < 10\}$

★ (b) $\left\{\dfrac{1}{3^x}: x \in \mathbb{N}\right\}$

    (c) $\left\{x \in \mathbb{R}: x + \dfrac{1}{x} < 5\right\}$

      (d) $\{x \in \mathbb{R}: 7x^2 + 14x + 2 < 23\}$

      (e) $\{x \in \mathbb{R}: x < 0 \text{ and } x - x^2 \le -2\}$

      (f) $\{2^{-x}: x \in \mathbb{R}\}$

      (g) $\{x \in \mathbb{R}: x - 10 < \log x\}$

2. Find a lower bound in $\mathbb{R}$ (if one exists) for each of the sets in Exercise 1.

3. Find the supremum and infimum, if they exist, of each of the following sets.

  ★ **(a)** $\left\{\dfrac{1}{n}: n \in \mathbb{N}\right\}$ **(b)** $\left\{\dfrac{n+1}{n}: n \in \mathbb{N}\right\}$

  ★ **(c)** $\{2^x: x \in \mathbb{Z}\}$ **(d)** $\left\{(-1)^n\left(1 + \dfrac{1}{n}\right): n \in \mathbb{N}\right\}$

  ★ **(e)** $\left\{\dfrac{n}{n+2}: n \in \mathbb{N}\right\}$ **(f)** $\{x \in \mathbb{Q}: x^2 < 10\}$

  ★ **(g)** $[-1, 1] \cup \{5\}$ **(h)** $[-1, 1] - \{0\}$

      (i) $\left\{\dfrac{x}{2^y}: x, y \in \mathbb{N}\right\}$ **(j)** $\{x: |x| > 2\}$

4. Let A and B be subsets of $\mathbb{R}$. Prove that

  ☆ **(a)** if A is bounded above and $B \subseteq A$, then B is bounded above.

      (b) if A is bounded below and $B \subseteq A$, then B is bounded below.

      (c) if A and B are bounded above, then $A \cup B$ is bounded above.

      (d) if A and B are bounded below, then $A \cup B$ is bounded below.

5. Let x be an upper bound for $A \subseteq \mathbb{R}$. Prove that

      (a) if $x < y$, then y is an upper bound for A.

      (b) if $x \in A$, then $x = \sup(A)$.

6. Let $A \subseteq \mathbb{R}$. Prove that

  ☆ **(a)** if A is bounded above, then A^c is not bounded above.

      (b) if A is bounded below, then A^c is not bounded below.

7. Give an example of a set $A \subseteq \mathbb{R}$ for which both A and A^c are unbounded above and below.

8. Let $A \subseteq \mathbb{R}$. Prove that

  ★ **(a)** if $\sup(A)$ exists, then it is unique. That is, if x and y are both least upper bounds for A, then $x = y$.

      (b) if $\inf(A)$ exists, then it is unique.

9. Let $A \subseteq B \subseteq \mathbb{R}$. Prove that

      (a) if $\sup(A)$ and $\sup(B)$ both exist, then $\sup(A) \le \sup(B)$.

      (b) if $\inf(A)$ and $\inf(B)$ both exist, then $\inf(A) \ge \inf(B)$.

10. Formulate and prove a characterization of greatest lower bounds similar to that in Theorem 7.1.1 for least upper bounds.

11. If possible, give an example of

      (a) a set $A \subseteq \mathbb{R}$ such that $\sup(A) = 4$ and $4 \notin A$.

      (b) a set $A \subseteq \mathbb{Q}$ such that $\sup(A) = 4$ and $4 \notin A$.

(c) a set $A \subseteq \mathbb{N}$ such that sup $(A) = 4$ and $4 \notin A$.

(d) a set $A \subseteq \mathbb{N}$ such that sup $(A) > 4$ and $4 \notin A$.

12. Give an example of a set of rational numbers that has a rational lower bound but no rational greatest lower bound.

13. Let $A \subseteq \mathbb{R}$. Prove that

★ **(a)** if sup (A) exists, then sup $(A) = \inf \{u: u$ is an upper bound of $A\}$.

(b) if inf (A) exists, then inf $(A) = \sup \{l: l$ is a lower bound of $A\}$.

14. Let A and B be subsets of $\mathbb{R}$.

★ **(a)** Prove that if sup (A) and sup (B) exist, then sup $(A \cup B)$ exists and sup $(A \cup B) = \max \{\sup (A), \sup (B)\}$.

(b) State and prove a similar result for inf $(A \cup B)$.

15. **(a)** Give an example of sets A and B of real numbers such that $A \cap B \neq \emptyset$, sup $(A \cap B) < \sup (A)$, and sup $(A \cap B) < \sup (B)$.

(b) For sets A and B such that $A \cap B \neq \emptyset$, state and prove a relationship between sup (A), sup (B), and sup $(A \cap B)$.

16. **(a)** Give an example of sets A and B of real numbers such that $A \cap B \neq \emptyset$, inf $(A \cap B) > \inf (A)$, and inf $(A \cap B) > \inf (B)$.

(b) For sets A and B such that $A \cap B \neq \emptyset$, state and prove a relationship between inf (A), inf (B), and inf $(A \cap B)$.

17. Use the completeness property of $\mathbb{R}$ to prove the Archimedean Principle for the real numbers: For every positive real number r there is an integer K such that $\frac{1}{K} < r$. *Hint:* Suppose the assertion is false for some real number r. Verify that the set $W = \{nr: n \in \mathbb{N}\}$ is nonempty and bounded above by 1. Let t be the supremum of W. Observe that $t - r < t$, so $t - r$ is not an upper bound for W. Then, by Theorem 7.1.1, there is a natural number m such that $t - r < mr$. It follows that $t < (m + 1)r$, contradicting the fact that t is an upper bound for W.

18. This exercise shows that every irrational number is "missing" from $\mathbb{Q}$. Let x be an irrational number. Find a subset A of $\mathbb{Q}$ such that A is bounded above in $\mathbb{Q}$ and sup (A) does not exist in $\mathbb{Q}$, but when A is considered a subset of $\mathbb{R}$, sup $(A) = x$.

☆ **19.** Prove that an ordered field F is complete iff every nonempty subset of F that has a lower bound in F has an infimum in F.

20. Let F be an ordered field and $x, y, z \in F$. Prove that

(a) exactly one of $x < y$, $x = y$, or $y < x$ is true.

(b) if $x < 0$, then $-x > 0$.

(c) $0 < 1$.

(d) $-1 < 0$.

(e) if $x < y$, then $-y < -x$.

(f) if $x < y$ and $z < 0$, then $x \cdot z > y \cdot z$.

(g) $(-1) \cdot (-1) = 1$.

(h) $0 \cdot x = 0$.

(i) $(-1) \cdot x = -x$.

Proofs to Grade   **21.** Assign a grade of A (correct), C (partially correct), or F (failure) to each. Justify assignments of grades other than A.

★   **(a)**   **Claim.**   Let $A \subseteq \mathbb{R}$. If $i = \inf(A)$ and $\varepsilon > 0$, then there is $y \in A$ such that $y < i + \varepsilon$.

"*Proof.*"   Let $y = i + \frac{\varepsilon}{2}$. Then $i < y$ so $y \in A$. By construction of y, $y < i + \varepsilon$.   ∎

(b)   **Claim.**   Let $A \subseteq \mathbb{R}$. If A is bounded above, then A^c is bounded below.

"*Proof.*"   If A is bounded above, then $\sup(A)$ exists (because $\mathbb{R}$ is complete). Since $\sup(A) = \inf(A^c)$ (see the figure), $\inf(A^c)$ exists. Thus A^c is bounded below.   ∎

$$\sup(A) = \inf(A^c)$$

(c)   **Claim.**   If $A \subseteq B \subseteq \mathbb{R}$, and $\sup(A)$ and $\sup(B)$ both exist, then $\sup(A) \leq \sup(B)$.

"*Proof.*"   Assume $A \subseteq B$ and $\sup(A) > \sup(B)$. We choose $\varepsilon = \frac{1}{2}(\sup(A) - \sup(B))$. Then $\varepsilon > 0$ and $\sup(B) < \sup(A) - \varepsilon < \sup(A)$. By part (ii) of Theorem 7.1.1, there is $y \in A$ such that $y > \sup(A) - \varepsilon$. Then $y \in B$ and $y > \sup(B)$. This is impossible. Therefore, $\sup(A) \leq \sup(B)$.   ∎

(d)   **Claim.**   If $f: \mathbb{R} \to \mathbb{R}$ and A is a bounded subset of $\mathbb{R}$, then $\text{Rng}(A)$ is bounded.

"*Proof.*"   Let m be an upper bound for A. Then $a \leq m$ for all $a \in A$. Therefore, $f(a) \leq f(m)$ for all $a \in A$. Thus $f(m)$ is an upper bound for $\text{Rng}(A)$.   ∎

(e)   **Claim.**   Let $(F, +, \cdot)$ be an ordered field and $x \in F$. If $-x < 0$, then $x > 0$.

"*Proof.*"   Suppose $-x < 0$. Then by property (4) of ordered fields $0 = -x + x < 0 + x = x$, so $0 < x$. Thus $x > 0$.   ∎

7.2   The Heine–Borel Theorem

In this section we begin by introducing some concepts used to describe sets of real numbers and then use the completeness of $\mathbb{R}$ to establish our first major result, the Heine–Borel Theorem. We state these results in terms of the real numbers, but the results apply more generally to any complete ordered field.

> **DEFINITION**   Let a and δ be real numbers with $\delta > 0$. The **δ-neighborhood of a** is the set
>
> $$\mathcal{N}(a, \delta) = \{x \in \mathbb{R}: |x - a| < \delta\}.$$

The δ-neighborhood of a consists of all points x whose distance from a is less than δ. Since $|x - a| < \delta$ is equivalent to $a - \delta < x < a + \delta$, $\mathcal{N}(a, \delta)$ is the open interval $(a - \delta, a + \delta)$. For example, $\mathcal{N}(3, 0.4) = (2.6, 3.4)$ and $\mathcal{N}(1, 0.01) = (0.99, 1.01)$. Both the 1-neighborhood of 0.7 and the 0.2 neighborhood of 0.7 are shown in Figure 7.2.1. Your intuition is best served by thinking of δ as a small positive number and $\mathcal{N}(a, \delta)$ as a small open interval of radius δ centered about a.

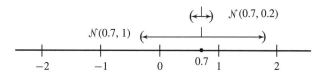

$\mathcal{N}(0.7, 1) = (-0.3, 1.7)$ and $\mathcal{N}(0.7, 0.2) = (0.5, 0.9)$.

Figure 7.2.1

Many concepts in mathematics may be expressed using neighborhood terminology. Recall that a function $f : \mathbb{R} \to \mathbb{R}$ is *continuous* at a point a in its domain iff

for all $\varepsilon > 0$, there exists $\delta > 0$ such that if $|x - a| < \delta$, then $|f(x) - f(a)| < \varepsilon$.

This means that whenever x is close to a, then $f(x)$ must be close to $f(a)$. In terms of neighborhoods, the definition is:

for all $\varepsilon > 0$, there exists $\delta > 0$ such that if $x \in \mathcal{N}(a, \delta)$, then $f(x) \in \mathcal{N}(f(a), \varepsilon)$.

This version of the definition has the advantage of specifying "closeness" in terms of sets (neighborhoods) instead of distances. Because there are systems other than the real numbers for which the concept of neighborhood may be introduced, the neighborhood version of the definition is a way to define continuity for those systems.

DEFINITION   For a set $A \subseteq \mathbb{R}$, a point x is an **interior point of** A iff there exists $\delta > 0$ such that $\mathcal{N}(x, \delta) \subseteq A$.

If x is an interior point of A, then not only is x contained in A, but all elements of some neighborhood around x are also contained in A.

For the interval $[2, 5)$, 3 is an interior point since $\mathcal{N}(3, 0.5) \subseteq [2, 5)$. Also, 4.9998 is an interior point because $\mathcal{N}(4.9998, 0.0001) \subseteq [2, 5)$. See Figure 7.2.2. In fact, every point in $(2, 5)$ is an interior point of $[2, 5)$. The point 2 is not interior to $[2, 5)$, since every δ-neighborhood of 2 contains points that are less than 2 and hence not in $[2, 5)$.

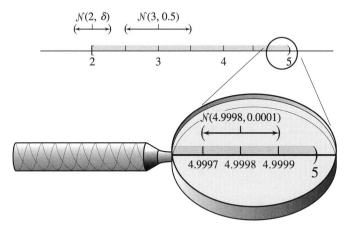

Figure 7.2.2

DEFINITIONS The set $A \subseteq \mathbb{R}$ is **open** iff every point of A is an interior point of A. The set A is **closed** iff its complement A^c is open.

The interval $(2, 5)$ is open because every point in $(2, 5)$ is an interior point of $(2, 5)$. Thus its complement, $(-\infty, 2] \cup [5, \infty)$, is a closed set. On the other hand, 2 is not an interior point of $[2, 5)$. Therefore, $[2, 5)$ is not open.

The interval $[2, 5)$ is not closed either, since its complement $[2, 5)^c = (-\infty, 2) \cup [5, \infty)$ contains 5, but 5 is not an interior point of $(-\infty, 2) \cup [5, \infty)$. In ordinary conversation, with references to objects like doors and eyes, the concepts of open and closed are opposites, but the interval $[2, 5)$ is an example of a subset of $\mathbb{R}$ that is neither open nor closed.

Examples. The set $\mathbb{R}$ is open since for every $x \in \mathbb{R}$, $\mathcal{N}(x, 1) \subseteq \mathbb{R}$. The empty set $\varnothing$ is also open since the statement "for all x, if $x \in \varnothing$, x is an interior point of $\varnothing$" is true because the antecedent is false. Since $\mathbb{R}$ and $\varnothing$ are complements, they are also closed sets. It can be shown that there are no other subsets of the reals that are both open and closed.

A set is open if about each element in the set there is a δ-neighborhood that lies entirely within the set. This means no point of the set can be on the "boundary" or outer edges of the set (see Exercise 11). The next two theorems will help you recognize open sets.

Theorem 7.2.1   Every open interval of real numbers is an open set.

Proof. Let (a, b) be an open interval and let $x \in (a, b)$. ⟨*To show (a, b) is open, we show that x is an interior point of (a, b). That is, we show $\mathcal{N}(x, \delta) \subseteq (a, b)$ for some $\delta > 0$.*⟩ We choose $\delta = \min \{x - a, b - x\}$. ⟨*This minimum is the largest possible δ we can use. See Figure 7.2.3.*⟩ Then $\delta > 0$. To show that $\mathcal{N}(x, \delta) \subseteq (a, b)$, let $y \in \mathcal{N}(x, \delta)$. Then $a = x - (x - a) \leq x - \delta < y < x + \delta \leq x + (b - x) = b$; and so $y \in (a, b)$. ∎

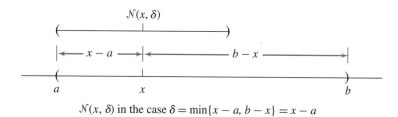

$\mathcal{N}(x, \delta)$ in the case $\delta = \min\{x - a, b - x\} = x - a$

Figure 7.2.3

Theorem 7.2.2   Let $\mathcal{A}$ be a nonempty collection of open subsets of $\mathbb{R}$. Then

(a) $\displaystyle\bigcup_{A \in \mathcal{A}} A$ is an open set.

(b)   If $\mathcal{A}$ is finite, then $\displaystyle\bigcap_{A \in \mathcal{A}} A$ is an open set.

Proof. Let $\mathcal{A}$ be a nonempty collection of open sets.

(a)   Suppose $x \in \displaystyle\bigcup_{A \in \mathcal{A}} A$. ⟨*We must show that x is an interior point of $\bigcup_{A \in \mathcal{A}} A$.*⟩ Since x is in the union over the collection, there exists $B \in \mathcal{A}$ such that $x \in B$. Since B is in the collection $\mathcal{A}$, B is open. Thus x is an interior point of B. Therefore there exists $\delta > 0$ such that $\mathcal{N}(x, \delta) \subseteq B$. Since $B \subseteq \displaystyle\bigcup_{A \in \mathcal{A}} A$, $\mathcal{N}(x, \delta) \subseteq \displaystyle\bigcup_{A \in \mathcal{A}} A$. Therefore x is an interior point of $\displaystyle\bigcup_{A \in \mathcal{A}} A$.

(b)   Suppose $\mathcal{A}$ is finite and $x \in \displaystyle\bigcap_{A \in \mathcal{A}} A$. Then $x \in A$ for all $A \in \mathcal{A}$, and so for each open set $A \in \mathcal{A}$ there corresponds $\delta_A > 0$ such that $\mathcal{N}(x, \delta_A) \subseteq A$. Let $\delta = \min \{\delta_A : A \in \mathcal{A}\}$. ⟨*Note that the minimum of a finite set of positive numbers must be positive.*⟩ Then $\delta > 0$ and $\mathcal{N}(x, \delta) \subseteq \mathcal{N}(x, \delta_A) \subseteq A$ for all $A \in \mathcal{A}$. Thus $\mathcal{N}(x, \delta) \subseteq \displaystyle\bigcap_{A \in \mathcal{A}} A$. Therefore x is an interior point of $\displaystyle\bigcap_{A \in \mathcal{A}} A$. ∎

For the proof of part (b) we chose $\delta = \min \{\delta_A : A \in \mathcal{A}\}$ and relied on the fact that it is always possible to find the minimum of a finite set of real numbers. However, if the set $\{\delta_A : A \in \mathcal{A}\}$ is infinite, we can not be sure of finding a minimum. The set $\left\{\frac{1}{n} : n \in \mathbb{N}\right\}$, for example, does not have a minimum element. The statement of Theorem 7.2.2 (b) is false if we omit the word "finite."

Example.   The family $\mathscr{A} = \left\{\left(2-\dfrac{1}{n},5\right): n \in \mathbb{N}\right\}$ is an infinite collection of open intervals. The intersection is

$$\bigcap_{n\in\mathbb{N}}\left(2-\dfrac{1}{n},5\right) = [2,5),$$

which we have seen is not an open set.

Theorems 7.2.1 and 7.2.2 can be used to produce many examples of open subsets of $\mathbb{R}$. For example, the following are open sets:

$$(5, 7) \cup (-3, 4) \cup (10, 20)$$

$$(2, \infty) = \bigcup_{A\in\mathscr{A}} A, \text{ where } \mathscr{A} = \{(2, x): x > 2\}$$

$$(-\infty, 2) = \bigcup_{A\in\mathscr{A}} A, \text{ where } \mathscr{A} = \{(x, 2): x < 2\}$$

$$(-5, 0) \cup (2, \infty)$$

$$\mathbb{R} - \{2\} = (-\infty, 2) \cup (2, \infty).$$

Corresponding to Theorem 7.2.1, it can be shown that every closed interval of real numbers is a closed set. See Exercise 7(b). The analog of Theorem 7.2.2 for closed sets is Exercise 8.

Every finite subset of $\mathbb{R}$ is closed because the complement of a finite set is the union of two open rays and a number of open intervals. Finite sets are good examples of sets that are both closed and bounded. We shall see that infinite closed and bounded sets have other properties in common with finite sets. For example, if A is a finite set, then the infimum and supremum of A exist and are elements of A. The next theorem shows this is true for all closed and bounded sets.

Theorem 7.2.3   If A is a nonempty closed and bounded subset of $\mathbb{R}$, then $\sup(A) \in A$ and $\inf(A) \in A$.

Proof.   Suppose A is a nonempty closed and bounded set. Let $s = \sup(A)$ and suppose $s \notin A$. Then $s \in A^c$, which is open since A is closed. Thus $\mathscr{N}(s, \delta) \subseteq A^c$ for some positive δ. This implies $s - \delta$ is an upper bound for A, since the interval $(s - \delta, s + \delta)$ is a subset of A^c. ⟨*No element of A is greater than s, or equal to s, or between $s - \delta$ and s.*⟩ This contradicts Theorem 7.1.1. Therefore, $s \in A$. The proof that $\inf(A) \in A$ is similar.   ∎

Examples.   Let $A = [2, 5]$,
$$B = [-2, 2] \cup [4, 10] \cup \{12\} \text{ and}$$
$$C = \{2^{-n}: n \in \mathbb{N}\} \cup \{0\}.$$

Each of these sets is closed and bounded, so by Theorem 7.2.3 each contains its supremum and infimum: $\inf(A) = 2$ and $\sup(A) = 5$ are elements of A; $\inf(B) = -2$ and $\sup(B) = 12$ are elements of B; and $\inf(C) = 0$ and $\sup(C) = \dfrac{1}{2}$ are elements of C.

The bounded set $[0, 1)$ does not contain its supremum; it does not satisfy the conditions of Theorem 7.2.3 because it is not closed. The closed set $[2, \infty)$ does not have a supremum; it does not satisfy the conditions of Theorem 7.2.3 because it is not bounded.

To understand how the completeness property of $\mathbb{R}$ is related to properties of closed and bounded sets, we need the concept of a cover for a set. A cover for a set A is a collection of open sets whose union includes A.

DEFINITIONS   Let A be a set of real numbers. A collection $\mathscr{C}$ of open subsets of $\mathbb{R}$ is a **cover** for A iff $A \subseteq \bigcup_{C \in \mathscr{C}} C$.

If $\mathscr{B} \subseteq \mathscr{C}$ and $\mathscr{B}$ is also a cover for A, we say $\mathscr{B}$ is a **subcover** of $\mathscr{C}$.

Example.   For the set $A = \{2, 4, 5, 6, 8, 9\}$, the collection of open intervals

$$\mathscr{C} = \{(-1, 2), (1, 4), (2, 5), (3, 6), (4, 7), (5, 9), (6, 10), (7, 11)\}$$

is a cover for A because $\mathscr{C}$ consists of open sets and $A \subseteq \bigcup_{C \in \mathscr{C}} C = (-1, 11)$. The collection

$$\mathscr{B} = \{(1, 4), (3, 6), (4, 7), (6, 10)\}$$

is a subset of $\mathscr{C}$ and is also cover of A since $\bigcup_{C \in \mathscr{B}} C = (1, 10)$. Thus $\mathscr{B}$ is a subcover of $\mathscr{C}$ for A.

Example.   Let $A_n = \left(n - \dfrac{1}{n}, n + \dfrac{1}{n} \right)$ for each $n \in \mathbb{N}$. The collection $\mathscr{A} = \{A_n \colon n \in \mathbb{N}\}$ is a cover for $\mathbb{N}$ that has no subcover other than itself (Figure 7.2.4).

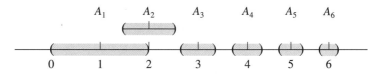

Figure 7.2.4

Example.   Since $\bigcup\limits_{n=1}^{\infty}(-\infty, n) = \mathbb{R}$, the collection $\mathscr{H} = \{(-\infty, n) \colon n \in \mathbb{N}\}$ is a cover for $\mathbb{R}$. The collection $\mathscr{K} = \{(-\infty, n) \colon n \in 2\mathbb{N}\}$, where $2\mathbb{N}$ is the set of even natural numbers, is a subcover of $\mathscr{H}$ for $\mathbb{R}$. We note that there are many subcovers of $\mathscr{H}$ for $\mathbb{R}$, but there is no finite subset of $\mathscr{H}$ that is a cover for $\mathbb{R}$.

A cover $\mathscr{C}$ of a set A may be imagined by thinking of the covering sets of $\mathscr{C}$ as providing shade for the set A. In Figure 7.2.5 we have the sun directly over the set A. The covering sets $C_\alpha, C_\beta, C_\delta, \ldots$ are drawn slightly above A. For this analogy, think of the sun's rays as parallel beams of light. As the sun shines straight down, $\mathscr{C}$ is a cover for A iff A is within the region shaded by the sets in $\mathscr{C}$—that is, iff $A \subseteq \bigcup\limits_{C \in \mathscr{C}} C$.

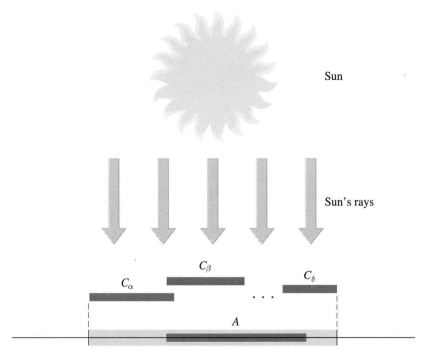

Figure 7.2.5

Let $A = \{x_1, x_2, x_3, \ldots x_n\}$ be a nonempty finite set of real numbers and let $\mathscr{C} = \{O_\alpha : \alpha \in \Delta\}$ be a cover for A. We may not need all the sets in $\mathscr{C}$ to make a cover for A, so we look for a subcover. For each $i = 1, 2, 3, \ldots, n$ there is α_i in Δ such that $x_i \in O_{\alpha_i}$. Then the collection $\mathscr{B} = \{O_{\alpha_i} : i = 1, 2, 3, \ldots n\}$ is a finite subset of $\mathscr{C}$ whose union includes A. In other words, given any cover for the finite set A, we can always construct a finite subcover for A. Sets of real numbers that have this property are called compact sets.

DEFINITION   A subset A of $\mathbb{R}$ is **compact** iff for every cover $\mathscr{C}$ for A, there is a finite subcover of $\mathscr{C}$ for A.

By the discussion above, every nonempty finite set is compact. Next is an infinite set that is compact.

Example. The set $A = \left\{ \dfrac{n+1}{n} : n \in \mathbb{N} \right\} \cup \{1\}$ is compact.

Proof. Let $\{O_\alpha : \alpha \in \Delta\}$ be any cover for A. One of the covering sets, call it O_{α^*}, contains the element 1. Since O_{α^*} is open, there is a δ-neighborhood $\mathcal{N}(1, \delta) \subseteq O_{\alpha^*}$. See Figure 7.2.6. ⟨*We will show that all but a finite number of elements of A are in $\mathcal{N}(1, \delta)$ and therefore in O_{α^*}.*⟩ Choose N such that $N > \frac{1}{\delta}$. If $n > N$, then $\frac{1}{\delta} < n$ and so $1 < n\delta$. This implies $n + 1 < n + n\delta$, which means $\dfrac{n+1}{n} < 1 + \delta$. Therefore, if $n > N$, then $\dfrac{n+1}{n} \in \mathcal{N}(1, \delta)$ and $\dfrac{n+1}{n} \in O_{\alpha^*}$.

Now choose $O_{\alpha_1}, O_{\alpha_2}, O_{\alpha_3}, \ldots, O_{\alpha_N}$ such that $2 \in O_{\alpha_1}, \frac{3}{2} \in O_{\alpha_2}, \frac{4}{3} \in O_{\alpha_3}, \ldots,$ and $\dfrac{N+1}{N} \in O_{\alpha_N}$. Then $A \subseteq O_{\alpha^*} \cup O_{\alpha_1} \cup O_{\alpha_2} \cup O_{\alpha_3} \cup \ldots \cup O_{\alpha_N}$. ⟨*The first N elements of A are in $O_{\alpha_1}, \ldots, O_{\alpha_N}$, and everything else is in O_{α^*}.*⟩ We have succeeded in finding a finite subcover of $N + 1$ sets for the cover $\{O_\alpha : \alpha \in \Delta\}$. Therefore, A is compact. ∎

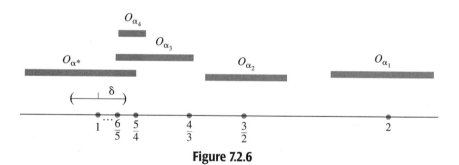

Figure 7.2.6

We have seen that $\mathcal{H} = \{(-\infty, n) : n \in \mathbb{N}\}$ is a cover for $\mathbb{R}$ with no finite subcover. Thus $\mathbb{R}$ is *not* compact. Neither is the open interval $(0, 1)$ compact because the collection $\mathcal{C} = \left\{ \left(\frac{1}{x}, 1 \right) : x \in (1, \infty) \right\}$ is a cover for $(0, 1)$ that has no finite subcover.

You may have noticed that all of our examples of compact sets have been closed and bounded sets and the two non-compact examples are either not closed or not bounded. This is no coincidence. The next theorem is a beautiful characterization of compact sets based on the work of Edward Heine* and Emile

---

* Edward Heine (1821–1881) was a German mathematician at the University of Halle (the same university where Georg Cantor spent his entire career) who made several contributions to analysis, especially to the descriptions and solutions of equations involving infinite series. As a senior professor, Heine gave the young Cantor a problem in analysis whose solution and generalization demonstrated the need to define the term "set" precisely.

Borel.* The proof uses the following lemma. Watch for the places where the proof of the Heine–Borel Theorem depends on the completeness of $\mathbb{R}$.

Lemma 7.2.4   Let A be a closed set and $x \in \mathbb{R}$. If $A \cap \mathcal{N}(x, \delta) \neq \varnothing$ for all $\delta > 0$, then $x \in A$.

Proof.   Exercise 12.   ∎

Theorem 7.2.5   ### The Heine–Borel Theorem
A subset A of $\mathbb{R}$ is compact iff A is closed and bounded.

Proof.

(i)   Suppose A is compact. We first show that A is bounded. We note that $A \subseteq \mathbb{R} = \bigcup_{n \in \mathbb{N}} (-n, n)$. Therefore, $\mathscr{H} = \{(-n, n) : n \in \mathbb{N}\}$ is a cover for A. By compactness, $\mathscr{H}$ has a finite subcover $\{(-n, n) : n \in \{n_1, n_2, \dots, n_k\}\}$. If we choose $N = \max\{n_1, n_2, \dots, n_k\}$, then $A \subseteq \bigcup_{i=1}^{k} (-n_i, n_i) = (-N, N)$. Therefore, A is bounded above by N and below by $-N$.

We next show that A is closed by proving A^c is open. Suppose $y \in A^c$. ⟨*We must show y is an interior point of A^c.*⟩ For each $x \in A$, $x \neq y$ and thus $\delta_x = \frac{1}{2}|x - y|$ is a positive number. The collection $\{\mathcal{N}(x, \delta_x) : x \in A\}$ is a family of open sets that covers A. Hence by the compactness of A,

$$A \subseteq \mathcal{N}(x_1, \delta_{x_1}) \cup \mathcal{N}(x_2, \delta_{x_2}) \cup \ldots \cup \mathcal{N}(x_k, \delta_{x_k})$$

for some $x_1, x_2, \dots, x_k \in A$. By choosing $\delta = \min\{\delta_{x_1}, \delta_{x_2}, \dots, \delta_{x_k}\}$, we have $\delta > 0$ and $\mathcal{N}(y, \delta) \subseteq A^c$. See Figure 7.2.7. ⟨*If $z \in A$, then $|z - x_i| < \delta_{x_i}$ for some i. Thus if $z \in \mathcal{N}(y, \delta)$, then $|z - y| < \delta \leq \delta_{x_i}$ and $|x_i - y| \leq |x_i - z| + |z - y| < 2\delta_{x_i} = |x_i - y|$.*⟩ Thus A^c is open. Hence A is closed.

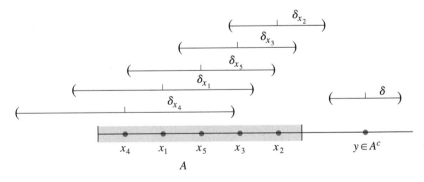

Figure 7.2.7

---

* Emile Borel (1871–1956) was a French mathematician and politician who contributed substantially to probability and game theory and the creation of the branch of mathematics called measure theory. He stated and proved the Heine–Borel Theorem for countable sets. He served many years in French politics and was a member of the French resistance in World War II.

(ii)   Conversely, suppose A is a closed and bounded set and $\mathscr{C}$ is a cover for A. For each $x \in \mathbb{R}$, let $A_x = \{a \in A: a \le x\}$. Also, let $D = \{x \in \mathbb{R}: A_x$ is included in a union of finitely many sets from $\mathscr{C}\}$.

Since A is bounded, $\inf(A)$ exists $\langle by\ the\ completeness\ of\ \mathbb{R}\rangle$. Thus if $x < \inf(A)$, $A_x = \varnothing$ and it follows that $x \in D$. Therefore $(-\infty, \inf(A)) \subseteq D$ and so D is nonempty.

We claim D has no upper bound. $\langle The\ following\ proof\ of\ this\ fact\ involves$ $showing\ that\ if\ D\ is\ bounded\ above,\ then\ \sup(D)\ is\ in\ A,\ and\ then\ using\ the$ $nature\ of\ D\ to\ build\ a\ contradiction.\rangle$ Suppose D is bounded above. Then $x_0 = \sup(D)$ exists $\langle by\ the\ completeness\ of\ \mathbb{R}\rangle$. Let $\delta > 0$ and choose $t \in D$ such that $x_0 - \delta < t \le x_0$ $\langle applying\ Theorem\ 7.1.1\rangle$. If $A \cap \mathscr{N}(x_0, \delta) = \varnothing$, then $A_t = \{a \in A: a \le t\} = \left\{a \in A: a \le x_0 + \dfrac{\delta}{2}\right\} = A_{x_0 + (\delta/2)}$. But t is in D, so $x_0 + \dfrac{\delta}{2}$ is in D. This is a contradiction to $x_0 = \sup(D)$. Therefore, for all $\delta > 0$, we have $A \cap \mathscr{N}(x_0, \delta) \ne \varnothing$. By Lemma 7.2.4, x_0 is in the closed set A.

Let C^* be an element of $\mathscr{C}$ such that $x_0 \in C^*$. Since C^* is open, there exists $\varepsilon > 0$ such that $\mathscr{N}(x_0, \varepsilon) \subseteq C^*$. Choose $x_1 \in D$ such that $x_0 - \varepsilon < x_1 \le x_0$. Since $x_1 \in D$, there are open sets $C_1, C_2, \ldots, C_n$ in $\mathscr{C}$ such that $A_{x_1} \subseteq C_1 \cup C_2 \cup \cdots \cup C_n$. Now let $x_2 = x_0 + \dfrac{\varepsilon}{2}$. Then $x_2 \in C^*$ and $A_{x_2} \subseteq C_1 \cup C_2 \cup \cdots \cup C_n \cup C^*$. Thus $x_2 \in D$, a contradiction, since $x_2 > x_0$ and $x_0 = \sup(D)$. We conclude that D has no upper bound.

Finally, since D has no upper bound, choose $x \in D$ such that $x > \sup(A)$ $\langle \sup(A)\ exists\ because\ \mathbb{R}\ is\ complete\rangle$. Thus $A_x = A$ and since $x \in D$, A is included in a union of finitely many sets from $\mathscr{C}$. Therefore, A is compact. ∎

Exercises 7.2

1. Find x and $\delta \in \mathbb{R}$ such that
    (a) $\mathscr{N}(x, \delta) = (7, 12)$.   ★ **(b)** $\mathscr{N}(x, \delta) = (3.8, 3.85)$.
    (c) $\mathscr{N}(x, \delta) = (6.023, 6.024)$.

2. For $x_1, x_2 \in \mathbb{R}$, $\delta_1 > 0$ and $\delta_2 > 0$, describe
    (a) $\mathscr{N}(x_1, \delta_1) \cap \mathscr{N}(x_1, \delta_2)$.   ☆ **(b)** $\mathscr{N}(x_1, \delta_1) \cap \mathscr{N}(x_2, \delta_1)$.
    (c) $\mathscr{N}(x_1, \delta_1) \cap \mathscr{N}(x_2, \delta_2)$.

★ 3. Write the definition of $\lim\limits_{x \to a} f(x) = L$ in terms of neighborhoods.

4. Find the set of interior points for each of these subsets of $\mathbb{R}$.
    (a) $(-1, 1)$ **(b)** $(-1, 1]$
    (c) $\mathbb{Q}$ **(d)** $\mathbb{R} - \mathbb{Q}$
    ★ **(e)** $\left\{\dfrac{1}{3k}: k \in \mathbb{N}\right\}$ **(f)** $\left\{\dfrac{1}{3k}: k \in \mathbb{N}\right\} \cup \{0\}$
    (g) $\mathbb{R} - \mathbb{N}$ **(h)** $\mathbb{R} - \left\{\dfrac{1}{3k}: k \in \mathbb{N}\right\}$
    ★ **(i)** $\bigcup\limits_{n \in \mathbb{N}} (n + 0.1, n + 0.2)$

5. Classify each of the following subsets of $\mathbb{R}$ as open, closed, or neither open nor closed.

   (a) $(-\infty, -3)$ ⋆ **(b)** $\mathcal{N}(a, \delta) - \{a\}$ for $a \in \mathbb{R}$ and $\delta > 0$

   (c) $(5, 8) \cup \{9\}$ **(d)** $\mathbb{Q}$

   ⋆ **(e)** $\mathbb{R} - \mathbb{N}$ **(f)** $\{x \colon |x - 5| = 7\}$

   (g) $\{x \colon |x - 5| > 7\}$ **(h)** $\{x \colon |x - 5| \neq 7\}$

   ⋆ **(i)** $\{x \colon |x - 5| \leq 7\}$ **(j)** $\{x \colon 0 < |x - 5| \leq 7\}$

6. Let $a \in \mathbb{R}$. Prove that every open ray, either (a, ∞) or $(-\infty, a)$, is an open set.

7. Let $a, b \in \mathbb{R}$. Prove that

   (a) every closed ray, either $[a, \infty)$ or $(-\infty, a]$, is a closed set.

   (b) every closed interval $[a, b]$ is a closed set.

8. Let $\mathcal{A}$ be a nonempty collection of closed subsets of $\mathbb{R}$.

   (a) Prove that $\bigcap_{A \in \mathcal{A}} A$ is a closed set.

   (b) If $\mathcal{A}$ is a finite collection, prove that $\bigcup_{A \in \mathcal{A}} A$ is a closed set.

   (c) Show by example that part (b) is false if we do not assume that $\mathcal{A}$ is finite.

9. Let A and B be subsets of $\mathbb{R}$ and $x \in \mathbb{R}$. Prove that

   (a) if A is open, then $A - \{x\}$ is open.

   (b) if A is open and B is closed, then $A - B$ is open.

   (c) if A is open and B is closed, then $B - A$ is closed.

10. Let A be a subset of $\mathbb{R}$. Prove that the set of all interior points of A is an open set.

11. A point x is a **boundary point** of the set A iff for all $\delta > 0$, $\mathcal{N}(x, \delta) \cap A \neq \varnothing$ and $\mathcal{N}(x, \delta) \cap A^c \neq \varnothing$.

    (a) Find all boundary points of $(2, 5]$, $(0, 1)$, $[3, 5] \cup \{6\}$, and $\mathbb{Q}$.

    (b) Prove that x is a boundary point of A iff x is not an interior point of A and not an interior point of A^c.

    (c) Prove that A is open iff A contains none of its boundary points.

    (d) Prove that A is closed iff A contains all of its boundary points.

12. Prove Lemma 7.2.4.

13. Which of the following subsets of $\mathbb{R}$ are compact?

    ⋆ **(a)** $\mathbb{Z}$ **(b)** $[0, 10] \cup [20, 30]$

    (c) $[\pi, \sqrt{10}]$ ⋆ **(d)** $\mathbb{R} - A$, where A is finite set

    (e) $\{1, 2, 3, 4, 9, 12, 18\}$ **(f)** $\{0\} \cup \left\{ \dfrac{1}{n} \colon n \in \mathbb{N} \right\}$

    ⋆ **(g)** $(-3, 5]$ **(h)** $[0, 1] \cap \mathbb{Q}$

14. Give an example of
    (a) a bounded subset of $\mathbb{R}$ and a cover of that set that has no finite sub-cover.
    (b) a closed subset of $\mathbb{R}$ and a cover of that set that has no finite sub-cover.
    (c) Sets A, B, C, and D of real numbers such that $A \subseteq B \subseteq C \subseteq D$, A is open, B is closed, C is neither open nor closed, and D is compact.

15. Let A and B be compact subsets of $\mathbb{R}$.
    ☆ (a) Use the definition of compact to prove that $A \cup B$ is compact.
    (b) Apply the Heine–Borel Theorem to prove that $A \cap B$ is compact.
    (c) Apply the Heine–Borel Theorem to prove that $A \cup B$ is compact.

16. Let $S = (0, 1]$ and let $\mathscr{C} = \left\{ \left(\dfrac{n+2}{2^n}, 2^{(1/n)} \right) : n \in \mathbb{N} \right\}$.
    (a) Prove that $\mathscr{C}$ is a cover for S.
    (b) Is there a finite subcover of $\mathscr{C}$ for S?
    (c) What does the Heine–Borel Theorem say about S?

17. Use the Heine–Borel Theorem to prove that if $\{A_\alpha : \alpha \in \Delta\}$ is a collection of compact sets, then $\bigcap_{\alpha \in \Delta} A_\alpha$ is compact.

18. Give an example of a collection $\{A_\alpha : \alpha \in \Delta\}$ of compact sets such that $\bigcup_{\alpha \in \Delta} A_\alpha$ is not compact.

Proofs to Grade

19. Assign a grade of A (correct), C (partially correct), or F (failure) to each. Justify assignments of grades other than A.
    (a) **Claim.** Let $a \in \mathbb{R}$. The open ray (a, ∞) is an open set.
    "Proof." Let $x \in (a, \infty)$. Let $\delta = a - x$. If $y \in \mathcal{N}(x, \delta)$, then $y > x - \delta$. Therefore, $y > a$ and so $y \in (a, \infty)$. Thus $\mathcal{N}(x, \delta) \subseteq (a, \infty)$. This proves x is an interior point of (a, ∞). Since every point of (a, ∞) is an interior point, (a, ∞) is open. ∎
    (b) **Claim.** If A and B are compact, then $A \cup B$ is compact.
    "Proof." If A and B are compact, then for any cover $\{O_\alpha : \alpha \in \Delta\}$ for A, there exists a finite subcover $O_{\alpha_1}, O_{\alpha_2}, \ldots, O_{\alpha_n}$, and for any open cover $\{U_\beta : \beta \in \Gamma\}$ for B, there exists a finite subcover $U_{\beta_1}, U_{\beta_2}, \ldots, U_{\beta_m}$. Thus $A \subseteq O_{\alpha_1} \cup O_{\alpha_2} \cup \cdots \cup O_{\alpha_n}$ and $B \subseteq U_{\beta_1} \cup U_{\beta_2} \cup \cdots \cup U_{\beta_m}$. Therefore, $A \cup B \subseteq O_{\alpha_1} \cup O_{\alpha_2} \cup \cdots \cup O_{\alpha_n} \cup U_{\beta_1} \cup U_{\beta_2} \cup \cdots \cup U_{\beta_m}$, a union of a finite number of open sets. Thus $A \cup B$ is compact. ∎
    ★ (c) **Claim.** If A is compact, $B \subseteq A$, and B is closed, then B is compact.
    "Proof." Let $\{O_\alpha : \alpha \in \Delta\}$ be a cover for B. If $\{O_\alpha : \alpha \in \Delta\}$ is a cover for A, then there is a finite subcover of $\{O_\alpha : \alpha \in \Delta\}$ that covers A and hence covers B. If $\{O_\alpha : \alpha \in \Delta\}$ is not a cover for A, add one more open set $O^* = \mathbb{R} - B$ to the collection to obtain a cover for A. This cover for A has a finite subcover of A that is a cover for B. In either case B is covered by a finite number of open sets. Therefore B is compact. ∎

(d) **Claim.** If A is compact, $B \subseteq A$, and B is closed, then B is compact.
    "Proof." B is closed by assumption. Since A is compact, A is bounded. Since $B \subseteq A$, B is also bounded. Thus B is closed and bounded. Therefore, B is compact. ■

(e) **Claim.** The set $(5, \infty)$ is compact.
    "Proof." The set $\mathscr{C} = \{(4, 12), (10, \infty)\}$ is a cover for $(5, \infty)$. Then $\mathscr{C}$ is a finite subcover of $\mathscr{C}$ for $(5, \infty)$, so $(5, \infty)$ is compact.

7.3   The Bolzano–Weierstrass Theorem

In the previous section we used the completeness of $\mathbb{R}$ to prove the Heine–Borel Theorem. In this section we use the Heine–Borel Theorem to prove another classical result of analysis, the Bolzano–Weierstrass* Theorem.

Let's begin with a closed interval $[a, b]$ and imagine that we must build a subset A of $[a, b]$ by selecting elements for A, one at a time, from $[a, b]$. Since $A \subseteq [a, b]$, A will necessarily be a bounded set. If A is finite, we could choose elements that are spread out across the interval. In other words, if A is finite, there need not be any point in the interval where the elements of A pile up or "accumulate." What the Bolzano–Weierstrass Theorem says is that if a set A is infinite and bounded, there *must* be at least one point in $[a, b]$ around which an infinite number of elements of A will be congregated. Before we get to that result, we give an example and define carefully what it means for elements of a set to accumulate around a point.

Example. For the bounded infinite set $A = \left\{ 1, -\frac{1}{2}, \frac{1}{3}, -\frac{1}{4}, \frac{1}{5}, \dots \right\}$, each element of A is contained in the interval $[-1, 1]$. Figure 7.3.1 shows that the number 0 is a

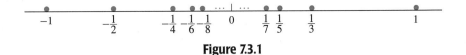

Figure 7.3.1

---

* Bernard Bolzano (1781–1848) was a Bohemian mathematician, philosopher, and logician. He pioneered several modern mathematics concepts (such as the rigorous definition of limit) but because of his strong antimilitary beliefs most of his work appeared in obscure publications. The Bolzano–Weierstrass Theorem was for several years called simply the Weierstrass Theorem until Bolzano's independent proof was discovered many years after his death.

Karl Weierstrass (1815–1897) was a German mathematician and the foremost leader in the 1800s in developing highly rigorous definitions and theorems that characterize much of modern mathematics. He was the first to develop rigorous proofs for the Intermediate Value Theorem, the Heine–Borel Theorem and several theorems whose titles now bear his name.

point around which the elements of A gather. We say 0 is an accumulation point for the set A.

DEFINITION   Let A be a set of real numbers. The number x is an **accumulation point for A** iff for all $\delta > 0$, $\mathcal{N}(x, \delta)$ contains at least one point of A distinct from x.

The definition says that for x to be an accumulation point for set A, it must be that for every $\delta > 0$, $(\mathcal{N}(x, \delta) - \{x\}) \cap A \neq \varnothing$. To verify that 0 satisfies the definition of an accumulation point for the set $A = \left\{1, -\frac{1}{2}, \frac{1}{3}, -\frac{1}{4}, \frac{1}{5}, \cdots\right\}$, we start by choosing some δ-neighborhood $\mathcal{N}(0, \delta)$ about 0. By the Archimedean Principle, there exists an odd natural number k such that $k > \frac{1}{\delta}$. Then $\frac{1}{k} < \delta$. Therefore $\frac{1}{k} \in \mathcal{N}(0, \delta)$ and since k is odd, $\frac{1}{k} \in A$. Thus every neighborhood of 0 contains a point of A that is distinct from 0.

Example.   Let $A = [3, 7)$. Prove that the set of accumulation points for A is $[3, 7]$.

Proof.   We consider cases: elements of $(3, 7)$, the endpoints, other numbers.

(i)   Let $x \in (3, 7)$. Suppose $\delta > 0$. The set $(3, 7)$ is open so there exists $\beta > 0$ such that $\mathcal{N}(x, \beta) \subseteq (3, 7)$. Let γ be the smaller of δ and β. Then $\mathcal{N}(x, \gamma) \subseteq (3, 7)$ and $x + \frac{\gamma}{2}$ is a point in $\mathcal{N}(x, \gamma)$ that is in A and distinct from x. Thus if x is in $(3, 7)$, x is an accumulation point for A.

(ii)   Let $x = 3$ and $\delta > 0$. If $\delta \geq 4$, then 5 is a point of $\mathcal{N}(x, \delta)$ that is in $[3, 7)$ and distinct from 3. If $\delta < 4$, then $3 + \frac{\delta}{2}$ is a point of $\mathcal{N}(x, \delta)$ that is in $[3, 7)$ and distinct from 3. Thus 3 is an accumulation point for A.

(iii)   Let $x = 7$. By an argument very similar to part (ii), 7 is also an accumulation point for A. (See Exercise 1.)

(iv)   Let $x < 3$ and $\delta = 3 - x$. Then $\mathcal{N}(x, \delta)$ and A are disjoint, so x is not an accumulation point for A.

(v)   Let $x > 7$. Using reasoning similar to that in part (iv), x is not an accumulation point for A.

We conclude that the set of all accumulation points for $[3, 7)$ is $[3, 7]$.   ∎

Example.   Let $B = (2, 6) \cup \{9\}$. The number 9 is not an accumulation point for B because, for example, $\mathcal{N}(9, 0.5)$ contains no points of B other than 9. The accumulation points of B are all the elements of $[2, 6]$.

We see from our examples that an accumulation point of a set is not necessarily an element of the set, and, conversely, an element of a set is not necessarily an accumulation point of the set.

We also see that for x to be an accumulation point of the set A, much more is required than just that there is an element of A distinct from x that is within δ of x. The next theorem shows that there must be infinitely many points of A that are within δ of x.

Theorem 7.3.1

A number x is an accumulation point for a set A iff for all $\delta > 0$, $\mathcal{N}(x, \delta)$ contains an infinite number of points of A.

Proof. If every neighborhood of x contains an infinite number of points of A, then each neighborhood certainly contains at least one point of A distinct from x. Therefore, x is an accumulation point.

Now suppose that x is an accumulation point for A. Suppose that $\mathcal{N}(x, \delta) \cap A$ is finite for some $\delta > 0$. Let $\delta_1 = \min\{|x - y| : y \in \mathcal{N}(x, \delta) \cap A, x \neq y\}$. ⟨*Our choice of δ_1 is so small that $\mathcal{N}(x, \delta_1)$ will have no points of A other than perhaps x itself.*⟩ Then $\mathcal{N}(x, \delta_1) \cap A = \{x\}$, which contradicts the initial assumption that x is an accumulation point for A. Therefore, every neighborhood of x must contain an infinite number of points of A. ∎

From Theorem 7.3.1 it follows that no finite set can have any accumulation points.

Examples. There are rational numbers between any two distinct real numbers, so it follows that for real number x and $\delta > 0$, the interval $(x - \delta, x + \delta)$ contains infinitely many rationals. Therefore, every real number is an accumulation point for $\mathbb{Q}$. Using similar reasoning, we can conclude that every real number is an accumulation point for the set of irrationals.

DEFINITION Let A be a set of real numbers. The set of accumulation points for A is called the **derived set** of A, and is denoted by A'.

Let $B = \{1.4, 1.41, 1.414, 1.4142, 1.41421, \ldots\}$ be the set of successive decimal approximations to $\sqrt{2}$. Then $B' = \{\sqrt{2}\}$. Other examples of derived sets of subsets of $\mathbb{R}$ are:

$$(3, 5)' = [3, 5)' = (3, 5]' = [3, 5]' = [3, 5]$$
$$((-1, 6] \cup (7, 8))' = [-1, 6] \cup [7, 8]$$
$$((-1, 6) \cup (6, 8))' = [-1, 8]$$
$$\mathbb{Q}' = \mathbb{R}$$
$$\mathbb{N}' = \varnothing.$$

The following theorem relates derived sets and closed sets.

Theorem 7.3.2   A set A is closed iff $A' \subseteq A$.

Proof. Suppose A is closed and $x \in A'$. If $x \notin A$, then $x \in A^c$, an open set. Thus $\mathcal{N}(x, \delta) \subseteq A^c$ for some positive δ. But then $\mathcal{N}(x, \delta)$ can contain no points of A. Thus x is not an accumulation point of A and so $x \notin A'$, a contradiction. We conclude that $x \in A$. Therefore $A' \subseteq A$.

Now suppose $A' \subseteq A$. To show that A is closed, we show A^c is open. If A^c is not open, there is at least one $x \in A^c$ that is not an interior point of A^c. Therefore, no δ-neighborhood of x is a subset of A^c; that is, each δ-neighborhood of x contains a point of A. This point must be different from x, since $x \in A^c$. Thus $x \in A'$. But $A' \subseteq A$, so $x \in A$. This is a contradiction. We conclude that A is closed. ■

At the beginning of this section we suggested that every bounded infinite set of real numbers, such as $A = \left\{ 1, -\frac{1}{2}, \frac{1}{3}, -\frac{1}{4}, \frac{1}{5}, \dots \right\}$, must have at least one accumulation point in $\mathbb{R}$. We are now in a position to prove this is so. The proof uses the Heine–Borel Theorem.

Theorem 7.3.3   **The Bolzano–Weierstrass Theorem**
Every bounded infinite set of real numbers has an accumulation point in $\mathbb{R}$.

Proof. Suppose the set A is bounded and infinite but has no accumulation points. Then $A' = \varnothing$. Since $A' \subseteq A$, A is closed $\langle$ *by Theorem 7.3.2* $\rangle$. Then by the Heine–Borel Theorem, A is compact.

Since A has no accumulation points, for each $x \in A$ there exists $\delta_x > 0$ such that $\mathcal{N}(x, \delta_x) \cap A = \{x\}$. Thus if $y \in A$ and $x \neq y$, then $y \notin \mathcal{N}(x, \delta_x)$. But this means the family $\{\mathcal{N}(x, \delta_x) : x \in A\}$ is an infinite collection of open sets that covers A and has no subcover other than itself. Hence A has no finite subcover. This contradicts the fact that A is compact. Therefore, A must have an accumulation point. ■

Exercises 7.3

1. Prove that
   (a) 7 is an accumulation point for $[3, 7)$.
   (b) 0 is an accumulation point for $\left\{ \dfrac{1 + (-1)^n}{n} : n \in \mathbb{N} \right\}$.
   ☆ (c) e is an accumulation point for $\left\{ \left(1 + \dfrac{1}{n}\right)^n : n \in \mathbb{N} \right\}$.

2. Find an example of an infinite subset of $\mathbb{R}$ that has
   (a) no accumulation points.
   (b) exactly one accumulation point.
   (c) exactly two accumulation points.
   (d) denumerably many accumulation points.
   (e) an uncountable number of accumulation points.

3. Find the derived set of each of the following sets.

★ (a) $\left\{\dfrac{n+1}{2n} : n \in \mathbb{N}\right\}$

(b) $\{2^n : n \in \mathbb{N}\}$

★ (c) $\{6n : n \in \mathbb{N}\}$

(d) $\left\{\dfrac{7}{2^n} : n \in \mathbb{N}\right\}$

(e) $(0, 1]$

(f) $(3, 7) \cup \{4, 6, 8\}$

★ (g) $\left\{1 + \dfrac{(-1)^n n}{n+1} : n \in \mathbb{N}\right\}$

(h) $\mathbb{Z}$

(i) $\mathbb{Q} \cap (0, 1)$

(j) $\left\{\dfrac{1 + n^2(1 + (-1)^n)}{n} : n \in \mathbb{N}\right\}$

(k) $\left\{\sin x : x \in \left(\dfrac{-\pi}{2}, \dfrac{\pi}{2}\right)\right\}$

(l) $\left\{\dfrac{\sin x}{x} : x \in (0, \pi)\right\}$

★ (m) $\left\{k + \dfrac{1}{n} : k, n \in \mathbb{N}\right\}$

(n) $\left\{\dfrac{x}{2^y} : x, y \in \mathbb{Z}\right\}$

★ 4. Let $S = (0, 1]$. Find $S' \cap (S^c)'$.

☆ 5. Prove that if $A \subseteq \mathbb{R}$, $z = \sup(A)$, and $z \notin A$, then z is an accumulation point of A.

6. (a) Prove that if $A \subseteq B \subseteq \mathbb{R}$, then $A' \subseteq B'$.
   (b) Is the converse of part (a) true? Explain.

7. Let A and B be subsets of $\mathbb{R}$.
   (a) Prove that $(A \cup B)' = A' \cup B'$. (The operation of finding the derived set preserves unions.)
   ☆ (b) Prove that $(A \cap B)' \subseteq A' \cap B'$.
   (c) Find a counterexample for $(A \cap B)' = A' \cap B'$.

8. Let A and B be sets of real numbers. Prove that
   (a) if B is closed and $A \subseteq B$, then $A' \subseteq B$.
   (b) $A \cup A'$ is closed.

9. (a) Prove that if x is an interior point of the set A, then x is an accumulation point for A.
   (b) Is the converse of part (a) true? Explain.
   (c) Prove that if $S \subseteq \mathbb{R}$ is open, then every point of S is an accumulation point of S.
   (d) Is the converse of part (c) true? Explain.

10. Which of the following must have at least one accumulation point?
   ★ (a) an infinite subset of $\mathbb{N}$
   (b) an infinite subset of $(-10, 10)$
   ★ (c) an infinite subset of $[0, 100]$
   (d) an infinite subset of $\mathbb{Q}$
   (e) $\left\{\dfrac{1}{2^k} : k \in \mathbb{N}\right\}$
   (f) $\left\{\dfrac{p}{q} : p, q \in \mathbb{N}, p < q\right\}$
   (g) an infinite subset of $\mathbb{Q} \cap [0, 1]$

11. Let A be a set of real numbers. Prove that $(A')^c \subseteq (A^c)'$.

12. Let A and F be sets of real numbers and let F be finite. Prove that if x is an accumulation point of A, then x is an accumulation point of $A - F$.

Proofs to Grade

13. Assign a grade of A (correct), C (partially correct), or F (failure) to each. Justify assignments of grades other than A.

☆ **(a)** **Claim.** For $A, B \subseteq \mathbb{R}$, $(A \cup B)' = A' \cup B'$.

"Proof."

(i) Since $A \subseteq A \cup B$, $A' \subseteq (A \cup B)'$ by exercise 6(a). Likewise, $B' \subseteq (A \cup B)'$. Therefore, $A' \cup B' \subseteq (A \cup B)'$.

(ii) To show that $(A \cup B)' \subseteq A' \cup B'$, let $x \in (A \cup B)'$. Then for all $\delta > 0$, $\mathcal{N}(x, \delta)$ contains a point of $A \cup B$ distinct from x. Restating this, we have, for all $\delta > 0$, that $\mathcal{N}(x, \delta)$ contains a point of A distinct from x or a point of B distinct from x. Thus for all $\delta > 0$, $\mathcal{N}(x, \delta)$ contains a point of A distinct from x, or, for all $\delta > 0$, $\mathcal{N}(x, \delta)$ contains a point of B distinct from x. But this means $x \in A'$ or $x \in B'$. Therefore, $x \in A' \cup B'$. ∎

(b) **Claim.** For $A \subseteq \mathbb{R}$, $(A^c)' = (A')^c$.

"Proof." $x \in (A')^c$ iff $x \notin A'$

         iff x is not an accumulation point for A

         iff x is an accumulation point for A^c

         iff $x \in (A^c)'$. ∎

★ **(c)** **Claim.** For $A, B \subseteq \mathbb{R}$, $(A - B)' \subseteq A' - B'$.

"Proof." $\begin{aligned}(A - B)' &= (A \cap B^c)' &&\langle\textit{definition of } A - B\rangle \\ &\subseteq A' \cap (B^c)' &&\langle\textit{Exercise 7(b)}\rangle \\ &\subseteq A' \cap (B')^c &&\langle\textit{since } (B^c)' \subseteq (B')^c\rangle \\ &= A' - B'. \end{aligned}$ ∎

(d) **Claim.** If A is closed, then $A' \subseteq A$.

"Proof." Suppose A is closed. Then A^c is open. Let $x \in A^c$. Then x is an interior point of A^c. Therefore, there exists $\delta > 0$ so that $\mathcal{N}(x, \delta) \subseteq A^c$. Hence $\mathcal{N}(x, \delta) \cap A = \varnothing$. Thus x is not an accumulation point for A. Since $x \in A^c$ implies $x \notin A'$, we conclude $A' \subseteq A$. ∎

(e) **Claim.** If A is a set with an accumulation point, $B \subseteq A$, and B is infinite, then B has an accumulation point.

"Proof." First, A is infinite because $B \subseteq A$ and B is infinite. Since A has an accumulation point, by the Bolzano–Weierstrass Theorem A must be bounded. Since $B \subseteq A$, this means B is bounded. Hence by the Bolzano–Weierstrass Theorem again, B has an accumulation point. ∎

7.4   The Bounded Monotone Sequence Theorem

Recall that a sequence of real numbers is a function x from $\mathbb{N}$ to $\mathbb{R}$ and the notation x_n is used to represent the nth term of the sequence. We showed in Section 4.6 that when a sequence converges (has a limit) then the limit is unique. For instance, the first few terms of the sequence a given by $a_n = \dfrac{1}{n^2}$ are $1, \dfrac{1}{4}, \dfrac{1}{9}, \dfrac{1}{16}, \ldots$. This sequence converges to 0 and we write $\lim_{n \to \infty} \dfrac{1}{n^2} = 0$ In this section we prove that the

Bolzano–Weierstrass Theorem implies that certain kinds of sequences of real numbers (bounded monotone sequences) must converge. We begin by defining bounded sequences and monotone sequences.

DEFINITIONS   For a sequence x of real numbers, if there exists a real number B such that $x_n \leq B$ for all $n \in \mathbb{N}$, we say x is **bounded above** (by B).

Similarly, if there exists a real number B such that $x_n \geq B$ for all $n \in \mathbb{N}$, we say x is **bounded below** (by B).

The sequence x is **bounded** iff x is bounded above and bounded below.

A sequence is bounded iff the terms of the sequence are never larger than some fixed number and never less than some other (smaller) fixed number. This is equivalent to saying that the set $\{x_n: n \in \mathbb{N}\}$ is a bounded set—that is, the range of the function x is a bounded subset of $\mathbb{R}$. Boundedness may also be described by the absolute values of the terms of the sequences.

Theorem 7.4.1   A sequence x of real numbers is bounded iff there exists a real number B such that $|x_n| \leq B$ for all $n \in \mathbb{N}$.

Proof.   Exercise 4. ∎

The sequence y given by $y_n = 2^n$ is unbounded. Intuitively it seems that y must diverge because its terms never approach any possible limit L. Our next theorem confirms that every unbounded sequence diverges.

Theorem 7.4.2   If a sequence of real numbers converges, then it is bounded.

Proof.   Suppose x is a sequence convergent to the real number L. For $\varepsilon = 1$, there is a natural number N such that if $n > N$, then $|x_n - L| < 1$. Since $||x_n| - |L|| \leq |x_n - L|$, we have for all $n > N$, $|x_n| - |L| < 1$. Thus for all $n > N$, $|x_n| < |L| + 1$. ⟨*All but the first N terms are bounded by $|L| + 1$. We now find an upper bound for those terms as well.*⟩ Let $B = \max\{|x_1|, |x_2|, \ldots, |x_N|, |L| + 1\}$. Then $|x_n| \leq B$ for all $n \in \mathbb{N}$, and x is bounded above. A similar argument proves that x is bounded below. Therefore x is bounded. ∎

The proof shows that after the first few terms (that is, after N terms), the remaining terms of the sequence must be close to the limit. Let x be the sequence

$$x_n = \begin{cases} (-2)^n & 1 \leq n \leq 1000 \\ \dfrac{15n}{n+1} & n > 1000 \end{cases}.$$

This sequence converges to 15 and therefore must be bounded. In this case the first "few" (1000) terms jump around before the terms settle in close to 15. The sequence is bounded above by 2^{1000} and below by -2^{999}.

> **DEFINITIONS**   Let x be a sequence of real numbers. The sequence x is **increasing** iff for all $n, m \in \mathbb{N}$, if $n < m$ then $x_n \leq x_m$.
>   We say x is **decreasing** iff for all $n, m \in \mathbb{N}$, if $n < m$ then $x_n \geq x_m$.
>   The sequence x is **monotone** iff x is either increasing or decreasing.

The sequence y given by $y_n = 2^n$ is increasing since $n < m$ implies $2^n \leq 2^m$. The sequence whose terms are $z_n = e^{-n}$ is decreasing since the value of e^{-n} gets smaller as n gets larger. A constant sequence k, such that every term k_n is the number c for some $c \in \mathbb{R}$, is both increasing and decreasing. The alternating sequence $1, -1, 1, -1, 1, \ldots$ is not monotone, because its terms are neither in increasing order nor in decreasing order.

A proof that a given sequence x is increasing (or decreasing) is similar to the proof that a real valued function is increasing (or decreasing) on an interval I. (See Section 4.2.)

Example.   Prove that the sequence x given by $x_n = \dfrac{n}{n + 1}$ is increasing.

Proof.   Suppose that m and n are natural numbers and $n < m$. Then

$$mn + n < mn + m,$$
$$n(m + 1) < m(n + 1),$$
$$\frac{n}{n + 1} < \frac{m}{m + 1} \quad \langle\text{since } n + 1 \text{ and } m + 1 \text{ are positive}\rangle.$$

Therefore, $x_n < x_m$. Hence the sequence x is increasing.   ∎

The next theorem relates all the concepts of this section: A sequence of real numbers that is both bounded and monotone must converge. The proof makes use of the Bolzano–Weierstrass Theorem.

Theorem 7.4.3

The Bounded Monotone Sequence Theorem
For every bounded monotone sequence x, there is a real number L such that $\lim\limits_{n \to \infty} x_n = L$.

Proof.   Assume x is a bounded and increasing sequence. (The proof in the case where x is decreasing is similar.)

If $\{x_n : n \in \mathbb{N}\}$ is finite, then let $L = \max\{x_n : n \in \mathbb{N}\}$. For some $N \in \mathbb{N}$, $x_N = L$ and, since x is increasing, $x_n = L$ for all $n > N$. Therefore, $\lim\limits_{n \to \infty} x_n = L$.

Suppose $\{x_n : n \in \mathbb{N}\}$ is infinite. Then by the Bolzano–Weierstrass Theorem, the bounded infinite set $\{x_n : n \in \mathbb{N}\}$ must have at least one accumulation point. Let

L be an accumulation point. We claim $x_n \leq L$ for all $n \in \mathbb{N}$. If there exists N such that $x_N > L$, then $x_n > L$ for all $n \geq N$. Since L is an accumulation point of $\{x_n : n \in \mathbb{N}\}$, and $\{x_n : n \leq N\}$ is finite, L is an accumulation point of $\{x_n : n > N\}$, by Exercise 12 of Section 7.3. Let $\delta = |x_N - L|$. Then $\mathcal{N}(L, \delta)$ contains no points of $\{x_n : n \geq N\}$. This is a contradiction. Thus $x_n \leq L$ for all n.

We claim that the sequence x converges to L. Let $\varepsilon > 0$. Since L is an accumulation point of $\{x_n : n \in \mathbb{N}\}$, there exists $M \in \mathbb{N}$ such that $x_M \in \mathcal{N}(L, \varepsilon)$. Thus $L - \varepsilon < x_M$, and so, for $n > M$, $L - \varepsilon < x_M \leq x_n \leq L < L + \varepsilon$. Therefore, for $n > M$, $|x_n - L| < \varepsilon$. Thus $\lim_{n \to \infty} x_n = L$. ∎

The Bounded Monotone Sequence Theorem can be used to prove the existence of several important real numbers. For example, the constant e, the base of the natural logarithm function, can be defined as

$$e = \lim_{n \to \infty} \left(1 + \frac{1}{n}\right)^n.$$

To show that the sequence whose nth term is $x_n = \left(1 + \frac{1}{n}\right)^n$ has a limit, we will show that the sequence is bounded above and increasing.

By the Binomial Expansion Theorem (Theorem 2.6.9), for any n,

$$
\begin{aligned}
x_n = \left(1 + \frac{1}{n}\right)^n &= 1 + \frac{n}{1!}\frac{1}{n} + \frac{n(n-1)}{2!}\frac{1}{n^2} + \frac{n(n-1)(n-2)}{3!}\frac{1}{n^3} + \cdots + \frac{1}{n^n} \\
&= 1 + 1 + \frac{1}{2!}\frac{n(n-1)}{n^2} + \frac{1}{3!}\frac{n(n-1)(n-2)}{n^3} + \cdots + \frac{1}{n!}\frac{n!}{n^n} \\
&\leq 1 + 1 + \frac{1}{2!} + \frac{1}{3!} + \cdots + \frac{1}{n!} \\
&\leq 1 + 1 + \frac{1}{2} + \frac{1}{4} + \cdots + \frac{1}{2^{n-1}} \\
&= 1 + \frac{2^n - 1}{2^{n-1}} \\
&< 1 + 2 \\
&= 3.
\end{aligned}
$$

Thus the sequence $x_n = \left(1 + \frac{1}{n}\right)^n$ is bounded above by 3.

We next show that x is an increasing sequence. We again use the Binomial Theorem to compare x_n and x_{n+1}:

$$
\begin{aligned}
x_n &= \left(1 + \frac{1}{n}\right)^n \\
&= 1 + 1 + \frac{1}{2!}\frac{n(n-1)}{n^2} + \frac{1}{3!}\frac{n(n-1)(n-2)}{n^3} + \cdots + \frac{1}{n!}\frac{n!}{n^n}
\end{aligned}
$$

and

$$x_{n+1} = \left(1 + \frac{1}{n+1}\right)^{n+1}$$

$$= 1 + 1 + \frac{1}{2!}\frac{(n+1)n}{(n+1)^2} + \frac{1}{3!}\frac{(n+1)(n)(n-1)}{(n+1)^3} + \cdots$$

$$+ \frac{1}{n!}\frac{(n+1)(n)(n-1)\cdot\cdots\cdot 3\cdot 2}{(n+1)^n} + \frac{1}{(n+1)!}\frac{(n+1)!}{(n+1)^{n+1}}.$$

We leave it as Exercise 7 to show that each term in the expansion of x_n is less than or equal to the corresponding term in the expansion of x_{n+1}. Additionally, the binomial expansion of x_{n+1} has one more positive term, $\dfrac{1}{(n+1)!}\dfrac{(n+1)!}{(n+1)^{n+1}}$, than the binomial expansion of x_n. Thus $x_n \le x_{n+1}$ for all $n \in \mathbb{N}$.

Because x is increasing, it is bounded below by its first term $x_1 = \left(1 + \dfrac{1}{1}\right)^{\frac{1}{1}} = 2$. Thus x is bounded. By the Bounded Monotone Sequence Theorem, x must converge. The limit of the sequence x is, by definition, the number e.

Exercises 7.4

1. For each sequence x, determine whether x is bounded, bounded above, or bounded below.

   ★ (a) $x_n = 10n$

   (b) $x_n = \dfrac{10}{n}$

   ★ (c) $x_n = 10^{-n}$

   (d) $x_n = \log_{10} n$

   ★ (e) $x_n = \dfrac{10}{n!}$

   (f) $x_n = \dfrac{\cos n}{n}$

   ★ (g) $x_n = n\tan\left(\dfrac{3n\pi}{4}\right)$

   (h) $x_n = (-1)^n n$

   ★ (i) $x_n = (-0.9)^n$

   (j) $x_n = (-1.1)^n$

   (k) $x_n = \dfrac{(-1)^n + 1}{n}$

   (l) $x_n = \dfrac{((-1)^n - 1)((-1)^n + 1)}{n}$

2. Give an example of
   (a) a bounded sequence that is not convergent.
   (b) an increasing sequence that is not convergent.
   (c) a convergent sequence that is not monotone.
   (d) a divergent sequence x such that the sequence whose nth term is $|x_n|$ converges.
   (e) an increasing sequence that converges to $\dfrac{\pi}{2}$.

   ☆ 3. Prove that if $x_n \to 0$ and y is a bounded sequence, then $x_n y_n \to 0$.

4. Prove Theorem 7.4.1.

5. For each sequence, determine whether the sequence is increasing, decreasing, or neither. Prove your answer.

   (a) $x_n = \dfrac{n+2}{n}$

   (b) $y_n = 2^{-n}$

   (c) $x_n = (n-2)(n-5)^2$

   (d) $y_n = \dfrac{10}{n!}$

   (e) $x_n = \dfrac{2n-5}{n+3}$

   ☆ (f) $y_n = \dfrac{n!}{n^n}$

   (g) $x_n = \sqrt{n+1}$

6. Give a proof of the Bounded Monotone Sequence Theorem for the case in which the sequence x is bounded and decreasing.

7. Complete the proof that $x_n = \left(1 + \dfrac{1}{n}\right)^n$ is an increasing sequence by showing that for all $k \le n$, $\dfrac{1}{k!} \dfrac{n(n-1)(n-2) \cdots \cdot [n-(k-1)]}{n^k}$ is less than or equal to $\dfrac{1}{k!} \dfrac{(n+1)(n)(n-1) \cdots \cdot [n-(k-2)]}{(n+1)^k}$.

8. Let x be a bounded increasing sequence. Use the completeness property of the reals and properties of suprema and limits to prove directly (without reference to the Bolzano–Weierstrass Theorem) that x converges. (*Hint:* Consider the supremum of the set of terms of x.)

9. Recall from Exercise 8 of Section 4.6 that the sequence y_n is a **subsequence** of x_n if and only if there is an increasing function $f: \mathbb{N} \to \mathbb{N}$ such that $y_n = x_{f(n)}$. Prove that if x is bounded, then every subsequence of x is bounded.

10. A sequence x of real numbers is a **Cauchy*** **sequence** iff for every $\varepsilon > 0$, there exists an integer M such that if $m, n > M$, then $|x_n - x_m| < \varepsilon$. That is, terms in the sequence are arbitrarily close together if the terms are chosen far enough along the sequence.
    (a) Prove that if x is a Cauchy sequence, then x is bounded. (It can also be shown that every Cauchy sequence converges.)
    (b) Prove that if x is a convergent sequence, then x is a Cauchy sequence.

11. Let x and y be positive real numbers with $x > y$. Let $a_1 = \dfrac{x+y}{2}$ and $b_1 = \sqrt{xy}$. The numbers a_1 and b_1 are called the *arithmetic* and *geometric means*, respectively. In general, for $n > 1$, let $a_{n+1} = \dfrac{a_n + b_n}{2}$ and $b_{n+1} = \sqrt{a_n b_n}$.
    (a) Use induction to show that for all $n \in \mathbb{N}$, $a_n > a_{n+1} > b_{n+1} > b_n$.

---

* Augustin Louis Cauchy (1789–1857) was a creative French mathematician and pioneer in the efforts to bring rigor to the infinitesimal calculus. He was the first to define complex numbers as a pair of real numbers. Cauchy's name is associated with concepts and results in many fields of mathematics, including geometry, analysis, and mathematical physics.

(b) Let a and b be the sequences whose terms are a_n and b_n, respectively. Show that both sequences a and b converge.

(c) Show that $\lim\limits_{n \to \infty} a_n = \lim\limits_{n \to \infty} b_n$. This number is called the *arithmetic-geometric mean*.

Proofs to Grade

12. Assign a grade of A (correct), C (partially correct), or F (failure) to each. Justify assignments of grades other than A.

(a) Claim. Every bounded decreasing sequence converges.

"Proof." Let x be a bounded decreasing sequence. Then $y_n = -x_n$ defines a bounded increasing sequence. By the proof of Theorem 7.4.3, $\lim\limits_{n \to \infty} y_n = L$ for some L. Thus $\lim\limits_{n \to \infty} x_n = -L$. ∎

(b) Claim. The sequence x, where $x_n = \dfrac{\ln n}{n}$, converges.

"Proof." Since $\ln n \le n$ for all natural numbers n, $\dfrac{\ln n}{n} \le 1$. Therefore, x is a bounded sequence. The derivative of $\dfrac{\ln n}{n}$ is $\dfrac{1 - \ln n}{n^2}$, which is less than 0 for every natural number n greater than e. Therefore, except for the first two terms, x is a decreasing sequence. Since x is bounded and decreasing, x converges. ∎

7.5 Equivalents of Completeness

We began the chapter by stating without proof that the real numbers are a complete ordered field. In subsequent sections we (1) used completeness to prove that a set of real numbers is compact iff it is closed and bounded, then (2) used that property to prove that every bounded infinite set of reals has an accumulation point, and then (3) used that property to prove that every bounded monotone sequence of real numbers converges. In this section we use the bounded monotone sequence property to prove that $\mathbb{R}$ is complete. This result completes a cycle of implications about the real numbers. See Figure 7.5.1.

When we finish this section, we will not have proved that the real numbers have any of the four properties we have studied. Rather, we will have shown that the completeness of $\mathbb{R}$ is equivalent to each of the other properties, so that we have a deeper understanding of the meaning and importance of completeness.

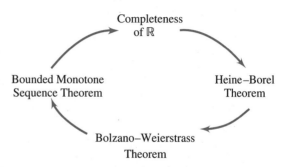

Figure 7.5.1

Before we get to the main theorem we need two lemmas about the convergence of sequences. Their proofs are Exercises 1 and 2.

Lemma 7.5.1   If x and y are two sequences such that $\lim\limits_{n\to\infty} y_n = s$ and $\lim\limits_{n\to\infty} (x_n - y_n) = 0$, then $\lim\limits_{n\to\infty} x_n = s$.

Lemma 7.5.2   If x is a sequence with $\lim\limits_{n\to\infty} x_n = s$ and t is a real number such that $t < s$, then there exists $N \in \mathbb{N}$ such that $x_n > t$ for all $n \geq N$.

Theorem 7.5.3   Suppose $\mathbb{R}$ has the property that every bounded monotone sequence must converge. Then $\mathbb{R}$ is complete.

Proof.   Let A be a nonempty subset of $\mathbb{R}$ that is bounded above by a real number b. To prove completeness, we must show sup (A) exists and is a real number. Since $A \neq \varnothing$, we may choose $a \in A$. If a is an upper bound for A, then $a = \sup (A)$ ⟨*see Exercise 5(b) of Section 7.1*⟩ and we are done. Assume that a is not an upper bound for A. If $(a + b)/2$ is an upper bound for A, let $x_1 = a$ and $y_1 = (a + b)/2$; if not, let $x_1 = (a + b)/2$ and $y_1 = b$. In either case $y_1 - x_1 = (b - a)/2$, x_1 is not an upper bound, and y_1 is an upper bound for A.

Now if $(x_1 + y_1)/2$ is an upper bound for A, let $x_2 = x_1$ and $y_2 = (x_1 + y_1)/2$; otherwise, let $x_2 = (x_1 + y_1)/2$ and $y_2 = y_1$. In either case the result is that $y_2 - x_2 = (b - a)/4$, $x_2 \geq x_1$, x_2 is not an upper bound for A, while $y_2 \leq y_1$, and y_2 is an upper bound for A.

Continuing in this manner, we inductively define an increasing sequence x such that no x_n is an upper bound for A, and a decreasing sequence y such that every y_n is an upper bound for A. By the hypothesis, since y is bounded below, y converges to a point $s \in \mathbb{R}$. In addition $y_n - x_n = (b - a)/2^n$, so $\lim\limits_{n\to\infty} (y_n - x_n) = \lim\limits_{n\to\infty} (b - a)/2^n = 0$. Therefore by Lemma 7.5.1, $\lim\limits_{n\to\infty} x_n = s$.

We claim that s is an upper bound for A. If $z > s$ for some $z \in A$, then $z > y_N$ for some N ⟨*because* $\lim\limits_{n\to\infty} y_n = s$⟩. This contradicts the fact that y_N is an upper bound for A.

Finally, if t is a real number and $t < s$, then $t < x_N$ for some $N \in \mathbb{N}$ by Lemma 7.5.2. Since x_N is not an upper bound for A, t is not an upper bound.

Thus s is a real number that is an upper bound for A, and no number less than s is an upper bound; that is, $s = \sup (A)$. Therefore, $\mathbb{R}$ is complete.   ∎

We saw in Section 7.1 that $(\mathbb{Q}, +, \cdot)$ is an ordered field that is not complete. Thus, all the properties described by the main theorems of this chapter must fail for the rational numbers. We give examples:

Example.   The set $A = \{x \in \mathbb{Q} : x^2 \leq 2\}$ is a closed and bounded subset of $\mathbb{Q}$. A is not compact because $\{(-x, x) : x \in A \text{ and } x \neq 0\}$ is a cover for A with no finite subcover. This example shows that the Heine–Borel Theorem fails for $\mathbb{Q}$.

Example. The set $B = \{1.4, 1.41, 1.414, 1.4142, \ldots\}$ of (rational) decimal approximations of $\sqrt{2}$ is a bounded and infinite subset of $\mathbb{Q}$ with no accumulation point in $\mathbb{Q}$. Thus the Bolzano–Weierstrass Theorem fails in $\mathbb{Q}$.

Example. A counterexample to the Bounded Monotone Sequence Theorem in $\mathbb{Q}$ is the sequence whose terms are the successive decimal expansions of $\sqrt{2}$. This bounded and increasing sequence fails to converge to any rational number.

Why is completeness such a crucial property of the real number system? In Section 7.4 we saw that the sequence whose nth term is $x_n = \left(1 + \dfrac{1}{n}\right)^n$ is bounded and increasing, so by the completeness of $\mathbb{R}$ (via the Bounded Monotone Sequence Theorem), the terms x_n must approach a unique real number, which is the number e. The fact that important constants such as e must exist in $\mathbb{R}$ is a consequence of completeness.

Not only must the limit of a bounded monotone sequence of rational numbers be in $\mathbb{R}$, but the same is true for a bounded monotone sequence of irrational numbers, or of rationals and irrationals. The completeness property and its equivalents assure us that every number that is a limit of a sequence of reals is in fact a real number.

A proof that $\mathbb{R}$ is a complete ordered field requires a much more rigorous definition of a real number than we gave in the *Preface to the Student*. Such a definition requires construction of the reals from the rationals in such a fashion that the completeness property holds. What this means is that we must identify some set of objects based on the rational numbers, tell how to add, multiply, and order these objects, and then show that all the properties of a complete ordered field hold for these objects.

One approach considers Cauchy sequences of rational numbers (see Exercise 10 of Section 7.4). Two Cauchy sequences $\{x_n\}$ and $\{y_n\}$ are equivalent iff the sequence $|x_n - y_n|$ converges to zero. In this approach, $\mathbb{R}$ is the set of equivalence classes of Cauchy sequences. For example, the real number $\sqrt{2}$ is represented by the equivalence class containing the rational sequence 1, 1.4, 1.41, 1.414, 1.4142, 1.41421, 1.414213, 1.4142135, ... and all other Cauchy sequences equivalent to this one. After carefully crafting the definitions of addition, multiplication, and the order properties for equivalence classes of Cauchy sequences, one can show that the resulting system is a complete ordered field. See Charles Chapman Pugh's *Real Mathematical Analysis* (Springer, 2002) for an explanation of how to define addition and multiplication and for why this system forms a complete ordered field.

A different approach to constructing the reals from the rationals sets up two-element partitions of $\mathbb{Q}$, called *Dedekind** cuts as a method for defining irrational numbers. For example, the pair $\{A_1, A_2\}$, where $A_1 = \{x \in \mathbb{Q} : x \le 0 \text{ or } x^2 < 2\}$

---

* Richard Dedekind (1831–1916) was Gauss' last graduate student at the University of Göttingen. A strong supporter of Cantor, he was noted for his work with infinite sets and axiomatic definitions of number systems. He was the first to show that a set is infinite iff it is equivalent to one of its proper subsets and among the first to point out the importance of groups in algebra.

and $A_2 = \{x \in \mathbb{Q}: x > 0 \text{ and } x^2 > 2\}$, is a cut that partitions $\mathbb{Q}$ into two sets: A_1 is all rational numbers less than $\sqrt{2}$ and A_2 is all rational numbers greater than $\sqrt{2}$. The cut $\{A_1, A_2\}$ represents the real number $\sqrt{2}$. Again, one must carefully define addition and multiplication of cuts and the ordering of the set of all cuts. It can be shown that the set of Dedekind cuts forms a complete ordered field. See Walter Rudin's *Principles of Mathematical Analysis*, 3rd ed. (McGraw-Hill, New York, 1976).

With much work we could show that $\mathbb{R}$ is essentially the *only* complete ordered field. In Chapter 6 we discussed the concept of isomorphisms of algebraic structures: one-to-one correspondences that preserve the algebraic structure. We could apply this concept to ordered fields, and the end result would be that every complete ordered field is isomorphic to the field of real numbers.

Equivalence classes of Cauchy sequences and Dedekind cuts give us vastly different mental images of the real numbers, but the study of these approaches is the means to explain why the real numbers possess the powerful properties described in this chapter.

Exercises 7.5

1. Prove Lemma 7.5.1.

2. Prove Lemma 7.5.2.

3. Give an example of
   ☆ (a) a closed subset A of $\mathbb{Q}$ such that $A \subseteq [7, 8]$ and A is not compact.
   (b) a bounded infinite subset of $\mathbb{Q} \cap [7, 8]$ that has no accumulation point in $\mathbb{Q}$.
   (c) a bounded increasing sequence x of rational numbers such that $\{x_n: n \in \mathbb{N}\} \subseteq [7, 8]$ and x has no limit in $\mathbb{Q}$.

4. For the set $\mathbb{R} - \mathbb{Q}$ of irrational numbers, give an example of
   (a) a closed subset A of $\mathbb{R} - \mathbb{Q}$ such that $A \subseteq [3, 4]$ and A is not compact.
   (b) a bounded infinite subset of $\mathbb{R} - \mathbb{Q} \cap [3, 4]$ that has no accumulation point in $\mathbb{R} - \mathbb{Q}$.
   (c) a bounded increasing sequence x of irrational numbers such that $\{x_n: n \in \mathbb{N}\} \subseteq [3, 4]$ and x has no limit in $\mathbb{R} - \mathbb{Q}$.

5. (a) Find $\bigcap_{n \in \mathbb{N}} A_n$ where for all $n \in \mathbb{N}$, A_n is defined as follows:

   (i) $A_n = \left[-\dfrac{1}{n}, \dfrac{1}{n}\right]$ (ii) $A_n = \left[2 - \dfrac{1}{n^2}, 4 + \dfrac{1}{n^2}\right]$

   (iii) $A_n = [n^3, \infty)$ (iv) $A_n = \left(0, \dfrac{1}{n}\right)$

   (b) **(The Nested Interval Theorem)** Show that if $A_n = [a_n, b_n]$ is a sequence of closed intervals such that $A_{n+1} \subseteq A_n$ for all $n \in \mathbb{N}$, then $\bigcap_{n \in \mathbb{N}} A_n \neq \varnothing$.

Proofs to Grade   **6.** Assign a grade of A (correct), C (partially correct), or F (failure) to each. Justify assignments of grades other than A.

☆ **(a)** **Claim.**   If every bounded monotone sequence in the reals is convergent, then the reals are complete.

"Proof."   Suppose the reals are not complete. Then there is a bounded infinite subset A of $\mathbb{R}$ such that A has no supremum in $\mathbb{R}$. Let $x_1 \in A$. Then x_1 is not an upper bound for A, or else x_1 would be the least upper bound $\langle$*since* $x_1 \in A\rangle$. Thus there is an $x_2 \in A$ such that $x_1 < x_2$. Likewise, $x_2 \in A$ and x_2 is not an upper bound, so there exists $x_3 \in A$ with $x_2 < x_3$. Continuing in this fashion, we build an increasing sequence $x_1, x_2, x_3, \ldots$. This sequence is bounded since it is a subset of A. Therefore, $L = \lim\limits_{n \to \infty} x_n$ exists. Since $L > x_n$ for all $n \in \mathbb{N}$, L is the supremum of A. Therefore $\sup(A)$ exists, which is a contradiction. Thus $\mathbb{R}$ is complete. ∎

★ **(b)** **Claim.**   The Bolzano–Weierstrass Theorem implies the completeness of $\mathbb{R}$.

"Proof."   Suppose that every bounded infinite subset of the reals has an accumulation point. Let A be an infinite subset of $\mathbb{R}$ with an upper bound a_0. Then $B = [0, a_0] \cap A$ is a bounded set. If B is finite, then B has a least upper bound, which is a least upper bound for A. If B is infinite, then by the Bolzano–Weierstrass Theorem, B has an accumulation point a_1 which, by construction, is the least upper bound of A. ∎

Answers to Selected Exercises

Exercises 1.1

1. **(b)** false

    (g) true

2. **(c)** not a proposition; the symbol x acts as a variable

    (f) a true proposition

3. **(a)**

P	$\sim P$	$P \wedge \sim P$
T	F	F
F	T	F

    (c)

P	Q	$\sim Q$	$P \wedge \sim Q$
T	T	F	F
F	T	F	F
T	F	T	T
F	F	T	F

    (e)

P	Q	$\sim Q$	$P \wedge Q$	$(P \wedge Q) \vee \sim Q$
T	T	F	T	T
F	T	F	F	F
T	F	T	F	T
F	F	T	F	T

    (i)

P	Q	R	$Q \vee R$	$P \wedge (Q \vee R)$
T	T	T	T	T
F	T	T	T	F
T	F	T	T	T
F	F	T	T	F
T	T	F	T	T
F	T	F	T	F
T	F	F	F	F
F	F	F	F	F

4. (a) false
    (c) true
    (f) false
    (g) true
6. (a) equivalent
    (c) equivalent
    (e) not equivalent
    (g) not equivalent
8. (a) Since P is equivalent to Q, P has the same truth table as Q. Therefore, Q has the same truth table as P, so Q is equivalent to P.
9. (c) tautology

P	Q	$P \wedge Q$	$\sim P \vee \sim Q$	$(P \wedge Q) \vee (\sim P \vee \sim Q)$
T	T	T	F	T
F	T	F	T	T
T	F	F	T	T
F	F	F	T	T

10. (a) contradiction
     (c) tautology
11. (a) x is not positive.
     (c) $5 < 3$
     (e) Roses are not red or violets are not blue.
13. (a) (i)

P	Q	$P \oslash Q$
T	T	F
F	T	T
T	F	T
F	F	F

Exercises 1.2

1. (a) Antecedent: squares have three sides.
         Consequent: triangles have four sides.
    (d) Antecedent: f is differentiable.
         Consequent: f is continuous.
    (f) Antecedent: f is integrable.
         Consequent: f is bounded.
    (i) Antecedent: An athlete qualifies for the Olympic team.
         Consequent: The athlete has a time of 3 minutes, 48 seconds or less.
2. (a) Converse: If triangles have four sides, then squares have three sides.
         Contrapositive: If triangles do not have four sides, then squares do not have three sides.
    (d) Converse: If f is continuous, then f is differentiable.
         Contrapositive: If f is not continuous, then f is not differentiable.
    (f) Converse: If f is bounded, then f is integrable.
         Contrapositive: If f is not bounded, then f is not integrable.

(j) Converse: A time of 3 minutes, 48 seconds or less is sufficient to qualify for the Olympic team.

Contrapositive: If an athlete records a time that is not 3 minutes and 48 seconds or less, then that athlete does not qualify for the Olympic team.

5. **(a)** true

(c) true

(e) true

6. **(a)** true

(c) true

7. **(b)**

P	Q	$\sim P$	$\sim P \Rightarrow Q$	$Q \Leftrightarrow P$	$(\sim P \Rightarrow Q) \vee (Q \Leftrightarrow P)$
T	T	F	T	T	T
F	T	T	T	F	T
T	F	F	T	F	T
F	F	T	F	T	T

(c)

P	Q	$\sim Q$	$Q \Leftrightarrow P$	$(\sim Q) \Rightarrow (Q \Leftrightarrow P)$
T	T	F	T	T
F	T	F	F	T
T	F	T	F	F
F	F	T	T	T

10. **(a)** f has a relative minimum at $x_0 \wedge f$ is differentiable at $x_0 \Rightarrow f'(x_0) = 0$.

(d) $x = 1 \vee x = -1 \Rightarrow |x| = 1$.

(e) x_0 is a critical point for $f \Leftrightarrow f'(x_0) = 0 \vee f'(x_0)$ does not exist.

11. **(b)** There are three nonequivalent ways to translate the sentence, using the symbols D "The Dolphins make the playoffs" and B "The Bears win all the rest of their games."

$$D \Rightarrow B \text{ or } (\sim B) \Rightarrow (\sim D)$$
$$B \Rightarrow D \text{ or } (\sim D) \Rightarrow (\sim B)$$
$$D \Leftrightarrow B \text{ or } (\sim B) \Leftrightarrow (\sim D)$$

The conditional meaning of *unless* (the first translation) is preferred, but the speaker may have intended any of the three.

12. **(b)**

P	Q	R	$P \wedge Q$	$P \wedge Q \Rightarrow R$	$\sim R$	$\sim Q$	$P \wedge \sim R$	$(P \wedge \sim R) \Rightarrow \sim Q$
T	T	T	T	T	F	F	F	T
F	T	T	F	T	F	F	F	T
T	F	T	F	T	F	T	F	T
F	F	T	F	T	F	T	F	T
T	T	F	T	F	T	F	T	F
F	T	F	F	T	T	F	F	T
T	F	F	F	T	T	T	T	T
F	F	F	F	T	T	T	F	T

Since the fifth and ninth columns are the same, the propositions $P \wedge Q \Rightarrow R$ and $(P \wedge \sim R) \Rightarrow \sim Q$ are equivalent.

13. **(a)** If 6 is an even integer, then 7 is an odd integer.

(c) not possible

16. **(a)** tautology

(d) neither

Exercises 1.3

1. (a) $\sim(\forall x)(x$ is precious $\Rightarrow x$ is beautiful).
Or, $(\exists x)(x$ is precious and x is not beautiful).

(b) *Hint:* This exercise is not the same as 1(a).

(h) $(\forall x)(x \in \mathbb{Z} \Rightarrow x > -4 \lor x < 6)$ or $(\forall x \in \mathbb{Z})(x > -4 \lor x < 6)$

(j) $\sim(\exists x)(\forall y)(x \geq y)$ or $(\forall x)(\exists y)(x < y)$

(l) $(\exists x)(x$ is a positive integer and x is smaller than all other positive integers).
Or, $(\exists x)(x$ is a positive integer and $(\forall y)(y$ is a positive integer $\Rightarrow x \leq y))$.

(m) $(\forall x)(\sim(\forall y)(x$ loves $y))$. Or, $\sim(\exists x)(\forall y)(x$ loves $y)$.

2. (a) $(\forall x)(x$ is precious $\Rightarrow x$ is beautiful). All precious stones are beautiful.

(h) $(\exists x \in \mathbb{Z})(x \leq -4 \land x < 6)$ There is an integer that is both less than or equal to -4 and greater than or equal to 6.

(j) $(\exists x)(\forall y)(x \geq y)$ There is an integer that is greater than or equal to every integer.

(l) $(\forall x)(x$ is a positive integer $\Rightarrow (\exists y)(y$ is a positive integer$) \land x > y))$.
For every positive integer there is a smaller positive integer. Or,
$\sim(\exists x)(x$ is a positive integer $\land (\forall y)(y$ is a positive integer $\Rightarrow x \leq y))$.
There is no smallest positive integer.

(m) $(\exists x)(\forall y)(x$ loves $y)$ Someone loves everyone.

5. The first interpretation may be translated as
$$(\forall x)[x \text{ is a person} \Rightarrow (\forall y)(y \text{ is a tax} \Rightarrow x \text{ dislikes } y)].$$

6. (a) $T, U, V,$ and W

7. (b) *Hint:* Every sentence of the form $P(x)$ is equivalent to $\sim\sim P(x)$. Use this fact to rewrite $(\forall x)(\sim A(x))$ and then simplify by using part (a).

8. (b) true
(e) false
(h) true

9. (b) Only one real number is both nonnegative and nonpositive.
(d) There is exactly one real number whose natural logarithm is 1.

10. (a) true
(d) false
(f) false
(i) false

11. (a) *Hint:* Begin by supposing that U is any universe and $A(x)$ is an open sentence.

(b) *Hint:* You must name a specific universe and a specific open sentence such that the converse is false.

(e) $(\forall x)(\sim A(x)) \lor (\exists y)(\exists z)(A(y) \land A(z) \land y \neq z)$

13. (d) This statement is not a denial. It implies the negation of $(\exists ! x)P(x)$, but if $(\forall x) \sim P(x)$, then both the statement and $(\exists ! x)P(x)$ are false.

14. For every backwards E, there exists an upside down A!

Exercises 1.4

1. (a) Suppose $(G, *)$ is a cyclic group.
$$\vdots$$
Thus, $(G, *)$ is abelian.
Therefore, if $(G, *)$ is a cyclic group, then $(G, *)$ is abelian.

4. (a) The crime took place in the library, not the kitchen. By fact (i), if the crime did not take place in the kitchen, then Professor Plum is guilty. Therefore, Professor Plum is guilty.

5. (h) **Proof.** Suppose x is even and y is odd. Then $x = 2k$ for some integer k, and $y = 2j + 1$ for some integer j. Therefore, $x + y = 2k + (2j + 1) = 2(k + j) + 1$, which is odd. (We use the fact that $k + j$ is an integer.)

6. (d) *Hint:* The four cases to consider are: case 1, in which $a \geq 0$ and $b \geq 0$; case 2, in which $a < 0$ and $b < 0$; case 3, in which $a \geq 0$ and $b < 0$; and case 4, in which $a < 0$ and $b \geq 0$. In case 3 it is worthwhile to consider two subcases: In subcase (i), $a + b \geq 0$, so that $|a + b| = a + b$; in subcase (ii), $a + b < 0$, so that $|a + b| = -(a + b)$. Now in subcase (i) we have $|a + b| = a + b < a$ (from $b < 0$) and $a < a + (-b)$ (from $0 < -b$). Thus, $|a + b| < a + (-b) = |a| + |b|$. Subcase (ii) is similar. Case 4 is the same as case 3 except for the names of the variables a and b.

7. (b) **Proof.** Let a be an integer. Assume that a is even. Then $a = 2k$ for some integer k. Therefore, $a + 1 = 2k + 1$, so $a + 1$ is odd.

(d) **Proof.** *Hint*: Let a be an integer. Use the fact that a is either even or odd to give a proof by cases. It is acceptable, but not necessary, to use the definitions of even and odd in proving these cases: previous exercises have laid the foundations we need. For the case when a is even, use Exercise 7(c) and 5(i). For the case when a is odd, we may use Exercise 5 (e), and then use Exercise 5(i) again.

(g) **Proof.** Suppose a and b are positive integers and a divides b. Then for some integer k, $b = ka$. ⟨*We must show that $a \leq b$, which is the same as $a \leq ka$. To show $a \leq ka$, we could multiply both sides of $1 \leq k$ by a, using the fact that a is positive. To do that, we must first show $1 \leq k$.*⟩ Since b and a are positive, k must also be positive. Since k is also an integer, $1 \leq k$. Therefore, $a = a \cdot 1 \leq a \cdot k = b$, so $a \leq b$.

(i) **Proof.** Suppose a and b are positive integers and $ab = 1$. Then a divides 1 and b divides 1. By part (g) $a \leq 1$ and $b \leq 1$. But a and b are positive integers, so $a = 1$ and $b = 1$.

10. (a) **Proof.** Suppose $A > C > B > 0$. Multiplying by the positive numbers C and B we have $AC > C^2 > BC$ and $BC > B^2$, so $AC > B^2$. AC is positive, so $4AC > AC$. Therefore, $4AC > B^2$, so $B^2 - 4AC < 0$. Thus, the graph must be an ellipse.

11. (a) F. This proof, while it appears to have the essence of the correct reasoning, has too many gaps. The first "sentence" is incomplete and the steps are not justified. The steps could be justified either by using the definitions or by referring to previous examples and exercises.

(c)   C. The order in which the steps are written makes it look as if the author of this "proof" assumed that $x + \dfrac{1}{x} \geq 2$. The proof could be fixed by beginning with the (true) statement that $(x - 1)^2 \geq 0$ and ending with the conclusion that $x + \dfrac{1}{x} \geq 2$.

(d)   F. This is not a proof of the statement. It is a proof of the converse of the statement.

Exercises 1.5

1. (a)   Suppose $(G, *)$ is not abelian.

$\vdots$

Thus $(G, *)$ is not a cyclic group.
Therefore, if $(G, *)$ is a cyclic group, then $(G, *)$ is abelian.

(c)   Suppose the set of natural numbers is finite.

$\vdots$

Therefore statement Q.

$\vdots$

Therefore statement $\sim Q$.
This is a contradiction.
Therefore the set of natural numbers is not finite.

(e)   (i)   Suppose that the inverse of the function f from A to B is a function from B to A.

$\vdots$

Therefore f is one-to-one.

$\vdots$

Therefore f is onto B.
Therefore f is one-to-one and onto B.

(ii)   Suppose that f is one-to-one and onto B.

$\vdots$

Therefore, the inverse of the function f from A to B is a function from B to A.

3. (a)   **Proof.** Suppose that the integer $x + 1$ is not odd. Then $x + 1$ is an even integer. Thus, there exists an integer k such that $x + 1 = 2k$. Then $x = 2k - 1 = 2(k - 1) + 1$, so x is not even. We have shown that if $x + 1$ is not odd, then x is not even. Therefore, if x is even, then $x + 1$ is odd.

(e)   *Hint:* The contrapositive statement for "if $x + y$ is even, then either x and y are odd or x and y are even" is "if it is not the case that either x and y are odd or x and y are even then $x + y$ is not even." This is equivalent to "if either x is even and y is odd or x is odd and y is even, then $x + y$ is odd."

4. (b)   **Proof.** Suppose it is not true that $2 < x < 3$. Then either $x \leq 2$ or $x \geq 3$. If $x \leq 2$, then $x - 2 \leq 0$ and $x - 3 \leq 0$ $\langle since\ x - 3 < x - 2 \rangle$.

Because the product of two nonpositive numbers is nonnegative, $(x - 2)(x - 3) = x^2 - 5x + 6 \geq 0$. In the other case, if $x \geq 3$, then $x - 3 \geq 0$ and $x - 2 \geq 0$. Therefore $(x-2)(x-3) = x^2 - 5x + 6 \geq 0$. In either case $x^2 - 5x + 6 \geq 0$. We have shown that if $x \leq 2$ or $x \geq 3$, then $x^2 - 5x + 6 \geq 0$. Therefore, if $x^2 - 5x + 6 < 0$ then $2 < x < 3$.

6. **(b)** **Proof.** Suppose a and b are positive integers. Suppose ab is odd and suppose a and b are not both odd. Then either a is even or b is even. If a is even, then $a = 2k$ for some integer k. Thus $ab = (2k)b = 2(kb)$ is even. Likewise, if b is even, then $b = 2m$ for some integer m and, again, $ab = a(2m) = 2(am)$ is even. Either case leads to a contradiction to the hypothesis that ab is odd. Therefore, if ab is odd then both a and b are odd.

12. **(b)** A.

Exercises 1.6

1. **(a)** Choose $m = -3$ and $n = 1$. Then $2m + 7n = 1$.
   (c) Suppose m and n are integers and $2m + 4n = 7$. Then 2 divides $2m$ and 2 divides $4n$, so 2 divides their sum $2m + 4n$. But 2 does not divide 7, so this is impossible.
   (f) *Hint:* See the statement of part (d). Can you prove that m and n are both negative whenever the antecedent is true?

2. **(b)** **Proof.** Assume a divides $b - 1$ and $c - 1$. ⟨*The proof involves writing $bc - 1$ as a sum of multiples of a, using the fact that if a divides a number, it divides any multiple of that number (Exercise 7(h) of Section* 1.4*) or, more generally, the fact that if a divides two numbers, it divides any sum of multiples of the two numbers (Exercise 2(a)).*⟩ Then a divides the product $(b - 1)(c - 1) = bc - b - c + 1$. By Exercise 2(a), a divides $(bc - b - c + 1) + (b - 1) = bc - c$. Then also by Exercise 2(a), a divides the sum $(bc - c) + (c - 1) = bc - 1$.

4. **(h)** *Hint:* For a counterexample, choose $x = 1$. Explain.
   (i) *Hint:* For a proof, choose $y = x$. For a different proof, choose $y = 1$.

6. **(c)** **Proof.** Let n be a natural number. Then both $2n$ and $2n + 1$ are natural numbers. Let $M = 2n + 1$. Then M is a natural number greater than $2n$.
   (g) **Proof.** Let $\varepsilon > 0$ be a real number. Then $\frac{1}{\varepsilon}$ is a positive real number and so has a decimal expression as an integer part plus a decimal part. Let M be the integer part of $\frac{1}{\varepsilon}$, plus 1. Then M is an integer and $M > \frac{1}{\varepsilon}$. To prove for all natural numbers $n > M$ that $\frac{1}{n} < \varepsilon$, let n be a natural number and $n > M$. Since $M > \frac{1}{\varepsilon}$, we have $n > \frac{1}{\varepsilon}$. Thus $\frac{1}{n} < \varepsilon$. Therefore, for every real number $\varepsilon > 0$, there is a natural number M such that for all natural numbers $n > M$, $\frac{1}{n} < \varepsilon$.
   (h) *Hint:* Because m is positive, the statement $\frac{1}{n} - \frac{1}{m} < \varepsilon$ follows from $\frac{1}{n} < \varepsilon$.

7. **(a)** F. The false statement referred to is not the opposite (denial) of the claim.

(b) C. The "proof" shows that there is a polynomial with the required properties, but must also show that there is no other polynomial with these properties.

(d) A.

(i) *Hint*: The grade should be C. What error must be corrected?

Exercises 1.7

1. **(a)** **Proof.** ⟨*We work both forwards and backwards: From the hypothesis that $3n + 1$ is odd we can deduce that $3n$ is even, from which we can deduce that n is even. We could reach the conclusion that $2n + 8$ is divisible by 4 if we knew that 4 divides $2n$ (since 8 is divisible by 4). In turn, the statement $2n$ is divisible by 4 may be derived from the statement that n is divisible by 2. We combine these steps in the proper order to create the proof.*⟩

 Suppose n is an integer and $3n + 1$ is odd. Therefore $3n$ is even, which implies that n is even. ⟨*We are now using properties of even and odd integers that we proved earlier, without referencing specific examples or exercises.*⟩ Since n is even, n is divisible by 2. Therefore $2n$ is divisible by 4. Finally since 8 is also divisible by 4, $2n + 8$ is divisible by 4. ∎

(b) **Proof.** Let a be a real number, $a \neq 3$. ⟨*The key to the proof is to use the idea of "solution" and then work with the resulting equation.*⟩

 Assume that a is a solution to $x^2 - x - 6 = 0$.

 Then a makes the equation true ⟨*by the definition of a solution to an equation*⟩.

 Thus $a^2 - a - 6 = (a - 3)(a + 2) = 0$.

 Then $a + 2 = 0$, because $a - 3 \neq 0$.

 Then $(a^2 + 1)(a + 2) = a^3 + 2a^2 + a + 3 = 0$.

 Therefore a is a solution to $x^3 + 2x^2 + x + 3 = 0$. ∎

(c) **Proof.** Assume that $a \neq 3$. ⟨*Observe that in the proof above, each step implies its predecessor. Thus we can modify the given proof to create an iff proof.*

 a is a solution to $x^2 - x - 6 = 0$

 iff $a^2 - a - 6 = (a - 3)(a + 2) = 0$.

 iff $a + 2 = 0$.   ⟨*Because $a - 3 \neq 0$.*⟩

 iff $(a^2 + 1)(a + 2) = a^3 + 2a^2 + a + 3 = 0$.   ⟨*Because $a^2 + 1 \neq 0$.*⟩

 iff a is a solution to $x^3 + 2x^2 + x + 3 = 0$. ∎

(d) **Proof.** Suppose $x^2 = 2x + 15$ and $x > 2$. Then $(x - 5)(x + 3) = 0$. Since $x > 2$, x must be 5. Then $x - 4$ and $x - 3$ are positive, so $(x - 4)/(x - 3) > 0$. ∎

(e) **Proof.** Let x and y be real numbers. ⟨*The statement has the form $P \Rightarrow (Q \vee R)$, so it might be proved by assuming P and $\sim Q$ and deducing R. In this case a proof by contrapositive works well.*⟩ Assume that neither x nor y is irrational. Then both x and y are rational, so they can be written in the form $x = \frac{p}{q}$ and $y = \frac{r}{s}$, where p, q, r, and s are integers,

$q \neq 0$, and $s \neq 0$. Therefore, $x + y = \dfrac{p}{q} + \dfrac{r}{s} = \dfrac{ps + rq}{qs}$. Since $ps + rq$ and qs are integers and $qs \neq 0$, $x + y$ is a rational number. We have shown that if x and y are rational, then $x + y$ is rational. We conclude that if $x + y$ is irrational, then either x or y is irrational. ∎

(f)   **Proof.** ⟨*If we let S be the set of all nonvertical lines in the xy-plane, we can simplify the symbolic form of the theorem as follows:*

$(\forall L_1 \in S)(\forall L_2 \in S)(L_1$ *and* L_2 *are perpendicular* $\Rightarrow$
(slope of $L_1) \cdot$ *(slope of* $L_2) = -1).$⟩

Let L_1 and L_2 be nonvertical lines. Suppose L_1 and L_2 are perpendicular. ⟨*We now use the fact that the slope of a nonvertical line is* tan(α), *where* α *is the angle of inclination of the line.*⟩ Let α_1 and α_2 be the angles of inclinations of L_1 and L_2, respectively. See the figure. We may assume that $\alpha_1 > \alpha_2$. ⟨*We can make this assumption because the two lines are arbitrary; if* $\alpha_1 < \alpha_2$ *simply interchange the labels of the lines.*⟩ Since L_1 and L_2 are perpendicular, $\alpha_1 = \alpha_2 + \frac{\pi}{2}$. Therefore,

$$\tan(\alpha_1) = \tan(\alpha_2 + \tfrac{\pi}{2}) = -\cot(\alpha_2) = -\dfrac{1}{\tan(\alpha_2)}.$$

⟨*We use trigonometric identities to rewrite* $\tan(\alpha_1)$.⟩ Thus, $\tan(\alpha_1) \cdot \tan(\alpha_2) = -1$. Since $\tan(\alpha_1)$ is the slope of L_1 and $\tan(\alpha_2)$ is the slope of L_2, the product of the slopes is -1. ∎

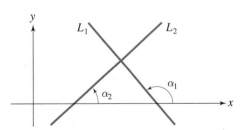

(g)   **Proof.** ⟨*This is a "non-existence" proof. We could restate the result as "Every point inside the circle is not on the line" and begin a direct proof by assuming that (x, y) is a point inside the circle. We would then have to prove that (x, y) is not on the line. In this instance, a better approach is to use a proof by contradiction. The statement has the form*

$\sim(\exists x)(\exists y)((x,y)$ *is inside the circle* $\wedge$ (x, y) *is on the line*).⟩

Suppose there is a point (a, b) that is inside the circle and on the line. Then $(a - 3)^2 + b^2 < 6$ and $b = a + 1$. ⟨*We now have two expressions to use.*⟩ Therefore,

$$(a - 3)^2 + (a + 1)^2 < 6$$
$$2a^2 - 4a + 10 < 6$$
$$a^2 - 2a + 5 < 3$$
$$a^2 - 2a + 1 < -1$$
$$(a - 1)^2 < -1.$$

This is a contradiction since $(a - 1)^2 \geq 0$. Thus, no point inside the circle is on the line. ∎

(h)   **Proof.** ⟨*Proofs that verify equalities or inequalities containing absolute value expressions usually involve cases, because of the two-part definition of* $|x|$. *The two cases are* $x - 2 \geq 0$ *and* $x - 2 < 0$. *The proof in each case is discovered by working backwards from the desired conclusion. The key steps are to note that, in the first case, if* $x \geq 2$, *then* $-6 \leq x$, *and, in the second case, that if* $x \geq 1$, *then* $\frac{6}{7} \leq x$.⟩

Let x be a real number greater than 1.

Case 1. Suppose $x - 2 \geq 0$. Then $|x - 2| = x - 2$. Since $x \geq 2$,

$$-6 \leq x$$
$$3x - 6 \leq 4x$$
$$\frac{3(x - 2)}{x} \leq 4. \quad \langle \textit{Remember that } x \textit{ is positive.}\rangle$$

Therefore, $\dfrac{3|x - 2|}{x} \leq 4$.

Case 2. Suppose $x - 2 < 0$. Then $|x - 2| = -(x - 2)$. By hypothesis, $x \geq 1$. Therefore,

$$\frac{6}{7} \leq x$$
$$6 \leq 7x$$
$$6 - 3x \leq 4x$$
$$3[-(x - 2)] \leq 4x$$

$$\frac{3[-(x - 2)]}{x} \leq 4. \quad \langle \textit{Remember that } x \textit{ is positive.}\rangle$$

Therefore, $\dfrac{3|x - 2|}{x} \leq 4$.   ∎

2. (e) *Hint:* Write $n^3 - n$ as the product of 3 consecutive integers.

4. (b) *Hint:* Is it possible that for all irrational numbers x and y, $x + y$ is irrational? Or could $x + y = 0$?

8. (a) **Proof.** Suppose (x, y) is inside the circle. Then from the distance formula, $(x - 3)^2 + (y - 2)^2 < 4$. Therefore, $|x - 3|^2 < 4$ and $|y - 2|^2 < 4$. It follows that $|x - 3| < 2$ and $|y - 2| < 2$, so $-2 < x - 3 < 2$ and $-2 < y - 2 < 2$. Thus $1 < x < 5$ and $0 < y < 4$. Therefore, $x^2 < 25$ and $y^2 < 16$, so $x^2 + y^2 < 41$.

9. (d) $-36 = (-8)5 + 4$. The quotient is -8 and remainder is 4.

14. (a) *Hint:* Use Theorem 1.7.3 and then Theorem 1.7.1.

(c) *Hint:* Assume that $d = \gcd(a, b) = 1$ and a divides bc. Write 1 as a linear combination of a and b, and then multiply by c.

16. (a) **Proof.** Suppose p is prime and a is any natural number. The only divisors of p are 1 and p, and $\gcd(p, a)$ divides p, so $\gcd(p, a) = 1$ or p. (i) Assume $\gcd(p, a) = p$. Then p divides a by definition of gcd. (ii) Suppose p divides a. Then p is a common divisor of p and a. Since p is the largest divisor of p it is the largest common divisor of p and a, so $\gcd(p, a) = p$.

20. 42

21. (a) *Hint:* Use a two-part proof. For the part of the proof that assumes a divides b, show both conditions (i) and (ii) for the lcm are satisfied by b.

(c) *Hint:* Assume that gcd $(a, b) = 1$. Since b divides $m = \text{lcm}(a, b)$, $m = kb$ for some integer k. Use part (b) to show that $k \leq a$. Then use part (c) of Exercise 14 to show that a divides k. Conclude that $a = k$, so $m = kb = ab$.

(f) *Hint:* By Exercise 14(d), $\gcd\left(\frac{a}{d}, \frac{b}{d}\right) = 1$. Use part (c) to find an expression for $\text{lcm}\left(\frac{a}{d}, \frac{b}{d}\right)$. Then use part (e) to obtain another expression for $\text{lcm}\left(\frac{a}{d}, \frac{b}{d}\right)$. Equate the two expressions and simplify.

23. (b) A.

Exercises 2.1

1. (a) $\{x \in \mathbb{N} : x < 6\}$ or $\{x : x \in \mathbb{N} \text{ and } x < 6\}$.

  (c) $\{x \in \mathbb{R} : 2 \leq x \leq 6\}$ or $\{x : x \in \mathbb{R} \text{ and } 2 \leq x \leq 6\}$.

3. (a) Suppose that X is a set. If $X \in X$ then X is not an ordinary set, so $X \notin X$. On the other hand, if $X \notin X$, then X is an ordinary set, so $X \in X$. Both $X \in X$ and $X \notin X$ lead to a contradiction. We conclude that the collection of ordinary sets is not a set.

4. (a) true

  (c) true

  (e) false

  (g) true

  (i) false

5. (a) true

  (c) true

  (e) false

  (g) false

  (i) false

  (k) true

6. (a) $A = \{1, 2\}, B = \{1, 2, 4\}, C = \{1, 2, 5\}$

  (c) $A = \{1, 2, 3\}, B = \{1, 4\}, C = \{1, 2, 3, 5\}$

8. *Hint:* To prove that if $A \subseteq B$ and $B \subseteq C$, then $A \subseteq C$, begin by assuming that $A \subseteq B$ and $B \subseteq C$. To show that $A \subseteq C$, we recall that $A \subseteq C$ means $(\forall x)(x \in A \Rightarrow x \in C)$. The first steps are: Let x be any object. Suppose $x \in A$. Now use the fact that $A \subseteq B$ and $B \subseteq C$.

9. *Hint:* To prove $A = B$, use the hypothesis $A \subseteq B$ and show $B \subseteq A$ by using Theorem 2.1.1(c).

14. (a) $\{\{0\}, \{\triangle\}, \{\square\}, \{0, \triangle\}, \{0, \square\}, \{\triangle, \square\}, X, \varnothing\}$

  (c) $\{\{\varnothing\}, \{\{a\}\}, \{\{b\}\}, \{\{a, b\}\}, \{\varnothing, \{a\}\}, \{\varnothing, \{b\}\}, \{\varnothing, \{a, b\}\}, \{\{a\}, \{b\}\},$
    $\{\{a\}, \{a, b\}\}, \{\{b\}, \{a, b\}\}, \{\varnothing, \{a\}, \{b\}\}, \{\varnothing, \{a\}, \{a, b\}\},$
    $\{\varnothing, \{b\}, \{a, b\}\}, \{\{a\}, \{b\}, \{a, b\}\}, X, \varnothing\}$.

15. (a) false

  (e) false

  (g) true

16. **(a)** no proper subsets
    (b) $\varnothing, \{1\}, \{2\}$
17. **(a)** true
    (c) true
    (e) true
    (g) true
    (i) true
    (k) true
19. **(c)** C. The "proof" asserts that $x \in C$, but fails to justify this assertion with a definite statement that $x \in A$ or that $x \in B$. This problem could be corrected by inserting a second sentence "Suppose $x \in A$" and a fourth sentence "Then $x \in B$."
    (e) F. The error repeatedly committed in this proof is to say $A \subseteq B$ means $x \in A$ and $x \in B$. The correct meaning of $A \subseteq B$ is that for every x, if $x \in A$, then $x \in B$.
    (h) C. The proof could be considered correct, but it lacks a statement of the hypothesis, helpful explanations and connecting words. How much explanation you include depends on the presumed level of the reader's knowledge. We prefer the use of words, not just symbols.
    (i) F. The claim is false. (For example, let $A = \{1, 2\}$, $B = \{1, 2, 4\}$, $C = \{1, 2, 5, 6, 7\}$.) The statement "Since $x \in B$, $x \in A \ldots$," would be correct if we knew that $B \subseteq A$.

Exercises 2.2

1. **(a)** $\{0, 1, 2, 3, 4, 5, 6, 7, 8, 9\}$
   (c) $\{1, 3, 5, 7, 9\}$
   (e) $\{3, 9\}$
   (g) $\{1, 5, 7\}$
   (i) $\{1, 5, 7\}$
2. **(b)** $(1, 8)$
   (d) $[2, 4)$
   (h) $(-\infty, 3) \cup [8, \infty)$
3. **(a)** $\{0, -2, -4, -6, -8, -10, \ldots\}$
   (d) $\mathbb{Z}^- \cup \{0\}$
   (g) $\{0, 2, 4, 6, 8, \ldots\}$
4. A and B are disjoint.
6. **(a)** A Venn diagram is helpful.

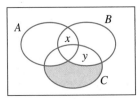

Since $C \subseteq A \cup B$, every element of C is either in A or B. In the diagram, the shaded area must be empty. Since $A \cap B$ is not a subset of C, there is some element x that is in $A \cap B$, but not in C. To ensure that C is non-empty, we must place an element y in any one of the three available regions of C. Our solution is $A = \{x\}$, $B = \{x, y\}$, $C = \{y\}$. For other correct examples, we could place other elements anywhere in the diagram (except in the shaded region).

9. (a) *Hint:* To prove that $A \subseteq B$ implies $A - B = \varnothing$, assume that $A \subseteq B$ and show that $x \in A - B$ is false for every object x. To prove the converse, assume $A - B = \varnothing$ and there is some object x in A.

   (c) **Proof.**
   (i) Assume $C \subseteq A \cap B$. Suppose $x \in C$. Then $x \in A \cap B$, so $x \in A$ and $x \in B$. Since $x \in C$ implies $x \in A$, $C \subseteq A$. Similarly, $C \subseteq B$.
   (ii) Assume $C \subseteq A$ and $C \subseteq B$. Suppose $x \in C$. Then from $C \subseteq A$ we have $x \in A$ and from $C \subseteq B$ we have $x \in B$. Therefore, $x \in A \cap B$. We conclude that $C \subseteq A \cap B$.

10. (c) **Proof.** Suppose $C \subseteq A$ and $D \subseteq B$. Assume that C and D are not disjoint. Then there is an object $x \in C \cap D$. But then $x \in C$ and $x \in D$. Since $C \subseteq A$ and $D \subseteq B$, $x \in A$ and $x \in B$. Therefore, $x \in A \cap B$, so A and B are not disjoint.

11. (a) $A = \{1, 2\}, B = \{1, 3\}, C = \{2, 3, 4\}$
    (c) $A = \{1, 2\}, B = \{1, 3\}, C = \{2, 3\}$
    (e) $A = \{1, 2\}, B = \{1, 3\}, C = \{1\}$

12. (a) **Proof.** $S \in \mathscr{P}(A \cap B)$
    iff $S \subseteq A \cap B$
    iff $\langle$by Exercise 9(c)$\rangle$ $S \subseteq A$ and $S \subseteq B$
    iff $S \in \mathscr{P}(A)$ and $S \in \mathscr{P}(B)$
    iff $S \in \mathscr{P}(A) \cap \mathscr{P}(B)$.

    (d) **Proof.** Let A and B be any sets. Since $\varnothing$ is a subset of every set we have $\varnothing \in \mathscr{P}(A - B)$. Also $\varnothing \in \mathscr{P}(A)$ and $\varnothing \in \mathscr{P}(B)$. Therefore $\varnothing \notin \mathscr{P}(A) - \mathscr{P}(B)$. This shows that $\mathscr{P}(A - B) \not\subseteq \mathscr{P}(A) - \mathscr{P}(B)$.

13. (b) $A \times B = \{(1, q), (1, \{t\}), (1, \pi), (2, q), (2, \{t\}), (2, \pi),$
    $(\{1, 2\}, q), (\{1, 2\}, \{t\}), (\{1, 2\}, \pi)\}$

    $B \times A = \{(q, 1), (\{t\}, 1), (\pi, 1), (q, 2), (\{t\}, 2), (\pi, 2),$
    $(q, \{1, 2\}), (\{t\}, \{1, 2\}), (\pi, \{1, 2\})\}$

15. (a) **Proof.** $(a, b) \in A \times (B \cap C)$
    iff $a \in A$ and $b \in B \cap C$
    iff $a \in A$ and $b \in B$ and $b \in C$
    iff $a \in A$ and $b \in B$ and $a \in A$ and $b \in C$
    iff $(a, b) \in A \times B$ and $(a, b) \in A \times C$
    iff $(a, b) \in (A \times B) \cap (A \times C)$.

19. (a) F. One serious error is the assertion that $x \notin C$, which has no justification. The author of this "proof" was misled by supposing $x \in A$, which is an acceptable step but not useful in proving $A - C \subseteq B - C$. After assuming that $A \subseteq B$, the natural first step for proving that $A - C \subseteq B - C$ is to suppose that $x \in A - C$.

(b) *Hint:* The second sentence of this proof says "Suppose $A - C$," which doesn't make good sense when you think about it. (What should we suppose about the set $A - C$?) Consider that the author of this proof may have meant to say "Suppose $x \in A - C$."

(d) C. The proof that $A \cap B = A$ is incomplete.

(e) F. The claim is false. The statement "$x \in A$ and $x \in \emptyset$ iff $x \in A$" is false.

(i) F. Although a picture may help by suggesting ideas around which a correct proof can be made, a picture alone is rarely sufficient for a proof. Thus a proof that consists only of Venn diagrams will usually have a grade of F. This "proof" is made better because of the explanation that is included, but the only way to give a complete proof is to show that $A \cup B \subseteq B$ and $B \subseteq A \cup B$.

Exercises 2.3

1. (a) $\displaystyle\bigcup_{A \in \mathscr{A}} A = \{1, 2, 3, 4, 5, 6, 7, 8\}$; $\displaystyle\bigcap_{A \in \mathscr{A}} A = \{4, 5\}$

(c) $\displaystyle\bigcup_{n \in \mathbb{N}} A_n = \{5, 6, 10, 11, 12, 15, 16, 17, 18\} \cup \{n \in \mathbb{N}: n \geq 20\}$; $\displaystyle\bigcap_{n \in \mathbb{N}} A_n = \emptyset$

(e) $\displaystyle\bigcup_{A \in \mathscr{A}} A = \mathbb{Z}$; $\displaystyle\bigcap_{A \in \mathscr{A}} A = \{10\}$

(g) $\displaystyle\bigcup_{n \in \mathbb{N}} A_n = (0, 1)$; $\displaystyle\bigcap_{n \in \mathbb{N}} A_n = \emptyset$

(i) $\displaystyle\bigcup_{r \in \mathbb{R}} A_r = [0, \infty)$; $\displaystyle\bigcap_{r \in \mathbb{R}} A_r = \emptyset$

(p) The union is the triangular region bounded by $y = 0$, $x = 1$, $y = x$. The intersection consists of the sides of the triangle that lie on the axes.

2. The family in Exercise 1(a) is not pairwise disjoint. The family in 1(b) is pairwise disjoint.

4. (a) *Hint:* For what real numbers is $(\forall A)(A \in \mathscr{A} \Rightarrow x \in A)$ true?

(b) *Hint:* For what real numbers is $(\exists A)(A \in \mathscr{A} \wedge x \in A)$ true?

5. (a) Let $\beta \in \Delta$. Suppose $x \in \displaystyle\bigcap_{\alpha \in \Delta} A_\alpha$. Then $x \in A_\alpha$ for each $\alpha \in \Delta$. Since $\beta \in \Delta$, $x \in A_\beta$. Therefore $\displaystyle\bigcap_{\alpha \in \Delta} A \subseteq A_\beta$.

6. (a) $x \in B \cap \displaystyle\bigcup_{\alpha \in \Delta} A_\alpha$ iff $x \in B$ and $x \in \displaystyle\bigcup_{\alpha \in \Delta} A_\alpha$

iff $x \in B$ and $x \in A_\alpha$ for some $\alpha \in \Delta$

iff $x \in B \cap A_\alpha$ for some $\alpha \in \Delta$

iff $x \in \displaystyle\bigcup_{\alpha \in \Delta} (B \cap A_\alpha)$.

7. **(a)** $\left(\bigcup_{\alpha \in \Delta} A_\alpha\right) \cap \left(\bigcup_{\beta \in \Gamma} B_\beta\right) = \bigcup_{\beta \in \Gamma}\left(\left(\bigcup_{\alpha \in \Delta} A_\alpha\right) \cap B_\beta\right)$

$$= \bigcup_{\beta \in \Gamma}\left(\bigcup_{\alpha \in \Delta}(A_\alpha \cap B_\beta)\right).$$

8. **(c)** *Hint:* The statement is correct.

9. **(a)** Suppose $x \in \bigcup_{\alpha \in \Gamma} A_\alpha$. Then $x \in A_\alpha$ for some $\alpha \in \Gamma$. Since $\Gamma \subseteq \Delta$, $\alpha \in \Delta$. Thus $x \in A_\alpha$ for some $\alpha \in \Delta$. Therefore, $x \in \bigcup_{\alpha \in \Delta} A_\alpha$.

10. **(a)** **Proof.** Let $x \in B$. For each $A \in \mathcal{A}$, $B \subseteq A$. Thus for each $A \in \mathcal{A}$, $x \in A$. Therefore $x \in \bigcap_{A \in \mathcal{A}} A$.

(b) $X = \bigcap_{A \in \mathcal{A}} A$.

15. **(c)** **Proof.** Let $x \in \bigcup_{i=k}^{m} A_i$. Then there exists $j \in \mathbb{N}$ such that $k \leq j \leq m$ and $x \in A_j$. Since $j \in \mathbb{N}$ and $x \in A_j$, $x \in \bigcup_{i=1}^{\infty} A_i$.

Alternate Proof: The set $\Gamma = \{k, k+1, k+2, \ldots, m\}$ is a subset of $\Delta = \{1, 2, 3, \ldots\}$. Therefore, by Exercise 9(a), $\bigcup_{i=k}^{m} A_i \subseteq \bigcup_{i=1}^{\infty} A_i$.

17. **(a)** Let $A_k = \left[-\frac{1}{k}, 1 + \frac{1}{k}\right)$ for each $k \in \mathbb{N}$.

18. **(a)** C. This proof omits an explanation of why there is some β in Δ such that $A_\beta \in \{A_\alpha : \alpha \in \Delta\}$. The explanation is that by definition of an indexed family, $\Delta \neq \varnothing$. If we allowed $\Delta = \varnothing$, the claim would be false.

(b) C. No connection is made between the first and second sentences. The connection that needs to be made is that if $x \in \bigcup_{\alpha \in \Delta} A_\alpha$ then $x \in A_\alpha$ for some $\alpha \in \Delta$, so $x \in B$ because $A_\alpha \subseteq B$.

(e) F. The claim is false. $\bigcup_{n=1}^{\infty} [n, n+1) = [1, \infty)$.

Exercises 2.4

1. **(b)** not inductive

(d) not inductive

(f) not inductive

2. **(b)** true

(e) false

4. **(e)** $(n+2)(n+1)$

5. **(c)** $A = \{n : n = 2^k \text{ for some } k \in \mathbb{N}\}$ may be defined as

(i) $2 \in A$.

(ii) if $x \in A$, then $2x \in A$.

6. (a) Proof.

    (i) The statement is true for $n = 1$ because $\sum_{i=1}^{1}(3i - 2) = 1$ and $\frac{1}{2}(3(1) - 1) = 1$.

    (ii) Assume that for some $n \in \mathbb{N}$, $\sum_{i=1}^{n}(3i - 2) = \frac{n}{2}(3n - 1)$. We must show $\sum_{i=1}^{n+1}(3i - 2) = \frac{n+1}{2}(3(n + 1) - 1)$. The equation's left-hand side is $\sum_{i=1}^{n}(3i - 2) + [3(n + 1) - 2] = \frac{n}{2}(3n - 1) + (3n + 1) = \frac{3}{2}n^2 + \frac{5}{2}n + 1$ and the equation's right-hand side is $\frac{1}{2}(n + 1)(3n + 2)$, which also simplifies to $\frac{3}{2}n^2 + \frac{5}{2}n + 1$. Thus the statement is true for $n + 1$.

    (iii) By the PMI, the statement is true for every $n \in \mathbb{N}$.

7. (b) Proof.

    (i) For $n = 1$, $4^1 - 1 = 3$, which is divisible by 3.

    (ii) Suppose for some $k \in \mathbb{N}$ that $4^k - 1$ is divisible by 3. Then
$$4^{k+1} - 1 = 4(4^k) - 1$$
$$= 4(4^k - 1) - 1 + 4$$
$$= 4(4^k - 1) + 3.$$
Both 3 and $4^k - 1$ are divisible by 3, so $4^{k+1} - 1$ is divisible by 3.

    (iii) By the PMI, $4^n - 1$ is divisible by 3 for all $n \in \mathbb{N}$.

(i) Proof.

    (i) $3^{3+1} = 3^4 = 81 > 64 = (1 + 3)^3$, so the statement is true for $n = 1$.

    (ii) Assume that $3^{n+3} > (n + 3)^3$ for some $n \in \mathbb{N}$. Then
$$3^{(n+1)+3} = 3^{n+4} = 3 \cdot 3^{n+3}$$
$$> 3(n + 3)^3 = 3(n^3 + 9n^2 + 27n + 27)$$
$$= 3n^3 + 27n^2 + 81n + 81$$
$$> n^3 + 12n^2 + 48n + 64 = (n + 4)^3 = ((n + 1) + 3)^3.$$
That is, $3^{(n+1)+3} > ((n + 1) + 3)^3$.

    (iii) By the PMI, $3^{n+3} > (n + 3)^3$ for every $n \in \mathbb{N}$.

8. (a) Proof.

    (i) $6^3 = 216 < 720 = 6!$, so the statement is true for $n = 6$.

    (ii) Assume $n^3 < n!$ for some $n \geq 6$. Then
$$(n + 1)^3 = n^3 + 3n^2 + 3n + 1$$
$$< n^3 + 3n^2 + 3n + n = n^3 + 3n^2 + 4n$$
$$< n^3 + 3n^2 + n^2 = n^3 + 4n^2$$
$$< n^3 + n^3 = 2n^3$$
$$< (n + 1)n^3$$
$$< (n + 1)n! = (n + 1)!.$$

(iii) By the PMI, $n^3 < n!$ for all $n \geq 6$.

(c) *Hint:* $(n + 2)[(n + 1)!] > (n + 2)2^{n+3} > 2(2^{n+3})$.

(g) *Hint:* For a convex polygon, any line segment drawn from an edge point to another edge point lies inside the polygon. For the inductive step, when you consider a polygon of $n + 1$ sides, draw a line as shown between two vertices. (Only part of the polygon is shown.)

The new line segment separates the upper left triangle from a convex polygon that has exactly n sides, so we can apply the hypothesis of induction to the n-sided polygon. Then use the result to compute the sum of the interior angles for the polygon of $n + 1$ sides.

10. *Hint:* Use $n = 3$ for the basis step. For the inductive step, assume the statement is correct for any collection of n points with no three points collinear. Consider a collection of $n + 1$ points, but apply the hypothesis of induction to only n of those points. Then calculate the total number of line segments determined by all $n + 1$ points.

11. *Hint:* For the induction step, visualize the starting position with $n + 1$ disks. When you think about the moves you would make to transfer all $n + 1$ disks from one peg to another, try to break down the task into three separate tasks, so you can use the assumption about how many moves are required to move n disks. The first task is to move the top n disks from the stack to another peg.

13. (a) F. The claim is obviously false, but this example of incorrect reasoning is well known because it's fun and the flaw is not easy to spot.

Let $S = \{n \in \mathbb{N}:$ all horses in every set of n horses have the same color$\}$. It is true that $1 \in S$. It is also true, **for $n \geq 2$**, that if $n \in S$, then $n + 1 \in S$. The "proof" fails in the case when $n = 1$ because $1 \in S$ but $2 \notin S$. In this case, when either horse is removed from the set, the remaining horse has the same color (as itself), [because there is only one horse left] but the two horses may have different colors. We conclude that the set S is not inductive, and in fact $S = \{1\}$.

(b) F. The basis step and the assumption that the statement is true for some n are correct. Perhaps the author hopes that just saying the statement is true for $n + 1$ is good enough. For a correct proof, one must use the statement about n to prove the statement about $n + 1$.

(e) F. The factorization of $xy + 1$ is wrong, and there is no reason to believe $x + 1$ or $y + 1$ is prime.

Exercises 2.5

1. **(a)** *Hint*: Let $S = \{n \in \mathbb{N}: n > 22 \text{ and } n = 3s + 4t \text{ for some integers } s \geq 3 \text{ and } t \geq 2\}$. Show that 23, 24, and 25 are in S. Then let $m > 22$ be a natural number and assume that for all $k \in \{23, 24, \ldots m - 1\}$, $k \in S$. To show that $m \in S$, proceed as follows: If $m = 23$, 24, or 25 we already know m is in S. Otherwise $m \geq 26$, so $m - 3 \geq 23$. By the hypothesis of induction, $m - 3 \in S$, so $m - 3 = 3s + 4t$ for some integers s and t, where $s \geq 3$ and $t \geq 2$. Then $m = 3(s + 1) + 4t$.

3. **(c)** *Hint:* Let $A = \{b \in \mathbb{N}: \text{there exists } a \in \mathbb{N} \text{ such that } a^2 = 2b^2\}$, assume A is nonempty, and use the WOP to reach a contradiction.

5. **(a)** *Hint:* The induction hypothesis is "Suppose f_{3k} is even and both f_{3k+1} and f_{3k+2} are odd for some natural number k." From this use the definition of Fibonacci numbers to show that $f_{3(k+1)}$ is even and both $f_{3(k+1)+1}$ and $f_{3(k+1)+2}$ are odd.

   (d) **Proof.**
   (i) In the case of $n = 1$, the formula is $f_1 = f_{1+2} - 1$, which is $1 = 2 - 1$. Thus the statement is true for $n = 1$.
   (ii) Suppose for some k that $f_1 + f_2 + f_3 + \cdots + f_k = f_{k+2} - 1$. Then

   $$\begin{aligned} f_1 + f_2 + f_3 + \cdots + f_k + f_{k+1} &= (f_1 + f_2 + f_3 + \cdots + f_k) + f_{k+1} \\ &= (f_{k+2} - 1) + f_{k+1} \\ &= (f_{k+2} + f_{k+1}) - 1 \\ &= f_{k+3} - 1. \end{aligned}$$

   Therefore the statement is true for $n + 1$.
   (iii) Thus, by the PMI, $f_1 + f_2 + f_3 + \cdots + f_n = f_{n+2} - 1$ for all natural numbers n.

6. **(d)** *Hint*: Consider the cases $n = 1$ and $n = 2$ separately. For $n > 2$, you will find it useful to multiply the equation $\alpha^2 = \alpha + 1$ by α^{n-2} and $\beta^2 = \beta + 1$ by β^{n-2}.

7. *Hint*: Modify the proof given for the case $a > 0$. Slightly different arguments are needed to show (i) that S is nonempty and (ii) that $r < |a| = -a$.

9. *Hint*: Suppose n is the smallest positive integer greater than 1 that is not prime and such that n can be expressed in two different ways as a product of primes (where we ignore the order in which the prime factors appear). Then n may be written as: $n = p_1 p_2 p_3 \ldots p_n$ and $n = q_1 q_2 q_3 \ldots q_m$. Apply Euclid's Lemma to show that $p_1 = q_j$ for some $1 \leq j \leq m$. Then find a contradiction.

12. **(b)** **Proof.** Let S be a subset of $\mathbb{N}$ such that $1 \in S$ and S is inductive. We wish to show that $S = \mathbb{N}$. Assume that $S \neq \mathbb{N}$ and let $T = \mathbb{N} - S$. By the WOP, the nonempty set T has a least element. This least element is not 1, because $1 \in S$. If the least element is n, then $n \in T$ and $n - 1 \in S$. But by the inductive property of S, $n - 1 \in S$ implies that $n \in S$. This is a contradiction. Therefore, $S = \mathbb{N}$.

13. **(b)** F. The claim is false. The flaw in the "proof" is the incorrect assumption that $m - 1$ is a natural number. In fact, 1 *is* the smallest natural number n

such that 3 does not divide $n^3 + 2n + 1$. There is no contradiction about a smaller natural number because there is no smaller natural number.

Exercises 2.6

2. **(b)** 16
3. *Hint:* Since $1{,}000{,}000 = (10^3)^2 = (10^2)^3 = 10^6$, there are 10^3 squares less than or equal to $1{,}000{,}000$; 10^2 cubes less than or equal to $1{,}000{,}000$; and 10 natural numbers that are both squares and cubes (sixth powers) less than or equal to $1{,}000{,}000$.
5. *Hint:* Complete this formula:
$$\overline{A \cup B \cup C \cup D} =$$
$$\overline{\overline{A}} + \cdots - \overline{\overline{A \cap B}} - \cdots + \overline{\overline{A \cap B \cap C}} + \cdots - \overline{\overline{A \cap B \cap C \cap D}}.$$
10. *Hint:* The answer is not $20 \cdot 19 \cdot 19 \cdot 18$; it is $130{,}340$. Consider the cases where the bottom right is colored the same as or differently from the upper left corner, and give two products that yield the correct sum.
21. **(a)** *Hint:* The algebra in the inductive step is:

$$\sum_{r=0}^{n+1} \binom{n+1}{r} a^r b^{n+1-r}$$

$$= b^{n+1} + \sum_{r=1}^{n} \binom{n+1}{r} a^r b^{n+1-r} + a^{n+1}$$

$$= b^{n+1} + \sum_{r=1}^{n} \left(\binom{n}{r} + \binom{n}{r-1} \right) a^r b^{n+1-r} + a^{n+1}$$

$$= b^{n+1} + \sum_{r=1}^{n} \binom{n}{r} a^r b^{n+1-r} + \sum_{r=1}^{n} \binom{n}{r-1} a^r b^{n+1-r} + a^{n+1}$$

$$= b \left(b^n + \sum_{r=1}^{n} \binom{n}{r} a^r b^{n-r} \right) + a \left(\sum_{r=1}^{n} \binom{n}{r-1} a^{r-1} b^{n+r-1} + a^n \right)$$

$$= b \left(\sum_{r=0}^{n} \binom{n}{r} a^r b^{n-r} \right) + a \left(\sum_{r=0}^{n-1} \binom{n}{r} a^r b^{n-r} + a^n \right)$$

$$= b \left(\sum_{r=0}^{n} \binom{n}{r} a^r b^{n-r} \right) + a \left(\sum_{r=0}^{n} \binom{n}{r} a^r b^{n-r} \right)$$

$$= (a + b) \left(\sum_{r=0}^{n} \binom{n}{r} a^r b^{n-r} \right).$$

(c) **Proof.** Choose one particular element x from a set A of n elements. The number of subsets of A with r elements is $\binom{n}{r}$. The collection of r-element subsets may be divided into two disjoint collections: those subsets containing x and those subsets not containing x. We count the number of subsets in each collection and add the results. First, there are $\binom{n-1}{r}$ r-element subsets of A that do not contain x, since each is a subset of $A - \{x\}$. Second, there are $\binom{n-1}{r-1}$ r-element subsets of A that do contain x, because each of these corresponds to the $(r-1)$-element sub-set of $A - \{x\}$ obtained by removing x from the subset. Thus the sum of the number of subsets in the two collections is

$$\binom{n-1}{r} + \binom{n-1}{r-1} = \binom{n}{r}.$$

23. **(b)** *Hint:* Consider two disjoint sets containing n and m elements.

Exercises 3.1

2. **(a)** domain $\mathbb{R}$, range $\mathbb{R}$
   (c) domain $[1, \infty)$, range $[0, \infty)$
   (e) domain $\mathbb{R}$, range $\mathbb{R}$
3. **(a)** 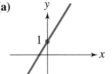   **(c)**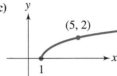

   (e)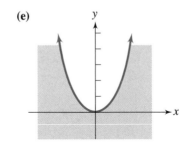

4. **(a)** $R_1^{-1} = R_1$
   (c) $R_3^{-1} = \left\{(x, y) \in \mathbb{R} \times \mathbb{R}: y = \frac{1}{7}(x + 10)\right\}$
   (e) $R_5^{-1} = \left\{(x, y) \in \mathbb{R} \times \mathbb{R}: y = \pm\sqrt{\dfrac{5-x}{4}}\right\}$
   (g) $R_7^{-1} = \left\{(x, y) \in \mathbb{R} \times \mathbb{R}: y > \dfrac{x+4}{3}\right\}$
   (i) $R_9^{-1} = \{(x, y) \in P \times P: y \text{ is a child of } x \text{ and } x \text{ is male}\}$
5. **(b)** $R \circ T = \{(3, 2), (4, 5)\}$
   (d) $R \circ R = \{(1, 2), (2, 2), (5, 2)\}$
6. **(a)** $R_1 \circ R_1 = \{(x, y): x = y\} = R_1$
   (d) $R_2 \circ R_3 = \{(x, y) \in \mathbb{R} \times \mathbb{R}: y = -35x + 52\}$
   (g) $R_4 \circ R_5 = \{(x, y) \in \mathbb{R} \times \mathbb{R}: y = 16x^4 - 40x^2 + 27\}$

(j) $R_6 \circ R_6 = \{(x, y) \in \mathbb{R} \times \mathbb{R} : y < x + 2\}$

(n) $R_3 \circ R_8 = \left\{(x, y) \in \mathbb{R} \times \mathbb{R} : y = \dfrac{14x}{x - 2} - 10\right\}$

(p) *Hint:* $R_9 \circ R_9$ is *not* $\{(x, y) : y$ is a grandfather of $x\}$.

15. **(a)** F. The statements "$x \in A \times B$" and "$x \in A$ and $x \in B$" are not equivalent. It's wise to avoid writing something like "Suppose $x \in A \times B$." Think of an element of a product as an ordered pair, and write "Suppose $(x, y) \in A \times B$."

(b) C. The only correction required is that $(a, c) \notin B \times D$ implies $a \notin B$ or $c \notin D$.

(d) A.

Exercises 3.2

1. **(a)** not reflexive, not symmetric, transitive

(e) reflexive, not symmetric, transitive

(l) not reflexive, symmetric, not transitive (*Note:* Sibling means "a brother or sister.")

2. **(a)** $\{(1, 1), (2, 2), (2, 3), (3, 1)\}$

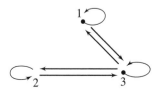

(d) $\{(1, 1), (2, 2), (3, 3), (1, 3), (2, 3), (3, 1), (3, 2)\}$

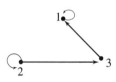

3. **(a)** Sketch the graph $y = 2x$. This relation is not reflexive on $\mathbb{R}$ because it does not contain $(1, 1)$, not symmetric because it contains $(1, 2)$ but not $(2, 1)$, and not transitive because it contains $(1, 2)$ and $(2, 4)$ but not $(1, 4)$.

(d) *Hint:* Sketch the line $y = x$ and the unit circle. This relation is not transitive because it contains $(1, 0)$ and $(0, -1)$ but not $(1, -1)$.

(f) 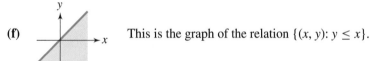 This is the graph of the relation $\{(x, y) : y \leq x\}$.

5. **(d)** *Hint:* To show that R is reflexive, let a be a natural number. All prime factorizations of a have the same number of 2's. Thus $a\,R\,a$. It must also

be shown that R is symmetric and transitive. Three elements of $4/R$ are $4 = 2 \cdot 2$, $28 = 2 \cdot 2 \cdot 7$, and $300 = 2 \cdot 2 \cdot 3 \cdot 5 \cdot 5$.

(g)   *Hint*: First show P is reflexive on $\mathbb{R} \times \mathbb{R}$ and symmetric. To show transitivity begin by supposing that $(x, y) \, P \, (z, w)$ and $(z, w) \, P \, (u, v)$. Then $|x - y| = |z - w|$ and $|z - w| = |u - v|$.

The equivalence class of $(0, 0)$ is the line $y = x$.

7. **(a)**   transitive, but not reflexive and not symmetric

   (c)   reflexive, symmetric, and transitive

8. **(a)** $\bar{0} = \{\ldots, -15, -10, -5, 0, 5, 10, \ldots\}$
   $\bar{1} = \{\ldots, -9, -4, 1, 6, 11, \ldots\}$
   $\bar{2} = \{\ldots, -8, -3, 2, 7, 12, \ldots\}$
   $\bar{3} = \{\ldots, -7, -2, 3, 8, 13, \ldots\}$
   $\bar{4} = \{\ldots, -6, -1, 4, 9, 14, \ldots\}$

10. **(b)**   *Hint:* See part (a).

    (c)   *Hint:* Assume $x \equiv_m y$. Part 1: Suppose $z \in \bar{x}$. Then $x \equiv_m z$. By symmetry $y \equiv_m x$ and by transitivity $y \equiv_m z$. Thus $z \in \bar{y}$. This shows $\bar{x} \subseteq \bar{y}$. In part 2, we must show $\bar{y} \subseteq \bar{x}$.

13. **(b)**   *Proof.*   Assume R is symmetric. Then $(x, y) \in R$ iff $(y, x) \in R$ iff $(x, y) \in R^{-1}$. Thus $R = R^{-1}$. Now, suppose $R = R^{-1}$. Then $(x, y) \in R$ implies $(x, y) \in R^{-1}$, which implies $(y, x) \in R$. Thus R is symmetric.

15. **(a)**   *Proof.*   Suppose $(x, y) \in R \cup R^{-1}$. Then $(x, y) \in R$ or $(x, y) \in R^{-1}$. If $(x, y) \in R$, then $(y, x) \in R^{-1}$. Likewise, if $(x, y) \in R^{-1}$, then $(y, x) \in R$. In either case, $(y, x) \in R \cup R^{-1}$. Thus, $R \cup R^{-1}$ is symmetric.

18. *Hint*: One part of the proof is to show that R is symmetric. Suppose $x \, R \, y$. Then $x \, L \, y$ and $y \, L \, x$, so $y \, L \, x$ and $x \, L \, y$ Therefore, $y \, R \, x$.

19. **(f)**   F. The last sentence confuses $R \cap S$ with $R \circ S$. A correct proof requires a more complete second sentence.

Exercises 3.3

2. **(d)**   The elements of $\mathscr{A}$ are natural numbers, not subsets of $\mathbb{N}$, so $\mathscr{A}$ is not a partition of $\mathbb{N}$. Note: $\{\{1, 2, 3, 4\}, \{n \in \mathbb{N}: n > 5\}\}$ *is* a partition of $\mathbb{N}$.

3. **(b)**   There are 10 equivalence classes. The class $0/R$ contains $0, 1, 2, \ldots, 9$, $100, 101, \ldots 109, 200, 201, \ldots, 209, \ldots$ and all the negatives of these numbers. The class of 10 modulo R contains all integers that have 1 as the tens digit, and so forth.

5. *Hint:* There are four subsets in the partition. One of them is the set $\{(1, -1), (-1, 1), (i, i), (-i, -i)\}$.

6. **(b)** $x \, R \, y$ iff $x = y$ or both $x > 2$ and $y > 2$.

   (d) $x \, R \, y$ iff (i) $x = y$ and $x \in \mathbb{Z}$ or (ii) $\text{int}(x) = \text{int}(y)$ and $x, y \notin \mathbb{Z}$. (Recall that $\text{int}(x)$ denotes the greatest integer function.)

8. **(a)** $\{(1, 1), (1, 2), (2, 1), (2, 2), (3, 3), (3, 4), (3, 5), (4, 3), (4, 4), (4, 5), (5, 3), (5, 4), (5, 5)\}$

11. No. Let R be the relation $\{(1, 1), (2, 2), (3, 3), (1, 2), (1, 3), (2, 1), (3, 1)\}$ on the set $A = \{1, 2, 3\}$. Then $R(1) = \{1, 2, 3\}$, $R_2 = \{1, 2\}$, and $R(3) = \{1, 3\}$. The set $\mathscr{A} = \{\{1, 2, 3\}, \{1, 2\}, \{1, 3\}\}$ is not a partition of A.

14. **(a)** Yes $\{B_1^c, B_2^c\}$ of a partition of A when $B_1 \neq B_2$, because $\{B_1^c, B_2^c\} = \{B_2, B_1\}$. If $B_1 = B_2$, then $B_1 = B_2 = A$ and $B_1^c = B_2^c = \varnothing$, so $\{B_1^c, B_2^c\}$ is not a partition of A.

15. **(d)** A (or C). The proof is correct because the ideas are all there, and every statement is true. You may give it a C if you feel the ideas are not well connected.

Exercises 3.4

1. **(a)** No, since $(2, 4)$ and $(4, 2)$ are in R.
   (c) No, since $(2, -2)$ and $(-2, 2)$ are in R.
   (f) No, since $(1, 3)$ and $(3, 1)$ are in R.
10. **(a)** $\{(a, a), (b, b), (c, c), (c, a), (c, b)\}$
11. **(a)** There are multiple correct answers for this question, depending on preferences. One answer is:

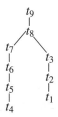

12. **(a)** **Proof.** Suppose $B - \{x\} \subseteq C \subseteq B$. For any $y \in B - \{x\}$, we have $y \in C$. If $x \in C$, then for all $y \in B$ we have $y \in C$. Therefore, $B \subseteq C$, which shows that $C = B$. On the other hand, if $x \notin C$, then for all $y \in C$ we have $y \in B$ and $y \neq x$. Thus, $C \subseteq B - \{x\}$, which shows that $C = B - \{x\}$. Therefore, there is no C different from B and $B - \{x\}$ such that $B - \{x\} \subseteq C \subseteq B$. Hence $B - \{x\}$ is an immediate predecessor of B.

13. **(b)** No. For example, consider a set of two squares where the squares are side by side within the rectangle.
    (c) A set containing two disjoint squares does not have a lower bound.

14. **(a)** *Hint:* Use parts of Theorem 2.2.1 and Exercise 9 of Section 2.2.
    (b) *Hint:* Use parts of Theorem 2.3.1 and Exercise 10 of Section 2.3.

20. **(b)** F. This proof does not show that $\sup(B)$ exists. All it shows is that *if* $\sup(B)$ *exists*, then $u = \sup(B)$. A correct proof would show that $u = \sup(B)$ by showing u has the two supremum properties (u is an upper bound and $u \, R \, v$ for all other upper bounds v.)

Exercises 3.5

2. **(d)** not possible
3. **(c)** not possible
9. *Hint:* Suppose the graph G has order $n \geq 2$, and all vertices have different degrees. Consider what these degrees must be. Can one vertex have degree $n - 1$ and another have degree 0?

Exercises 4.1

1. **(a)** This is a function with domain and range $\{0, \triangle, \square, \cup, \cap\}$. Other possible codomains are $\{0, \triangle, \square, \cup, \cap, \$\}$, $\{0, \triangle, \square, \cup, \cap, +, \#, \%\}$ and $\{\triangle, \square, \cup, \cap, 0, 1, 2, 3\}$.

3. **(a)** Domain $= \mathbb{R} - \{-1\}$. Range $= \{y \in \mathbb{R}: y \neq 0\}$. A possible codomain is $\mathbb{C}$.

   (d) Domain $= \mathbb{R} - \{\frac{\pi}{2} + k\pi: k \in \mathbb{Z}\}$. Range $= \mathbb{R}$. A possible codomain is $\mathbb{C}$.

4. **(a)** $\text{Dom}(f) = \mathbb{R} - \{3\}$; $\text{Rng}(f) = \mathbb{R} - \{-1\}$.

5. **(a)** If $x = 1$, then $2x + y = 2 + y$. For $y = 1, 2, 3$, or 4, $2 + y$ is prime and not 5 iff $y = 1$.

   If $x = 2$ then $2x + y = 4 + y$. For $y = 1, 2, 3$, or 4, $4 + y$ is prime and not 5 iff $y = 3$.

   If $x = 3$ then $2x + y = 6 + y$. For $y = 1, 2, 3$, or 4, $6 + y$ is prime and not 5 iff $y = 1$.

   If $x = 4$ then $2x + y = 8 + y$. For $y = 1, 2, 3$, or 4, $8 + y$ is prime and not 5 iff $y = 3$.

   For each $x \in A$, there is a unique y in A such that $(x, y) \in R$.

8. **(a)** A.

9. **(a)** Let $x_n = -n$. Then x is the sequence $-1, -2, -3, -4, \ldots$.

   (c) *Hint:* Find a rule so that $x_1 = 3\frac{1}{2}$ and $x_2 = 3\frac{2}{3}$.

10. **(a)** $f(3) = \overline{3} = \{\ldots, -9, -3, 3, 9, \ldots\}$.

11. **(b)** This rule is a function.

    (e) Not a function. For example in $\mathbb{Z}_3$ we have $\overline{0} = \overline{3} = \overline{6} = \overline{9}$. The rule assigns to $\overline{0}$ these different images: $[0], [3], [2]$, and $[1]$ in $\mathbb{Z}_4$.

15. **(a)** $\text{Dom}(S)$

16. **(a)** ***Proof.*** Let $x, y, z \in \mathbb{N}$.
    (i) By definition of absolute value, $d(x, y) = |x - y| \geq 0$ for all $x, y \in \mathbb{N}$.
    (ii) $d(x, y) = |x - y| = 0$ iff $x - y = 0$ iff $x = y$.
    (iii) $d(x, y) = |x - y| = |y - x| = d(y, x)$.
    (iv) By the triangle property of absolute value, $|x - y| + |y - z| \geq |x - z|$. Thus $d(x, y) + d(y, z) \geq d(x, z)$.

17. **(c)** $\binom{m}{2} n^2$ We choose 2 elements from A for first coordinates; each may then be assigned any element of B as its image.

Exercises 4.2

1. **(a)** $(f \circ g)(x) = 17 - 14x$, $(g \circ f)(x) = -29 - 14x$

   (c) $(f \circ g)(x) = \sin(2x^2 + 1)$, $(g \circ f)(x) = 2\sin^2 x + 1$

   (e) $f \circ g = \{(k, r), (t, r), (s, l)\}$, $g \circ f = \emptyset$

   (i) Observe that $(f \circ g)(x) = \begin{cases} f(2x) & \text{if } x \leq -1 \\ f(-x) & \text{if } x > -1 \end{cases}$.

We must consider cases. If $x \leq -1, f(g(x)) = f(2x)$. Since in this case $2x \leq 0$, $f(2x) = 2x + 1$. If $x > -1$, $f(g(x)) = f(-x)$. There are two sub-cases:

If $x \geq 0$, then $-x \leq 0$, so $f(-x) = -x + 1$.
If $-1 < x < 0$, then $1 > -x > 0$, so $f(-x) = -2x$.

Therefore,

$$(f \circ g)(x) = \begin{cases} 2x + 1 & \text{if } x \leq -1 \\ -2x & \text{if } -1 < x < 0. \\ -x + 1 & \text{if } x \geq 0 \end{cases}$$

Similarly,

$$(g \circ f)(x) = \begin{cases} g(x + 1) & \text{if } x \leq 0 \\ g(2x) & \text{if } x > 0 \end{cases}.$$

If $x \leq 0$, $g(f(x)) = g(x + 1)$. There are two subcases:

If $x \leq -2$, then $x + 1 \leq -1$, so $g(x + 1) = 2x + 2$.
If $x > -2$, then $x + 1 > -1$, so $g(x + 1) = -x - 1$.

If $x > 0$, $g(f(x)) = g(2x)$. Since $x > 0$, $2x > 0$, so $2x > -1$. Thus $g(2x) = -2x$.

$$\text{Therefore, } (g \circ f)(x) = \begin{cases} 2x + 2 & \text{if } x \leq -2 \\ -x - 1 & \text{if } -2 < x \leq 0. \\ -2x & \text{if } x > 0 \end{cases}$$

2. **(a)** $\text{Dom}(f \circ g) = \mathbb{R} = \text{Rng}(f \circ g) = \text{Dom}(g \circ f) = \text{Rng}(g \circ f)$
   (c) $\text{Dom}(f \circ g) = \mathbb{R}, \text{Rng}(f \circ g) = [-1, 1]$
   $\text{Dom}(g \circ f) = \mathbb{R}, \text{Rng}(g \circ f) = [1, 3]$
   (e) $\text{Dom}(f \circ g) = \{k, t, s\}, \text{Rng}(f \circ g) = \{r, l\}$
   $\text{Dom}(g \circ f) = \varnothing = \text{Rng}(g \circ f)$
   (i) $\text{Dom}(f \circ g) = \mathbb{R}, \text{Rng}(f \circ g) = (-\infty, 2)$
   $\text{Dom}(g \circ f) = \mathbb{R}, \text{Rng}(g \circ f) = (-\infty, 1)$
3. **(a)** Example 1: $f(x) = x^2$, $g(x) = 3x + 7$.
   Example 2: $f(x) = (x + 7)^2$, $g(x) = 3x$.
5. **(a)** $f^{-1}(x) = \dfrac{x - 2}{5}$
   (c) $f^{-1}(x) = \dfrac{1 - 2x}{x - 1}$
   (e) $f^{-1}(x) = -3 + \ln x$
9. **(a)** $\{(x, y) \in \mathbb{R} \times \mathbb{R}: y = 0 \text{ if } x < 0 \text{ and } y = x^2 \text{ if } x \geq 0\}$
   $\{(x, y) \in \mathbb{R} \times \mathbb{R}: y = x^2\}$
13. *Hint:* Write $A \cup C$ as $A \cup (C - E)$. Then show $h \cup g = h \cup (g \mid_{C-E})$ and use Theorem 4.2.5.

14. **(a)** $h \cup g$ is a function.

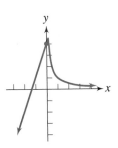

16. **(a)** ***Proof.*** Let $x, y \in \mathbb{R}$. Suppose. $x < y$. Then $3x < 3y$ and $3x - 7 < 3y - 7$. There-fore, $f(x) < f(y)$.

(d) ***Proof.*** Suppose x and y are in $(-3, \infty)$ and $x < y$. Then $4x < 4y$, so $3x - y < 3y - x$. Thus $xy + 3x - y - 3 < xy + 3y - x - 3$ or $(x - 1)(y + 3) < (y - 1)(x + 3)$. Using the fact that x and y are in $(-3, \infty)$, we know $x + 3$ and $y + 3$ are positive. Dividing both sides of the last inequality by $x + 3$ and $y + 3$, we conclude that

$$\frac{x - 1}{x + 3} < \frac{y - 1}{y + 3}.$$

Thus $f(x) < f(y)$.
⟨*Note: This proof was found by working backward from the conclusion.*⟩

17. **(d)** The function f given by

$$f(x) = \begin{cases} x + 1 & \text{if } x \le 0 \\ 1 - x & \text{if } 0 < x < 1 \\ x - 1 & \text{if } x \ge 1 \end{cases}$$

is a counterexample. Another counterexample is f given by $f(x) = (x + 1)(x - 1)^2$.

18. **(a)** ***Proof.*** We show that $f_1 + f_2$ is a function with a domain $\mathbb{R}$. First $f_1 + f_2$ is by definition a relation. For all $x \in \mathbb{R}$ there is some $u \in \mathbb{R}$ such that $(x, u) \in f_1$ because $f_1 : \mathbb{R} \to \mathbb{R}$ and there exists $v \in \mathbb{R}$ such that $(x, v) \in f_2$ because $f_2 : \mathbb{R} \to \mathbb{R}$. Then $(x, u + v) \in f_1 + f_2$, so $x \in \text{Dom}(f_1 + f_2)$. It is clear from the definition of $f_1 + f_2$ that $x \in \text{Dom}(f_1 + f_2)$ implies $x \in \mathbb{R}$, so $\text{Dom}(f_1 + f_2) = \mathbb{R}$.
    Let $x \in \mathbb{R}$. Suppose (x, c) and (x, d) are in $f_1 + f_2$. Then $c = f_1(x) + f_2(x) = d$. Therefore, $f_1 + f_2$ is a function.

20. **(a)** A. On line 3 the author of this proof chose not to mention that $x \in A$ (because A is the domain of f). The reader is expected to observe this to verify that $(x, y) \in I_A$.

Exercises 4.3

1. **(a)** Onto $\mathbb{R}$. ***Proof.*** Let $w \in \mathbb{R}$. Then for $x = 2(w - 6)$, $x \in \mathbb{R}$, and $f(x) = \frac{1}{2}[2(w - 6)] + 6 = w$. Thus $w \in \text{Rng}(f)$. Therefore, f maps onto $\mathbb{R}$.

(c) Not onto $\mathbb{N} \times \mathbb{N}$. Since $(5, 8) \in \mathbb{N} \times \mathbb{N}$ and $(5, 8) \notin \mathrm{Rng}(f)$, f does not map onto $\mathbb{N} \times \mathbb{N}$.

(k) ***Proof.*** First, if $x \in [2, 3)$, then $x - 2 \geq 0$ and $3 - x > 0$, so $f(x) = \dfrac{x - 2}{3 - x} \geq 0$. Therefore $\mathrm{Rng}(f) \subseteq [0, \infty)$. Now let $w \in [0, \infty)$.

Choose $x = \dfrac{3w + 2}{w + 1}$. Then $w \geq 0$, so $2w + 2 \leq 3w + 2 < 3w + 3$. Dividing by $w + 1$, we have $2 \leq x < 3$. Thus $x \in [2, 3)$, and

$$f(x) = \left[\frac{3w + 2}{w + 1} - 2\right] \div \left[3 - \frac{3w + 2}{w + 1}\right]$$

$$= \frac{3w + 2 - 2(w + 1)}{3(w + 1) - (3w + 2)} = w.$$

⟨*This value for x was found by working backward from the desired result.*⟩ Therefore, f maps onto $[0, \infty)$.

2. (a) One-to-one. ***Proof.*** Suppose $f(x) = f(y)$ for some $x, y \in \mathbb{R}$. Then $\frac{1}{2}x + 6 = \frac{1}{2}y + 6$. Then $\frac{1}{2}x = \frac{1}{2}y$, so $x = y$.

(c) One-to-one. ***Proof.*** Suppose $m, n \in \mathbb{N}$ and $f(m) = f(n)$. Then $(m, m) = (n, n)$, so $m = n$.

(k) f is one-to-one. ***Proof.*** Suppose $x, z \in [2, 3)$ and $f(x) = f(z)$. Then $\dfrac{x - 2}{3 - x} = \dfrac{z - 2}{3 - z}$, so $3x - xz - 6 + 2z = 3z - xz - 6 + 2x$. Thus $x = z$.

4. (a) $B = \{0, 3\}$, $f = \{(1, 0), (2, 3), (3, 0), (4, 0)\}$

8. (a) Let $f: \mathbb{R} \to \mathbb{R}$ be given by $f(x) = 2x$ and $g: \mathbb{R} \to \mathbb{R}$ be given by $g(x) = x^2$. Then f maps onto $\mathbb{R}$ but $g \circ f$ is not onto $\mathbb{R}$.

(e) Let $A = \{a, b, c\}$, $B = \{1, 2, 3\}$, and $C = \{x, y, z\}$. The function f must not be one-to-one. Let $f = \{(a, 2), (b, 2), (c, 3)\}$ and $g = \{(1, x), (2, y), (3, z)\}$. Then g is one-to-one, but the composite $g \circ f = \{(a, y), (b, y), (c, z)\}$ is not.

9. (c) ***Proof.*** We verify that f maps onto $\mathbb{R}$ as follows: $1 \in \mathrm{Rng}(f)$, because $f(4) = 1$. For $w \neq 1$, choose $x = \dfrac{4w + 2}{1 - w}$. Then $x \neq -4$ and

$$f(x) = \left[\frac{4w + 2}{1 - w} - 2\right] \div \left[\frac{4w + 2}{1 - w} + 4\right] = \frac{6w}{6} = w.$$

Therefore, f maps onto $\mathbb{R}$.

To show that f is one-to-one, suppose $f(x) = f(z)$. Then for $x \neq -4$ and $z \neq -4$, $\dfrac{x - 2}{x + 4} = \dfrac{z - 2}{z + 4}$. Therefore, $xz - 2z + 4x - 8 = xz - 2x + 4z - 8$, so $6x = 6z$ and $x = z$. We must also consider whether $f(x)$ might be 1, for $x \neq -4$. But if $x \neq -4$ and $f(x) = \dfrac{x - 2}{x + 4} = 1$, then $x - 2 = x + 4$, so $-2 = 4$. This is impossible. Therefore f is one-to-one.

11. (a) ***Proof.*** f is not a surjection because $[1]$ has no pre-image in $\mathbb{Z}_4$. This is because if $f(\bar{x}) = [1]$, then $[2x] = [1]$. But then 8 divides the odd number $2x - 1$, which is impossible. To show f is an injection, suppose $f(\bar{x}) = f(\bar{z})$. Then $[2x] = [2z]$, so $2x = 2z \pmod 8$. Therefore, 8 divides $2x - 2z$, so 4 divides $x - z$, and thus $\bar{x} = \bar{z}$.

12. (b) Example 1: $x_n = n$ for all $n \in \mathbb{N}$ (the identity function).

$$\text{Example 2: } x_n = \begin{cases} 50 - n & \text{if } n < 50 \\ n & \text{if } n \geq 50 \end{cases}.$$

$$\text{Example 3: } x_n = \begin{cases} n + 1 & \text{if } n \text{ is odd} \\ n - 1 & \text{if } n \text{ is even} \end{cases}.$$

The sequence of example 3 is $2, 1, 4, 3, 6, 5, 8, 7, \ldots$.

13. (c) None.

   (f) Since $m = n + 1$, one element of B has two pre-images. This element of B can be selected in n ways, and the two pre-images in $\binom{n+1}{2}$ ways. We can assign each of the remaining $n - 1$ elements of B as the image of exactly one of the $n - 1$ remaining elements of A in $(n - 1)!$ ways. Thus there are $n\binom{n+1}{2}(n-1)! = n!\binom{n+1}{2}$ functions from A onto B.

14. (a) F. To show that f is onto $\mathbb{R}$ we must prove that $\mathbb{R} \subseteq \text{Rng}(f)$. The proof shows only that $\text{Rng}(f) \subseteq \mathbb{R}$.

   (b) *Hint:* What additional information should be included in this proof?

   (d) A. Notice that a direct proof would have been a little easier to follow.

Exercises 4.4

---

3. (b) Let h be the inverse of g. Then

$$(x, y) \in h \text{ iff } (y, x) \in g$$

$$\text{iff } x = \frac{4y}{y + 2}$$
$$\text{iff } xy + 2x = 4y$$
$$\text{iff } y(x - 4) = -2x$$
$$\text{iff } y = \frac{2x}{4 - x}, \text{ for } x < 4.$$

Therefore $h(x) = \dfrac{2x}{4 - x}$. To verfiy that this formula is correct, suppose $x > -2$. Then

$$(h \circ g)(x) = h(g(x))$$
$$= h\left(\frac{4x}{x + 2}\right)$$
$$= \frac{2\left(\frac{4x}{x+2}\right)}{4 - \frac{4x}{x+2}}$$

$$= \frac{\dfrac{8x}{x+2}}{\dfrac{4(x+2)-4x}{x+2}}$$

$$= x.$$

Since the composite is the identity function on the domain of g, we conclude that $h = g^{-1}$.

9. **(c)** [2 5 4 6 1 7 3]

    (e) [6 1 4 7 5 3 2]

    (i) [5 3 7 4 1 2 6]

Exercises 4.5

1. **(a)** $h(\varnothing)=\varnothing, h(\{1\})=\{4\}, h(\{2\})=\{4\}, h(\{3\})=\{5\}, h(\{1,2\})=\{4\},$
   $h(\{1,3\}) = \{4, 5\}, h(\{2, 3\}) = \{4, 5\}, h(A) = \{4, 5\}.$

2. **(a)** $[2, 10]$

    (c) $\{0\}$

4. **(b)** $[2, 5.2]$

    (c) $\left[2 - \sqrt{3}, \dfrac{3 - \sqrt{5}}{2}\right) \cup \left(\dfrac{3 + \sqrt{5}}{2}, 2 + \sqrt{3}\right]$

8. **(a)** Suppose $b \in f\left(\bigcap_{\alpha \in \Delta} D_\alpha\right)$. Then $b = f(a)$ for some $a \in \bigcap_{\alpha \in \Delta} D_\alpha$. Thus $a \in D_\alpha$ for every $\alpha \in \Delta$. Since $b = f(a)$ we conclude that $b \in f(D_\alpha)$ for every $\alpha \in \Delta$. Therefore, $b \in \bigcap_{\alpha \in \Delta} f(D_\alpha)$. This proves that $f\left(\bigcap_{\alpha \in \Delta} D_\alpha\right) \subseteq \bigcap_{\alpha \in \Delta} f(D_\alpha)$.

9. *Hint:* There must be at least two sets D_1 and D_2 in the family, and $b \in f(D_\alpha)$ for all $\alpha \in \Delta$, but no element a in $\bigcap_{\alpha \in \Delta} D_\alpha$ such that $f(a) = b$. The function f cannot be one-to-one.

10. **(a)** Suppose $b \in f(f^{-1}(E))$. Then there is $a \in f^{-1}(E)$ such that $f(a) = b$. Since $a \in f^{-1}(E)$, $f(a) \in E$. But $f(a) = b$, so $b \in E$. Therefore, $f(f^{-1}(E)) \subseteq E$.

    (d) First, suppose $E = f(f^{-1}(E))$. Suppose $b \in E$. Then $b \in f(f^{-1}(E))$. Thus there is $a \in f^{-1}(E)$ such that $b = f(a)$, so $b \in \text{Rng}(f)$. Therefore, $E \subseteq \text{Rng}(f)$.

        Now assume $E \subseteq \text{Rng}(f)$. We know by part (a) that $f(f^{-1}(E)) \subseteq E$, so to prove equality, we must show $E \subseteq f(f^{-1}(E))$. Suppose $b \in E$. Then $b \in \text{Rng}(f)$, so $b = f(a)$ for some $a \in A$. Since $b = f(a) \in E$, $a \in f^{-1}(E)$. Thus $b = f(a)$ and $a \in f^{-1}(E)$, so $b \in f(f^{-1}(E))$. Therefore, $E \subseteq f(f^{-1}(E))$.

11. **(b)** **Proof.** Suppose $t \in f(X) - f(Y)$. Then $t \in f(X)$, so there exists $x \in X$ such that $f(x) = t$. We note $x \notin Y$ since $t = f(x) \notin f(Y)$. Thus $x \in X - Y$ and therefore $t = f(x) \in f(X - Y)$.

12. The converse is true. To prove f is one-to-one, suppose $x, y \in A$ and $x \neq y$. Then $\{x\} \cap \{y\} = \emptyset$ and thus $f(\{x\} \cap \{y\}) = \emptyset$. By hypothesis, $f(\{x\} \cap \{y\}) = f(\{x\}) \cap f(\{y\}) = \{f(x)\} \cap \{f(y)\}$. Thus $f(x) \neq f(y)$.

15. (a) If f is one-to-one, then the induced function is one-to-one.

18. (a) F. The claim is not true. We cannot conclude $x \in X$ from $f(x) \in f(X)$.

Exercises 4.6

3. (a) does not exist

(c) $\frac{4}{5}$

(e) 0

(g) e^2 (Recall that $\lim\limits_{n \to \infty} \left(1 + \dfrac{1}{n}\right)^n = e$.)

(i) 0

5. (a) *Hint:* To show x diverges, suppose the limit is L and let $\varepsilon = 1$.

(c) *Hint:* x diverges; use $\varepsilon = 1$.

(f) *Hint:* $x_n \to 0$; for $\varepsilon > 0$, use $N > (2\varepsilon)^{-2}$ and

$$\sqrt{n+1} - \sqrt{n} = \left(\sqrt{n+1} - \sqrt{n}\right)\left(\frac{\sqrt{n+1} + \sqrt{n}}{\sqrt{n+1} + \sqrt{n}}\right).$$

6. (a) **Proof.** Let $\varepsilon > 0$. Then $\frac{\varepsilon}{2} > 0$. Since $x_n \to L$, there exists $N_1 \in \mathbb{N}$ such that if $n > N_1$, then $|x_n - L| < \frac{\varepsilon}{2}$. Likewise, there exists $N_2 \in \mathbb{N}$ such that $n > N_2$ implies $|y_n - M| < \frac{\varepsilon}{2}$. Let $N_3 = \max\{N_1, N_2\}$, and assume $n > N_3$. Therefore we have $|(x_n + y_n) - (L + M)| = |(x_n - L) + (y_n - M)| \leq |x_n - L| + |y_n - M| < \frac{\varepsilon}{2} + \frac{\varepsilon}{2} = \varepsilon$. Therefore, $x_n + y_n \to L + M$.

(f) *Hint:* $|x_n| - |L| \leq |x_n - L|$.

7. (a) *Hint:* Since $x_n \to L$, $|x_n| \to |L|$, by Exercise 6(f). Now apply the definition of $|x_n| \to |L|$ with $\varepsilon = \frac{|L|}{2}$.

10. (b) A. The proof uses Exercise 6(b).

Exercises 5.1

2. (a) finite

(c) finite

(e) infinite

(f) finite

6. (a) Suppose A is finite. Since $A \cap B$ is a subset of A, $A \cap B$ is finite.

7. *Hint:* Write $A \cup B = (A - B) \cup B$ and $A = (A - B) \cup (A \cap B)$ and apply Theorem 5.1.7(a).

9. (a) *Hint:* Define $f: A \to A \cup \{x\}$ by $f(a) = (a, x)$, for each $a \in A$. Now show that f is one-to-one and onto $A \cup \{x\}$.

10. *Hint:* First show $\mathcal{P}(A \times B)$ is finite.

11. **(c)**   not possible

13. Obviously $\mathbb{N}_r - \{x\} \approx \mathbb{N}_{r-1}$ when $x = r$. If $x \neq r$, define a function on $\mathbb{N}_r - \{x\}$ by considering first the images of elements that are less than x and then images of elements greater than x.

15. *Hint:* Use the argument that if $n < m$, then the finite set $\mathbb{N}_m$ is equivalent to one of its proper subsets.

18. **(a)**   *Hint:* Suppose f is not onto B and consider the range of f.

   (b)   *Hint:* Suppose f is not one-to-one. Then A is not empty and since A and B are finite and $A \approx B$, there is some $n \in \mathbb{N}$ such that $\mathbb{N}_n \approx A$ and $B \approx \mathbb{N}_n$. Use these facts to construct a function F from $\mathbb{N}_n$ onto $\mathbb{N}_n$ that is not one-to-one. Then for some $x, y, z \in \mathbb{N}_n, f(x) = f(y) = z$. Removing (y, z) from F produces a function from a proper subset of $\mathbb{N}_n$ onto $\mathbb{N}_n$. Now apply Exercise 17.

19. *Hint:* Use induction on the number of elements in the domain.

20. ***Proof.***   Suppose A and B are finite, $\overline{\overline{A}} = m$, $\overline{\overline{B}} = n$, $m > n$, and $f: A \to B$ is one-to-one. Then there exists a one-to-one (and onto) function $g: \mathbb{N}_m \to A$ and a one-to-one (and onto) function $h: B \to \mathbb{N}_n$. See the diagram below. Then $h \circ f \circ g$ is a one-to-one function from $\mathbb{N}_m$ to $\mathbb{N}_n$ by Theorem 4.3.4. But $m > n$ so the conclusion that $h \circ f \circ g$ is a one-to-one contradicts Theorem 5.1.9.

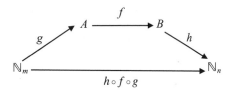

21. **(b)**   *Hint:* The largest possible sum of 10 elements of $\mathbb{N}_{99}$ is $90 + 91 + \cdots + 99 = 945$. Thus there are no more than 945 possible sums. However, there are $2^{10} - 1 = 1{,}023$ nonempty subsets of S. Apply the Pigeonhole Principle, and delete any common elements from two subsets that have the same sum to form disjoint subsets.

22. **(b)**   C. In Case 2, it is not correct that $\mathbb{N}_k \cup \mathbb{N}_1 \approx \mathbb{N}_{k+1}$. In fact, $\mathbb{N}_k \cup \mathbb{N}_1 = \mathbb{N}_k$.

Exercises 5.2

3. **(a)**   ***Proof.***   Let $f: \mathbb{N} \to D^+$ be given by $f(n) = 2n - 1$ for each $n \in \mathbb{N}$. We show that f is one-to-one and maps onto D^+. First, to show f is one-to-one, suppose $f(x) = f(y)$. Thus $2x - 1 = 2y - 1$, which implies $x = y$. Also, f maps onto D^+ since if d is an odd positive integer, then d has the form $d = 2r - 1$ for some $r \in \mathbb{N}$. But then $f(r) = d$.

   (e)   *Hint:* Consider $f(x) = -(x + 12)$ with domain $\mathbb{N}$.

4. (a) *Hint:* Let $f(0, 1) \to (1, \infty)$ be given by $f(x) = \dfrac{1}{x}$.

(b) *Hint:* Let $f(0, 1) \to (a, \infty)$ be given by $f(x) = (a - 1) + \dfrac{1}{x}$.

(d) *Hint:* Let $f(0, 1) \to [1, 2) \cup (5, 6)$ be given by

$$f(x) = \begin{cases} 2 - 2x & \text{if } 0 < x \le \dfrac{1}{2} \\ 2x + 4 & \text{if } \dfrac{1}{2} < x < 1 \end{cases}.$$

6. (a) Define $g \colon \mathbb{N} \to E^+$ by

$$g(x) = \begin{cases} 20 & \text{if } x = 1 \\ 2 & \text{if } x = 10 \\ 2x & \text{if } x \ne 1, x \ne 10 \end{cases}.$$

7. (a) c

(c) $\aleph_0$

(e) c

12. (a) F. W is certainly an infinite subset of $\mathbb{N}$, and D^+ is denumerable, but this "proof" claims without justification that every infinite subset of $\mathbb{N}$ is denumerable. To show W is denumerable, we need to use another theorem or a bijection between W and a denumerable set.

(c) F. The claim is false. Also "A and B are finite" is not a denial of "A and B are infinite."

(d) F. Writing an infinite set A as $\{x_1, x_2, \ldots\}$ is the same as assuming A is denumerable.

Exercises 5.3

6. *Hint:* Use a proof by induction on n, the number of sets in the family.

7. *Hint:* This theorem has been proved in the cases where A and B are finite (Theorem 5.1.7(a)), where one set is denumerable and the other is finite (Theorem 5.3.4), and where A and B are denumerable and disjoint (Theorem 5.3.5). The only remaining case is where A and B are denumerable and not disjoint. Write $A \cup B$ as $A \cup (B - A)$, a union of disjoint sets. Now explain why $B - A$ is countable, and then apply Theorems 5.3.4 and 5.3.5.

8. *Hint:* Give a proof by induction.

11. *Hint:* For each $m \in \mathbb{N}$, let $\overline{\overline{B}}_m = k_m$. Then there is a bijection $f_m \colon B_m \to \mathbb{N}_{k_m}$. Define $h \colon \bigcup_{i \in \mathbb{N}} B_i \to \mathbb{N}$ by $h(x) = \left(\sum_{i=1}^{m-1} k_i \right) + f_m(x)$, for $x \in B_m$.

13. (b) *Hint:* First prove that each set T_n is infinite. Then for $a \in T_n$ define $f(a) = 2^{a_1} \cdot 3^{a_2} \cdot 5^{a_3} \cdot \cdots p_k^{a_k}$, where p_k is the kth prime and $a_i = 0$ for all $i > k$. Explain why T_n is equivalent to $\mathrm{Rng}(f)$ and why $\mathrm{Rng}(f)$ is countable.

15. (a) C. The proof is valid only when $f(1) = x$. In the case when $f(1) \ne x$, we need a new function g that is almost the same as f except that the image of 1 will be x. This involves removing the two ordered pairs with second coordinates x and $f(1)$ and replacing them with two other ordered pairs. Let t be the unique element of $\mathbb{N}$ such that $f(t) = x$ and define $f^* = (f - \{(1, f(1)), (t, x)\}) \cup \{(1, x), (t, f(1))\}$. Now let $g(n) = f^*(n + 1)$ for all $n \in \mathbb{N}$.

Exercises 5.4

2. *Hint:* Recall that $\overline{\overline{\mathcal{P}(\mathbb{N})}} = \overline{\overline{\mathbb{R}}}$.

5. (b) true

8. (a) $\overline{\overline{\varnothing}} < \overline{\overline{\{0\}}} < \overline{\overline{\{0, 1\}}} < \overline{\overline{\mathbb{Q}}} < \overline{\overline{(0, 1)}} = \overline{\overline{[0, 1]}} = \overline{\overline{\mathbb{R} - \mathbb{N}}}$
$$= \overline{\overline{\mathbb{R}}} < \overline{\overline{\mathcal{P}(\mathbb{R})}} < \overline{\overline{\mathcal{P}(\mathcal{P}(\mathbb{R}))}}.$$

11. (b) not possible

16. (a) *Hint:* Assume that there is a bijection. Associate each function f in $\mathcal{F}$ with the corresponding real number a such that $0 \le a \le 1$, and write f as f_a. Define $g: [0, 1] \to [0, 1]$ by

$$g(x) = \begin{cases} 0 & \text{if } f_x(x) \ne 0 \\ 1 & \text{if } f_x(x) = 0 \end{cases}.$$

Then $g = f_b$ for some $b \in [0, 1]$. Compute $g(b)$ to obtain a contradiction. [Ref.: J. Robertson, "A Student Exercise on Cardinality," *Mathematics and Computer Education* 32 (1998), 17–18.]

(b) *Hint:* Consider the set of constant functions in $\mathcal{F}$.

17. (b) F. The claim is false. We have not defined or discussed properties of operations such as addition for infinite cardinal numbers. Thus the equation $\overline{\overline{C}} = \overline{\overline{B}} + \overline{\overline{(C - B)}}$ applies only to finite sets.

(d) F. The "proof" assumes that every element of B is in the range of f.

Exercises 5.5

1. (b) The Axiom of Choice is not necessary since there are only a finite number of sets in the collection.

(h) The Axiom of Choice is necessary.

5. *Proof.* Let $B \subseteq A$ with B infinite and A denumerable. Since $B \subseteq A$, $\overline{\overline{B}} \le \overline{\overline{A}}$. Since A is denumerable, $\overline{\overline{A}} = \overline{\overline{\mathbb{N}}}$. Since B is infinite, B has a denumerable subset D by Theorem 5.5.4. Thus $\overline{\overline{A}} = \overline{\overline{\mathbb{N}}} = \overline{\overline{D}} \le \overline{\overline{B}}$. By the Cantor–Schröder–Bernstein Theorem, $\overline{\overline{B}} = \overline{\overline{A}}$. Thus $B \approx A$.

Alternate Proof. The set B is an infinite subset of the countable set A, so B is countable and denumerable. Since both A and B are equivalent to $\mathbb{N}$, A is equivalent to B.

8. *Hint:* Let $x \in A$. By Theorem 5.5.4, $A - \{x\}$ has a denumerable subset $\{a_n: n \in \mathbb{N}\}$. Construct a one-to-one correspondence between A and $A - \{x\}$.

10. (a) A. Note that the range of every sequence is countable.

(c) F. The idea of this "proof" is to take out countably many elements, one at a time, until denumerably many elements are left. But if A is uncountable and C is countable, then the set $B = A - C$ of leftover elements will always be uncountable. (See Exercise 9(b) of Section 5.3.)

Exercises 6.1

1. (a) yes

(e) no

2. (a) not commutative

(e) not an operation

3. (a) not associative

(e) not an operation

4. (a) a is the identity element

(b) Yes. This is tedious to verify because one must verify that 64 equations of the form $(x \circ y) \circ z = x \circ (y \circ z)$ are all true. It helps to observe that if $x = a$ (the identity) then $(x \circ y) \circ z = y \circ z = x \circ (y \circ z)$. Similarly, if $y = a$ or $z = a$, the equation is easily seen to be true. This leaves only 27 cases to verify when none of x, y, or z is a. For example, $(b \circ c) \circ b = d \circ b = c$, while $b \circ (c \circ b) = (b \circ d) = c$, so the equation is true when $x = b$, $y = c$, and $z = b$.

(c) Yes, because the table is symmetric about its main diagonal. To verify by cases that the equation $x \circ y = y \circ x$ is true for every choice of x and y, consider first that case that one of x or y is a, then the case that $x = y$, and finally the other 3 cases.

(d) The inverses of a, b, c, d, are a, b, c, d, respectively.

(e) No. The product $b \circ c = d$ is not in B_1, so B_1 is not closed under $\circ$.

(f) Yes. $a \circ a = a$, $a \circ c = c$, $c \circ a = c$, $c \circ c = a$.

(g) $\{a\}$, $\{a, b\}$, $\{a, c\}$, $\{a, d\}$, and $\{a, b, c, d\}$.

(h) True. In fact, for all $x \in A$, $x \circ x = a$.

8. *Hint:* Compute $(ac)(db)$ and $(db)(ac)$.

9. (a) *Hint:* Compute $x \circ (a \circ y)$ and $(x \circ a) \circ y$.

10. (b) *Hint:* Assume that for some natural number n, every product of t elements $a_1, a_2, \ldots, a_t$ of A is equal to $(\ldots((a_1 * a_2) * a_3) \ldots) * a_t$ for every $t \le n$. Now consider a product of $n + 1$ factors $a_1, a_2, \ldots, a_{n+1}$ in that order. This product has the form $b_1 * b_2$, where b_1 is a product of some k factors $(k \le n)$ $a_1, a_2, \ldots, a_k$ in that order and b_2 is a product of the remaining factors $a_{k+1}, \ldots, a_{n+1}$ in that order. First use the induction hypothesis to write b_1 and b_2 in left-associated form.

   Now consider two cases. If $k > 1$, then there are at least two factors a_1 and a_2 in the product b_1. Denote the product $a_1 * a_2$ by c, which is an element of A. Replace $a_1 * a_2$ by c and use the hypothesis of induction to write $b_1 * b_2$ as a left-associated product.

If $k = 1$ and b_2 has only one factor, then the product $b_1 * b_2$ is $a_1 * a_2$ which is already in left-associated form. Otherwise, b_2 has at least 2 factors, and we may denote by d the product of the first two factors a_2 and a_3 of b_2. Replace $a_2 * a_3$ by d in the product $b_1 * b_2$ and apply the hypothesis of induction. Finally, apply the associative property to $a_1 * (a_2 * a_3)$ to write the entire product in left-associated form.

12. *Hint:* Assume that $a = c \pmod m$ and $b = d \pmod m$. Then there exist integers x and y such that $mx = a - c$ and $my = b - d$. To show that $a + b = c + d \pmod m$, compute the sum of mx and my. To show that $a \cdot b = c \cdot d \pmod m$, simplify $bmx + cmy$.

15. **(a)**   2, 4, and 6.

    (c)   There are no divisors of zero.

16. **(b)**   F. The claim is false. One may premultiply (multiply on the left) or postmultiply both sides of an equation by equal quantities. Multiplying one side on the left and the other on the right does not always preserve equality.

    (d)   F. The proof makes the assumption that $xy \neq 0$, which may be false.

Exercises 6.2

1. **(c)**

$\cdot$	1	-1	i	$-i$
1	1	-1	i	$-i$
-1	-1	1	$-i$	i
i	i	$-i$	-1	1
$-i$	$-i$	i	1	-1

We see from the table that the set is closed under $\cdot$, 1 is the identity, and each element has an inverse. Also, $\cdot$ is associative.

    (d)   *Hint:* $\varnothing$ is the identity.

2.

	e	u	v	w
e	e	u	v	w
u	u	v	w	e
v	v	w	e	u
w	w	e	u	v

4. **(c)**   The group is abelian.

8. **(b)**   [3 2 1 4], [1 2 3 4], [2 4 1 3].

12. *Hint:* For a, $b \in G$, compute $a^2 b^2$ and $(ab)^2$ two ways.

13. *Hint:* To have both cancellation properties, every element must occur in every row and in every column of the table.

16. **(a)** v, w, e, and u

    (b)   *Hint:* Let $a, b \in G$. For an element x such that $a * x = b$, try $a^{-1} * b$.

19. Since $(p - 1)(p - 1) = p^2 - 2p + 1 = p(p - 2) + 1$, $(p - 1)^2 = 1 \pmod p$. Therefore $(p - 1)(p - 1) = 1$ in $\mathbb{Z}_p$, and hence $(p - 1)^{-1} = p - 1$.

20. **(a)** $x = 0, 4, 8, 12, 16$

21. **(c)** $x = 6$
    (e) No solution
22. **(a)** $(x - 1)(x + 1) = x^2 - 1 = 0$
23. **(b)** F. A minor criticism is that no special case is needed for e. The fatal flaw is the use of the undefined division notation.
    (c) A. The proof is correct, but provides minimal explanation.

Exercises 6.3

1. **(a)** $\{0\}, \mathbb{Z}_8, \{0, 4\}, \{0, 2, 4, 6\}$
   (d) *Hint:* There are six subgroups.
7. **(a)** Yes. Assume G is abelian and H is a subgroup of G. Suppose $x, y \in H$. Then $x, y \in G$. Therefore, $xy = yx$.
9. **(c)** The order of 0 is 1. The elements 1, 3, 5, and 7 have order 8. The elements 2 and 6 have order 4. The order of 4 is 2.
11. **Proof.** The set C_a is not empty because $ea = a = ae$, so $e \in C_a$. Let $x, y \in C_a$. Then $xa = ax$ and $ya = ay$. Multiplying both sides of the last equation by y^{-1}, we have $y^{-1}(ya)y^{-1} = y^{-1}(ay)y^{-1}$. Thus $(y^{-1}y)(ay^{-1}) = (y^{-1}a)(yy^{-1})$, or $ay^{-1} = y^{-1}a$. Therefore, $(xy^{-1})a = x(y^{-1}a) = x(ay^{-1}) = (xa)y^{-1} = (ax)y^{-1} = a(xy^{-1})$. This shows $xy^{-1} \in C_a$. Therefore, C_a is a subgroup of G, by Theorem 6.3.3.
14. **Proof.** The identity $e \in H$ because H is a group, and thus $a^{-1}ea \in K$. Thus K is not empty. Suppose $b, c \in K$. Then $b = a^{-1}h_1a$ and $c = a^{-1}h_2a$ for some $h_1, h_2 \in H$. Thus $bc^{-1} = (a^{-1}h_1a)(a^{-1}h_2a)^{-1} = (a^{-1}h_1a)(a^{-1}h_2^{-1}a) = a^{-1}h_1(aa^{-1})h_2^{-1}a = a^{-1}h_1h_2^{-1}a$. But H is a group, so $h_1h_2^{-1} \in H$. Thus $bc^{-1} \in K$. Therefore, K is a subgroup of G.
17. *Hint:* Let H be a subgroup of (a). If H is the trivial subgroup it is clearly cyclic. Otherwise show that there is a positive integer t such that a^t is in H. Now use the Well Ordering Principle to find the smallest such t and use the Division Algorithm to show that this power of a is a generator for H.
18. **(c)** a^{10}, a^{20}
19. **(a)** C. The proof omits the step of verifying that $H \cap G$ is nonempty (because it contains the identity).

Exercises 6.4

4. Let $f, g \in F$. Then $I(f + g) = \int_a^b (f + g)(x)\,dx = \int_a^b f(x)\,dx + \int_a^b g(x)\,dx = I(f) + I(g)$.
8. **(a)** Let $C, D \in \mathscr{P}(A)$. Then $f(C \cup D) = f(C) \cup f(D)$ by Theorem 4.5.1(b). Therefore, f is an OP mapping.
10. **(a)** *Hint:* Suppose $G = (\{e, a\}, \circ)$ and $H = (\{i, b\}, *)$ are two groups with identity elements e and i. To define an isomorphism from G to H, first determine the image of e.
    (c) *Hint:* Use Theorem 6.4.3(c) to show that the algebraic system in Exercise 2 of Section 6.2 is not isomorphic to $(\mathbb{Z}_4, +)$.

14. **(a)** Suppose $\bar{y} = \bar{x}$ in $\mathbb{Z}_{18}$. Then 18 divides $x - y$. Therefore, 6 divides $x - y$, and so 24 divides $4(x - y) = 4x - 4y$. Thus $f(\bar{x}) = [4x] = [4y] = f(\bar{y})$ in $\mathbb{Z}_{24}$, so f is well defined. Now let $\bar{x}, \bar{y} \in \mathbb{Z}_{18}$. Then $f(\bar{x} +_{18} \bar{y}) = [4(x + y)] = [4x + 4y] = [4x] +_{24} [4y] = f(\bar{x}) +_{24} f(\bar{y})$.

(b) Rng $(f) = \{[0], [4], [8], [12], [16], [20]\}$. The table is:

	[0]	[4]	[8]	[12]	[16]	[20]
[0]	[0]	[4]	[8]	[12]	[16]	[20]
[4]	[4]	[8]	[12]	[16]	[20]	[0]
[8]	[8]	[12]	[16]	[20]	[0]	[4]
[12]	[12]	[16]	[20]	[0]	[4]	[8]
[16]	[16]	[20]	[0]	[4]	[8]	[12]
[20]	[20]	[0]	[4]	[8]	[12]	[16]

Exercises 6.5

1. **(b)** The interval $[-1, 1]$ is not a ring because it is not closed under addition. (For example, $\frac{3}{4} + \frac{3}{4}$ in not an element of $[-1, 1]$).

4. *Hint:* First show that $\mathbb{Z} \times \mathbb{Z}$ is closed under $\oplus$ and $\otimes$ and that the additive identity is $(0, 0)$. The inverse of (a, b) is $(-a, -b)$ because

$$(a, b) \oplus (-a, -b) = (a + (-a), b + (-b)) = (0, 0), \text{ and}$$
$$(-a, -b) \oplus (a, b) = (-a + a, -b + b) = (0, 0).$$

Explain why the remaining axioms hold for $\mathbb{Z} \times \mathbb{Z}$.

13. **(a)** *Hint:* Let $m \in \mathbb{N}$. First, show that if m is composite, then m has divisors of 0. Now suppose m is prime and there are nonzero elements a and b in $\mathbb{Z}_m$ such that $ab = 0$. Apply Euclid's Lemma to reach a contradiction.

Exercises 7.1

1. **(b)** $\frac{1}{3}, 1, 2, 3$

2. **(b)** 0 and all negative real numbers are lower bounds.

3. **(a)** supremum: 1; infimum: 0
   (c) supremum: does not exist; infimum: 0
   (e) supremum: 1; infimum: $\frac{1}{3}$
   (g) supremum: 5; infimum: -1

4. **(a)** *Hint:* Show that an upper bound for A is an upper bound for B.

6. *Hint:* Suppose b is an upper bound for A. Therefore, for all x, if $x \in A$ then $x \le b$. This means that for all x, if $x > b$, then $x \notin A$. Explain why A^c is not bounded above.

8. **(a)** **Proof.** Let x and y be least upper bounds for A. Then x and y are upper bounds for A. Since y is an upper bound and x is a least upper bound,

$x \leq y$. Since x is an upper bound and y is a least upper bound, $y \leq x$. Thus $x = y$.

13. **(a)** **Proof.** Let $s = \sup(A)$, $B = \{u : u$ is an upper bound for $A\}$. Then B is bounded below (by elements of A) so $\inf(B)$ exists. Let $t = \inf(B)$. We must show $s = t$.

(i) To show $t \leq s$ we note that since $s = \sup(A)$, s is an upper bound for A. Thus $s \in B$. Therefore, $t \leq s$.

(ii) To show $s \leq t$ we will show t is an upper bound for A. If t is not an upper bound for A, then there exists $a \in A$ with $a > t$. Let $\varepsilon = \frac{a - t}{2}$. Since $t = \inf(B)$ and $t < t + \varepsilon$, there exists $u \in B$ such that $u < t + \varepsilon$. But $t + \varepsilon < a$. Therefore, $u < a$, a contradiction, since $u \in B$ and $a \in A$.

14. **(a)** **Proof.** Suppose $\sup(A)$ and $\sup(B)$ exist. Then $A \cup B$ is bounded above by $m = \max\{\sup(A), \sup(B)\}$ (see Exercise 4(c)). By the completeness property, $\sup(A \cup B)$ exists. We show that $\sup(A \cup B) = m$.

(i) Since $A \subseteq A \cup B$, we have $\sup(A) \leq \sup(A \cup B)$. Also, $B \subseteq A \cup B$ implies that $\sup(B) \leq \sup(A \cup B)$. It follows that $m = \max\{\sup(A), \sup(B)\} \leq \sup(A \cup B)$.

(ii) It suffices to show m is an upper bound for $A \cup B$. Let $x \in A \cup B$. If $x \in A$, then $x \leq \sup(A) \leq m$. If $x \in B$, then $x \leq \sup(B) \leq m$. Thus m is an upper bound for $A \cup B$. Hence $\sup(A \cup B) \leq m$.

19. *Hint:* Let F be an ordered field. Assume that F is complete and let A be a nonempty subset of F that has a lower bound in F. To show that A has an infimum in F, begin by defining the set $A^- = \{-x : x \in A\}$. Prove that A^- has a supremum and then find an infimum for A.

21. **(a)** F. The claim is true but $y = i + \frac{\varepsilon}{2}$ might not be in A.

Exercises 7.2

1. **(b)** $x = 3.825$, $\delta = 0.025$
2. **(b)** *Hint:* In the case where $|x_2 - x_1| < 2\delta_1$,

$$\mathcal{N}(x_1, \delta_1) \cap \mathcal{N}(x_2, \delta_1) = \mathcal{N}\left(\frac{x_1 + x_2}{2}, \delta_1 - \frac{|x_2 - x_1|}{2}\right).$$

3. $\lim_{x \to a} f(x) = L$ iff for all $\varepsilon > 0$ there exists $\delta > 0$ such that if $x \in \mathcal{N}(a, \delta)$, then $f(x) \in \mathcal{N}(L, \varepsilon)$.

4. **(e)** $\varnothing$

(i) $\bigcup_{n \in \mathbb{N}} (n + 0.1, n + 0.2)$

5. **(b)** open

(e) open

(i) closed

13. **(a)** not compact (not bounded)
    (d) not compact (neither closed nor bounded)
    (g) not compact (not closed)
15. **(a)** *Hint:* First show that a cover for $A \cup B$ is a cover for A and a cover for B.
19. **(c)** C. With the addition of O^* to the cover $\{O_\alpha : \alpha \in \Delta\}$ we are assured that there is a finite subcover of $\{O^*\} \cup \{O_\alpha : \alpha \in \Delta\}$, but not necessarily a subcover of $\{O_\alpha : \alpha \in \Delta\}$. Since $O^* = A - B$ is useless in a cover for B, it can be deleted from the subcover.

Exercises 7.3

1. **(c)** *Hint:* Use the fact the $\lim_{n \to \infty} \left(1 + \dfrac{1}{n}\right)^n = e$.
3. **(a)** $\left\{\dfrac{1}{2}\right\}$
   (c) $\varnothing$
   (g) $\{0, 2\}$
   (m) $\mathbb{N}$
4. $\{0, 1\}$
5. *Hint:* Let $a \in A$. Then $z > a$. Show z is an accumulation point of A by using Theorem 7.1.1.
7. **(b)** *Hint:* Use Exercise 6(a).
10. **(a)** The set has no accumulation points.
    (c) The set has at least one accumulation point.
13. **(a)** F. *Hint:* Identify the misuse of quantifiers.
    (c) F. The claim is false. $(B^c)'$ need not be a subset of $(B')^c$.

Exercises 7.4

1. **(a)** bounded below by 10, not bounded above
   (c) bounded; bounded above by $\dfrac{1}{10}$, bounded below by 0
   (e) bounded; bounded above by 10, bounded below by 0
   (g) not bounded above, not bounded below
   (i) bounded; bounded above by 0.81, bounded below by -0.9
3. *Hint:* Let y be bounded, and B a number such that $|y_n| \le B$ for all $n \in \mathbb{N}$. Use the definition of $x_n \to 0$ with $\dfrac{\varepsilon}{B}$.
5. **(f)** *Hint:* It suffices to show that for $n \in \mathbb{N}$, $\dfrac{(n+1)!}{(n+1)^{n+1}} < \dfrac{n!}{n^n}$.

Exercises 7.5

3. **(a)** *Hint:* Consider rational numbers in $[7, 8]$ whose square is less than 50.
6. **(a)** *Hint:* Consider a set A that includes $[0, 2]$ and the sequence $x_n = n/(n+1)$.

(b)   F. The claim is correct but there is little that is correct in this proof. For instance, the upper bound a_0 for A may be negative, in which case B would have to be defined differently. The most serious error is that there is no connection between being an accumulation point and being an upper bound for a set.

INDEX

Note: Page numbers followed by n indicate items appearing in footnotes.

A

Abel, Niels, 285
Abelian groups, 285, 309
Accumulation point, 337–339
Aha! Gotcha: Paradoxes to Puzzle and Delight
     (Gardner), 254n
Algebraic proofs, 130
Algebraic structures, 275–280
Algebraic system
   definition of, 276
   order of, 276
   properties of, 277–280
Antecedent, 9
Antisymmetric property of relations, 164–165
Appel, Kenneth, 36
Archimedean Principle, 108–109, 319, 337
Arcs, 138
Arithmetic mean, 347
Associative Laws, 5
Associative property, 277–280
Assumptions
   identification of, 60
   statement of, 28
Axiomatic set theory, 72
Axiom of Choice, 255, 262, 267–272
Axioms
   consistent systems of, 46
   definition of, 27
   of natural numbers, 100
   statement of, 28

B

Banach-Tarski paradox, 270
Bernstein, Felix, 262
Biconditional sentences
   definition of, 12
   proof of, 43–44

Bijections. *See also* One-to-one correspondence
   construction of, 214
   definition of, 212
   permutation as, 217
Binary operations, 79, 275–277
Binomial coefficients, 127, 129
Binomial Theorem, 344–345
Bolzano, Bernard, 336n
Bolzano-Weierstrass Theorem, 336–339
Borel, Emile, 331–332
Bound, upper and lower, 168, 318
Bounded Monotone Sequence Theorem, 341–345
Bounded sequences, 342–345, 347
Bounded sets, 336–339

C

Canonical map, 191
Cantor, Georg, 251, 259, 261, 262
Cantor-Schröder-Bernstein Theorem, 262–266
Cantor's Theorem, 261–262
Cardinality
   comparability of cardinal numbers and, 267–272
   countable sets and, 251–257
   equivalent sets and, 234–236
   finite sets and, 236–240
   infinite sets and, 236, 242–248
   ordering of cardinal numbers and, 259–266
   symbol for, 243
Cardinal numbers
   comparability of, 267–272
   definition of, 259–260
   finite, 243
   infinite, 242, 243, 247
   ordering of, 259–266
Cartesian products
   definition of, 85
Cauchy, Augustin Louis, 346n

Cauchy sequences, 346, 349–350
Cayley, Arthur, 277n
Cayley's Theorem, 303
Cayley tables, 277, 302
Characteristic function, 190
Choice function, 268
Codomain, 186
Cohen, Paul, 272
Combination Rule, 128–131
Combinatorial proofs, 130
Commutative Laws, 5
Commutative property, 277, 279, 285
Commutative ring, 310, 311–312
Compact sets, 330–331
Comparability property, 164
Comparability Theorem, 267–268, 270
Complement
    of digraph, 156
    of set, 83
Complete graphs, 176
Complete induction, 114
Completeness
    of an ordered field, 320
    equivalents of, 347–350
Component of a graph, 179
Composition
    of function, 156
    of relation, 83
Conditionals, 9–15
Conditional sentences
    converse and contrapositive of, 11–12
    definition of, 9
    direct proof of, 33
Congruence, 151
Congruence modulo, 278
Conjunction
    definition of, 2
    negation of, 6
Connected graph, 180
Consequent, 9
Consistent axiom systems, 46
Continuum hypothesis, 247, 271–272
Contradiction
    definition of, 4
    negation of, 4
    proof by, 41–43, 46, 50–52, 61,
        263–264
Contraposition, proof by, 40–41, 45
Contrapositive, of sentence, 11
Convergence, of sequence, 225–229, 348
Converse, of sentence, 11
Correspondence
    rules of, 191
    single-valued, 187
Countable sets, 245–246, 251–257, 269

Counterexamples, 56
Counting, two-way, 128
Counting principles
    Combination Rule, 128–131
    Generalized Product Rule, 125–127
    Generalized Sum Rule, 122–124
    Permutation Rule, 127–128
    Product Rule, 125
    Sum Rule, 122
Cyclic group, 295, 296
Cyclic subgroup, 295, 296

D

Dedekind, Richard, 349n
Dedekind cuts, 349–350
Deductive reasoning, 1
Definitions, 28
Degree of a vertex, 175
DeMorgan, Augustus, 5n
De Morgan's Laws, 5, 6, 29, 93
Denumerable sets, 251, 252, 255, 256, 269
Denumerable subsets, 271
Derived sets, 338
Descartes, René, 85
Difference, set operations of, 79–81
Digraphs
    complement of, 156
    definition of, 138
Direct proof
    examples of, 33–35, 48
    explanation of, 31–33
    form of, 49–50
    use of, 61
Disjunction, 2
Distributive Laws, 5
Division Algorithm, 62–65, 117–118
Domain
    of function, 186
    of relation, 137
Double Negation Law, 5

E

Edge, 174
Element-chasing proof, 77
Equivalence relations
    algebraic structures based on, 278–279
    definition of, 147
    partitions and, 157–161, 179
Equivalent sets, 233–236
Equivalents of completeness, 347–350
Euclidean axioms, 28
Euclid of Alexandria, 28n, 42, 46
Euclid's Lemma, 45, 65
Euler, Leonard, 51

Exhaustion, proof by, 36
Existence theorems, 51, 52, 54
Existential quantifier, 19

F
Family of sets
   definition of, 89
   indexed, 92–95
Fibonacci, Leonardo, 115n
Fibonacci numbers, 115, 116
Fields
   algebraic properties of, 320–321
   complete, 320
   discussion of, 311–312, 315
   ordered, 316–321, 347, 349, 350
Fifth postulate (Euclid), 46
Finite sets
   definition of, 236
   Pigeonhole Principle and, 239–240
Four-Color Theorem, 36
Fraenkel, Abraham, 72
Functions
   characteristic, 190
   choice, 268
   codomain of, 186, 188, 205, 206
   composite, 198
   constructions of, 195–202
   decreasing, 201
   definition of, 185
   greatest integer, 190
   identity, 189, 217
   image of sets and, 220–223
   inclusion, 189
   increasing, 201
   induced, 223
   inverse of, 196
   one-to-one, 208–210
   one-to-one correspondence and inverse, 213–218
   onto, 205–208
   piecewise defined, 201
   range of, 186
   real, 188
   as relations, 186–191
   sequences and, 225–230
   step, 190
Fundamental Theorem of Arithmetic, 118–119

G
Galois, Evariste, 283
Gardner, Martin, 239
Gauss, Carl Friedrich, 151
Generalized Principle of Mathematical Induction, 109
Generalized Product Rule, 125–127
Generalized Sum Rule, 122–124
Generator, 295

Geometric mean, 347
Gödel, Kurt, 272n
Goldbach, Christian, 57n
Goldbach Conjecture, 57n
Graphs
   complete, 176
   connected, 180
   definition of, 174
   to describe relations, 136
   directed, 138
   discussion of, 177–181
   of functions, 187, 188, 190, 200–202
   isomorphic, 175
   null, 176
   subgraphs, 177
Graph theory, 174
Greatest common divisor, 62, 63
Greatest integer function, 190
Groups. *See also* Subgroups
   abelian, 285, 308, 309
   axiomatic approach to define, 283, 289
   commutative, 285
   cyclic, 295
   definition of, 283–285
   homomorphism, 300
   Law of Exponents and, 288
   permutation, 286–287

H
Haken, Wolfgang, 36
Handshaking Lemma, 176–177
Hasse diagram, 167, 169
Heine, Edward, 331–332
Heine-Borel Theorem, 332–333, 336
Hierarchy of connectives, 6, 15
Hilbert, David, 254
Homomorphic image, 300, 301, 303
Homomorphism
   discussion of, 300–303
   group, 300
   ring, 309, 310
Horizontal Line Test, 206

I
Identity element, 277, 278
Identity function, 189, 217
Identity permutation, 217, 218
Identity subgroup, 293
Images of sets, 220–223
Immediate predecessor, 166
Inclusion function, 189
Index, 92
Indexing set, 92–93
Indirect proofs, 40–41
Induced function, 223

Induction
  complete, 114
  Generalized Principle of Mathematical, 109
  hypothesis of, 102
  Principle of Complete, 114–116
  Principle of Mathematical, 100–109, 114
  proofs by, 102–103
  Well-Ordering Principle, 116–119
Inductive set, 100
Infimum, 168, 318
Infinite Hotel, 254
Infinite order, 295
Infinite sequence, 190
Infinite sets, 236, 242–248
Integral domain, 310, 311
Intermediate Value Theorem, 20
Intersection
  over family of sets, 90
  set operation of, 80, 81
Inverse element, 277, 278
Inverse functions, 215–218
Inverse permutation, 218
Isomorphic
  graphs, 175
  groups, 302
Isomorphism, 302

J
Join operation, 144

L
The Last Recreations (Gardner), 239
Law of Cosines, 45
Law of Excluded Middle, 4
Law of Exponents, 288
Leibnitz, G. W., 185
Leonardo of Pisa (Leonardo Fibonacci), 115
Linear combination, 63
Linear order, 169, 170
Loop, 138, 148
Lower bound, 168, 318

M
Mean Value Theorem, 51
Modulus, of congruence, 151, 152
Modus ponens, 29–30
Monotone sequences, 342–345, 347

N
Natural numbers
  infinite sets and, 243–248
  property of sets of, 100
Negations
  of proposition, 2
  simplified form of, 23, 24
Neighborhood, 325–326, 338

Normalized form, 246
Null graphs, 176

O
One-to-one correspondence. *See also* bijections
  elements in set and, 233, 234
  explanation of, 213–218, 234
  infinite sets and, 244
One-to-one functions, 208–210
Open sets, 326–329
Operation preserving (OP), 298
Operation tables, 277
Order
  of an algebraic system, 276
  of an element, 295
  of a graph, 175
Ordered field properties of real numbers, 316–321
Ordered fields, 316–321, 347, 349, 350
Ordered *n*-tuples, 84
Ordered pairs
  definition of, 84
  as relation, 137
Ordering relations, 163–170

P
Pairwise disjoint families, 94, 95
Paradox, 2
Parentheses, 6–7
Partially ordered set (poset), 165–169
Partitions
  definition of, 157–159
  equivalence relation and, 159–161, 179
Pascal, Blaise, 130n
Pascal's triangle, 130–131
Path, 177–180
Peano, Giuseppe, 100
Permutation groups, 286–287
Permutation Rule, 127–128
Permutations
  composites of, 218
  counting number of, 128–129
  definition of, 217, 285
  identity, 218
  inverses of, 217
  of sets, 126, 285–286
Piecewise-defined functions, 201
Pigeonhole Principle, 239–240
Poset, 165
Postulates, 27
Power set, 75
Principle of Complete Induction (PCI), 114–116
Principle of Inclusion and Exclusion, 124
Principle of Mathematical Induction (PMI)
  discussion and use of, 100–109, 114
  generalized, 109
Product Rule, 125

Products, 275
Proofs
    algebraic, 130
    basic methods I, 27–36
    basic methods II, 40–46
    of biconditional sentences, 43–44
    combinatorial, 130
    by contradiction, 41–43, 46, 50–53, 61
    by contraposition, 40–41, 45
    direct, 31–35, 48–50
    element-chasing, 77
    by exhaustion, 36
    explanation of, 1, 27
    indirect, 40–41
    induction, 102–103
    involving quantifiers, 48–56
    strategies for writing, 29, 60–62
Proper subset, 74
Propositions
    ambiguity and, 6–7
    antecedent, 9
    compound, 3–4
    conditionals and biconditionals and, 9–15
    consequent, 9
    contrapositive and, 11
    converse and, 11
    definition of, 1–2
    denial of, 6
    equivalent, 4–6
    examples of, 2
    formation of, 2
    negation of, 2
Pugh, Charles Chapman, 349
Pythagoras, 28n
Pythagorean Theorem, 28

Q
Quantified sentences
    explanation of, 22, 23
    strategies for dealing with, 61
Quantifiers
    existential, 19
    explanation of, 18–25
    hidden, 20–21
    incorrect deductions and, 55–56
    proofs involving, 48–56
    unique existential, 24
    universal, 20

R
Range
    of function, 186
    of relation, 137
Real functions, 188
Real Mathematical Analysis (Pugh), 349
Reasoning, deductive, 1

Reflexive property of relations, 147–149, 164
Relational databases, 143
Relations
    antisymmetric property of, 164–165
    construction of, 139–144
    definition of, 135
    equivalence, 147–153, 159–161, 191
    functions as, 185–191
    graphs to represent, 174–181
    inverse of, 139
    n-tuple, 143
    ordering, 163–170
    reflexivity property of, 147–149, 164
    symmetric property of, 147–149
    transitive property of, 147–149, 164
Replacement rule, 29
Restrictions, of function, 199
Ring homomorphism, 309, 310
Ring isomorphism, 310
Rings
    commutative, 310, 311
    definition of, 307
    with unity, 310
Rolle's Theorem, 51
Russell, Bertrand, 2n, 72

S
Sandwich theorem, 229
Schröder, Ernst, 262
Sentences
    conditional, 9–15
    contrapositive of, 11
    converse of, 11
    quantified, 22, 23, 61
Sequences
    bounded, 342–345, 347
    Cauchy, 346, 349–350
    convergence of, 225
    decreasing, 343
    definition of, 190, 225
    divergence of, 225
    increasing, 343
    limit of, 226–229
    monotone, 342–345, 347
    nth term of, 190, 225
    properties of, 225–226
Set operations
    discussion of, 79–86
    extended, 89–95
Sets
    bounded, 318
    compact, 330–331
    countable, 245–246, 251–257, 269
    counting elements in, 233, 234
    denumerable, 251, 252, 255, 256, 269
    derived, 338

Sets (*continued*)
 equivalent, 234–236
 family of, 89–95
 finite, 236–240
 images of, 220–223
 infinite, 236, 242–248
 open, 326–329
 partially ordered, 165–169
 permutations of, 126, 285–286
 supremum of, 318, 319
 truth, 18–20
 uncountable, 245–246
Set theory
 axiomatic, 72
 basic concepts of, 71–77
 counting principles and, 122–131
 equivalent forms of induction and, 114–119
 extended set operations and indexed families of sets and, 89–95
 set operations and, 79–86
 Zermelo-Fraenkel, 72, 268, 272
Single-valued correspondence, 187
Squeeze theorem, 229
Step function, 190
Subgroups, 292–296. *See also* Groups
Subsequence, 346
Subsets
 of algebraic system, 276
 of countable sets, 253
 definition of, 72
 denumerable, 271
 of finite sets, 237
 of groups, 294
 proper, 74
Sum Rule, 122, 238
Supremum, 168, 318
Surjection, 205, 206, 208, 209
Symmetric closure, 156
Symmetric group on *n* symbols, 286
Symmetric property of relations, 147–149

T
Tautologies, 4, 29
Tautology rule, 29
Terminal vertex, 177
Ternary operations, 276
Theorems
 existence, 51, 52, 54
 explanation of, 27

Transitive closure, 156
Transitive property of relations, 147–149, 164
Triangles, Pascal's, 130–131
Trichotomy property, 267–268
Trigonometric functions, 200, 215, 247
Trivial subgroup, 293
Truth set, 18–20
Truth table, 10–11
Two-way counting, 128

U
Unary operations, 276
Undecidable statements, 46
Undefined terms, 28
Union
 of finite numbers, 238
 over family of sets, 90
 set operation of, 80, 81
Unique existential quantifiers, 24
Universal quantifier, 20
Universe of discourse, 18, 83
Upper bound, 168, 318

V
Venn, John, 75n
Venn diagrams, 75, 80, 81, 91
Vertex
 adjacent, 175
 definition of, 138, 174
 initial, 177
 isolated, 175
 terminal, 177
Vertical Line Test, 188

W
Walk, 177
Weierstrass, Karl, 336n
Well ordering, 170
Well-Ordering Principle (WOP), 116–117, 170, 253
Well-Ordering Theorem, 170

Z
Zermelo, Ernst, 72
Zermelo-Fraenkel set theory, 72, 268, 272